OPTIMIZATION TECHNIQUES FOR SOLVING COMPLEX PROBLEMS

WILEY SERIES ON PARALLEL AND DISTRIBUTED COMPUTING

Editor: Albert Y. Zomaya

A complete list of titles in this series appears at the end of this volume.

OPTIMIZATION TECHNIQUES FOR SOLVING COMPLEX PROBLEMS

Edited by

Enrique Alba
University of Málaga

Christian Blum
Technical University of Catalonia

Pedro Isasi
University Carlos III of Madrid

Coromoto León
University of La Laguna

Juan Antonio Gómez
University of Extremadura

WILEY

A JOHN WILEY & SONS, INC., PUBLICATION

Published by John Wiley & Sons, Inc., Hoboken, New Jersey.
Published simultaneously in Canada.

Library of Congress Cataloging-in-Publication Data:

Optimization techniques for solving complex problems / [edited by] Enrique Alba, Christian Blum, Pedro Isasi, Coromoto León, Juan Antonio Gómez

Includes bibliographic references and index.

ISBN 978-0-470-29332-4 (cloth)

Printed in the United States of America

10 9 8 7 6 5 4 3 2 1

Enrique Alba, *To my family*

Christian Blum, *To María and Marc*

Pedro Isasi, *To my family*

Coromoto León, *To Juana*

Juan Antonio Gómez, *To my family*

CONTENTS

CONTRIBUTORS

E. Alba, Universidad de Málaga, Dpto. de Lenguajes y Ciencias de la Computación, Málaga (Spain)

R. Aler, Universidad Carlos III de Madrid, Dpto. de Informática, Escuela Politécnica Superior, Madrid (Spain)

M. J. Blesa, Universitat Politècnia de Catalunya, Dpto. de Llenguatges i Sistemes Informàtics, Barcelona (Spain)

C. Blum, Universitat Politècnica de Catalunya, Dpto. de Llenguatges i Sistemes Informàtics, Barcelona (Spain)

J. Carretero, Universitat Politècnica de Catalunya, Dpto. d'Arquitectura de Computadors, Barcelona (Spain)

F. Chicano, Universidad de Málaga, Dpto. de Lenguajes y Ciencias de la Computación, Málaga (Spain)

C. Cotta, Universidad de Málaga, Dpto. de Lenguajes y Ciencias de la Computación, Málaga (Spain)

B. Dorronsoro, Université de Luxembourg (Luxembourg)

J. J. Durillo, Universidad de Málaga, Dpto. dé Lenguajes y Ciencias de la Computación, Málaga (Spain)

C. Estébanez, Universidad Carlos III de Madrid, Dpto. de Informática, Escuela Politécnica Superior, Madrid (Spain)

A. J. Fernández, Universidad de Málaga, Dpto. de Lenguajes y Ciencias de la Computación, Málaga (Spain)

F. Fernández, Universidad de Extremadura, Dpto. de Tecnologías de Computadores y Comunicaciones, Centro Universitario de Mérida, Mérida (Spain)

J. E. Gallardo, Universidad de Málaga, Dpto. de Lenguajes y Ciencias de la Computación, Málaga (Spain)

I. M. Galván, Universidad Carlos III de Madrid, Dpto. de Informática, Escuela Politécnica Superior, Madrid (Spain)

J. García-Nieto, Universidad de Málaga, Dpto. de Lenguajes y Ciencias de la Computación, Málaga (Spain)

A. Gómez, Centro de Investigaciones Energéticas, Medioambientales y Tecnológicas (CIEMAT), Centro Extremeño de Tecnologías Avanzadas, Trujillo (Spain)

J. A. Gómez, Universidad de Extremadura, Dpto. de Tecnologías de Computadores y Comunicaciones, Escuela Politécnica, Cáceres (Spain)

J. M. Granado, Universidad de Extremadura, Dpto. de Ingeniería de Sistemas Informáticos y Telemáticos, Escuela Politécnica, Cáceres (Spain)

J. M. Guerra, Universidad Complutense de Madrid, Dpto. de Optica, Madrid (Spain)

J. L. Guisado, Universidad de Sevilla, Dpto. de Arquitectura y Tecnología de Computadors, Sevilla (Spain)

P. Isasi, Universidad Carlos III de Madrid, Dpto. de Informática, Escuela Politécnica Superior, Madrid (Spain)

M. D. Jaraiz, Universidad de Extremadura, Dpto. de Tecnologías de Computadores y Comunicaciones, Escuela Politécnica, Cáceres (Spain)

F. Jiménez-Morales, Universidad de Sevilla, Dpto. de Física de la Materia Condensada, Sevilla (Spain)

G. Leguizamón, Universidad Nacional de San Luis, Laboratorio de Investigación y Desarrollo en Inteligencia Computacional (LIDIC), San Luis (Argentina)

C. León, Universidad de La Laguna, Dpto. de Estadística, I.O. y Computación, La Laguna (Spain)

D. Lombraña, Universidad de Extremadura, Cátedra CETA-CIEMAT, Centro Universitario de Mérida, Mérida (Spain)

F. Luna, Universidad de Málaga, Dpto. de Lenguajes y Ciencias de la Computación, Málaga (Spain)

C. Luque, Universidad Carlos III de Madrid, Dpto. de Informática, Escuela Politécnica Superior, Madrid (Spain)

G. Luque, Universidad de Málaga, Dpto. de Lenguajes y Ciencias de la Computación, Málaga (Spain)

E. Lutton, Institut National de Recherche en Informatique et en Automatique (INRIA), Orsay (France)

G. Miranda, Universidad de La Laguna, Dpto. de Estadística, I.O. y Computación, La Laguna (Spain)

G. Molina, Universidad de Málaga, Dpto. de Lenguajes y Ciencias de la Computación, Málaga (Spain)

J. M. Molina, Universidad de Málaga, Dpto. de Lenguajes y Ciencias de la Computación, Málaga (Spain)

S. Molina, Universided de Nacional San Luis, Laboratorio de Investigación y Desarrollo en Inteligencia Computacional (LIDIC), San Luis (Argentina)

A. Mochón, Universidad Nacional de Educación a Distancia (UNED), Dpto. de Economía Aplicada e Historia Económica, Madrid (Spain)

A. J. Nebro, Universidad de Málaga, Dpto. de Lenguajes y Ciencias de la Computación, Málaga (Spain)

G. Olague, Centro de Investigación Cientifica y de Educación Superior de Ensenada (CICESE), Dpto. de Ciencias Informáticas, Ensenada (México)

G. Ordóñez, Universidad Nacional de San Luis, Laboratorio de Investigación y Desarrollo en Inteligencia Computacional (LIDIC), San Luis (Argentina)

C. B. Pérez, Centro de Investigación Cientifica y de Educación Superior de Ensenada (CICESE), Dpto. de Ciencias Informáticas, Ensenada (México)

D. Quintana, Universidad Carlos III de Madrid, Dpto. de Informática, Escuela Politécnica Superior, Madrid (Spain)

C. Rodríguez, Universidad de La Laguna, Dpto. de Estadística, I. O. y Computación, La Laguna (Spain)

M. Rubio, Centro de Investigaciones Energéticas, Medioambientales y Tecnológicas (CIEMAT), Centro Extremeño de Tecnologías Avanzadas, Trujillo (Spain)

Y. Sáez, Universidad Carlos III de Madrid, Dpto. de Informática, Escuela Politécnica Superior, Madrid (Spain)

C. Salto, Universidad Nacional de La Pampa, Facultad de Ingeniería, General Pico (Argentina)

J. M. Sánchez, Universidad de Extremadura, Dpto. de Tecnologías de Computadores y Comunicaciones, Escuela Politécnica, Cáceres (Spain)

J. M. Valls, Universidad Carlos III de Madrid, Dpto. de Informática, Escuela Politécnica Superior, Madrid (Spain)

M. A. Vega, Universidad de Extremadura, Dpto. de Tecnologías de Computadores y Comunicaciones, Escuela Politécnica, Cáceres (Spain)

F. Xhafa, Universitat Politècnia de Catalunya, Dpto. de Llenguatges i Sistemes Informàtics, Barcelona (Spain)

FOREWORD

The topic of optimization, especially in the context of solving complex problems, is of utmost importance to most practitioners who deal with a variety of optimization tasks in real-world settings. These practitioners need a set of new tools for extending existing algorithms and developing new algorithms to address a variety of real-world problems. This book addresses these very issues.

The first part of the book covers many new ideas, algorithms, and techniques. These include modern heuristic methods such as genetic programming, neural networks, genetic algorithms, and hybrid evolutionary algorithms, as well as classic methods such as divide and conquer, branch and bound, dynamic programming, and cryptographic algorithms. Many of these are extended by new paradigms (e.g., new metaheuristics for multiobjective optimization, dynamic optimization) and they address many important and practical issues (e.g., constrained optimization, optimization of time series).

The second part of the book concentrates on various applications and indicates the applicability of these new tools for solving complex real-world problems. These applications include DNA sequencing and reconstruction, location of antennas in telecommunication networks, job scheduling, cutting and packing problems, multidimensional knapsack problems, and image processing, to name a few.

The third and final part of the book includes information on the possibility of remote optimization through use of the Internet. This is definitely an interesting option, as there is a growing need for such services.

I am sure that you will find this book useful and interesting, as it presents a variety of available techniques and some areas of potential applications.

ZBIGNIEW MICHALEWICZ

University of Adelaide, Australia
February 2008

PREFACE

This book is the result of an ambitious project to bring together various visions of many researchers in both fundamental and applied issues of computational methods, with a main focus on optimization. The large number of such techniques and their wide applicability make it worthwhile (although difficult) to present in a single volume some core ideas leading to the creation of new algorithms and their application to new real-world tasks.

In addition to researchers interested mainly in algorithmic aspects of computational methods, there are many researchers whose daily work is rather application-driven, with the requirement to apply existing techniques efficiently but having neither the time, the resources, nor the interest in algorithmic aspects. This book is intended to serve all of them, since these two points of view are addressed in most of the chapters. Since the book has these two parts (fundamentals and applications), readers may use chapters of either part to enhance their understanding of modern applications and of optimization techniques simultaneously.

Since this is an edited volume, we were able to profit from a large number of researchers as well as from new research lines on related topics that have begun recently; this is an important added value that an authored book would probably not provide to such an extent. This can easily be understood by listing the diverse domains considered: telecommunications, bioinformatics, economy, cutting, packing, cryptography, hardware, laser industry, scheduling, and many more.

We express our profound appreciation to all who have contributed a chapter to this book, since any merit the work deserves must be credited to them. Also, we thank the research groups that contributed to the book for their efforts and for their help in making this project successful. We also appreciate the support received from Wiley during the entire editing process, as well as the decisive endorsement by Professor A. Zomaya that made this idea a reality. To all, thank you very much.

Enrique Alba
Christian Blum
Pedro Isasi
Coromoto León
Juan Antonio Gómez

February 2008

PART I

METHODOLOGIES FOR COMPLEX PROBLEM SOLVING

CHAPTER 1

Generating Automatic Projections by Means of Genetic Programming

C. ESTÉBANEZ and R. ALER

Universidad Carlos III de Madrid, Spain

1.1 INTRODUCTION

The aim of inductive machine learning (ML) is to generate models that can make predictions from analysis of data sets. These data sets consist of a number of instances or examples, each example described by a set of attributes. It is known that the quality or relevance of the attributes of a data set is a key issue when trying to obtain models with a satisfactory level of generalization. There are many techniques of feature extraction, construction, and selection [1] that try to improve the representation of data sets, thus increasing the prediction capabilities of traditional ML algorithms. These techniques work by filtering nonrelevant attributes or by recombining the original attributes into higher-quality ones. Some of these techniques were created in an automatic way by means of genetic programming (GP).

GP is an evolutionary technique for evolving symbolic programs [2]. Most research has focused on evolving functional expressions, but the use of loops and recursion has also been considered [3]. Evolving circuits are also among the successes of GP [4]. In this work we present a method for attribute generation based on GP called the GPPE (genetic programming projection engine). Our aim is to evolve symbolic mathematical expressions that are able to transform data sets by representing data on a new space, with a new set of attributes created by GP. The goal of the transformation is to be able to obtain higher accuracy in the target space than in the original space. The dimensions of the new data space can be equal to, larger, or smaller than those of the original. Thus, we also intend that GPPE be used as a dimension reduction technique as

Optimization Techniques for Solving Complex Problems, Edited by Enrique Alba, Christian Blum, Pedro Isasi, Coromoto León, and Juan Antonio Gómez

well as creating highly predictive attributes. Although GPPE can either increase or reduce dimensionality, the work presented in this chapter focuses on reducing the number of dimensions dramatically while attempting to improve, or at least maintain, the accuracy obtained using the original data.

In the case of dimension reduction, the newly created attributes should contain all the information present in the original attributes, but in a more compact way. To force the creation of a few attributes with a high information content, we have established that the data in the projected space must follow a nearly linear path. To test GPPE for dimensionality reduction, we have applied it to two types of data mining domains: classification and regression. In classification, linear behavior will be measured by a fast classification algorithm based on selecting the nearest class centroid. In regression, linear behavior will be determined by simple linear regression in the projected space.

GP is very suitable for generating feature extractors, and some work has been done in this field. In the following section we overview briefly some approaches proposed in the literature. Then, in Section 1.4 we focus on GPPE, which can be used in both the classification and regression domains, and we show some experimental results in Section 1.5. We finish with our conclusions and some suggestions for future work.

1.2 BACKGROUND

There are many different constructive induction algorithms, using a wide variety of approaches. Liu et al. [1] provide a good starting point for the exploration of research into feature extraction, construction, and selection. Their book compiles contributions from researchers in this field and offers a very interesting general view. Here we discuss only works that use GP or any other evolutionary strategy, and we focus on those that are among the most interesting for us because they bear some resemblance to GPPE.

Otero et al. [5] use typed GP for building feature extractors. The functions are arithmetic and relational operators, and the terminals are the original (continuous) attributes of the original data set. Each individual is an attribute, and the fitness function uses the information gain ratio. Testing results using C4.5 show some improvements in some UCI domains. In Krawiec's work [6], each individual contains several subtrees, one per feature. C4.5 is used to classify in feature space. Their work allows us to cross over subtrees from different features.

Shafti and Pérez [7] discuss the importance of applying GA as a global search strategy for constructive induction (CI) methods and the advantages of using these strategies instead of using classic greedy methods. They also present MFE2/GA, a CI method that uses GA to search through the space of different combination of attribute subsets and functions defined over them. MFE2/GA uses a nonalgebraic form of representation to extract complex interactions between the original attributes of the problem.

Kuscu [8] introduced the GCI system. GCI is a CI method based on GP. It is similar to GPPE in the sense that it uses basic arithmetic operators and the fitness is computed measuring the performance of an ML algorithm (a quick-prop net) using the attributes generated. However, each individual represents a new attribute instead of a new attribute set. In this way, GCI can only generate new attributes that are added to the original ones, thus increasing the dimensionality of the problem. The possibility of reducing the number of attributes of the problem is mentioned only as possible and very interesting future work.

Hu [9] introduced another CI method based on GP: GPCI. As in GCI, in GPCI each individual represents a newly generated attribute. The fitness of an individual is evaluated by combining two functions: an absolute measure and a relative measure. The absolute measure evaluates the quality of a new attribute using a gain ratio. The relative measure evaluates the improvement of the attribute over its parents. A function set is formed by two Boolean operators: AND and NOT. GPCI is applied to 12 UCI domains and compared with two other CI methods, achieving some competitive results.

Howley and Madden [10] used GP to evolve kernels for support vector machines. Both scalar and vector operations are used in the function set. Fitness is computed from SVM performance using a GP-evolved kernel. The hyperplane margin is used as a tiebreaker to avoid overfitting. Although evolved kernels are not forced by the fitness function to satisfy standard properties (such as Mercer's property) and therefore the evolved individuals are not proper kernels, results in the testing data sets are very good compared to those of standard kernels. We believe that evolving proper distance functions or kernels is difficult because some properties (such as transitivity or Mercer's property) are not easy to impose on the fitness computation.

Eads et al. [11] used GP to construct features to classify time series. Individuals were made of several subtrees returning scalars (one per feature). The function set contained typical signal-processing primitives (e.g., convolution), together with statistical and arithmetic operations. SVM was then used for classification in feature space. Cross-validation on training data was used as a fitness function. The system did not outperform the SVM, but managed to reduce dimensionality. This means that it constructed good features to classify time series. However, only some specific time series domains have been tested. Similarly, Harvey et al. [12] and Szymanski et al. [13] assemble image-processing primitives (e.g., edge detectors) to extract multiple features from the same scene to classify terrains containing objects of interest (i.e., golf courses, forests, etc.). Linear fixed-length representations are used for the GP trees. A Fisher linear discriminant is used for fitness computation. Results are quite encouraging but are restricted to image-processing domains.

Results from the literature show that, in general, the GP projection approach has merit and obtains reasonable results, but that more research is needed. New variations of the idea and more domains should be tested. Regression problems are not considered in any of the works reviewed, and we believe that a lot more research on this topic is also needed.

1.3 DOMAINS

In this chapter we are interested in applying GPPE to two classical prediction tasks: classification and regression. We have used bankruptcy prediction as the classification domain and IPO underpricing prediction as the regression domain.

1.3.1 Bankruptcy Prediction

In general terms, the bankruptcy prediction problem attempts to determine the financial health of a company, and whether or not it will soon collapse. In this chapter we use a data set provided and described by Vieira et al. [14]. This data set studies the influence of several financial and economical variables on the financial health of a company. It includes data on 1158 companies, half of which are in a bankruptcy situation (class 0) and the rest of which have good financial health (class 1). Companies are characterized by 40 numerical attributes [14]. For validation purposes we have divided the data set into a training set and a test set, containing 766 (64%) and 400 (36%) instances, respectively.

1.3.2 IPO Underpricing Prediction

IPO underpricing is an interesting and important phenomenon in the stock market. The academic literature has long documented the existence of important price gains in the first trading day of initial public offerings (IPOs). That is, there is usually a big difference between the offering price and the closing price at the end of the first trading day. In this chapter we have used a data set composed of 1000 companies entering the U.S. stock market for the first time, between April 1996 and November 1999 [15]. Each company is characterized by seven explicative variables: underwriter prestige, price range width, price adjustment, offer price, retained stock, offer size, and relation to the tech sector. The target variable is a real number which measures the profits that could be obtained by purchasing the shares at the offering price and selling them soon after dealing begins. For validation purposes we have divided the data set into a training set and a test set, containing 800 (80%) and 200 (20%) instances, respectively.

1.4 ALGORITHMIC PROPOSAL

In this section we describe the genetic programming projection engine (GPPE). GPPE is based on GP. Only a brief summary of GP is provided here. The reader is encouraged to consult Koza's book [2] for more information.

GP has three main elements:

1. A population of individuals, in this case, computer programs
2. A fitness function, used to measure the goodness of the computer program represented by the individual

3. A set of genetic operators, the basic operators being reproduction, mutation, and crossover

The GP algorithm enters into a cycle of fitness evaluation and genetic operator application, producing consecutive generations of populations of computer programs, until a stopping criterion is satisfied. Usually, GP is stopped when an optimal individual is found or when a time limit is reached. Every genetic operator has the probability of being applied each time we need to generate an offspring individual for the next generation. Also, GP has many other parameters, the most important ones being the size of the population (M) and the maximum number of generations (G).

GPPE is a GP-based system for computing data projections with the aim of improving prediction accuracy in prediction tasks (classification and regression). The training data E belong to an N-dimensional space U, which will be projected into a new P-dimensional space V. In classification problems the goal of the projection is that classification becomes easier in the new space V. By "easier" we mean that data in the new space V are as close to linear separability as possible. Similarly, in regression problems the aim is to project data so that they can be approximated by a linear regression. To include both cases, we will talk about linear behavior. P can be larger, equal to, or smaller than N. In the latter case, in addition to improving prediction, we would also reduce the number of dimensions.

GPPE uses standard GP to evolve individuals made of P subtrees (as many as the number of dimensions of the projected space V). Fitness of individuals is computed by measuring the degree of linear behavior of data in the space projected by using the individual as a projection function from U to V. The system stops if 100% linear behavior has been achieved (i.e., a 100% classification rate or 0.0 error in regression) or if the maximum number of generations is reached. Otherwise, the system outputs the highest-fitness individual (i.e., the most accurate individual on the training data). Algorithm 1.1 displays pseudocode for GPPE operation. For the implementation of our application, we have used Lilgp 1.1, a software package for GP developed at Michigan State University by Douglas Zongker and Bill Punch.

Next, we describe the main GPPE elements found in all GP-based systems: the terminal and function set for the GP individuals, the structure of the GP individuals themselves, and the fitness function.

1.4.1 Terminal and Function Set

The terminal and function set is composed of the variables, constants, and functions required to represent mathematical projections of data. For instance, $(v_0, v_1) = (3 * u_0 + u_1, u_2^2)$ is a projection from three to two dimensions comprising the following functions and terminals:

$$\text{functions} = \{+, *, ^2\}$$

$$\text{terminals} = \{3, u_0, u_1, u_2, 2\}$$

Algorithm 1.1 Pseudocode for GPPE Operation

```
P = Initial population made of random projections;
generation = 0; stopCondition = FALSE;
while (generation < maxGenerations) AND (NOT stopCondition) do
 for each projection p ∈ P do
  f_p = fitness(p);
  if (f_p = perfectFitness) then stopCondition = TRUE;
 end for
 P' = geneticOperatorsApplication(P, f);
 P = P';
 generation = generation + 1;
end while
```

The set of functions and terminals is not easy to determine: It must be sufficient to express the problem solution. But if the set is too large, it will increase the search space unnecessarily. In practice, different domains will require different function and terminal sets, which have to be chosen by the programmer. We consider this to be an advantage of GP over other methods with fixed primitives because it allows us to insert some domain knowledge into the learning process. At this point in our research, we have tested some generic sets appropriate for numerical attributes. In the future we will analyze the domains to determine which terminals and functions are more suitable.

This generic terminal set contains the attributes of the problem expressed in coordinates of U $(u_0, u_1, \ldots, u_N)$, and the ephemeral random constants (ERCs). An ERC is equivalent to an infinite terminal set of real numbers. The generic functions we have used in GPPE so far are the basic arithmetical functions $+$, $-$, $*$, and /, and the square and square root. We have judged them to be sufficient to represent numerical projections, and experimental data have shown good results.

1.4.2 Individuals

Projections can be expressed as

$$(v_0 \cdots v_P) = (f_1(u_0 \cdots u_N), \ldots, f_P(u_0 \cdots u_N))$$

To represent them, individuals are made of P subtrees. Every subtree number i represents function f_i, which corresponds to coordinate v_i in the target space. All subtrees use the same set of functions and terminals.

1.4.3 Fitness Function

The fitness function evaluates an individual by projecting the original data and determining the degree of linear behavior in the target space. For this task we designed two different fitness functions: one for classification problems and one for regression problems.

Classification The classification fitness function uses a centroid-based classifier. This classifier takes the projected data and calculates a centroid for each class. Centroids are the centers of mass (baricenters) of the examples belonging to each class. Therefore, there will be as many centroids as classes. The centroid-based classifier assigns to every instance the class of the nearest centroid.

This function tries to exert a selective pressure on the projections that forces every instance belonging to the same class to get close. The great advantage of this function is that it is fast and simple. We call this function CENTROIDS.

To avoid overfitting, fitness is actually computed using cross-validation on the training data. That is, in every cross-validation cycle the centroids-based classifier is trained on some of the training data and tested on the remaining training data. Training the centroids means computing the baricenters, and testing them means using them to classify the part of the training data reserved for testing. The final fitness is the average of all the cross-validation testing results.

Regression In this case, linear behavior is defined as data fitting a hyperplane, so in this case, the goal is to adjust projected data to a hyperplane as closely as possible. For the regression tasks, a simple linear regression algorithm is used to compute fitness. More precisely, the fitness is the error produced by the linear regression on the projected data. This error is measured by the normalized mean-square error (NMSE).

1.5 EXPERIMENTAL ANALYSIS

In this section we test GPPE in the classification and regression domains, described in Section 1.3. We want to show that GPPE can help in both tasks. We have executed 100 GP runs in both the classification domain (bankruptcy prediction) and the regression domain (IPO underpricing prediction), each with the parameters displayed in Table 1.1. They were found empirically by running GPPE five times on each domain. The fitness function CENTROIDS was used in the classification domain, and REGRESSION in the regression domain. Tenfold cross-validation was used to compute the training fitness, to avoid overfitting the training data.

Every GP run involves running GPPE on the same problem but with a different random seed. However, typical use of GP involves running the GP engine several times and selecting the best individual according to its training fitness. We have simulated this procedure by using bootstrapping [16] on these 100 samples, as follows:

- Repeat B times:
  - Select R samples from the 100 elements data set.

TABLE 1.1 Common GPPE Parameters for Both Experimentation Domains

Parameter	Value
Max. generations (G)	100
Population size (M)	500 (bankruptcy)/5000 (IPO)
Max. nodes per tree	75
Initialization method	Half and half
Initialization depth	2–6
Number of genetic operators	3
Genetic operator 1	Reproduction rate = 15% Selection = fitness proportion
Genetic operator 2	Crossover rate = 80% Selection = tournament Tournament size = 4
Genetic operator 3	Mutation selection = tournament Tournament size = 2

- From the set of R samples, select the best sample i, according to its training fitness.
- Return the accuracy of i, according to the test set.

In this case, $B = 100$ and $R = 5$. This means that we have simulated 100 times the procedure of running GPPE five times and selecting the best individual according to its training fitness. In both the classification and regression domains, the resulting bootstrapping distributions follow a normal distribution, according to a Kolmogorov–Smirnov test, with $p < 0.01$. Table 1.2 displays the average, standard deviation, and median values for the bootstrapping distribution on the test set.

To determine whether GPPE projections generate descriptive attributes and improve results over those of the original data set, we have used several well-known machine learning algorithms (MLAs) for both domains before and after projecting the data. We have chosen the best-performing GPPE individuals according to their training fitness (individuals e95 from the classification domain and e22 from the regression domain). e95 obtained a training fitness of 81.33% and a test fitness of 80.00%; e22 obtained a training NMSE of 0.59868 and a

TABLE 1.2 Average and Median Values of the Bootstrapping Distribution for B = 100 and R = 5

	Average	Median
Classification domain	79.05 ± 1.01	78.88
Regression domain	0.864084 ± 0.02	0.861467

TABLE 1.3 Comparison of MLA Performance Using Original and Projected Data

	Classification Domain			
	MLP	SMO	Simple Logistics	RBF Network
Original	78.50	79.25	61.75	72.75
Proj. by e95	**80.75**	**80.25**	**80.25**	72.75
PCA var. = 0.95	76.25	**80.25**	79.75	**73.25**
	Regression Domain			
	Linear Reg.	Simp. Lin. Reg.	Least Med. Sq.	Additive Reg.
Original	0.878816	0.932780	1.056546	0.884140
Proj. by e22	**0.838715**	**0.837349**	**1.012531**	**0.851886**
PCA var. = 0.95	0.904745	0.932780	1.061460	0.899076

test NMSE of 0.844891. Multilayer perceptron [17], support vector machine (SMO) [18], simple logistics [19], and a radial basis function network [20] were applied to the projected and unprojected data in the classification domain. The same process was carried out for the regression data, with linear regression, simple linear regression [19], least median square [21], and additive regression [19]. Table 1.3 displays these results, which were computed using the training and test sets described in Section 1.3 (they are the same sets on which GPPE was run). To compare with a commonly used projection algorithm, Table 1.3 also shows the accuracy of the same algorithms with data preprocessed by principal component analysis (PCA) with a variance of 0.95.

Table 1.3 highlights those values where projecting the data yields better results than for the same algorithm working on unprojected data. In both domains, classification and regression accuracy improve by projecting the data, even when PCA is applied. Unfortunately, the differences in Table 1.3 are not statistically significant (according to a t-test with 0.05 confidence). Even so, it must be noted that GPPE reduced the number of attributes of the problem from 40 to only 3 in the classification domain, and from 8 to 3 in the regression domain. PCA generated 26 and 6 attributes on the classification and regression domains, respectively.

1.6 CONCLUSIONS

Inductive ML algorithms are able to learn from a data set of past cases of a problem. They try to extract the patterns that identify similar instances and use this experience to forecast unseen cases. It is clear that the prediction rate is affected enormously by the quality of the attributes of the data set. To obtain higher-quality attributes, researchers and practitioners often use feature extraction methods, which can filter, wrap, or transform the original attributes, facilitating the learning process and improving the prediction rate.

Some work has been dedicated to implementing a GP system that can produce ad hoc feature extractors for each problem, as explained in Section 1.2. In this chapter we present our own contribution: GPPE, a genetic programming–based method to project a data set into a new data space, with the aim of performing better in data mining tasks. More specifically, in this work we focus on reducing dimensionality while trying to maintain or increase prediction accuracy. GPPE has been tested in the classification and regression domains. Results show that:

- GPPE is able to reduce dimensionality dramatically (from 40 to 3 in one of the domains).
- Attributes created by GPPE enforce nearly linear behavior on the transformed space and therefore facilitate classification and regression in the new space.
- Different learning algorithms can also benefit from the new attributes generated by GPPE.

In the future we would like to test GPPE on more complex domains with a large number of low-level attributes. For instance, image recognition tasks where attributes are individual pixels would be very appropriate. Classifying biological data involving multiple time series, including, for example, heart rate, EEG, and galvanic skin response, may also be a candidate domain. Such specialized domains might require specialized primitives (e.g., averages, summations, derivatives, filters). GP has the advantage that any primitive whatsoever can be utilized by adding it into the function set. We believe that in these domains, our projection strategy will be able to improve accuracy in addition to reducing dimensionality.

We have also compared the results obtained by PCA with those obtained by GPPE, but these two techniques are in no way in competition. On the contrary, in the future we want to use PCA to filter irrelevant attributes, thus reducing the GPPE search space and possibly improving its results.

Automatically defined functions (ADFs) are considered as a good way to improve the performance and optimization capabilities of GP. They are independent subtrees that can be called from the main subtree of a GP individual, just like any other function. The main subtree plays the same part in a GP individual as in the main function of a C program. ADFs evolve separately and work as subroutines that main subtrees can call during execution of the individual. It could be very interesting in the future to study the impact of ADFs in GPPE performance. We think that this could improve the generalization power of our method.

Acknowledgments

The authors are partially supported by the Ministry of Science and Technology and FEDER under contract TIN2006-08818-C04-02 (the OPLINK project).

REFERENCES

1. H. Liu et al., eds. *Feature Extraction, Construction and Selection: A Data Mining Perspective*. Kluwer Academic, Norwell, MA, 1998.
2. J. R. Koza. *Genetic Programming: On the Programming of Computers by Means of Natural Selection*. MIT Press, Cambridge, MA, 1992.
3. J. R. Koza. *Genetic Programming II: Automatic Discovery of Reusable Programs*. MIT Press, Cambridge, MA, 1994.
4. J. R. Koza, M. A. Keane, M. J. Streeter, W. Mydlowec, J. Yu, and G. Lanza. *Genetic Programming IV: Routine Human–Competitive Machine Intelligence*. Kluwer Academic, Norwell, MA, 2003.
5. F. E. B. Otero, M. M. S. Silva, A. A. Freitas, and J. C. Nievola. Genetic programming for attribute construction in data mining. In C. Ryan et al., eds., *Genetic Programming (Proceedings of EuroGP'03)*, vol. 2610 of Lecture Notes in Computer Science. Springer-Verlag, New York, 2003, pp. 389–398.
6. K. Krawiec. Genetic programming-based construction of features for machine learning and knowledge discovery tasks. *Genetic Programming and Evolvable Machines*, 3(4):329–343, 2002.
7. L. S. Shafti and E. Pérez. Constructive induction and genetic algorithms for learning concepts with complex interaction. In H.-G. Beyer et al., eds., *Proceedings of the Genetic and Evolutionary Computation Conference (GECCO'05)*, Washington, DC. ACM Press, New York, 2005, pp. 1811–1818.
8. I. Kuscu. A genetic constructive induction model. In P. J. Angeline et al., eds., *Proceedings of the 1999 Congress on Evolutionary Computation*, vol. 1. IEEE Press, Piscataway, NJ, 1999, pp. 212–217.
9. Y.-J. Hu. A genetic programming approach to constructive induction. In J. R. Koza et al., eds., *Genetic Programming 1998 (Proceedings of the 3rd Annual Conference)*, Madison, WI. Morgan Kaufmann, San Francisco, CA, 1998, pp. 146–151.
10. T. Howley and M. G. Madden. The genetic kernel support vector machine: description and evaluation. *Artificial Intelligence Review*, 24(3–4):379–395, 2005.
11. D. Eads, D. Hill, S. Davis, S. Perkins, J. Ma, R. Porter, and J. Theiler. Genetic algorithms and support vector machines for time series classification. In *Proceedings of SPIE 4787 Conference on Visualization and Data Analysis*, 2002, pp. 74–85.
12. N. R. Harvey, J. Theiler, S. P. Brumby, S. Perkins, J. J. Szymanski, J. J. Bloch, R. B. Porter, M. Galassi, and A. C. Young. Comparison of GENIE and conventional supervised classifiers for multispectral image feature extraction. *IEEE Transactions on Geoscience and Remote Sensing*, 40(2):393–404, 2002.
13. J. J. Szymanski, S. P. Brumby, P. Pope, D. Eads, D. Esch-Mosher, M. Galassi, N. R. Harvey, H. D. W. McCulloch, S. J. Perkins, R. Porter, J. Theiler, A. C. Young, J. J. Bloch, and N. David. Feature extraction from multiple data sources using genetic programming. In S. S. Shen et al., eds., *Algorithms and Technologies for Multispectral, Hyperspectral, and Ultraspectral Imagery VIII*, vol. 4725 of *SPIE*, 2002, pp. 338–345.
14. A. Vieira, B. Ribeiro, and J. C. Neves. A method to improve generalization of neural networks: application to the problem of bankruptcy prediction. In *Proceedings of the 7th International Conference on Adaptive and Natural Computing Algorithms*

(*ICANNGA'05*), Coimbra, Portugal, vol. 1. *Springer-Verlag Series on Adaptative and Natural Computing Algorithms*. Springer-Verlag, New York, 2005, pp. 417.

15. D. Quintana, C. Luque, and P. Isasi. Evolutionary rule-based system for IPO underpricing prediction. In H.-G. Beyer et al., eds,. *Proceedings of the Genetic and Evolutionary Computation Conference* (*GECCO'05*), New York, vol. 1. ACM SIGEVO (formerly ISGEC). ACM Press, New York, 2005, pp. 983–989.
16. B. Efron and R. J. Tsibirani. *An Introduction to the Bootstrap*. Chapman & Hall, New York, 1994.
17. C. Bishop. *Neural Networks for Pattern Recognition*. Oxford University Press, New York, 1995.
18. C. J. Stone. Additive regression and other nonparametric models. *Annals of Statistics*, 13(2):689–705, 1985.
19. I. H. Witten and E. Frank. *Data Mining: Practical Machine Learning Tools and Techniques*, 2nd ed. Morgan Kaufmann, San Francisco, CA, 2005.
20. M. Buhmann and M. Albowitz. *Radial Basis Functions: Theory and Implementations*. Cambridge University Press, New York, 2003.
21. P. Rousseeuw. Least median squares of regression. *Journal of the American Statistical Association*, 49:871–880, 1984.

CHAPTER 2

Neural Lazy Local Learning

J. M. VALLS, I. M. GALVÁN, and P. ISASI

Universidad Carlos III de Madrid, Spain

2.1 INTRODUCTION

Lazy learning methods [1–3] are conceptually straightforward approaches to approximating real- or discrete-valued target functions. These learning algorithms defer the decision of how to generalize beyond the training data until a new query is encountered. When the query instance is received, a set of similar related patterns is retrieved from the available training patterns set and is used to approximate the new instance. Similar patterns are chosen by means of a distance metric in such a way that nearby points have higher relevance.

Lazy methods generally work by selecting the k nearest input patterns from the query points. Usually, the metric used is the Euclidean distance. Afterward, a local approximation using the samples selected is carried out with the purpose of generalizing the new instance. The most basic form is the *k-nearest neighbor method* [4]. In this case, the approximation of the new sample is just the most common output value among the k examples selected. A refinement of this method, called *weighted k-nearest neighbor* [4], can also be used, which consists of weighting the contribution of each of the k neighbors according to the distance to the new query, giving greater weight to closer neighbors. Another strategy to determine the approximation of the new sample is locally weighted linear regression [2], which constructs a linear approximation of the target function over a region around the new query instance. The regression coefficients are based on the k nearest input patterns to the query.

By contrast, *eager learning methods* construct approximations using the entire training data set, and the generalization is carried out beyond the training data before observing the new instance. Artificial neural networks can be considered as eager learning methods because they construct a global approximation that covers the entire input space and all future query instances. That approximation

Optimization Techniques for Solving Complex Problems, Edited by Enrique Alba, Christian Blum, Pedro Isasi, Coromoto León, and Juan Antonio Gómez

over the training data representing the domain could lead to poor generalization properties, especially if the target function is complex or when data are not evenly distributed in the input space. In these cases, the use of a lazy strategy to train artificial neural networks could be appropriate because the complex target function could be described by a collection of less complex local approximations.

Bottou and Vapnik [5] have introduced a lazy approach in the context of artificial neural networks. The approach is based on the selection, for each query pattern, of the k closest examples from the training set. With these examples, a linear neural network classifier is trained to predict the test patterns. However, the idea of selecting the k nearest patterns might not be the most appropriate, mainly because the network will always be trained with the same number of training examples.

In this work we present a lazy learning approach for artificial neural networks. This lazy learning method recognizes from the entire training data set the patterns most similar to each new query to be predicted. The most similar patterns are determined by using weighting kernel functions, which assign high weights to training patterns close (in terms of Euclidean distance) to the new query instance received. The number of patterns retrieved will depend on the new query point location in the input space and on the kernel function, but not on the k parameter as in classical lazy techniques. In this case, selection is not homogeneous as happened in ref. 5; rather, for each testing pattern, it is determined how many training patterns would be needed and the importance in the learning process of each of them.

The lazy learning strategy proposed is developed for radial basis neural networks (RBNNs) [6,7]. It means that once the most similar or relevant patterns are selected, the approximation for each query sample is constructed by an RBNN. For this reason the method is called lazy RBNN (LRBNN). Other types of artificial neural networks could be used (e.g., the multilayer perceptron). However, RBNNs seem to be more appropriate for the low computational cost of training.

The chapter is organized as follows. Section 2.2 includes a detailed description of the lazy learning method proposed. In this section, two kernel weighting functions are presented: the Gaussian and inverse functions. Issues about the influence of using those kernel functions for making the selection of training patterns given a query instance are presented and analyzed. This section also includes the sequential procedure of the LRBNN method and two important aspects that must be taken into account when an RBNN is trained in a lazy way: random initialization of the RBNN parameters and the possibility of selecting no training patterns when the query point is located in certain regions of the input space. Section 2.3 covers the experimental validation of LRBNN. It includes a brief description of the domains used in the experiments as well as the results obtained using both kernel functions. Also in this section, LRBNN is compared with classical lazy techniques and with eager or classical RBNN. Finally, in Section 2.4 we draw conclusions regarding this work.

2.2 LAZY RADIAL BASIS NEURAL NETWORKS

The learning method presented consists of training RBNNs with a lazy learning approach [8,9]. As mentioned in Section 2.1, this lazy strategy could be applied to other types of artificial neural networks [10], but RBNNs have been chosen for the low computational cost of their training process. The method, called LRBNN (the lazy RBNN method), is based on the selection for each new query instance of an appropriate subset of training patterns from the entire training data set. For each new pattern received, a new subset of training examples is selected and an RBNN is trained to predict the query instance.

Patterns are selected using a *weighting measure* (also called a *kernel function*), which assigns a weight to each training example. The patterns selected are included one or more times in the resulting training subset and the network is trained with the most useful information, discarding those patterns that not only do not provide any knowledge to the network but that might confuse the learning process. Next, either the weighting measures for selecting the training patterns or the complete procedure to train an RBNN in a lazy way are explained. Also, some aspects related to the learning procedure are analyzed and treated.

2.2.1 Weighting Measures for Training Pattern Selection

Let us consider $\mathbf{q}$, an arbitrary query instance, described by an n-dimensional vector and let $X = \{\mathbf{x_k}, \mathbf{y_k}\}_{k=1,\ldots,N}$ be the entire available training data set, where $\mathbf{x_k}$ is the input vector and $\mathbf{y_k}$ is the corresponding output. When a new instance $\mathbf{q}$ must be predicted, a subset of training patterns, named X_q, is selected from the entire training data set X.

In order to select X_q, which contains the patterns most similar to the query instance $\mathbf{q}$, Euclidean distances (d_k) from all the training samples $\mathbf{x_k}$ to $\mathbf{q}$ must be evaluated. Distances may have very different magnitude depending on the problem domains, due to their different data values and number of attributes. It may happen that for some domains the maximum distance between patterns is many times the maximum distance between patterns for other domains. To make the method independent of the distance's magnitude, relative values must be used. Thus, a relative distance, d_{rk}, is calculated for each training pattern:

$$d_{rk} = \frac{d_k}{d_{\max}} \qquad k - 1, \ldots, N \tag{2.1}$$

where $d_{\max}$ is the maximum distance to the query instance; this is $d_{\max} = \max(d_1, d_2, \ldots, d_N)$.

The selection of patterns is carried out by establishing a weight for each training pattern depending on its distance to the query instance $\mathbf{q}$. That weight is calculated using a kernel function $K(\cdot)$, and it is used to indicate how many times a training pattern will be replicated into the subset X_q.

Kernel functions must reach their maximum value when the distance to the query point is null, and it decreases smoothly as this distance increases. Here we present and compare two kernel functions that fulfill the conditions above: the Gaussian and inverse functions, which are described next.

Weighted Selection of Patterns Using the Gaussian Function The Gaussian function assigns to each training pattern $\mathbf{x_k}$ a real value or weight according to

$$K(\mathbf{x_k}) = \frac{1}{\sigma\sqrt{2\pi}} e^{-d_{rk}^2/2\sigma^2} \qquad k = 1, \ldots, N \tag{2.2}$$

where σ is a parameter that indicates the width of the Gaussian function and d_{rk} is the relative Euclidean distance from the query to the training input pattern $\mathbf{x_k}$ given by Equation 2.1.

The weight values $K(\mathbf{x_k})$, calculated in Equation 2.2, are used to indicate how many times the training pattern $(\mathbf{x_k}, \mathbf{y_k})$ will be included in the training subset associated with the new instance $\mathbf{q}$. Hence, those real values must be transformed into natural numbers. The most intuitive way consists of taking the integer part of $K(\mathbf{x_k})$. Thus, each training pattern will have an associated natural number given by

$$n_k = \text{int}(K(\mathbf{x_k})) \tag{2.3}$$

That value indicates how many times the pattern $(\mathbf{x_k}, \mathbf{y_k})$ is included in the subset X_q. If $n_k = 0$, the kth pattern is not selected and is not included in the set X_q.

Weighted Selection of Patterns Using the Inverse Function One problem that arises with the Gaussian function is its strong dependence on the parameter σ. For this reason, another weighting function, the inverse function, can be used:

$$K(\mathbf{x_k}) = \frac{1}{d_{rk}} \qquad k = 1, \ldots, N \tag{2.4}$$

This function does not depend on any parameter, but it is important to have some control over the number of training patterns selected. For this reason, an n-dimensional sphere centered at the test pattern is established in order to select only those patterns located in the sphere. Its radius, r, is a threshold distance, since all the training patterns whose distance to the novel sample is bigger than r will be discarded. To make it domain independent, the sphere radius will be relative with respect to the maximum distance to the test pattern. Thus, the relative radius, r_r, will be used to select the training patterns situated in the sphere centered at the test pattern, r_r being a parameter that must be established before employing the learning algorithm.

Due to the asymptotic nature of the inverse function, small distances could produce very large function values. For this reason, the values $K(\mathbf{x_k})$ are normalized such that their sum equals the number of training patterns in X. These normalized values, denoted as $K_n(\mathbf{x_k})$, are calculated in the following way:

$$K_n(\mathbf{x_k}) = VK(\mathbf{x_k}) \tag{2.5}$$

where

$$V = \frac{N}{\sum_{k=1}^{N} K(\mathbf{x_k})} \tag{2.6}$$

As in the case of the Gaussian function, the function values $K_n(\mathbf{x_k})$ calculated in (2.5) are used to weight the selected patterns that will be used to train the RBNN. The main difference is that now, all the patterns located in the sphere, and only those, will be selected. Thus, both the relative distance d_{rk} calculated previously and the normalized weight value $K_n(\mathbf{x_k})$ are used to decide whether the training pattern $(\mathbf{x}_k, \mathbf{y}_k)$ is selected and, in that case, how many times it will be included in the training subset X_q. Hence, they are used to generate a natural number, n_k, following the next rule:

$$\begin{array}{lll} \text{if} & d_{rk} < r_r & \text{then} \\ & n_k = \text{int}(K_n(\mathbf{x_k})) & \\ \text{else} & & \\ & n_k = 0 & \end{array} \tag{2.7}$$

At this point, each training pattern in X has an associated natural number n_k (see Equation 2.7), which indicates how many times the pattern $(\mathbf{x}_k, \mathbf{y}_k)$ will be used to train the RBNN for the new instance $\mathbf{q}$. If the pattern is selected, $n_k > 0$; otherwise, $n_k = 0$.

Examples of these kernel functions (Gaussian and inverse) are presented in Figure 2.1. The x-axis represents the relative distance from each training example to the query, and the y-axis represents the value of the kernel function. When the Gaussian kernel function is used [see Figure 2.1(a)], the values of the function for patterns close to $\mathbf{q}$ are high and decrease quickly when patterns are moved away. Moreover, as the width parameter decreases, the Gaussian function becomes tighter and higher. Therefore, if the width parameter is small, only patterns situated very close to the new sample $\mathbf{q}$ are selected, and are repeated many times. However, when the width parameter is large, more training patterns are selected but they are replicated fewer times. We can observe that the value of σ affects the number of training patterns selected as well as the number of times they will be replicated.

If the selection is made using the inverse kernel function [see Figure 2.1(b)], the number of patterns selected depends only on the relative radius, r_r. Patterns close to the query instance $\mathbf{q}$ will be selected and repeated many times. As the distance to the query $\mathbf{q}$ increases, the number of times that training patterns are replicated decreases, as long as this distance is lower than r_r [in Figure 2.1(b), r_r has a value of 0.5]. If the distance is greater than r_r, they will not be selected. Figure 2.1(b) shows that the closest patterns will always be selected and replicated

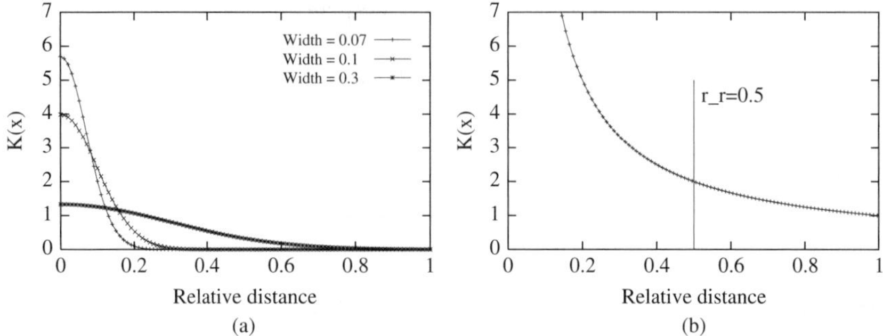

Figure 2.1 (a) Gaussian function with different widths; (b) inverse function and relative radius.

the same number of times, regardless of the radius value. This behavior does not happen with the Gaussian function, as we can see in Figure 2.1(a).

2.2.2 Training an RBNN in a Lazy Way

Here, we present the LRBNN procedure. For each query instance $\mathbf{q}$:

1. The standard Euclidean distances d_k from the pattern $\mathbf{q}$ to each input training pattern are calculated.
2. The relative distances d_{rk}, given by Equation 2.1, are calculated for each training pattern.
3. A weighting or kernel function is chosen and the n_k value for each training pattern is calculated. When the Gaussian function is used, the n_k values are given by Equation 2.3, and for the inverse function they are calculated using Equation 2.7.
4. The training subset X_q is obtained according to value n_k for each training pattern. Given a pattern $(\mathbf{x_k}, \mathbf{y_k})$ from the original training set X, that pattern is included in the new subset if the value n_k is higher than zero. In addition, the pattern $(\mathbf{x_k}, \mathbf{y_k})$ is placed in the training set X_q n_k times in random order.
5. The RBNN is trained using the new subset X_q. The training process of the RBNN implies the determination of the centers and dilations of the hidden neurons and the determination of the weights associated with those hidden neurons to the output neuron. Those parameters are calculated as follows:
   (a) The centers of neurons are calculated in an unsupervised way using the K-means algorithm in order to cluster the input space formed by all the training patterns included in the subset X_q, which contains the replicated patterns selected.
   (b) The neurons dilations or widths are evaluated as the geometric mean of the distances from each neuron center to its two nearest centers.

(c) The weights associated with the connections from the hidden neurons to the output neuron are obtained, in an iterative way, using the gradient descent method to minimize the mean-square error measured in the output of the RBNN over the training subset X_q.

2.2.3 Some Important Aspects of the LRBNN Method

To apply the lazy learning approach, two features must be taken into account. On the one hand, the results would depend on the random initialization of the K-means algorithm, which is used to determine the locations of the RBNN centers. Therefore, running the K-means algorithm with different random initialization for each test sample would have a high computational cost.

On the other hand, a problem arises when the test pattern belongs to regions of the input space with a low data density: It could happen that no training example would be selected. In this case, the method should offer some alternative in order to provide an answer for the test sample. We present solutions to both problems.

K-Means Initialization Given the objective of achieving the best performance, a deterministic initialization is proposed instead of the usual random initializations. The idea is to obtain a prediction of the network with a deterministic initialization of the centers whose accuracy is similar to the one obtained when several random initializations are done. The initial location of the centers will depend on the location of the closest training examples selected. The deterministic initialization is obtained as follows:

- Let $(\mathbf{x_1}, \mathbf{x_2}, \ldots, \mathbf{x_l})$ be the l training patterns selected, ordered by their values $(K(\mathbf{x_1}), K(\mathbf{x_2}), \ldots, K(\mathbf{x_l}))$ when the Gaussian function is used (Equation 2.2) or their normalized values when the selection is made using the inverse function calculated in Equation 2.5:

$$(K_n(\mathbf{x_1}), K_n(\mathbf{x_2}), \ldots, K_n(\mathbf{x_l}))$$

- Let m be the number of hidden neurons of the RBNN to be trained.
- The center of the ith neuron is initialized to the $\mathbf{x_i}$ position for $i = 1, 2, \ldots, m$.

It is necessary to avoid situations where $m > l$. The number of hidden neurons must be fixed to a number smaller than the patterns selected, since the opposite would not make any sense.

Empty Training Set It has been observed that when the input space data are highly dimensional, in certain regions of it the data density can be so small that any pattern is selected. When this situation occurs, an alternative method to select the training patterns must be used. Here we present two different approaches.

1. If the subset X_q associated with a test sample $\mathbf{q}$ is empty, we apply the method of selection to the closest training pattern as if it were the test pattern. In more detail: Let $\mathbf{x_c}$ be the closest training pattern to $\mathbf{q}$. Thus, we will consider $\mathbf{x_c}$ the new test pattern, being named $\mathbf{q}'$. We apply our lazy method to this pattern, generating the $X_{q'}$ training subset. Since $\mathbf{q}' \in X$, $X_{q'}$ will always have at least one element. At this point the network is trained with the set $X_{q'}$ to answer to the test point $\mathbf{q}$.
2. If the subset X_q associated with a test sample $\mathbf{q}$ is empty, the network is trained with X, the set formed by all the training patterns. In other words, the network is trained as usual, with all the patterns available.

As mentioned previously, if no training examples are selected, the method must provide some answer to the test pattern. Perhaps the most intuitive solution consists of using the entire training data set X; this alternative does not have disadvantages since the networks will be trained, in a fully global way, with the entire training set for only a few test samples. However, to maintain the coherence of the idea of training the networks with some selection of patterns, we suggest the first alternative. The experiments carried out show that this alternative behaves better.

2.3 EXPERIMENTAL ANALYSIS

The lazy strategy described above, with either the Gaussian or the inverse kernel function, has been applied to different RBNN architectures to measure the generalization capability of the networks in terms of the mean absolute error over the test data. We have used three domains to compare the results obtained by both kernel functions when tackling different types of problems. Besides, we have applied both eager RBNN and classical lazy methods to compare our lazy RBNN techniques with the classical techniques. The classical lazy methods we have used are k-nearest neighbor, weighted k-nearest neighbor, and weighted local regression.

2.3.1 Domain Descriptions

Three domains have been used to compare the different lazy strategies. Two of them are synthetic problems and the other is a real-world time series domain. The first problem consists of approximating the *Hermite polynomial*. This is a well-known problem widely used in the RBNN literature [11,12]. A uniform random sampling over the interval $[-4, 4]$ is used to obtain 40 input–output points for the training data. The test set consists of 200 input–output points generated in the same way as the points in the training set.

The next domain corresponds to the *Mackey–Glass time series* prediction and is, as the former one, widely regarded as a benchmark for comparing the generalization ability of an RBNN [6,12]. It is a chaotic time series created by the Mackey–Glass delay-difference equation [13]:

$$\frac{dx(t)}{dt} = -bx(t) + a\frac{x(t-\tau)}{1 + x(t-\tau)^{10}} \tag{2.8}$$

The series has been generated using the following values for the parameters: $a = 0.2, b = 0.1$, and $\tau = 17$. The task for the RBNN is to predict the value $x[t+1]$ from the earlier points $(x[t], x[t-6], x[t-12], x[t-18])$. Fixing $x(0) = 0$, 5000 values of the time series are generated using Equation (2.8). To avoid the initialization transients, the initial 3500 samples are discarded. The following 1000 data points have been chosen for the training set. The test set consists of the points corresponding to the time interval [4500, 5000].

Finally, the last problem, a real-world time series, consists of predicting the water level at Venice lagoon [14]. Sometimes, in Venice, unusually high tides are experienced. These unusual situations are called *high water*. They result from a combination of chaotic climatic elements in conjunction with the more normal, periodic tidal systems associated with that particular area. In this work, a training data set of 3000 points, corresponding to the level of water measured each hour, has been extracted from available data in such a way that both stable situations and high-water situations appear represented in the set. The test set has also been extracted from the data available and is formed of 50 samples, including the high-water phenomenon. A nonlinear model using the six previous sampling times seems appropriate because the goal is to predict only the next sampling time.

2.3.2 Experimental Results: Lazy Training of RBNNs

When RBNNs are trained using a lazy strategy based on either the Gaussian or the inverse kernel function to select the most appropriate training patterns, some conditions must be defined. Regarding the Gaussian kernel function, experiments varying the value of the width parameter from 0.05 to 0.3 have been carried out for all the domains. That parameter determines the shape of the Gaussian and therefore, the number of patterns selected to train the RBNN for each testing pattern. Those maximum and minimum values have been chosen such that the shape of the Gaussian allows selection of the same training patterns, although in some cases no training patterns might be selected. For the inverse selective learning method, and for reasons similar to those for the Gaussian, different values of the relative radius have been set, varying from 0.04 to 0.24 for all the domains. In addition, experiments varying the number of hidden neurons have been carried out to study the influence of that parameter in the performance of the method.

As noted in Section 2.2.3, two issues must be considered when employing the lazy learning approach. First, the results would depend on the random initialization of the RBNN centers used by the K-means algorithm. Thus, running the K-means algorithm with different random initializations for each testing pattern would be computationally expensive. To avoid this problem, we propose a deterministic initialization of the K-means algorithm. Second, when the testing pattern belongs to input space regions with a low data density, it might happen that no training example was selected. In this case, as we explained earlier, the method offers two alternatives for providing an answer to the query sample.

To evaluate the deterministic initialization proposed and compare it with the usual random initialization we have developed experiments for all the domains using the lazy learning approach when the neurons centers are randomly initialized. Ten runs have been made and we have calculated the mean values of the corresponding errors. We have also applied the proposed deterministic initialization.

The results show that for all the domains studied, the errors are similar or slightly better. That is, when we carry out the deterministic initialization, the errors are similar to the mean of the errors obtained when 10 different random initializations are made. For instance, in Table 2.1 we show the mean of these errors (10 random initializations) for the Mackey–Glass time series when the inverse selection function is used for different RBNN architectures and relative radius (the standard deviations for these 10 error values are shown in Table 2.2). Table 2.3 shows the results obtained for the Mackey–Glass time series when the deterministic initialization is used. Values lower than the corresponding errors in Table 2.1 are shown in boldface. We can observe that the error values are slightly better than those obtained when the neuron centers were located randomly. Thus, we can use the deterministic initialization instead of several random initializations, reducing the computational cost.

In Tables 2.1 and 2.3 the column NP displays the number of null patterns, that is, test patterns for which the number of training patterns selected is zero. As explained in Section 2.2.3, two alternative ways of treating these anomalous patterns are presented. Method (a) retains the local approach, and method (b)

TABLE 2.1 Mean Errors with Random Initialization of Centers: Mackey–Glass Time Series

	Hidden Neurons						
r_r	7	11	15	19	23	27	NP
0.04	0.02527	0.02641	0.02683	0.02743	0.02691	0.02722	45
0.08	0.02005	0.01891	0.01705	0.01571	0.01716	0.01585	0
0.12	0.02379	0.01954	0.01792	0.01935	0.01896	0.01940	0
0.16	0.02752	0.02223	0.01901	0.02106	0.02228	0.02263	0
0.2	0.03031	0.02427	0.02432	0.02287	0.02281	0.02244	0
0.24	0.03422	0.02668	0.02627	0.02482	0.02635	0.02798	0

TABLE 2.2 Standard Deviations of Errors Corresponding to 10 Random Initializations of Centers: Mackey–Glass Time Series

	Hidden Neurons					
r_r	7	11	15	19	23	27
0.04	0.001272	0.001653	0.002001	0.001148	0.001284	0.001280
0.08	0.001660	0.000503	0.001205	0.000900	0.002154	0.001906
0.12	0.001334	0.001387	0.001583	0.001703	0.000986	0.001324
0.16	0.001495	0.001847	0.002146	0.000844	0.000969	0.001327
0.2	0.001423	0.001160	0.001655	0.001712	0.001336	0.001218
0.24	0.001978	0.001322	0.001213	0.001775	0.001453	0.001992

TABLE 2.3 Errors with Deterministic Initialization of Centers: Mackey–Glass Time Series

	Hidden Neurons						
r_r	7	11	15	19	23	27	NP
0.04	0.02904	0.03086	0.03096	0.03109	0.03231	0.03295	45
0.08	**0.01944**	**0.01860**	**0.01666**	**0.01565**	**0.01551**	0.01585	0
0.12	**0.02131**	**0.01742**	**0.01644**	**0.01607**	**0.01628**	**0.01602**	0
0.16	**0.02424**	**0.02029**	**0.01812**	**0.01729**	**0.01783**	**0.01809**	0
0.2	**0.02837**	**0.02083**	**0.01927**	**0.01874**	**0.02006**	**0.02111**	0
0.24	**0.03082**	**0.02439**	**0.02256**	**0.02199**	**0.02205**	**0.02293**	0

renounces the local approach and follows a global one, assuming that the entire training set must be taken into account.

When null patterns are found, we have tested both approaches. After experiments in all domains, we have concluded that it is preferable to use method (a) [9], not only because it behaves better but also because it is more consistent with the idea of training the networks with a selection of patterns. For instance, for the Mackey–Glass domain with the inverse kernel function, 45 null patterns are found when $r_r = 0.04$ (see Table 2.3). Table 2.4 shows error values obtained for both methods, and we observe that errors are better when method (a) is used.

TABLE 2.4 Errors with Deterministic Initialization of Centers and Null Pattern Processing (r_r = 0.04): Mackey–Glass Time Series

	Hidden Neurons						
	7	11	15	19	23	27	NP
Method (a)	0.02974	0.03043	0.03132	0.03114	0.03309	0.03373	45
Method (b)	0.03385	0.03641	0.03545	0.03464	0.03568	0.03408	45

Next, we summarize the results obtained for all the domains, for both the inverse and the Gaussian selection functions, using the deterministic initialization in all cases and method (a) if null patterns are found. Figures 2.2, 2.3 and 2.4 show for the three application domains the behavior of the lazy strategy when the Gaussian and inverse kernel function are used to select the training patterns. In those cases, the mean error over the test data is evaluated for every value of the width for the Gaussian case and for every value of the relative radius for the inverse case. These figures show that the performance of the lazy learning method proposed to train RBNNs does not depend significantly on the parameters that determine the number of patterns selected when the selection is based on the inverse kernel function. With the inverse function and for all application domains, the general tendency is that there exists a wide interval of relative radius values, r_r, in which the errors are very similar for all architectures. Only when the r_r parameter is fixed to small values is generalization of the method poor in some domains. This is due to the small number of patterns selected, which is not sufficient to construct an approximation. However, as the relative radius increases, the mean error decreases and then does not change significantly.

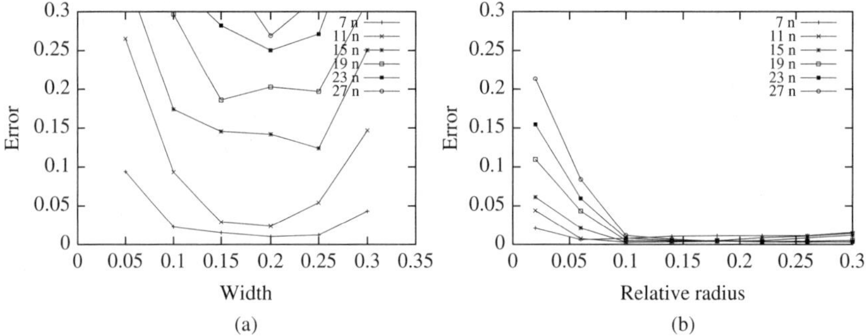

Figure 2.2 Errors with (a) Gaussian and (b) inverse selection for the Hermite polynomial.

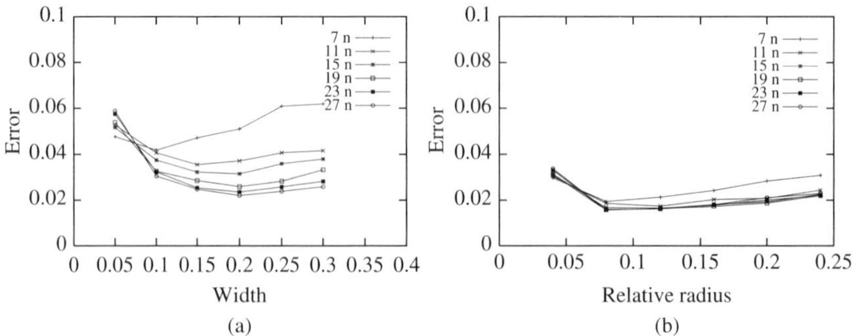

Figure 2.3 Errors with (a) Gaussian and (b) inverse selection for the Mackey–Glass time series.

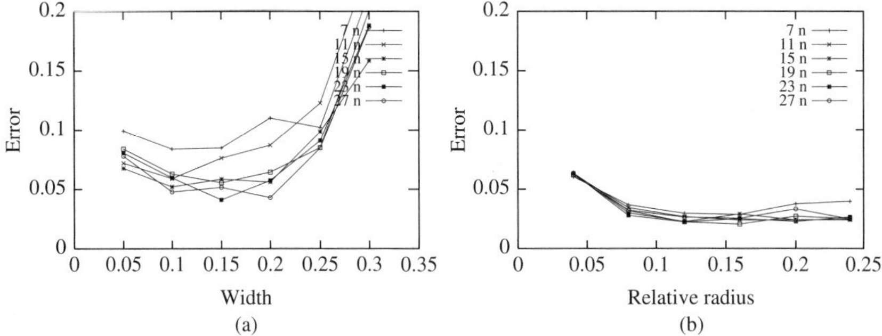

Figure 2.4 Errors with (a) Gaussian and (b) inverse selection for lagoon-level prediction.

Additionally, it is observed that the number of hidden neurons is not a critical parameter in the method.

When Gaussian selection is used, the performance of the method presents some differences. First, although there is also an interval of width values in which the errors are similar, if the width is fixed to high values, the error increases. For those values, the Gaussian is flatter and it could also happen that an insufficient number of patterns are selected. Second, in this case, the architecture of the RBNN is a more critical factor in the behavior of the method.

2.3.3 Comparative Analysis with Other Classical Techniques

To compare the proposed method with classical techniques related to our approach, we have performed two sets of experiments over the domains studied, in one case using eager RBNN and in the other, well-known lazy methods. In the first set we have used different RBNN architectures trained with all the available training patterns to make the predictions on the test sets; in the second set of experiments, we have applied to the domains studied the following classical lazy methods: k-nearest neighbor, weighted k-nearest neighbor, and weighted local regression. In all cases, we have measured the mean absolute error over the test data sets.

Regarding classical RBNN, different architectures, from 10 to 130 neurons, have been trained for all the domains in an eager way, using the entire training data set. The best mean errors over the test set for all the domains obtained by the various methods (lazy RBNN, eager RBNN) are shown in Table 2.5. The table shows that the generalization capability of RBNNs increases when they are trained using a lazy strategy instead of the eager or traditional training generally used in the context of neural networks. The results improve significantly when the selection is based on the input information contained in the test pattern and when this selection is carried out with the inverse function.

With respect to the second set of experiments using the classical lazy methods, k-nearest neighbor, weighted k-nearest neighbor, and weighted local regression

TABLE 2.5 Best Error for the Lazy Strategy to Train RBNNs Compared with Traditional Training

	Mean Error (Parameters)		
	Gaussian Lazy Method	Inverse Lazy Method	Traditional Method
Hermite polynomial	0.01040 $\sigma = 0.2$, 7 neurons	0.002994 $r_r = 0.1$, 11 neurons	0.01904 40 neurons
Mackey–Glass time series	0.02207 $\sigma = 0.2$, 27 neurons	0.01565 $r_r = 0.8$, 19 neurons	0.10273 110 neurons
Level at Venice lagoon	0.04107 $\sigma = 0.15$, 23 neurons	0.02059 $r_r = 0.16$, 19 neurons	0.09605 50 neurons

TABLE 2.6 Best Error for *k*-Nearest Neighbor, Weighted *k*-Nearest Neighbor, and Linear Local Regression Methods

	Mean Error (k Value)		
	k-Nearest Neighbor	Weighted k-Nearest Neighbor	Local Linear Regression
Hermite polynomial	0.01274 $k = 2$	0.00697 $k = 2$	0.02156 $k = 2$
Mackey–Glass time series	0.02419 $k = 2$	0.02404 $k = 6$	0.02579 $k = 2$
Level at Venice lagoon	0.05671 $k = 2$	0.05611 $k = 3$	0.04385 $k = 45$

methods [2] have been run for different values of the k parameter (number of patterns selected). Table 2.6 shows the best mean errors (and the corresponding k-parameter value) obtained by these methods for all the domains.

Tables 2.5 and 2.6 show that the lazy training of RBNNs using the inverse selection function improves the performance of traditional lazy methods significantly. This does not happen when the Gaussian selection function is used: The errors are better than those obtained by traditional lazy methods, but not in all cases. Therefore, the selection function becomes a key issue in the performance of the method.

2.4 CONCLUSIONS

Machine learning methods usually try to learn some implicit function from the set of examples provided by the domain. There are two ways of approaching this problem. On the one hand, eager or global methods build up an approximation using all available training examples. Afterward, this approximation will allow us to forecast any testing query. Artificial neural networks can be considered as eager learning methods because they construct a global approximation that covers the entire input space and all future query instances. However, in many domains with nonhomogeneous behavior where the examples are not evenly distributed in the input space, these global approximations could lead to poor generalization properties.

Lazy methods, on the contrary, build up a local approximation for each testing instance. These local approximations would adapt more easily to the specific characteristics of each input space region. Therefore, for those types of domains, lazy methods would be more appropriate. Usually, lazy methods work by selecting the k nearest input patterns (in terms of Euclidean distance) from the query points. Afterward, they build the local approximation using the samples selected in order to generalize the new query point. Lazy methods show good results, especially in domains with mostly linear behavior in the regions. When regions show more complex behavior, those techniques produce worse results.

We try to complement the good characteristics of each approach by using lazy learning for selecting the training set, but using artificial neural networks, which show good behavior for nonlinear predictions, to build the local approximation. Although any type of neural model can be used, we have chosen RBNNs, due to their good behavior in terms of computational cost.

In this work we present two ways of doing pattern selection by means of a kernel function: using the Gaussian and inverse functions. We have compared our approach with various lazy methods, such as k-nearest neighbor, weighted k-nearest neighbor, and local linear regression. We have also compared our results with eager or classical training using RBNNs.

One conclusion of this work is that the kernel function used for selecting training patterns is relevant for the success of the network since the two functions produce different results. The pattern selection is a crucial step in the success of the method: It is important to decide not only what patterns are going to be used in the training phase, but also how those patterns are going to be used and the importance of each pattern in the learning phase. In this work we see that the Gaussian function does not always produce good results. In the validation domains, the results are rather poor when the selection is made using the Gaussian function, although they are better than the results obtained using traditional methods. However, they are always worse than when the inverse function is used. This function has good results in all domains, also improving the results obtained using classical lazy techniques. Besides, the proposed deterministic initialization of the neuron centers produces results similar to those from the usual random initialization, thus being preferable because only one run is necessary. Moreover, the method is able to predict 100% of the test patterns, even in those extreme cases when no training examples are selected.

Summarizing, the results show that the combination of lazy learning and RBNN improves eager RBNN learning substantially and could reach better results than those of other lazy techniques. Two different aspects must be taken into consideration: first, we need a good method (compatible with lazy approaches) for selecting training patterns for each new query; and second, a nonlinear method must be used for prediction. The regions may have any type of structure, so a general method will be required.

Acknowledgments

This work has been financed by the Spanish-funded MCyT research project OPLINK, under contract TIN2005-08818-C04-02.

REFERENCES

1. D. W. Aha, D. Kibler, and M. Albert. Instance-based learning algorithms. *Machine Learning*, 6:37–66, 1991.
2. C. G. Atkeson, A. W. Moore, and S. Schaal. Locally weighted learning. *Artificial Intelligence Review*, 11:11–73, 1997.
3. D. Wettschereck, D. W. Aha, and T. Mohri. A review and empirical evaluation of feature weighting methods for a class of lazy learning algorithms. *Artificial Intelligence Review*, 11:273–314, 1997.
4. B.V. Dasarathy, ed. *Nearest Neighbor (NN) Norms: NN Pattern Classification Techniques*. IEEE Computer Society Press, Los Alamitos, CA, 1991.
5. L. Bottou and V. Vapnik. Local learning algorithms. *Neural Computation*, 4(6):888–900, 1992.
6. J. E. Moody and C. J. Darken. Fast learning in networks of locally-tuned processing units. *Neural Computation*, 1:281–294, 1989.
7. T. Poggio and F. Girosi. Networks for approximation and learning. *Proceedings of the IEEE*, 78:1481–1497, 1990.
8. J. M. Valls, I. M. Galván, and P. Isasi. Lazy learning in radial basis neural networks: a way of achieving more accurate models. *Neural Processing Letters*, 20(2):105–124, 2004.
9. J. M. Valls, I. M. Galván, and P. Isasi. LRBNN: a lazy RBNN model. *AI Communications*, 20(2):71–86, 2007.
10. I. M. Galván, P. Isasi, R. Aler, and J. M. Valls. A selective learning method to improve the generalization of multilayer feedforward neural networks. *International Journal of Neural Systems*, 11:167–157, 2001.
11. A. Leonardis and H. Bischof. An efficient MDL-based construction of RBF networks. *Neural Networks*, 11:963–973, 1998.
12. L. Yingwei, N. Sundararajan, and P. Saratchandran. A sequential learning scheme for function approximation using minimal radial basis function neural networks. *Neural Computation*, 9:461–478, 1997.
13. M. C. Mackey and L. Glass. Oscillation and chaos in physiological control systems. *Science*, 197:287–289, 1997.
14. J. M. Zaldívar, E. Gutiérrez, I. M. Galván, F. Strozzi, and A. Tomasin. Forecasting high waters at Venice lagoon using chaotic time series analysis and nonlinear neural networks. *Journal of Hydroinformatics*, 2:61–84, 2000.

CHAPTER 3

Optimization Using Genetic Algorithms with Micropopulations

Y. SÁEZ
Universidad Carlos III de Madrid, Spain

3.1 INTRODUCTION

Complex problems can be encountered in real-life domains such as finance and economics, telecommunications, and industrial environments. Solving these problems with evolutionary computation (EC) techniques requires deciding on what the best representations, operators, evaluation function, and parameters are. Furthermore, some complex problems need long computation times to evaluate the solutions proposed by the technique applied or even interaction with an expert. In those cases, it is necessary to decrease the number of evaluations made. One approach commonly used with population-based approaches is to reduce the number of individuals in the population. Micropopulations, which means at most 10 individuals, reduce the genetic diversity dramatically, diminishing the overall performance of the technique applied. However, there are some alternatives to addressing this problem, such as a fitness prediction genetic algorithm (FPGA) or the chromosome appearance probability matrix (CAPM) algorithm.

3.1.1 Solving Complex Problems with Micropopulations

EC encompasses computational models that follow a biological evolution metaphor. The success of these techniques is based on the maintenance of genetic diversity, for which it is necessary to work with large populations. The population size that guarantees an optimal solution in a short time has been a topic of intense research [2,3]. Large populations generally converge toward best solutions, but they require more computational cost and memory requirements. Goldberg and Rundnick [4] developed the first population-sizing equation

Optimization Techniques for Solving Complex Problems, Edited by Enrique Alba, Christian Blum, Pedro Isasi, Coromoto León, and Juan Antonio Gómez

based on the variance of fitness. They further enhanced the equation, which allows accurate statistical decision making among competing building blocks (BBs) [2]. Extending the decision model presented by Goldberg et al. [2], Harik et al. [3] tried to determine an adequate population size which guarantees a solution with the desired quality. To show the real importance of the population size in evolutionary algorithms (EAs), He and Yao [5] showed that the introduction of a nonrandom population decreases the convergence time. However, it is not always possible to deal with such large populations: for example, when the adequacy values must be estimated by a human being [interactive evolutionary computation (IEC)] or when problems involve high computational cost. Population size determines the probability that the algorithm will find the global optimum. Increasing the population size will increase the performance of the search, although it will also increase the number of evaluations required and therefore will involve extra processing cost. For example, let us suppose a genetic algorithm (GA) with an evaluation function that requires 1 hour to calculate how accurate the solution proposed is. If we use a GA size of 100 individuals, more than 4 days are required to evaluate all possible candidate solutions simply to complete one generation. To run 100 generations, more than one year would be needed ($\simeq$416 days). The problem of the computational cost can sometimes be tackled with techniques such as parallel processing, grid computing, or supercomputing machines. However, if the resources are not available, are not sufficient, or the problem time dependencies do not allow splitting the tasks, other techniques are required. In this example, if instead of using 100 individuals, the GA could work with only 10, the workload would be decreased dramatically: only 10 hours per generation (100 generations $\simeq$41 days). But in contrast, when working with small populations, two problems arise: low genetic diversity, which means bad performance, and a tendency to premature convergence (convergence to a nonoptimal solution). Therefore, those environments require methods capable of performing well with a small number of individuals (micropopulations).

The remainder of the chapter is organized as follows. First, a brief survey of the background is given. The algorithms selected for our proposal are then described in Section 3.2. Section 3.3 reports on the problem environment and the results of the comparative tests. Finally, some conclusions are drawn in Section 3.4.

3.1.2 Background

In this section we describe several techniques designed to deal with micropopulations:

1. *Predictive fitness*. To deal with micropopulations, some authors propose predicting fitness beginning with the study of a small subset of user-evaluated individuals. This prediction allows us to work with larger populations ($>$100). The new fitness values are calculated by applying an automatic fitness function to all the individuals before showing a small subset to the user. There are two principal learning methods for predicting fitness: Euclidean distances in the

search space [16,17] and rule systems trained by neural networks [14]. There are other proposals for predicting fitness based on fuzzy logic or hybrid techniques [27,32,33].

2. *Best individual function*. This technique tries to build the perfect individuals during each generation by applying a specific function that takes the best possible features. The aim of this technique is to speed up convergence by making generational jumps with the best individual selections [34].

3. *Initializing the first generation*. Another proposal to speed up the algorithm consists of initializing the first generation with the experts' main preferences, decreasing the convergence time [5,28,35].

4. *Searching masks*. The expert introduces his or her preferences, which are applied like a mask for the mutation and crossover operator (i.e., face search engine) [36].

5. *Estimation of distribution algorithms*. (EDAs). These are a type of GA, but EDAs use probabilistic models to guide exploration of the search space. Even though these techniques are not designed to deal with micropopulations, they have demonstrated very good abilities when working with small sets of individuals [37].

6. *Hybrid techniques*. Probabilistic approaches mixed with EC techniques are a possible combination for improving performance when working with small populations [39,40]. Among others, the CAPM algorithm has achieved very competitive results for IEC problems [1] and for numerical optimizations [31].

3.2 ALGORITHMIC PROPOSAL

In this section we describe four algorithms selected to deal with micropopulations: the classical GA [the simple genetic algorithm (SGA)], the fitness predictive genetic algorithm (FPGA), a probabilistic algorithm called population-based incremental learning (PBIL), and the chromosome appearance probability matrix (CAPM). The SGA is included because it is the best point of reference for the entire comparison. The FPGA is one of the most recent proposals for solving IEC problems [17]. The probabilistic approach offers another interesting point of view to be compared with, because it is the starting point for the estimation of distribution algorithms (EDAs) [18,19]. Finally, the CAPM [1,31] is selected because it achieved very competitive results in the studies carried out so far. In the following the algorithms selected will be explained briefly except for the SGA, which is broadly known.

3.2.1 Fitness Predictive Genetic Algorithm

As stated previously, the FPGA is one of the most recent proposals for solving IEC problems. In IEC environments the user must be present during the evaluation process; therefore, it is necessary to work with micropopulations that are easily evaluated. Nevertheless, a small number or a reduced population affects the productivity of GAs negatively, and as a result, improvements on the conventional GA are proposed. These improvements depend on the size of the population and on the fitness prediction for all the individuals not evaluated by the user [17,30]. Thus, they deal with an algorithm of M individuals, and only a subset of N is shown to the user. Then the user makes a personal evaluation. Finally, the fitness of the remainder $(M - N)$ is obtained according to the differences encountered with the selection made by the user in the previous iteration. There are several types of approaches to estimate the fitness of the rest, but only the simplest has been developed for this chapter: the Euclidean distance approach.

The FPGA follows the following steps (see Algorithm 3.1):

1. *Initiate the population with N possible random solutions*. During this process N individuals are initiated randomly from the population. The value of N depends on the quantity of individuals who want to be shown to the user per iteration and is also linked to the type of problem that is to be solved. The experiments done have been based on $N = 10$.

2. *Evaluate N candidates*. This process is responsible for the evaluation of N feasible candidates, just as the user would be. As a result, and if decoding were necessary, each genotype is converted to its phenotype. Then the N individuals are evaluated with an evaluation function that depends on the problem domain. At this point, all other individuals of the population $(M - N)$ remain to be evaluated. For the experiments made we have used populations of 100 individuals, out of which the user evaluates the best 10, $M = 100$ and $N = 10$.

3. *Forecast fitness values of candidates $M - N$*. This process is used if the condition that guarantees the number of iterations is greater than 2. Alternatively, it is used when an initial reference evaluation has been made which makes possible the forecasting of the fitness values of the rest of the population

Algorithm 3.1 Fitness Predictive Genetic Algorithm

```
t ← 0
Initialize(P_a)
Evaluate(Sub_Population(N))
PredictFitness(Sub_Population(M−N))
while not EndingCondition(t,P_t) do
  Parents ← SelectionParents(P_t)
  Offspring ← Crossover(Parents)
  Mutate(Offspring)
  Evaluate(Sub_Population(N))
  PredictFitness(Sub_Population(M−N))
  P_t ← Replacement(Offspring,P_t)
  t ← t+1
end while
```

$(M - N)$. The necessary modifications for adapting the algorithm to the selection method have been made carefully and correspond to the proposals regarding the algorithm put forward by Hsu and Chen [17]. Predictive fitness is obtained by calculating the Euclidean distance between the referenced chromosomes (those selected by the user and those not evaluated by the user). FPGAs are effective in typical IEC problems in which the parameters are coded such that the Euclidean distance gives an idea of how close or far an individual is from the one selected by the user. As the FPGA is not designed to be applied to numerical optimization problems and is not foreseen as being used with the coding proposed, it is possible that due to their genotype differences, two very close variables (as far as the phenotype is concerned) could produce great distances.

4. *Select two parents*. As in the SGA, the selection operator implemented for the experimentation is based on selection of the two best individuals according to their adaptation or fitness value. However, the selection can only be done with the best N individuals of the population. This limitation, mentioned previously, is typical of FPGAs and IEC techniques in which the user evaluates. Therefore, the selection forces the evaluation of a subset of the population (N individuals). During each iteration, the selections made by the user are stored in the ps_1 and ps_2 variables with the intention of making other calculations which will obtain the predictive fitness of the incoming iterations.

5. *Cross pairs with parents*. This process is exactly the same as in the SGA. Once the two best individuals are selected (i.e., those with the higher fitness values), the crossover operator is used to obtain a new generation of individuals. The replacement strategy is elitist, which means that the best always survives and the parents selected go directly to the following generation. The remainder are generated from the equally probable crossing genes of the parents.

6. *Mutate the descendants obtained*. As in the SGA, this process is responsible for mutating, with a certain probability (P_{mutation}), several genes of the individuals generated. The aim is to guarantee the appearance of new characteristics in the following populations and to maintain enough genetic diversity.

7. *Repeat until the final condition*. As in the SGA, the stop condition of the algorithm is imposed by a maximum limit on iterations.

3.2.2 Population-Based Incremental Learning

The basis for introducing learning in GAs was established with the population-based incremental learning (PBIL) algorithm proposed by Baluja et al. [25]. This algorithm is an alternative approach to the GAs that obtains excellent results in certain problems, such as those described in refs. 20–22. In addition, it has been the starting point for EDAs [18,19] and a good example for the CAPM. The PBIL algorithm [25] is based on the equilibrium genetic algorithm (EGA) [23] but with some improvements. The authors present it as a mix between an EC algorithm and a hillclimbing approach [20]. The main concept of the PBIL algorithm is the

Algorithm 3.2 Population-Based Incremental Learning

```
t ← 0
InitProbVector(P_a,0.5)
while not EndingCondition(t, P_t) do
  while Number_of_Samples do
    Sample_i ← GenerateSample()
    Evaluate(Sample_i)
  end while
  max ← FindSampleMaxEvaluation()
  while LengthProbVector(P_t) do
    P_t ← P_t * (1.0-α)+max * (α)
    mutate(P_t)
  end while
  t ← t+1
end while
```

substitution of the genetic population by a set of statistics representing information about the individuals. Therefore, the selection and crossover operators are no longer needed; instead, a probability distribution is responsible for changing the populations at each iteration. The success of this statistic approach opened a wide research topic with numerous works and applications [18,19,26,29 (among others)].

The PBIL algorithm comprises the following steps (see Algorithm 3.2):

1. *Initialize the probability matrix*. This procedure is responsible for randomly initializing all the individuals, a method called *solution vectors*. For this task the probability matrix is first initialized with uniform distribution, and all the alphabet elements will have the same likelihood of appearing:

$$P_{i=[0\cdots(k-1)],j=[0\cdots(L-1)]} = \frac{1.0}{k} \tag{3.1}$$

where k represents the alphabet size and L the chromosome size.

2. *Evaluate and generate N solution vectors*. This procedure deals with evaluation of the solution vectors. These vectors are generated after the first iteration through sampling of the probability matrix. Once the best solutions are selected, they are used to update the information stored in the probability matrix.

3. *Update the probability matrix*. This step uses the following updating rule:

$$P_i(X = j) = [P_i(X = j) \times (1.0 - \alpha)] + (\alpha \times M_{i,j}) \tag{3.2}$$

where $P_i(X = j)$ is the probability of generating any of the j values that belong to the alphabet in the ith position of the chromosome. α is the learning factor and $M_{i,j}$ is the probability matrix, column i, row j. The learning factor of the algorithm (α) can differ depending on the problem. Furthermore, it affects the final results [25]. Smaller learning rates imply wider searches, and higher values mean deeper search processes. However, its value should not be too high, because as the learning factor increases, the dependency between the solution and the initial population is higher.

4. *Mutate the probability matrix*. The mutation operator plays an important role during the search process to guarantee convergence, avoiding local optimums and maintaining the diversity through the iterations. The mutation operator in PBIL algorithms can be made at two levels: solution vector or probability matrix. Both of them are useful for maintaining genetic diversity, but after several experiments made in different studies, the probability matrix mutation appears to be slightly better [24,25].

5. *Repeat until the final condition*. As in the SGA and the FPGA, the stop condition is imposed by a maximum limit of iterations that depend on the problem to be solved.

3.2.3 Chromosome Appearance Probability Matrix

In PBIL algorithms, the recombination operator is replaced by an independent probability vector for each variable, and sampling this vector implies study of the selections made by the algorithm until that moment. This concept can be applied to IEC to speed up the evolution in regard to user needs. This was the key motivation for developing this new method based on the chromosome appearance probability matrix. The steps in this algorithm are explained in detail by Sáez et al. [1,31] (see Algorithm 3.3).

This method introduces the following new features, which differ from these of a canonical GA.

1. *Probability matrix for guiding mutation*. When the user selects an element of the population, his or her selection is usually based on the collective combination of features in each element of an individual. For example, if the user is searching for tables, he or she will appreciate the combination of several characteristics, such as the color of the legs of a table, the number, and the shape of the surface. Therefore, the information about the whole chromosome should be kept. To do this, a multidimensional array, with the same number of dimensions as genes, has been included. The bounds of the dimensions are determined by the number of alleles of the various genes. The idea is similar to the one presented by Zhang et al. [38] but employs the user selections instead of the similarities between parent and offspring. The probability matrix M is initialized by $M(\text{gene}_1, \text{gene}_2, \text{gene}_3, \text{gene}_m) = 1/T$, where m is the number of genes and gene_i could have values in $[\text{allele}^i_1, \text{allele}^i_2, \ldots, \text{allele}^i_{n_i}]$, and n_i is the number of alleles of gene i. The total possible combinations of chromosomes T is calculated

Algorithm 3.3 Chromosome Appearance Probability Matrix

```
t ← 0
Initialize(Pa)
InitProbMatrix(Mp,0.5)
Evaluate(Pa)
while not EndingCondition(t,Pt)do
  Parents ← SelectionParents(Pt)
  UpdateProbMatrix(Mt,α)
  Offspring ← Crossover(Parents)
  while not Clones(Pt)do
    OrientedMutation(Offspring,Mt)
    CloneRemover(Offspring)
  end while
  Evaluate(Offspring)
  t ← t+1
endwhile
```

by multiplying the maximum sizes of each gene ($T = \Pi_{i=1}^{m} n_i$). This array shows the probability of being chosen of each possible combination of alleles. Each iteration implies the selection of one or two individuals, and its chromosomes represent a position in the array above. After the selection, the corresponding position in the array is updated by a factor of α, with the increment factor of the update rule, Δ_M, calculated as

$$\Delta_M = [M_{\text{gene}_s^1,\ldots,\text{gene}_s^n} \times (1.0 + \alpha)] - M_{\text{gene}_s^1,\ldots,\text{gene}_s^n} \tag{3.3}$$

which can be simplified to

$$\Delta_M = M_{\text{gene}_s^1,\ldots,\text{gene}_s^n} \alpha \tag{3.4}$$

The example in Figure 3.1 shows how the update rule works for one chromosome with two different genes: gen_1 with four alleles {pos1, ..., pos4}, and gen$_{-2}$ with 10, {0, ..., 9}. It is clear how the probability matrix M is updated with $\alpha = 0.005$ and how it affects the remainder cells.

The update operations take care that the sum of all the elements of the array will be 1. This array is very useful for keeping information about the selection frequency of a determined chromosome and therefore for helping the mutation process to evolve toward the preferences of the user.

2. *Guided mutation operator*. The mutation operator is responsible for the mutation of the individuals. Once a gene has been selected for mutation, a specific chromosome is taken as the base of the mutation process (reference chromosome). This chromosome is selected from all possible chromosomes following

Gen_1 = [1..4]

Gen_2 = [0..9]

	Pos1	Pos2	Pos3	Pos4
0	0.045	0.025	0.025	0.02
1	0.015	0.025	0.025	0.015
2	0.025	0.01	0.025	0.025
3	0.025	0.04	0.025	0.04
4	0.025	0.025	0.025	0.025
5	0.025	0.025	0.025	0.025
6	0.015	0.025	0.02	0.025
7	0.025	0.025	0.02	0.025
8	0.025	0.025	0.035	0.025
9	0.025	0.025	0.025	0.025

1. Δm = [M[2,3]↓ (1.0 + α)] − M[2,3]
   Δm = [0.04↓ (1 + 0.005)] − M[2,3]
   Δm = ⌊0.0402 − 0.04 = 0.0002
2. [M[2,3] = M[2,3] + Δm = 0.0402
3. [M[1,0] = M[1,0] − (Δm / 40) = 0.044995
   M[1,1] = M[1,1] − (Δm / 40) = 0.014995
   (...)
   M[4,9] = M[2,9] − (Δm / 40) = 0.024995

Figure 3.1 Update rule example for the CAPM algorithm, $\alpha = 0.005$.

a uniform distribution fixed by the probability array. The higher the value of a chromosome's likelihood array, the better the probability of being chosen. In the mutation process, a position in the chromosome to be mutated is randomly selected. Then the gene in this position is substituted by a gene from the reference chromosome in that same position. Thus, the mutation operator is the result of a function from the chromosome to be mutated and the reference chromosome:

$$\text{newChrom}' = \text{mutateOneGen(OldChrom, referenceChrom)} \tag{3.5}$$

This approach has the ability to broadcast the preferences of the user toward all the chromosomes of the individuals.

3. *The inclusion of a clone remover operator*. The clone remover operator is responsible for mutating all those individuals that have exactly the same genetic structure as the other individuals in the same population.

4. *Replacement of all the population except parents with the new individuals*. The strategy proposed is elitist; however, as the user is not interested in evaluating the same individuals between iterations, the algorithm mutates the parents for the next generation, making them slightly different.

3.3 EXPERIMENTAL ANALYSIS: THE RASTRIGIN FUNCTION

In this section we test all the algorithms outside an IEC framework. The test function used as proposed by Rastrigin in 1974 [42] was limited to two variables. This use was extended years later by Mülhlenbein et al. [41] to allow increasing the number of variables to be optimized:

$$F(x) = 10n \sum_{i=1}^{n} x(i)^2 - 10 \cos 2\pi x(i) \tag{3.6}$$

which represents a compilation of peaks and troughs with numerous local minima (see Figure 3.2).

This characteristic makes the function very difficult to optimize, although it is an ideal benchmark for GAs. The generalized Rastrigin function is nonlinear, that is, multidimensional, multimodal, and quadratic, with a minimum of zero at its origin. In the tests done, the search space is limited to $-4.87 \leq x_i \leq 5.12$, with $i = 1.20$ and $\Delta x_i = 0.01$ on each axis. The search space of the original algorithm has been modified and discretized to adapt it to the encoding proposal. This means that the search becomes more complicated since it is done in an asymmetrical space of variables. Continuous solution would made necessary other techniques, such as evolutionary strategies or real codification for GAs. The proposed coding is based on a vector of three integers that can represent a total of 1000 different possible solutions; the total search space $n = 10$ is formed by $1000^{10} = 10^{30}$ possible solutions. The representation chosen for all the chromosomes is the same for all the contrasted algorithms (see Figure 3.3). Each individual

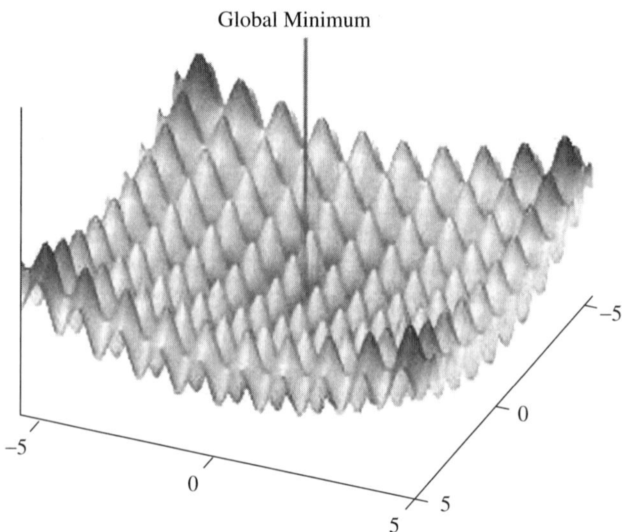

Figure 3.2 Rastrigin plot for two variables ($N = 2$).

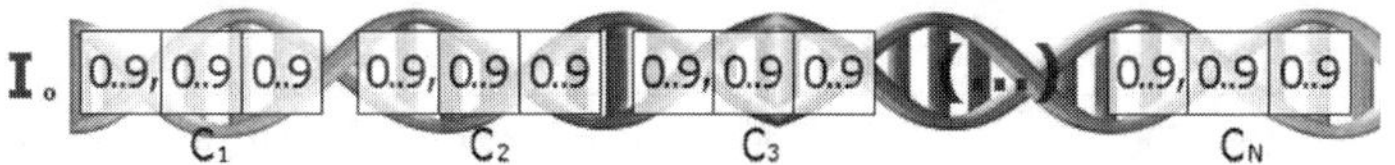

Figure 3.3 Codification for all the algorithms tested.

TABLE 3.1 Parameters Used for the Experiments

	SGA_100	SGA_10	FPGA_100	PBIL_10	CAPM_10
Population size	100	10	100 (10)	10	10
Experiments	100	100	100	100	100
Iterations	250	250	250	250	250
Selection	2	2	2	N/A	2
Crossover	50%	50%	50%	N/A	50%
Mutation (P_m)	5%	5%	5%	3%	20%
Learning rate (α)	—	—	—	0.005	0.001

is made up of various chromosomes, which are, in turn, made up of a vector of integers.

We have tested the algorithms proposed in this chapter for optimizing problems with micropopulations and we have decided to test the GA, also working with 100 individuals (as a well-known point of reference). The experiments have been run with the parameters shown in Table 3.1. All experiments (100 runs per algorithm) were programmed with .net and processed by an AMD Opteron 2.00-GHz server running Microsoft Windows 2003R2x64. The Rastrigin function has been selected as a benchmark. The algorithms proposed have been used to optimize the variables for minimizing the Rastrigin function. The mutation probability and the learning rate have been obtained following a detailed empirical study on each experiment for each algorithm. One hundred independent runs are performed for each algorithm tested.

3.3.1 Results

When comparing the five algorithms, during the first 100 iterations the SGA_100 algorithm converges fastest, followed very closely by the CAPM_10 algorithm. This difference is due mainly to the fact that the size of the population affects negatively the FPGA_100, PBIL_10, and CAPM_10 algorithms, which work with only 10 individuals and perform fewer evaluations per iteration (25,000 vs. 2500).

In fact, Table 3.2 shows that the reduced size of the population also affects the SGA_10 negatively, and as a result, occupies the last place in the ranking. The bad performance of the FPGA_100 is due to the fact that the automatic evaluation function prediction (Euclidean distances) is not good enough for the

TABLE 3.2 Average Fitness and Number of Solutions over 100 Runs and 250 Iterations

Iter.	SGA_100	SGA_10	FPGA_100	PBIL_10	CAPM_10
0	**105.08**(0)	136.65(0)	136.54(0)	153.52(0)	134.12(0)
25	**1.24**(0)	19.59(0)	21.77(0)	27.05(0)	16.30(0)
50	**0.53**(0)	8.89(0)	9.83(0)	11.09(0)	4.64(0)
100	**0.36**(2)	3.95(0)	3.32(0)	3.70(0)	0.43(2)
150	0.36(6)	2.39(0)	2.30(0)	1.15(0)	**0.18(28)**
200	0.18(14)	1.61(0)	1.63(0)	0.23(0)	**0.00(45)**
250	0.18(17)	1.63(0)	1.19(0)	0.14(0)	**0.00(57)**
Evals.	25,000	2500	2500	2500	2500

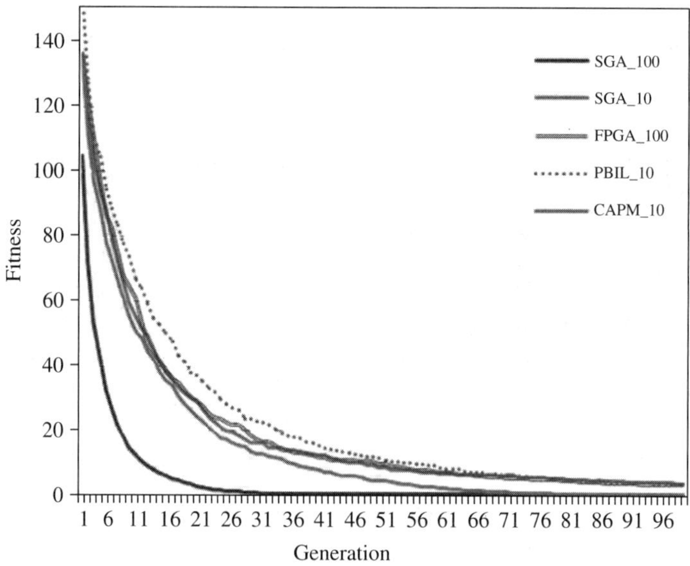

Figure 3.4 Evolution of average fitness (first 100 generations).

Rastrigin function values. However, results are very competitive compared to SGA_10. During the first 100 iterations (Figure 3.4), the SGA_100 is the algorithm that shows the best behavior, followed by the CAPM_10. The remainder converge in a very similar way, with the PBIL_10 being worst during the first 77 generations.

As can be seen in Table 3.3, the standard deviation is lower for the SGA_100. This is due to the population size, which gives more robustness to the results of the SGA_100. However, when working with micropopulations, the diversity of the population has to be maintained because it is a key factor in the

TABLE 3.3 Standard Deviation and Number of Solutions over 100 Runs and 250 Iterations

Iter.	SGA_100	SGA_10	FPGA_100	PBIL_10	CAPM_10
0	17.35(0)	27.08(0)	18.54(0)	26.96(0)	19.67(0)
25	3.70(0)	7.91(0)	8.56(0)	8.01(0)	7.75(0)
50	0.88(0)	4.63(0)	4.46(0)	5.01(0)	3.68(0)
100	0.28(2)	1.74(0)	2.12(0)	1.91(0)	0.66(2)
150	0.20(6)	0.60(0)	1.23(0)	1.12(0)	0.28(28)
200	0.12(14)	0.40(0)	0.71(0)	0.65(0)	0.19(45)
250	0.05(17)	0.26(0)	0.20(0)	0.19(0)	0.15(57)
Evals.	25,000	2500	2500	2500	2500

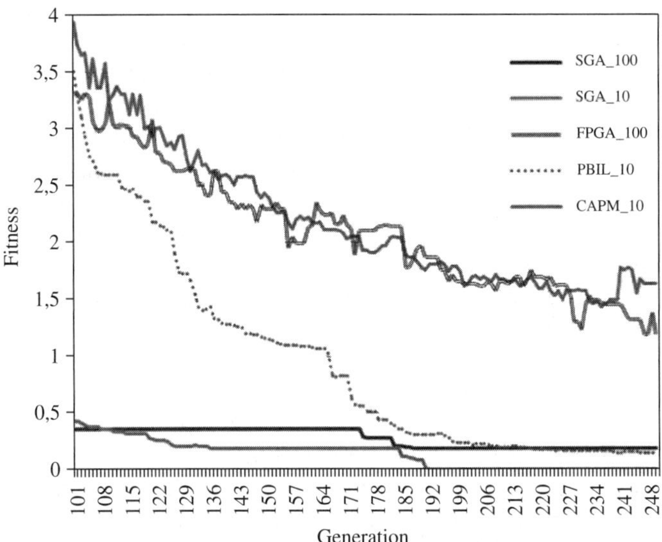

Figure 3.5 Evolution of average fitness (100 to 250 generations).

performance of the algorithm used. All the algorithms that work with 10 individuals in this experiment have higher standard deviations than the SGA_100, but all finish near zero at iteration 250. In the following generations (from 100 to 250; see Figure 3.5), the CAPM_10 surpasses the results obtained by SGA_100, even when working with 100 individuals. It is also important to note how PBIL_10 became the second-best approach, also surpassing the SGA_100 marks (generation 215).

It is worth highlighting the good efficiency of the CAPM_10 and PBIL_10 algorithms compared with their competitors. In addition, the CAPM_10 reaches the optimum solution in 57% of the experiments carried out, compared to the

TABLE 3.4 Comparing Groups: Kruskal–Wallis Test

Groups	n	Rank Sum	Mean Rank
SGA_100	250	92,518.0	370.07
SGA_10	250	210,737.5	842.95
FPGA_100	250	208,338.5	833.35
PBIL_10	250	166,641.0	666.56
CAPM_10	250	103,640.0	414.56
Kruskal–Wallis statistic	387.86		
χ^2 statistic	387.86		
DF	4		
p	<0.0001		

17% success rate seen in experiments done with SGA_100. The remaining methods never reached an optimum solution during the first 250 iterations and 100 experiments. During 100% of the experiments carried out, the CAPM_10 method outperforms all the algorithms tested that work with small populations (SGA_10, FPGA_100, and PBIL_10). In addition, the CAPM_10 is (from the 108th iteration) capable of surpassing the SGA_100 on average (an algorithm that works with 100 individuals and whose efficiency is sufficiently well known). To prove the significance of these results, a statistical test was needed. The Kolmogorov–Smirnov test was carried out to see if the results followed a Gaussian distribution. As the test failed for all approaches, a nonparametric Kruskal–Wallis test was chosen. The results for this test are shown in Table 3.4.

As the p-value is near zero, this casts doubt on the null hypothesis and suggests that at least one sample median is significantly different from the others. However, as can be seen in Table 3.4, SGA_100 and CAPM_10 are the samples that are most similar. Therefore, we decided to repeat the Kruskal–Wallis test again, but this time only to analyze these two techniques. The result of the test was also significant, with a p-value of 0.026. Thus, we can state that those samples are significantly different, with a confidence interval of 97.4%.

3.4 CONCLUSIONS

Frequently, complex problems need high computational cost. Within EC techniques the main problem is that the number of evaluations needed to guarantee an optimal solution is high. The computational cost for evaluation can be a constraint for these techniques. In this chapter we compare various optimization techniques that are capable of dealing with micropopulations. These methods dramatically reduce the number of evaluations needed in traditional EC approaches. However, as the micropopulations affect EC algorithms negatively, our goal is to study which are the most appropriate techniques for optimizing in complex problems with micro-populations. To develop an experimental comparative study, we have explained several techniques. Proceeding with the main goal of solving complex problems as fast as possible, we take the Rastrigin test function as a benchmark, and it is executed with the five algorithms described. The aim is to show the results of the algorithms in terms of the quality of solution and number of iterations. The results show clearly that the use of appropriate heuristics can reduce the number of evaluations and perform better, even working outside the IEC framework.

The CAPM method is the best of the methods tested. It performs even better than the GA with a population of 100 individuals. The CAPM is a hybrid GA with a probabilistic matrix for mutation. Through the mutation operator, it has the ability to propagate and transmit the objective function to the following generations of individuals. This ability is still based on natural genetic evolution, since the most spectacular improvements or changes in primitive humans happened without sexual reproduction, and thus as a result of mutation, and not only crossover. The results suggest application of the CAPM method to more problems to enable study of its behavior. It appears to be an ideal method for solving complex problems that require working with small populations.

Finally, keeping in mind that the test environment belongs to a numerical optimization environment (not typical of IEC, to which CAPM belongs), it should be highlighted that the results are very promising. However, in the future, more complex problems and other algorithmic proposals must also be studied.

Acknowledgments

The authors are supported partially by the Spanish MEC and FEDER under contract TIN2005-08818-C04-02 (the OPLINK project).

REFERENCES

1. Y. Sáez, P. Isasi, J. Segovia, and J. C. Hernández. Reference chromosome to overcome user fatigue in IEC. *New Generation Computing*, 23(2):129–142, 2005.
2. D. E. Goldberg, K. Deb, and J. H. Clark. Genetic algorithms, noise, and the sizing of populations. *Complex Systems*, 6(4):333–362, 1992.
3. G. Harik, E. Cantú-Paz, D. E. Goldberg, and B. L. Miller. The gambler's ruin problem, genetic algorithms, and the sizing of populations. *Transactions on Evolutionary Computation*, 7:231–253, 1999.
4. D. E. Goldberg and M. Rundnick. Genetic algorithms and the variance of fitness. *Complex Systems*, 5(3):265–278, 1991.
5. J. He and X. Yao. From an individual to a population: an analysis of the first hitting time of population-based evolutionary algorithms. *IEEE Transactions on Evolutionary Computation*, 6:495–511, Oct. 2002.
6. K. Sims. Artificial evolution for computer Graphics. *Computer Graphics*, 25(4): 319–328, 1991.
7. J. H. Moore. GAMusic: genetic algorithm to evolve musical melodies. Windows 3.1 Software available at http://www.cs.cmu.edu/, 1994.

8. J. Graf and W. Banzhaf. Interactive evolutionary algorithms in design. In *Proceedings of Artificial Neural Nets and Genetic Algorithms*, Ales, France, 1995, pp. 227–230.
9. F. J. Vico, F. J. Veredas, J. M. Bravo, and J. Almaraz. Automatic design synthesis with artificial intelligence techniques. *Artificial Intelligence in Engineering*, 13:251–256, 1999.
10. A. Santos, J. Dorado, J. Romero, B. Arcay, and J. Rodríguez. Artistic evolutionary computer systems. In *Proceedings of the GECCO Workshop*, Las Vegas, NV, 2000, pp. 149–150.
11. T. Unemi. SBART 2.4: an IEC tool for creating 2D images, movies and collage. In *Proceedings of the GECCO Conference*, Las Vegas, NV, 2000, pp. 153–157.
12. D. Rowland. Evolutionary co-operative design methodology: the genetic sculpture park. In *Proceedings of the GECCO Workshop*, Las Vegas, NV, 2000, pp. 146–148.
13. P. Bentley. From coffee tables to hospitals: generic evolutionary design. In Evolutionary Design by Computers. Morgan-Kauffman, San Francisco, CA, pp. 405–423, 1999.
14. J. Biles, P. G. Anderson, and L. W. Loggi. Neural network fitness functions for a musical IGA. In *Proceedings of the International ICSC Symposia on Intelligent Industrial Automation and Soft Computing*. ICSC Academic Press, Reading, UK, 1996, pp. B39–B44.
15. H. Takagi. Interactive evolutionary computation: fusion of the capabilities of EC optimization and human evaluation. *Proceedings of the IEEE*, 9(89):1275–1296, 2001.
16. J.-Y. Lee and S.-B. Cho. Sparse fitness evaluation for reducing user burden in interactive genetic algorithm. *Proceedings of FUZZ-IEEE'99*, 1999, pp. 998–1003.
17. F.-C. Hsu and J.-S. Chen. A study on a multicriteria decision making model: interactive genetic algorithms approach. *Proceedings of the IEEE International Conference on System, Man, and Cybernetics* (*SMC'99*), 1999, pp. 634–639.
18. P. Larrañaga and J. A. Lozano. Estimation of Distribution Algorithms: *A New Tool for Evolutionary Computation.* Academic, Norwell, MA, Kluwer, 2001.
19. P. Larrañaga J. A. Lozano, and E. Bengoetxea. Estimation of distribution algorithms based on multivariate normal and Gaussian networks. *Technical Report KZZA-IK-1-01.* Department of Computer Science and Artificial Intelligence, University of the Basque Country, Spain, 2001.
20. S. Baluja. An empirical comparison of seven iterative and evolutionary heuristics for static function optimization (extended abstract). In *Proceedings of the 11th International Conference on Systems Engineering* (*CMU-CS-95-193*), Las Vegas, NV, 1996, pp. 692–697.
21. M. Sebag and A. Ducoulombier. Extending population-based incremental learning to continuous search spaces. *Lecture Notes in Computer Science*, 1498:418–426, 1998.
22. N. Monmarché, E. Ramat, G. Dromel, M. Slimane, and G. Venturini. On the similarities between AS, BSC and PBIL: toward the birth of a new metaheuristic. *Internal Report 215*, Université de Tours, 1999.
23. A. Juels, S. Baluja, and A. Sinclair. The Equilibrium Genetic Algorithm and the Role of Crossover. Department of Computer Science, University of California at Berkeley, 1993.

24. A. Juels and M. Wattenberg. Stochastic Hillclimbing as a Baseline Method for Evaluating Genetic Algorithms. Department of Computer Science, University of California at Berkeley, 1995.
25. S. Baluja, D. Pomerleau, and T. Jochem. Towards automated artificial evolution for computer-generated images. *Connection Science*, 1994, pp. 325–354.
26. C. Gonzalez, J. Lozano, and P. Larranarraga. Analyzing the PBIL algorithm by means of discrete dynamical systems. *Complex Systems*, 12(4):465–479, 1997.
27. M. Ohsaki and H. Takagi. Application of interactive evolutionary computation to optimal tuning of digital hearing aids. In *Proceedings of the International Conference on Soft Computing* (*IIZUKA'98*), World Scientific, Fukuoka, Japan, 1998, pp. 849–852.
28. Y. Sáez, O. Sanjuán, J. Segovia, and P. Isasi. Genetic algorithms for the generation of models with micropopulations. Proceedings of EUROGP'03, University of Essex, UK, Apr., vol. 2611, 2003, pp. 547–582.
29. S. Kern, S. D. Muller, N. Hansen, D. Buche, J. Ocenasek, and P. Koumoutsakos. Learning Probability Distributions in Continuous Evolutionary Algorithms: A Comparative Review. Kluwer Academic, Norwell, MA, 2004, pp. 77–112.
30. K. Nishio, M. Murakami, E. Mizutani, and N. Honda. Efficient fuzzy fitness assignment strategies in an interactive genetic algorithm for cartoon face search. In *Proceedings of the 6th International Fuzzy Systems Association World Congress* (*IFSA'95*), 2005, pp. 173–176.
31. Y. Saez, P. Isasi, and J. Segovia. Interactive evolutionary computation algorithms applied to solve Rastrigin test functions. *Proceedings of the 4th IEEE International Workshop on Soft Computing as Transdisciplinary Science and Technology* (*WSTST 05*). Springer-Verlag, New York, 2005, pp. 682–691.
32. F. Sugimoto and M. Yoneyama. Robustness against instability of sensory judgment in a human interface to draw a facial image using a psychometrical space model. In *Proceedings of the IEEE International Conference on Multimedia and Expo*, 2000, pp. 635–638.
33. G. Dozier. Evolving robot behavior via interactive evolutionary computation: from real-world to simulation. In *Proceedings of the ACM Symposium on Applied Computing*, Las Vegas, NV, 2001, pp. 340–344.
34. T. Ingu and H. Takagi. Accelerating a GA convergence by fitting a single-peak function. In *Proceedings of the IEEE International Conference on Fuzzy Systems* (*FUZZ-IEEE'99*), Seoul, Korea, 1999, pp. 1415–1420.
35. Y. Sáez, O. Sanjuán, and J. Segovia. Algoritmos genéticos para la generacion de modelos con micropoblaciones. *Procedimiento Algoritmos Evolutivos y Bioinspirados* (*AEB'02*), 2002, pp. 547–582.
36. C. Caldwell and V. S. Johnston. Tracking a criminal suspect through face space with a genetic algorithm. In *Proceedings of ICGA-4*, 1991, pp. 416–421.
37. T. Gosling, N. Jin, and E. Tsang. Population based incremental learning with guided mutation versus genetic algorithms: iterated prisoners dilemma. In *Proceedings of the IEEE Congress on Evolutionary Computation*, 1(2–5):958–965, 2005.
38. Q. Zhang, J. Sun, E. Tsang, and J. Ford. Combination of guided local search and estimation of distribution algorithm for solving quadratic assignment problem. In *Proceedings of the Genetic and Evolutionary Computation Conference*, 2003, pp. 42–48.

39. Q. Zhang, J. Sun, and E. Tsang. An evolutionary algorithm with guided mutation for the maximum clique problem. *IEEE Transactions on Evolutionary Computation*, 2005, pp. 192–200.

40. Q. Zhang, J. Sun, and E. Tsang. Combinations of estimation of distribution algorithms and other techniques. *International Journal of Automation and Computing*, 2007, pp. 273–280.

41. H. Mühlenbein, D. Schomisch, and J. Born. The parallel genetic algorithm as function optimizer. *Parallel Computing*, 17(6–7):619–632, 1991.

42. L. A. Rastrigin. Extremal control systems. In *Theoretical Foundations of Engineering Cybernetics Series*, Moscow, Nauka, 1973.

CHAPTER 4

Analyzing Parallel Cellular Genetic Algorithms

G. LUQUE and E. ALBA
Universidad de Málaga, Spain

B. DORRONSORO
Université du Luxembourg, Luxembourg

4.1 INTRODUCTION

Genetic algorithms are population-based metaheuristics that are very suitable for parallelization because their main operations (i.e., crossover, mutation, local search, and function evaluation) can be performed independently on different individuals. As a consequence, the performance of these population-based algorithms is improved especially when run in parallel.

Two parallelizing strategies are especially relevant for population-based algorithms: (1) parallelization of computation, in which the operations commonly applied to each individual are performed in parallel, and (2) parallelization of population, in which the population is split into different parts, each evolving in semi-isolation (individuals can be exchanged between subpopulations). Among the most widely known types of structured GAs, the *distributed* (dGA) (or coarse-grained) and *cellular* (cGA) (or fine-grained) *algorithms* are very popular optimization procedures (see Figure 4.1). The parallelization of the population strategy is especially interesting, since it not only allows speeding up the computation, but also improving the search capabilities of GAs [1,2].

In this chapter we focus on cGAs, which have been demonstrated to be more accurate and efficient than dGAs for a large set of problems [3]. Our main contribution in this chapter is the implementation and comparison of several parallel models for cGAs. This model has typically been parallelized in massively parallel machines, but few studies propose parallel models for clusters of computers

Optimization Techniques for Solving Complex Problems, Edited by Enrique Alba, Christian Blum, Pedro Isasi, Coromoto León, and Juan Antonio Gómez

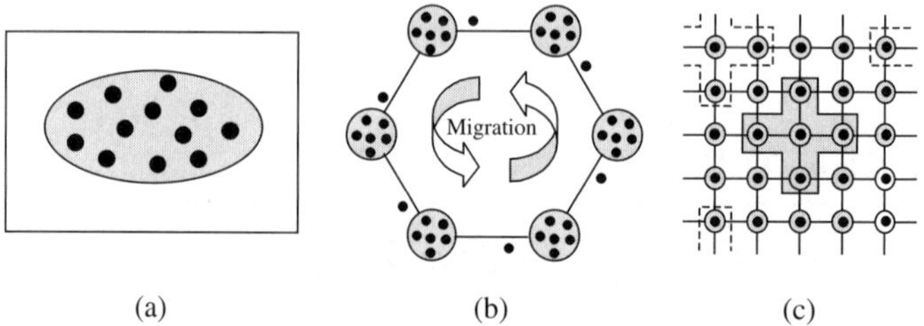

Figure 4.1 (a) Panmictic, (b) distributed, and (c) cellular GAs.

with distributed memory. We implement the main models of parallel cGAs that appear in the literature for clusters of computers, and compare their performance.

The chapter is organized as follows. In Section 4.2 we describe the standard model of a cGA, in which the individuals in the population are structured in a grid. Next, different implementations of parallel cGAs are presented, followed by parallel models for the cGAs used in the literature. In Section 4.5 we test and compare the behavior of several parallel models when solving instances of the well-known p-median problem. Finally, we summarize our most important conclusions.

4.2 CELLULAR GENETIC ALGORITHMS

A canonical cGA follows the pseudocode in Algorithm 4.1. In this basic cGA, the population is usually structured in a regular grid of d dimensions ($d = 1, 2, 3$), and a neighborhood is defined on it. The algorithm iteratively applies the variation operators to each individual in the grid (line 3). An individual may only interact with individuals belonging to its neighborhood (line 4), so its parents are chosen

Algorithm 4.1 Pseudocode for a Canonical cGA

```
 1: proc Steps_Up(cga)  //Algorithm parameters in 'cga'
 2: while not Termination_Condition() do
 3:   for individual ←1 to  cga.popSize do
 4:        n_list←Get_Neighborhood(cga,position(individual));
 5:        parents←Selection(n_list);
 6:        offspring←Recombination(cga.Pc,parents);
 7:        offspring←Mutation(cga.Pm,offspring);
 8:      Evaluate_Fitness(offspring);
 9:      Insert(position(individual),offspring,cga);
10:   end for
11: end while
12: end proc Steps_Up;
```

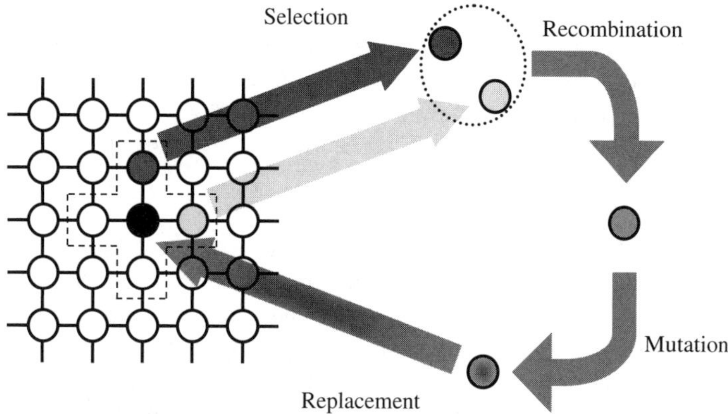

Figure 4.2 In cellular GAs, individuals are allowed to interact with their neighbors only during the breeding loop.

among its neighbors (line 5) with a given criterion. Crossover and mutation operators are applied to the individuals in lines 6 and 7 with probabilities P_c and P_m, respectively. Afterward, the algorithm computes the fitness value of the new offspring individual (or individuals) (line 8), and inserts it (or one of them) into the place of the current individual in the population (line 9) following a given replacement policy. This loop is repeated until a termination condition is met (line 2). The most usual termination conditions are to reach the optimal value (if known), to perform a maximum number of fitness function evaluations, or a combination.

In Figure 4.2 we show the steps performed during the breeding loop in cGAs for every individual, as explained above. There are two possible ways to update the individuals in the population [4]. The one shown in Figure 4.2 is called *asynchronous*, since the newly generated individuals are inserted into the population (following a given replacement policy) and can interact in the breeding loops of its neighbors. However, there is also the possibility of updating the population in a synchronous way, meaning that all the individuals in the population are updated at the same time. For that, an auxiliary population with the newly generated individuals is kept, and after applying the breeding loop to all the individuals, the current population is replaced by the auxiliary population.

4.3 PARALLEL MODELS FOR cGAs

As noted earlier, cGAs were developed initially in massively parallel machines, although other models more appropriate for the currently existing distributed architectures have emerged. In this section we describe briefly the primary conceptual models of the major parallel cGA paradigms that can be used.

Independent Runs Model This extremely simple method of doing simultaneous work can be very useful. In this case, no interaction exists among the independent runs. For example, it can be used to execute several times the same problem with different initial conditions, thus allowing gathering statistics on the problem. Since cGAs are stochastic in nature, the availability of this type of statistics is very important.

Master–Slave Model This method consists of distributing the objective function evaluations among several slave processors, while the main loop of the cGA is executed in a master processor. This parallel paradigm is quite simple to implement and its search space exploration is conceptually identical to that of a cGA executing on a serial processor. The master processor controls the parallelization of the objective function evaluation tasks (and possibly the fitness assignment and/or transformation) performed by the slaves. This model is generally more efficient as the objective evaluation becomes more expensive to compute, since the communication overhead is insignificant with respect to the fitness evaluation time.

Distributed Model In these models the overall cGA population is partitioned into a small number of subpopulations, and each subpopulation is assigned to a processor (island). In this scheme, the designer has two alternatives:

1. Each island executes a complete cGA using only its subpopulations, and individuals occasionally migrate between one particular island and its neighbors, although these islands usually evolve in isolation for the majority of the cGA run. The main idea is that each island can search in very different regions of the entire search space, and the migration helps to avoid the premature convergence of each island. This model is not considered in this chapter since it does not fit the canonical cGA model.

2. The global behavior of the parallel cGA is the same as that of a sequential (cellular) cGA, although the population is divided in separated processors. For that, at the beginning of each iteration, all the processors send the individuals of their first/last column/row to their neighboring islands. After receiving the individuals from the neighbors, a sequential cGA is executed in each subpopulation. The partition of the population can be made using different schemas. The two principal existing schemes are shown in Figure 4.3.

4.4 BRIEF SURVEY OF PARALLEL cGAs

In Table 4.1 the main existing parallel cEAs in the literature are summarized. Some examples of cGAs developed on SIMD machines are those studied by Manderick and Spiessens [5] (improved later in ref. 24), Mühlenbein [25,26], Gorges-Schleuter [27], Collins [28], and Davidor [6], where some individuals are located on a grid, restricting the selection and recombination to small

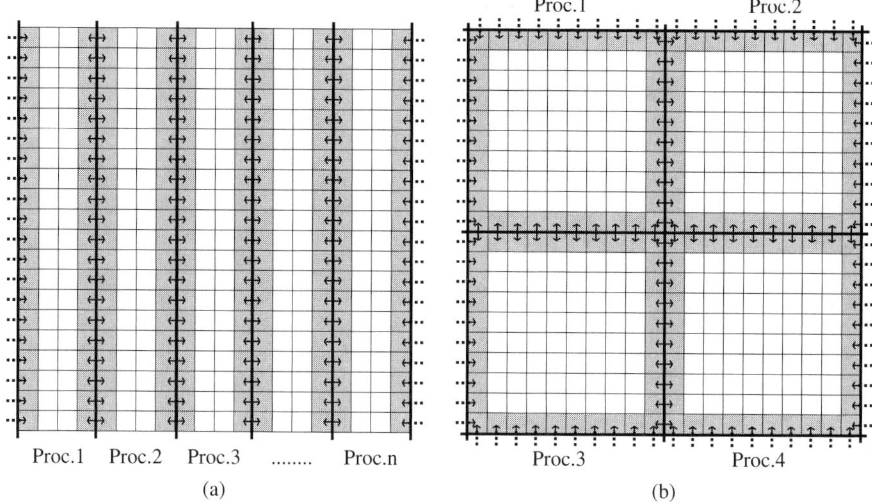

Figure 4.3 (a) CAGE and (b) combined parallel model of a cGA.

neighborhoods in the grid. ASPARAGOS, the model of Mühlenbein and Gorges-Schleuter, was implemented on a *transputers network*, with the population structured in a cyclic stair. Later it evolved, including new structures and matching mechanisms [14], until it was constituted as an effective optimization tool [29].

We also note the work of Talbi and Bessière [30], who studied the use of small neighborhoods, and that of Baluja [8], where three models of cGAs and a GA distributed in islands are analyzed on a MasPar MP-1, obtaining the best behavior of the cellular models, although Gordon and Whitley [31] compared a cGA to a coarse-grained GA, with the latter being slightly better. Kohlmorgen et al. [32] compared some cGAs and presented the equivalent sequential GA, showing clearly the advantages of using the cellular model. Alba and Troya [33] published a more exhaustive comparison than previous ones between a cGA, two panmictic GAs (steady-state and generational GAs), and a GA distributed in an island model in terms of the temporal complexity, the selection pressure, the efficacy, and the efficiency, among others issues. The authors conclude the existence of an important superiority of structured algorithms (cellular and island models) over nonstructured algorithms (the two panmictic GAs).

In 1993, Maruyama et al. [34] proposed a version of a cGA on a system of machines in a local area network. In this algorithm, called DPGA, an individual is located in each processor, and in order to reduce the communication to a minimum, in each generation each processor sends a copy of its individual to another randomly chosen processor. Each processor keeps a list of *suspended individuals* in which the individuals are located when they arrive from other processors. When applying the genetic operators in each processor, this list of suspended individuals behaves as the neighborhood. This model is compared to

TABLE 4.1 Brief Summary of the Main Existing Parallel cEAs

Algorithm	Reference	Model
Manderick & Spiessens	[5](1989)	Parallel cGA on SIMD machines
ECO-GA	[6](1991)	Neighborhood of eight individuals; two offsprings per step
HSDGA	[7](1992)	Fine- and coarse-grained hierarchical GA
fgpGA	[8](1993)	cGA with two individuals per processor
GAME	[9](1993)	Generic library for constructing parallel models
PEGAsuS	[10](1993)	Fine- and coarse-grained for MIMD
LICE	[11](1994)	Cellular model of evolutionary strategy
RPL2	[12](1994)	Fine- and coarse-grained; very flexible
Juille & Pollack	[13](1996)	Cellular model of genetic programming
ASPARAGOS	[14](1997)	Asynchronous; local search applied if no improvement
dcGA	[15](1998)	Cellular or steady-state islands models
Gorges-Schleuter	[16](1999)	Cellular model of evolutionary strategy
CAGE	[17](2001)	Cellular model of genetic programming
MALLBA	[18](2002)	Generic library for constructing parallel models in C++
Combined cGA	[19](2003)	Population composed of some cellular subpopulations
ParadisEO	[20](2004)	Generic library for constructing parallel models in C++
Weiner et al.	[21](2004)	Cellular ES with a variable neighborhood structure
Meta-cGA	[22](2005)	Parallel cGA for local area networks using MALLBA
PEGA	[23](2007)	Island-distributed cGA (for *grid computing*)

APGA, an asynchronous cGA proposed by Maruyama et al. [35], a sequential GA, and a specialized heuristic for the problem being studied. As a conclusion, the authors remark that DPGA shows an efficiency similar to that of the equivalent sequential algorithm.

There also exist more modern parallel cGA models, which work on connected computers in local area networks. These models should be designed to reduce communications to a minimum, as, due to their own characteristics, the cellular models need a high number of communications. In this frame, Nakashima et al. [36] propose a *combined cGA* in which there exist subpopulations with evolving cellular structure, interacting through their borders. A graph of this model can be seen on the right in Figure 4.3. In a later work [19], the authors proposed some parameterizations with different numbers of subpopulations, methods of replacement, and the topology of the subpopulation, and they analyze the results. The authors used this model in a sequential machine, but it is extrapolated directly to a parallel model, where each processor contains one of the subpopulations. This idea of splitting the population into smaller cellular subpopulations running in parallel was also studied by Luque et al. [22], who present the meta-cGA, which was developed using the MALLBA framework, and compared versus other sequential and parallel models in clusters of 2, 4, 8, and 16 computers.

Folino et al. [37] proposed CAGE, a parallel GP. In CAGE the population is structured in a bidimensional toroidal grid, and it is divided into groups of columns that constitute subpopulations (see the graph on the left in Figure 4.3). In this way, the number of messages to send is reduced according to other models, which divide the population in two dimensions (x and y axes). In CAGE, each processor contains a determined number of columns that evolve, and at the end of the generation the two columns at the borders are sent to neighboring processors so that they can use these individuals as neighbors of the individuals located in the limits of its subpopulation.

Dorronsoro et al. [23] proposed a new parallel cGA called PEGA (parallel cellular genetic algorithm), which adopts the two principal parallelization strategies commonly used for GAs: fine- and coarse-grained. The population is divided into several islands (dGA), the population of each island being structured by following the cellular model (cGA). Furthermore, each cGA implements a master–slave approach (parallelization of the computation) to compute the most costly operations applied to their individuals. Periodically, cGAs exchange information (migration) with the goal of inserting some diversity into their populations, thus avoiding their falling into local optima. By using such a structure in the population, the authors keep a good balance between exploration and exploitation, thus improving the capability of the algorithm to solve complex problems [1,38]. PEGA was applied successfully to the largest existing instances of the vehicle routing problem in a grid of about 150 computers.

Finally, there exist some generic programming frameworks of parallel algorithms that make it easy to implement any type of parallel algorithm, including the cellular models considered. These include GAME [9], ParadisEO [20], and MALLBA [18].

4.5 EXPERIMENTAL ANALYSIS

In this section we perform several experimental tests to study the behavior of the various parallel models described in previous sections. To test parallel algorithms we have used the well-known p-median problem [39]. We describe the p-median problem briefly in the next subsection and then discuss the results. First, we compare several well-known parallel models: a parallel cGA using independent runs (ircGA), a master–slave cGA (mscGA), and two different distributed cGAs, one that divides the populations into groups of columns [c-dcGA; see in Figure 4.3(a)], and another that partitions the population in squares [s-dcGA; see Figure 4.3(b)]. Later, we study some modifications of distributed versions.

Specifically, we test an asynchronous distributed variant and change the migration gap. This causes the behavior of the algorithm to change with respect to the canonical serial model.

4.5.1 *p*-Median Problem

The *p-median problem* (PMP) is a basic model from location theory [39]. Let us consider a set $I = \{1, \ldots, n\}$ of potential locations for p facilities and an $n \times n$ matrix $(g_{i,j})$ of transportation costs for satisfying the demands of customers. The p-median problem is to locate the p facilities at locations I to minimize the total transportation cost for satisfying customer demands. Each customer is supplied from the closest open facility.

This problem was shown to be NP-hard. The PMP has numerous applications in operations research, telecommunications, medicine, pattern recognition, and other fields. For the experiments we have used five instances (pmed16–21) provided by Beasley [40]. There are 400 customers and 5, 10, 40, 80, and 133 positions to be chosen, depending on the circumstance.

4.5.2 Analysis of Results

In this section we study the behavior of different parallel implementations of a cGA when solving the p-median problem. We begin with a description of the parameters of each algorithm. No special configuration analysis has been made for determining the optimum parameter values for each algorithm. The entire population comprises 400 individuals. In parallel implementations, each processor has a population of $400/n$, where n is the number of processors (in these experiments, $n = 8$). All the algorithms use the uniform crossover operator (with probability 0.7) and bit-flip mutation operator (with probability 0.2). All the experiments are performed on eight Pentium 4 processors using 2.4-GHz PCs linked by a fast Ethernet communication network. Because of the stochastic nature of genetic algorithms, we perform 30 independent runs of each test to gain sufficient experimental data.

Performance of Parallel cGA Models Next we analyze and compare the behavior of four parallel variants, which are also compared against the canonical sequential cGA model. Table 4.2 summarizes the results of applying various parallel schemas to solve the five PMP instances. In this table we show the mean error with respect to the optimal solution (% error columns) and the mean execution time to find the best solution in seconds (time columns). We do not show the standard deviation because the fluctuations in the accuracy of different runs are rather small, showing that the algorithms are very robust. For the mean speedup time of these algorithms, shown in Figure 4.4, we use the *weak* definition of speedup [1]; that is, we compare the parallel implementation runtime with respect to the serial implementation runtime.

When we interpret the results in Table 4.2, we notice several facts. On the one hand, the ircGA model allows us to reduce the search time and obtain

TABLE 4.2 Overall Results of Parallel cGA Models

	cGA		ircGA		mscGA		c-dcGA		s-dcGA	
Instance	% Error	Time	% Error	Time	% Error	Time	% Error	Time	% Error	Time
pmed16	0.00	4.89	3.54	0.62	0.00	3.37	0.00	0.71	0.00	0.70
pmed17	0.00	7.24	1.95	0.93	0.00	4.81	0.00	1.03	0.00	0.99
pmed18	12.87	26.53	25.81	3.67	12.79	15.60	13.01	3.79	12.93	3.77
pmed19	21.01	97.50	57.37	12.39	21.65	48.91	21.43	13.67	22.14	13.70
pmed20	43.18	156.91	103.01	20.17	44.18	74.01	43.02	22.77	42.79	22.60
Average	15.41	58.61	38.34	7.56	15.72	29.34	15.49	8.39	15.57	8.35

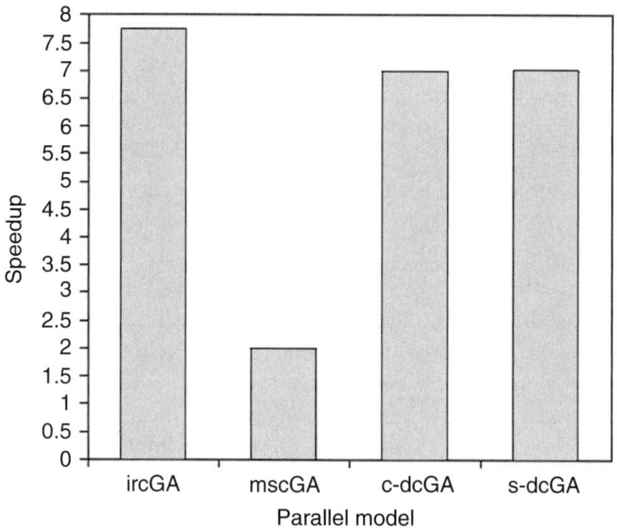

Figure 4.4 Mean weak speedup.

very good speedup (see Figure 4.4), nearly linear, but the results are worse than those of serial algorithms. This is not surprising, since when we use several islands, the population size of each is smaller than when using a single island, and the algorithm is not able to maintain enough diversity to provide a global solution. On the other hand, the behavior of the remainder of the variants is similar to that of the sequential version since they obtain the same numerical results (with statistical significance, p-value > 0.05). This is expected because these parallel models do not change the dynamics of the method with respect to the canonical version. Then we must analyze the execution time. The profit using the master–slave method is very low (see Figure 4.4) because the execution time of the fitness function does not compensate for the overhead cost of the communications.

The distributed versions of the cGA are better than the sequential algorithm in terms of search time. The execution time is quite similar for both versions, and in fact, there is no statistical difference between them. The speedup is quite good, although it is sublinear. These results are quite surprisingly since these algorithms perform a large number of exchanges. This is due to the fact that the communication time is quite insignificant with respect to the computation time.

Performance of Asynchronous Distributed cGA Since the results of the distributed versions of cGA are very promising, we have extended this model using asynchronous communication. In the previous (synchronous) experiments, each island computes the new population, sends the first/last columns/rows to its neighbors, and waits to receive the data from the other islands. In the asynchronous case, the last step is different; the island does not wait, and if pending data exist, they are incorporated into its subpopulation, but otherwise, it computes the next iteration using the latest data received from its neighborhood. This asynchronous model demonstrates behavior that differs from that of the canonical model, since the islands are not working with the latest version of the solutions. In this scheme we must define the *migration gap*, the number of steps in every subpopulation between two successive exchanges (steps of isolated evolution).

We have tested several values of this parameter: 1, 10, 20, 50, and 100, the results of this experiment being summarized in Table 4.3. In this case we used the pmed20 instance, the most complex. Several conclusions can be extracted from Table 4.3. First, we can observe that the asynchronous versions of the dcGA allows us to reduce the error when we use a small–medium-sized migration gap. However, for large values of the migration gap this model becomes less suitable and obtains poorer solutions. This is due to low–medium values of the migration gap that allow each island to search in a different space region, and is beneficial. However, if this parametric value is too large, subpopulations converge quickly to suboptimal solutions such as that which occurs in the ircGA variant. Second, we can notice that the c-dcGA version obtains a better numerical performance than the s-dcGA version, since the rectangular shape of each subpopulation makes it possible to maintain population diversity longer and is a key factor when dealing

TABLE 4.3 Overall Results of Asynchronous Distributed cGA Models

	c-dcGA		s-dcGA	
Migration Gap	% Error	Time	% Error	Time
1	40.63	22.66	41.85	22.71
10	36.86	22.57	38.94	22.74
20	35.12	22.70	42.74	22.61
50	42.04	22.31	44.59	22.53
100	45.28	21.65	47.83	22.06

with small populations. With respect to the execution time, it is slightly reduced when the migration gap increases, but the difference is quite small, because as we said before, the computational cost is significant higher than the communication time.

4.6 CONCLUSIONS

In this chapter we have discussed several parallel models and implementations of cGAs. This method appeared as a new algorithmic model intended to profit from the hardware features of massively parallel computers, but this type of parallel computer has become less and less popular. We then extended the basic concepts of canonical cGAs with the aim of offering the reader parallel models that can be used in modern parallel platforms. An overview of the most important up-to-date cEA systems was provided. The reference list can serve as a directory to provide the reader access to the valuable results that parallel cEAs offer the research community.

Finally, we performed an experimental test with the most common parallel models used in the literature. We use two distributed cellular, master–slave, and independent runs models to solve the well-known p-median problem. We note that the master–slave model is not suitable for this problem since the overhead provoked by the communications is not compensated by the execution time of the objective function. The independent runs model obtains very low execution times, but the solution quality gets worse. Use of distributed models improves the search time. Later, we tested an asynchronous version of distributed models, observing that the isolation computation of each island imposed for this scheme is beneficial for the algorithm when a low or medium-sized value of the migration gap is used, allowing us to reduce the error and the execution time.

Acknowledgments

The authors are partially supported by the Ministry of Science and Technology and FEDER under contract TIN2005-08818-C04-01 (the OPLINK project).

REFERENCES

1. E. Alba and M. Tomassini. Parallelism and evolutionary algorithms. *IEEE Transactions on Evolutionary Computation*, 6(5):443–462, 2002.
2. E. Cantú-Paz. *Efficient and Accurate Parallel Genetic Algorithms*, 2nd ed., vol. 1 of *Series on Genetic Algorithms and Evolutionary Computation*. Kluwer Academic, Norwell, MA, 2000.
3. E. Alba and B. Dorronsoro. *Cellular Genetic Algorithsm*, vol. 42 of *Operations Research/Computer Science Interfaces*. Springer-Verlag, Heidelberg, Germany 2008.

4. E. Alba, B. Dorronsoro, M. Giacobini, and M. Tomassini. Decentralized cellular evolutionary algorithms. In *Handbook of Bioinspired Algorithms and Applications*. CRC Press, Boca Raton, FL, 2006, pp. 103–120.
5. B. Manderick and P. Spiessens. Fine-grained parallel genetic algorithm. In J. D. Schaffer, ed., *Proceedings of the 3rd International Conference on Genetic Algorithms* (*ICGA*). Morgan Kaufmann, San Francisco, CA, 1989, pp. 428–433.
6. Y. Davidor. A naturally occurring niche and species phenomenon: the model and first results. In R. K. Belew and L. B. Booker, eds., *Proceedings of the 4th International Conference on Genetic Algorithms* (*ICGA*), San Diego, CA. Morgan Kaufmann, San Francisco, CA, 1991, pp. 257–263.
7. H. M. Voigt, I. Santibáñez-Koref, and J. Born. Hierarchically structured distributed genetic algorithms. In R. Männer and B. Manderick, eds., *Proceedings of the International Conference on Parallel Problem Solving from Nature II* (*PPSN-II*). North-Holland, Amsterdam, 1992, pp. 155–164.
8. S. Baluja. Structure and performance of fine-grain parallelism in genetic search. In S. Forrest, ed., *Proceedings of the 5th International Conference on Genetic Algorithms* (*ICGA*). Morgan Kaufmann, San Francisco, CA, 1993, pp. 155–162.
9. J. Stender, ed. *Parallel Genetic Algorithms: Theory and Applications*. IOS Press, Amsterdam, 1993.
10. J. L. Ribeiro-Filho, C. Alippi, and P. Treleaven. Genetic algorithm programming environments. In J. Stender, ed., *Parallel Genetic Algorithms: Theory and Applications*. IOS Press, Amsterdam, 1993, pp. 65–83.
11. J. Sprave. Linear neighborhood evolution strategies. In A. V. Sebald and L. J. Fogel, eds., *Proceedings of the Annual Conference on Evolutionary Programming*. World Scientific, Hackensack, NJ, 1994, pp. 42–51.
12. P. D. Surry and N. J. Radcliffe. RPL2: a language and parallel framework for evolutionary computing. In Y. Davidor, H.-P. Schwefel, and R. Männer, eds., *Proceedings of the International Conference on Parallel Problem Solving from Nature III* (*PPSN-III*). Springer-Verlag, Heidelberg, Germany, 1994, pp. 628–637.
13. H. Juille and J. B. Pollack. Massively parallel genetic programming. In *Advances in Genetic Programming*, vol. 2. MIT Press, Cambridge, MA, 1996, pp. 339–358.
14. M. Gorges-Schleuter. Asparagos96 and the traveling salesman problem. In *Proceedings of the IEEE International Conference on Evolutionary Computation* (*CEC*). IEEE Press, New York, 1997, pp. 171–174.
15. C. Cotta, E. Alba, and J. M. Troya. Un estudio de la robustez de los algoritmos genéticos paralelos. *Revista Iberoamericana de Inteligencia Artificial*, 98(5):6–13, 1998.
16. M. Gorges-Schleuter. An analysis of local selection in evolution strategies. In W. Banzhaf et al., eds., *Proceedings of the Genetic and Evolutionary Computation Conference* (*GECCO*), vol. 1. Morgan Kaufmann, San Francisco, CA, 1999, pp. 847–854.
17. G. Folino, C. Pizzuti, and G. Spezzano. Parallel hybrid method for SAT that couples genetic algorithms and local search. *IEEE Transactions on Evolutionary Computation*, 5(4):323–334, 2001.
18. E. Alba and the MALLBA Group. MALLBA: A library of skeletons for combinatorial optimization. In R. F. B. Monien, ed., *Proceedings of the Euro-Par*, Paderborn,

Germany, vol. 2400 of *Lecture Notes in Computer Science*. Springer-Verlag, Heidelberg, Germany, 2002, pp. 927–932.

19. T. Nakashima, T. Ariyama, T. Yoshida, and H. Ishibuchi. Performance evaluation of combined cellular genetic algorithms for function optimization problems. In *Proceedings of the IEEE International Symposium on Computational Intelligence in Robotics and Automation*, Kobe, Japan. IEEE Press, Piscataway, NJ, 2003, pp. 295–299.
20. S. Cahon, N. Melab, and E.-G. Talbi. ParadisEO: a framework for the reusable design of parallel and distributed metaheuristics. *Journal of Heuristics*, 10(3):357–380, 2004.
21. K. Weinert, J. Mehnen, and G. Rudolph. Dynamic neighborhood structures in parallel evolution strategies. *Complex Systems*, 13(3):227–243, 2001.
22. G. Luque, E. Alba, and B. Dorronsoro. Parallel metaheuristics: a new class of algorithms. In *Parallel Genetic Algorithms*. Wiley, Hoboken, NJ, 2005, pp. 107–125.
23. B. Dorronsoro, D. Arias, F. Luna, A. J. Nebro, and E. Alba. A grid-based hybrid cellular genetic algorithm for very large scale instances of the CVRP. In W. W. Smari, ed., *2007 High Performance Computing and Simulation Conference* (*HPCS'07*). IEEE Press, Piscataway, NJ, 2007, pp. 759–765.
24. P. Spiessens and B. Manderick. A massively parallel genetic algorithm: implementation and first analysis. In R. K. Belew and L. B. Booker, eds., *Proceedings of the 4th International Conference on Genetic Algorithms* (*ICGA*). Morgan Kaufmann, San Francisco, CA, 1991, pp. 279–286.
25. H. Mühlenbein. Parallel genetic algorithms, population genetic and combinatorial optimization. In J. D. Schaffer, ed., *Proceedings of the 3rd International Conference on Genetic Algorithms* (*ICGA*). Morgan Kaufmann, San Francisco, CA, 1989, pp. 416–421.
26. H. Mühlenbein, M. Gorges-Schleuter, and O. Krämer. Evolution algorithms in combinatorial optimization. *Parallel Computing*, 7:65–88, 1988.
27. M. Gorges-Schleuter. ASPARAGOS: an asynchronous parallel genetic optimization strategy. In J. D. Schaffer, ed., *Proceedings of the 3rd International Conference on Genetic Algorithms* (*ICGA*). Morgan Kaufmann, San Francisco, CA, 1989, pp. 422–428.
28. R. J. Collins and D. R. Jefferson. Selection in massively parallel genetic algorithms. In R. K. Belew and L. B. Booker, eds., *Proceedings of the 4th International Conference on Genetic Algorithms* (*ICGA*). Morgan Kaufmann, San Francisco, CA, 1991, pp. 249–256.
29. M. Gorges-Schleuter. A comparative study of global and local selection in evolution strategies. In A. E. Eiben et al., eds., *Proceedings of the International Conference on Parallel Problem Solving from Nature V* (*PPSN-V*), vol. 1498 of *Lecture Notes in Computer Science*. Springer-Verlag, Heidelberg, Germany, 1998, pp. 367–377.
30. E.-G. Talbi and P. Bessière. A parallel genetic algorithm for the graph partitioning problem. In E. S. Davidson and F. Hossfield, eds., *Proceedings of the International Conference on Supercomputing*. ACM Press, New York, 1991, pp. 312–320.
31. V. S. Gordon and D. Whitley. Serial and parallel genetic algorithms as function optimizers. In S. Forrest, ed., *Proceedings of the 5th International Conference on Genetic Algorithms* (*ICGA*). Morgan Kaufmann, San Francisco, CA, 1993, pp. 177–183.
32. U. Kohlmorgen, H. Schmeck, and K. Haase. Experiences with fine-grained parallel genetic algorithms. *Annals of Operations Research*, 90:203–219, 1999.

33. E. Alba and J. M. Troya. Improving flexibility and efficiency by adding parallelism to genetic algorithms. *Statistics and Computing*, 12(2):91–114, 2002.
34. T. Maruyama, T. Hirose, and A. Konagaya. A fine-grained parallel genetic algorithm for distributed parallel systems. In S. Forrest, ed., *Proceedings of the 5th International Conference on Genetic Algorithms* (*ICGA*). Morgan Kaufmann, San Francisco, CA, 1993, pp. 184–190.
35. T. Maruyama, A. Konagaya, and K. Konishi. An asynchronous fine-grained parallel genetic algorithm. In R. Männer and B. Manderick, eds., *Proceedings of the International Conference on Parallel Problem Solving from Nature II* (*PPSN-II*). North-Holland, Amsterdam, 1992, pp. 563–572.
36. T. Nakashima, T. Ariyama, and H. Ishibuchi. Combining multiple cellular genetic algorithms for efficient search. In L. Wang et al., eds., *Proceedings of the Asia-Pacific Conference on Simulated Evolution and Learning* (*SEAL*). IEEE Press, Piscataway, NJ, 2002, pp. 712–716.
37. G. Folino, C. Pizzuti, and G. Spezzano. A scalable cellular implementation of parallel genetic programming. *IEEE Transactions on Evolutionary Computation*, 7(1):37–53, 2003.
38. E. Alba and B. Dorronsoro. A hybrid cellular genetic algorithm for the capacitated vehicle routing problem. In *Engineering Evolutionary Intelligent Systems*, vol. 82 of *Studies in Computational Intelligence*. Springer-Verlag, Heidelberg, Germany, 2007, pp. 379–422.
39. C. ReVelle and R. Swain Central facilities location. *Geographical Analysis*, 2:30–42, 1970.
40. J. E. Beasley. A note on solving large-median problems. *European Journal on Operations Reserach*, 21:270–273, 1985.

CHAPTER 5

Evaluating New Advanced Multiobjective Metaheuristics

A. J. NEBRO, J. J. DURILLO, F. LUNA, and E. ALBA

Universidad de Málaga, Spain

5.1 INTRODUCTION

Many sectors of industry (e.g., mechanical, chemistry, telecommunication, environment, transport) are concerned with large and complex problems that must be optimized. Such problems seldom have a single objective; on the contrary, they frequently have several contradictory criteria or objectives that must be satisfied simultaneously. Multiobjective optimization is a discipline focused on the resolution of these types of problems.

As in single-objective optimization, the techniques to solve a *multiobjective optimization problem* (MOP) can be classified into *exact* and *approximate* (also named *heuristic*) *algorithms*. Exact methods such as *branch and bound* [18,22], the *A* algorithm* [20], and *dynamic programming* [2] are effective for problems of small size. When problems become more difficult, usually because of their NP-hard complexity, approximate algorithms are mandatory.

In recent years an approximate optimization technique known as *metaheuristics* has become an active research area [1,9]. Although there is not a commonly accepted definition of metaheuristics [1], they can be considered high-level strategies that guide a set of simpler techniques in the search for an optimum. Among these techniques, evolutionary algorithms for solving MOPs are very popular in multiobjective optimization, giving raise to a wide variety of algorithms, such as NSGA-II [5], SPEA2 [25], PAES [12], and many others [3,4].

Multiobjective optimization seeks to optimize several components of a cost function vector. Contrary to single-objective optimization, the solution of a MOP is not a single solution but a set of solutions known as a *Pareto optimal set*, called a Pareto border or *Pareto front* when plotted in the objective space. Any

Optimization Techniques for Solving Complex Problems, Edited by Enrique Alba, Christian Blum, Pedro Isasi, Coromoto León, and Juan Antonio Gómez

solution of this set is optimal in the sense that no improvement can be made on a component of the objective vector without worsening at least one other of its components. The main goal in the resolution of a multiobjective problem is to obtain a set of solutions within the Pareto optimal set and, consequently, the Pareto front.

In the current scenario of the multiobjective optimization field, many new metaheuristics are proposed continuously, and to ensure their efficacy, they used to be compared against reference techniques: that is, the NSGA-II algorithm and, to a lesser extent, SPEA2. The goal is, given a set of benchmark problems, to show that the new technique improves the results of the reference algorithms according to a number of quality indicators (see, e.g., refs. 11,16, and 17). The point here is that many of these studies frequently are not fully satisfactory, for a number of reasons. First, the algorithms compared are implemented by different authors, which makes a comparison unfair, because not only are the algorithms per se evaluated, but also the ability of the programmers, the performance of the programming language used, and so on. Second, the benchmark problems are not fully representative, although sometimes this is a consequence of space constraints in the length of the papers. Finally, there is poor experiment methodology: It is usual to find studies that carry out too few independent runs of the experiments, and accurate statistical analyses to test the significance of the results are often missing.

In this chapter we analyze the performance of new advanced multiobjective metaheuristics that attempt to avoid the aforementioned issues. Our contributions can be summarized as follows:

- We compare the following techniques: a scatter search, AbYSS [16]; a particle swarm optimization, OMOPSO [17]; two genetic algorithms, NSGA-II [5] and SPEA2 [25]; and a cellular genetic algorithm, MOCell [14].
- We use three benchmarks, which cover a broad range of problems having different features (e.g., concave, convex, disconnected, deceptive). The benchmarks are the well-known ZDT [24], DTLZ [6], and WFG [10] problem families. A total of 21 problems are used.
- To ensure the statistical confidence of the results, we perform 100 independent runs of each experiment, and we use parametric or nonparametric tests according to the normality of the distributions of the data.
- All the algorithms have been coded using jMetal [8], a Java-based framework for developing multiobjective optimization metaheuristics. Thus, the techniques share common internal structures (populations, individuals, etc.), operators (selection, mutation, crossover, etc.), and solution encodings. In this way we ensure a high degree of fairness when comparing the metaheuristics.

The remainder of the chapter is structured as follows. In Section 5.2 we provide some background of basic concepts related to multiobjective optimization. The metaheuristics that we compare are described in Section 5.3. The next section is devoted to detailing the experimentation methodology used to evaluate the algorithms, which includes quality indicators, benchmark problems, and statistical tests. In Section 5.5 we analyze the results obtained. Finally, Section 5.6 provides conclusions and lines of future work.

5.2 BACKGROUND

In this section we provide some background on multiobjective optimization. We first define basic concepts, such as Pareto optimality, Pareto dominance, Pareto optimal set, and Pareto front. In these definitions we are assuming, without loss of generality, the minimization of all the objectives.

A general multiobjective optimization problem (MOP) can be defined formally as follows:

Definition 5.1 To obtain a *multiobjective optimization problem* (MOP), find a vector $\vec{x}^* = \left[x_1^*, x_2^*, \ldots, x_n^*\right]$ that satisfies the m inequality constraints $g_i\,(\vec{x}) \geq 0, i = 1, 2, \ldots, m$, the p equality constraints $h_i\,(\vec{x}) = 0, i = 1, 2, \ldots, p$, and minimizes the vector function $\vec{f}\,(\vec{x}) = \left[f_1(\vec{x}), f_2(\vec{x}), \ldots, f_k(\vec{x})\right]^{\mathrm{T}}$, where $\vec{x} = [x_1, x_2, \ldots, x_n]^{\mathrm{T}}$ is a vector of decision variables.

The set of all the values satisfying the constraints defines the *feasible region* Ω, and any point $\vec{x} \in \Omega$ is a *feasible solution*. As mentioned earlier, we seek the Pareto optimum. Its formal definition is provided next.

Definition 5.2 A point $\vec{x}^* \in \Omega$ is *Pareto optimal* if for every $\vec{x} \in \Omega$ and $I = \{1, 2, \ldots, k\}$, either $\forall_{i \in I} \left[f_i\,(\vec{x}) = f_i(\vec{x}^*)\right]$ or there is at least one $i \in I$ such that $f_i\,(\vec{x}) > f_i\,(\vec{x}^*)$.

This definition states that $\vec{x}^*$ is Pareto optimal if no feasible vector $\vec{x}$ exists that would improve some criterion without causing simultaneous worsening in at least one other criterion. Other important definitions associated with Pareto optimality are the following:

Definition 5.3 A vector $\vec{u} = (u_1, \ldots, u_k)$ is said to *Pareto dominate* $\vec{v} = (v_1, \ldots, v_k)$ (denoted by $\vec{u} \prec \vec{v}$) if and only if $\vec{u}$ is partially less than $\vec{v}$ [i.e., $\forall i \in \{1, \ldots, k\}\,,\ u_i \leq v_i \ \wedge\ \exists i \in \{1, \ldots, k\} : \ u_i < v_i$].

Definition 5.4 For a given MOP $\vec{f}(\vec{x})$, the *Pareto optimal set* is defined as $\mathcal{P}^* = \{\vec{x} \in \Omega | \neg \exists \vec{x'} \in \Omega, \vec{f}(\vec{x'}) \prec \vec{f}(\vec{x})\}$.

Definition 5.5 For a given MOP $\vec{f}(\vec{x})$ and its Pareto optimal set $\mathcal{P}^*$, the *Pareto front* is defined as $\mathcal{PF}^* = \{\vec{f}(\vec{x}), \vec{x} \in \mathcal{P}^*\}$.

Obtaining the Pareto front of a MOP is the main goal of multiobjective optimization. Notwithstanding, when stochastic techniques such as metaheuristics are applied, the goal becomes obtaining a finite set of solutions having two properties: *convergence* to the true Pareto front and homogeneous *diversity*. The first property ensures that we are dealing with optimal solutions, while the second, which refers to obtaining a uniformly spaced set of solutions, indicates that we have carried out adequate exploration of the search space, so we are not losing valuable information.

To clarify these concepts, let us examine the three fronts included in Figure 5.1. Figure 5.1(a) shows a front having a uniform spread of solutions, but the points

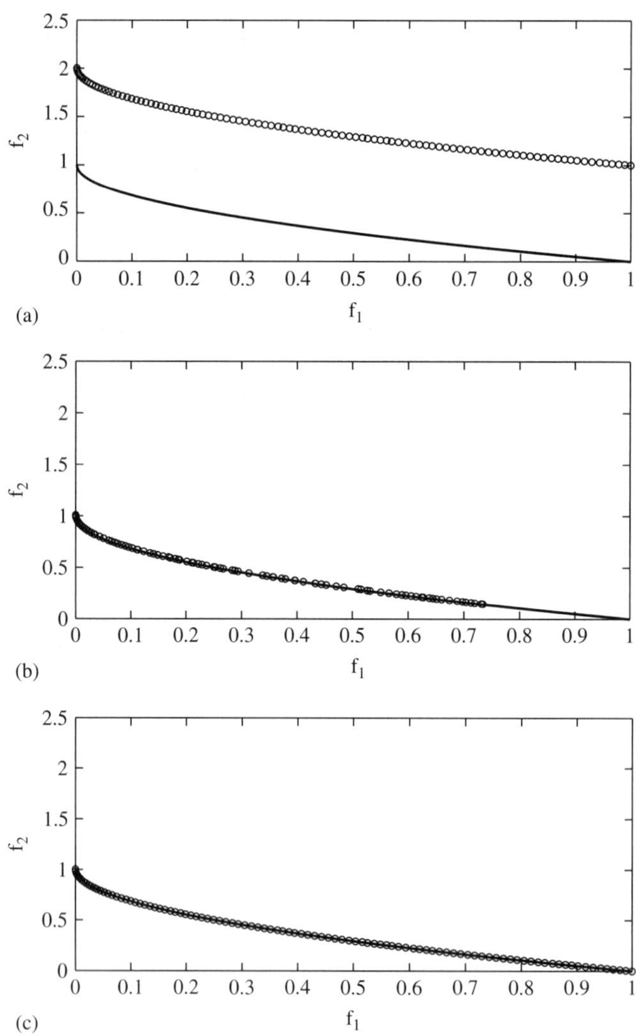

Figure 5.1 Examples of Pareto fronts: (a) poor convergence and good diversity; (b) good convergence and poor diversity; (c) good convergence and diversity.

are far from a true Pareto front; this front is not attractive because it does not provide Pareto optimal solutions. The second example [Figure 5.1(b)] contains a set of solutions that are very close to the true Pareto front, but some regions of the true Pareto front are not covered, so the decision maker could lose important trade-off solutions. Finally, the front depicted in Figure 5.1(c) has the two desirable properties of convergence and diversity.

5.3 DESCRIPTION OF THE METAHEURISTICS

In this section we detail briefly the main features of the techniques compared (for further information regarding the algorithms, see the references cited). We describe first NSGA-II and SPEA2, the two reference algorithms, to present three modern metaheuristics: AbYSS, OMOPSO, and MOCell. We include the settings of the most relevant parameters of the metaheuristics (summarized in Table 5.1).

Nondominated Sorting Genetic Algorithm II The NSGA-II algorithm proposed by Deb et al. [5] is based on obtaining a new population from the original one by applying typical genetic operators (i.e., selection, crossover, and mutation); then the individuals in the two populations are sorted according to their rank, and the best solutions are chosen to create a new population. In the case of having to select some individuals with the same rank, a density estimation based on measuring the crowding distance to the surrounding individuals belonging to the same rank is used to get the most promising solutions.

We have used a real-coded version of NSGA-II and the parameter settings used by Deb et al. [5]. The operators for crossover and mutation are simulated binary crossover (SBX) and polynomial mutation, with distribution indexes of $\mu_c = 20$ and $\mu_m = 20$, respectively. The crossover probability $p_c = 0.9$ and

TABLE 5.1 Parameter Setting Summary

Parameter	NSGA-II	SPEA2	OMOPSO	MOCell	AbYSS
Population size	100	100	100	100	20
Archive size	—	100	100	100	100
Crossover operator	SBX	SBX	—	SBX	SBX
Crossover probability	0.9	0.9	—	0.9	1.0
Mutation operator	Polynomial	Polynomial	Uniform, nonuniform	Polynomial	Polynomial
Mutation probability	$1/n$	$1/n$	$1/n$	$1/n$	$1/n$

mutation probability $p_m = 1/n$ (where n is the number of decision variables) are used. The population size is 100 individuals.

Strength Pareto Evolutionary Algorithm SPEA2 was proposed by Zitzler et al. [25]. In this algorithm, each individual is assigned a fitness value, which is the sum of its strength raw fitness and a density estimation. The algorithm applies the selection, crossover, and mutation operators to fill an archive of individuals; then individuals of both the original population and the archive are copied into a new population. If the number of nondominated individuals is greater than the population size, a truncation operator based on calculating distances to the kth nearest neighbor is used. In this way, individuals having the minimum distance to any other individual are chosen for removal.

We have used the following values for the parameters: both the population and the archive have a size of 100 individuals, and the crossover and mutation operators are the same as those used in NSGA-II, employing the same values for their application probabilities and distribution indexes.

Optimized MOPSO OMOPSO is a particle swarm optimization algorithm for solving MOPs [17]. Its main features include use of the NSGA-II crowding distance to filter out leader solutions, the use of mutation operators to accelerate the convergence of the swarm, and the concept of ϵ-dominance to limit the number of solutions produced by the algorithm. We consider here the leader population obtained after the algorithm has finished its result.

We have used a swarm size of 100 particles and a leader archive of 100 elements. Two mutation operators are used, uniform and nonuniform (for further details as to how the mutation operators are used, see ref. 17). The values of the mutation probability and ϵ are $1/n$ and 0.0075, respectively.

Multiobjective Cellular Genetic Algorithm MOCell is an adaptation of the canonical cellular genetic algorithm model to the multiobjective field [14]. MOCell uses an external archive to store the nondominated solutions found during execution of the algorithm, as many other multiobjective evolutionary algorithms do (e.g., PAES, SPEA2, OMOPSO). In this chapter we use the fourth asynchronous version of MOCell described by Nebro et al. [15].

The parameter settings of MOCell are the following. The size of the population and the archive is 100 (a grid of 10×10 individuals). The mutation and crossover operators are the same as those used in NSGA-II, with the same values as to probabilities and distribution indexes.

Archive-Based Hybrid Scatter Search AbYSS is an adaptation of the scatter search metaheuristic to the multiobjective domain [16]. It uses an external archive to maintain the diversity of the solutions found; the archive is similar to the one employed by PAES [12], but uses the crowding distance of NSGA-II instead of PAES's adaptive grid. The algorithm incorporates operators of the evolutionary algorithm domain, including polynomial mutation and simulated binary crossover in the improvement and solution combination methods, respectively.

We have used the parameter settings of Nebro et al. [16]. The size of the initial population P is equal to 20, and the reference sets RefSet_1 and RefSet_2 have 10 solutions. The mutation and crossover operators are the same as those used in NSGA-II, with the same values for their distribution indexes.

5.4 EXPERIMENTAL METHODOLOGY

This section is devoted to presenting the methodology used in the experiments carried out in this chapter. We first define the quality indicators used to measure the search capabilities of the metaheuristics. After that, we detail the set of MOPs used as a benchmark. Finally, we describe the statistical tests that we employed to ensure the confidence of the results obtained.

5.4.1 Quality Indicators

To assess the search capabilities of algorithms regarding the test problems, two different issues are normally taken into account: The distance from the Pareto front generated by the algorithm proposed to the exact Pareto front should be minimized, and the spread of solutions found should be maximized, to obtain as smooth and uniform a distribution of vectors as possible. To determine these issues it is necessary to know the exact location of the true Pareto front; in this work, we use a set of benchmark problems whose Pareto fronts are known (families ZDT, DTLZ, and WFG; see Section 5.4.2).

The quality indicators can be classified into three categories, depending on whether they evaluate the closeness to the Pareto front, the diversity in the solutions obtained, or both [4]. We have adopted one indicator of each type.

Inverted Generational Distance (IGD) This indicator was introduced by Van Veldhuizen and Lamont [21] to measure how far the elements are in the Pareto optimal set from those in the set of nondominated vectors, defined as

$$\text{IGD} = \frac{\sqrt{\sum_{i=1}^{n} d_i^2}}{n} \tag{5.1}$$

where n is the number of vectors in the Pareto optimal set and d_i is the Euclidean distance (measured in objective space) between each of these solutions and the nearest member of the set of nondominated vectors found. A value of $\text{IGD} = 0$ indicates that all the elements generated are in the Pareto front and they cover all the extension of the Pareto front. To get reliable results, nondominated sets are normalized before calculating this distance measure.

Spread The *spread* metric [5] is a diversity metric that measures the extent of spread achieved among the solutions obtained. This metric is defined as

$$\Delta = \frac{d_f + d_l + \sum_{i=1}^{N-1} \left| d_i - \bar{d} \right|}{d_f + d_l + (N-1)\bar{d}} \tag{5.2}$$

where d_i is the Euclidean distance between consecutive solutions, $\bar{d}$ the mean of these distances, and d_f and d_l the Euclidean distances to the *extreme* (bounding) solutions of the exact Pareto front in the objective space (see ref. 5 for details). This metric takes the value zero for an ideal distribution, pointing out a perfect spread out of the solutions in the Pareto front. We apply this metric after a normalization of the objective function values.

Hypervolume (HV) The HV indicator calculates the volume, in the objective space, covered by members of a nondominated set of solutions Q for problems where all objectives are to be minimized [26]. In the example depicted in Figure 5.2, HV is the region enclosed within the dashed line, where $Q = \{A, B, C\}$ (in the figure, the shaded area represents the objective space that has been explored). Mathematically, for each solution $i \in Q$, a hypercube v_i is constructed with a reference point W and the solution i as the diagonal corners of the hypercube. The reference point can be found simply by constructing a vector of the worst objective function values. Thereafter, a union of all hypercubes is found and its hypervolume is calculated:

$$\text{HV} = \text{volume}\left(\bigcup_{i=1}^{|Q|} v_i\right) \tag{5.3}$$

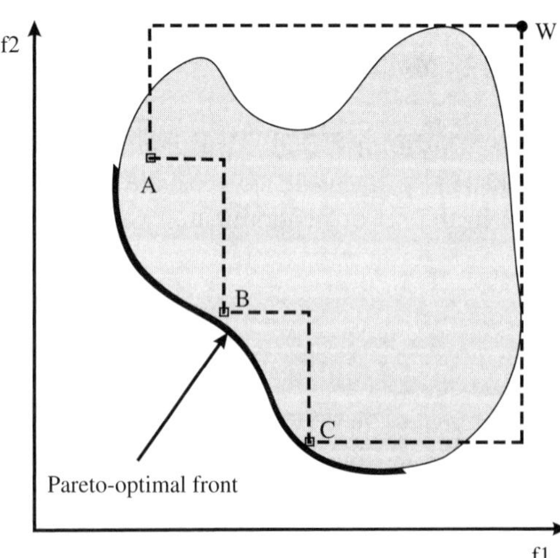

Figure 5.2 Hypervolume enclosed by the nondominated solutions.

Algorithms with higher HV values are desirable. Since this indicator is not free from arbitrary scaling of objectives, we have evaluated the metric using normalized objective function values.

HV is a Pareto-compliant quality indicator [13], while the IDG and the spread are Pareto noncompliant. A Pareto-compliant indicator assures that whenever a Pareto front A is preferable to a Pareto front B with respect to Pareto dominance, the value of the indicator for A should be at least as good as the indicator value for B. Conversely, Pareto-noncompliant indicators are those that do not fulfill this requirement. The use of only Pareto-compliant indicators should be preferable; however, many works in the field employ noncompliant indicators. The justification is that they are useful to assess only an isolated aspect of an approximation set's quality, such as proximity to the optimal front or its spread in the objective space. Thus, they allow us to refine the analysis of set of solutions having identical Pareto-compliant indicator values.

5.4.2 Test Problems

In this section we describe the various sets of problems solved in this work. These problems are well known, and they have been used in many studies in this area. The problem families are the following:

- *Zitzler–Deb–Thiele* (ZDT). This benchmark comprises five bi-objective problems [24]: ZDT1 (convex), ZDT2 (nonconvex), ZDT3 (nonconvex, disconnected), ZDT4 (convex, multimodal), and ZDT6 (nonconvex, nonuniformly spaced). These problems are scalable according to the number of decision variables.
- *Deb–Thiele–Laumanns–Zitzler* (DTLZ). The problems of this family are scalable in both number of variables and objectives [6]. The family consists of the following seven problems: DTLZ1 (linear), DTLZ2 to 4 (nonconvex), DTLZ5 and 6 (degenerate), and DTLZ7 (disconnected).
- *Walking Fish Group* (WFG). This set consists of nine problems, WFG1 to WFG9, constructed using the WFG toolkit [10]. The properties of these problems are detailed in Table 5.2. They are all scalable as to both number of variables and number of objectives.

In this chapter we use the bi-objective formulation of the DTLZ and WFG problem families. A total of 21 MOPs are used to evaluate the five metaheuristics.

5.4.3 Statistical Tests

Since we are dealing with stochastic algorithms and we want to provide the results with confidence, we have made 100 independent runs of each experiment, and the following statistical analysis has been performed throughout this work [7,19]. First, a Kolmogorov–Smirnov test is used to check whether or not the values of the results follow a normal (Gaussian) distribution. If the distribution

TABLE 5.2 Properties of the MOPs Created Using the WFG Toolkit

Problem	Separability	Modality	Bias	Geometry
WFG1	separable	uni	polynomial, flat	convex, mixed
WFG2	non-separable	f_1 uni, f_2 multi	no bias	convex, disconnected
WFG3	non-separable	uni	no bias	linear, degenerate
WFG4	non-separable	multi	no bias	concave
WFG5	separable	deceptive	no bias	concave
WFG6	non-separable	uni	no bias	concave
WFG7	separable	uni	parameter dependent	concave
WFG8	non-separable	uni	parameter dependent	concave
WFG9	non-separable	multi, deceptive	parameter dependent	concave

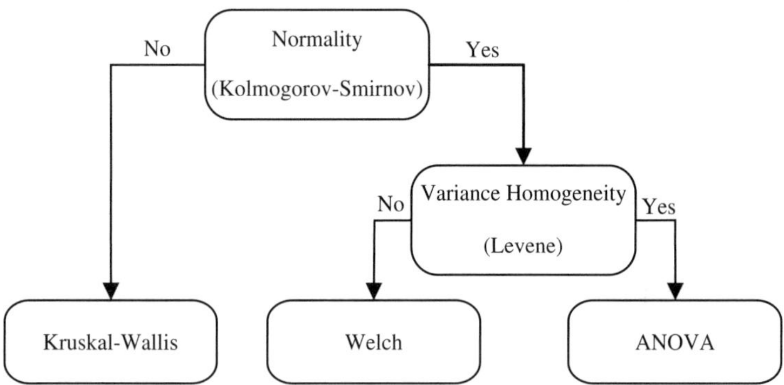

Figure 5.3 Statistical analysis performed in this work.

is normal, the Levene test checks for the homogeneity of the variances (homocedasticity). If the sample has equal variance (positive Levene test), an ANOVA test is carried out; otherwise, a Welch test is performed. For non-Gaussian distributions, the nonparametric Kruskal–Wallis test is used to compare the medians of the algorithms. Figure 5.3 summarizes the statistical analysis.

In this chapter we always use a confidence level of 95% (i.e., significance level of 5% or p-value under 0.05) in the statistical tests, which means that the differences are unlikely to have occurred by chance with a probability of 95%.

5.5 EXPERIMENTAL ANALYSIS

In this section we analyze the five metaheuristics according to the quality indicators measuring convergence, diversity, and both features (see Tables 5.3 to 5.5). Successful statistical tests are marked with + signs in the last column in all the tables in this section; conversely, a − sign indicates that no statistical confidence

TABLE 5.3 Median and Interquartile Range of the IGD Quality Indicator, $\tilde{x}_{\text{IQR}}$

Problem	NSGAII	SPEA2	OMOPSO	MOCell	AbYSS	Stat. Conf.
ZDT1	1.892e−04 1.1e−05	1.515e−04 4.4e−06	**1.360e−04** **1.7e−06**	*1.386e−04* *1.9e−06*	1.413e−04 3.9e−06	+
ZDT2	1.923e−04 1.4e−05	1.536e−04 8.9e−06	*1.411e−04* *2.1e−06*	**1.409e−04** **2.4e−06**	1.445e−04 3.2e−06	+
ZDT3	2.532e−04 1.6e−05	2.365e−04 1.3e−05	*2.018e−04* *7.4e−06*	**1.977e−04** **4.8e−06**	2.053e−04 4.9e−03	+
ZDT4	2.212e−04 6.5e−05	4.948e−04 1.2e−03	*2.243e−01* *1.4e−01*	**1.612e−04** **3.9e−05**	*2.148e−04* *1.0e−04*	+
ZDT6	3.604e−04 7.3e−05	6.432e−04 1.2e−04	*1.190e−04* *8.6e−06*	1.612e−04 2.1e−05	**1.178e−04** **5.4e−06**	+
DTLZ1	4.854e−04 1.5e−04	*4.692e−04* *3.0e−04*	2.554e+00 2.3e+00	**3.649e−04** **1.1e−04**	5.636e−04 7.0e−04	+
DTLZ2	4.426e−04 2.4e−05	3.470e−04 8.0e−06	3.393e−04 5.7e−06	**3.343e−04** **7.4e−06**	*3.351e−04* *6.8e−06*	+
DTLZ3	*7.993e−02* *1.2e−01*	1.572e−01 1.5e−01	7.494e+00 6.5e+00	**5.785e−02** **7.3e−02**	1.453e−01 1.4e−01	+
DTLZ4	1.081e−04 6.6e−03	**8.021e−05** **6.6e−03**	9.077e−05 1.2e−05	8.898e−05 6.6e−03	*8.590e−05* *6.4e−06*	+
DTLZ5	4.416e−04 2.4e−05	3.457e−04 7.3e−06	3.393e−04 7.8e−06	*3.352e−04* *6.5e−06*	**3.349e−04** **7.8e−06**	+
DTLZ6	*2.629e−03* *2.3e−03*	2.154e−02 4.5e−03	**3.327e−04** **6.2e−06**	*3.011e−03* *2.8e−03*	6.960e−03 3.6e−03	+
DTLZ7	2.146e−04 1.5e−05	1.908e−04 8.8e−06	*1.614e−04* *2.8e−06*	**1.609e−04** **4.7e−06**	1.627e−04 3.1e−06	+
WFG1	**9.527e−04** **2.0e−03**	3.487e−03 5.6e−04	3.683e−03 1.3e−03	1.271e−03 2.1e−03	3.849e−03 1.5e−03	+
WFG2	3.492e−04 2.8e−04	3.481e−04 2.9e−04	**4.860e−05** **3.4e−06**	3.480e−04 5.9e−07	*3.479e−04* *4.4e−07*	+
WFG3	6.841e−04 2.2e−07	6.840e−04 2.7e−07	6.840e−04 1.5e−08	6.838e−04 1.7e−07	**6.836e−04** **5.1e−07**	+
WFG4	1.220e−04 1.1e−05	1.138e−04 5.7e−06	2.070e−04 1.5e−05	**8.995e−05** **1.8e−06**	*9.014e−05* *1.9e−06*	+
WFG5	5.493e−04 3.5e−06	5.416e−04 1.9e−06	5.377e−04 7.8e−07	**5.374e−04** **3.4e−07**	*5.375e−04* *4.7e−07*	+
WFG6	1.609e−04 9.7e−05	*1.335e−04* *9.6e−05*	**8.712e−05** **1.2e−06**	2.480e−04 5.2e−04	6.573e−04 6.0e−04	+
WFG7	1.203e−04 1.1e−05	9.537e−05 3.9e−06	**8.265e−05** **1.3e−06**	*8.358e−05* *1.5e−06*	8.385e−05 2.0e−06	+
WFG8	*1.034e−03* *4.8e−05*	1.040e−03 6.5e−05	**9.020e−04** **1.5e−04**	1.036e−03 1.5e−04	1.042e−03 9.4e−06	+
WFG9	1.388e−04 1.3e−05	1.114e−04 1.3e−05	1.189e−04 5.5e−06	**1.017e−04** **1.6e−05**	1.089e−04 2.8e−05	+

TABLE 5.4 Median and Interquartile Range of the Δ Quality Indicator, $\tilde{x}_{\text{IQR}}$

Problem	NSGAII	SPEA2	OMOPSO	MOCell	AbYSS	Stat. Conf.
ZDT1	3.695e−01	1.523e−01	**7.342e−02**	*7.643e−02*	1.048e−01	+
	4.2e−02	2.2e−02	**1.7e−02**	*1.3e−02*	2.0e−02	
ZDT2	3.814e−01	1.549e−01	**7.286e−02**	*7.668e−02*	1.071e−01	+
	4.7e−02	2.7e−02	**1.5e−02**	*1.4e−02*	1.8e−02	
ZDT3	7.472e−01	7.100e−01	*7.082e−01*	**7.041e−01**	7.087e−01	+
	1.8e−02	7.5e−03	*6.4e−03*	**6.2e−03**	9.7e−03	
ZDT4	4.018e−01	2.721e−01	8.849e−01	**1.105e−01**	*1.275e−01*	+
	5.8e−02	1.6e−01	4.6e−02	**2.8e−02**	*3.5e−02*	
ZDT6	3.565e−01	2.279e−01	2.947e−01	*9.335e−02*	**8.990e−02**	+
	3.6e−02	2.5e−02	1.1e+00	*1.3e−02*	**1.4e−02**	
DTLZ1	4.027e−01	1.812e−01	7.738e−01	**1.053e−01**	*1.401e−01*	+
	6.1e−02	9.8e−02	1.3e−01	**3.6e−02**	*1.7e−01*	
DTLZ2	3.838e−01	1.478e−01	1.270e−01	**1.078e−01**	*1.088e−01*	+
	3.8e−02	1.6e−02	2.0e−02	**1.7e−02**	*1.9e−02*	
DTLZ3	9.534e−01	1.066e+00	7.675e−01	**7.452e−01**	*7.548e−01*	+
	1.6e−01	1.6e−01	9.3e−02	**5.5e−01**	*4.5e−01*	
DTLZ4	3.954e−01	1.484e−01	1.233e−01	*1.230e−01*	**1.080e−01**	+
	6.4e−01	8.6e−01	1.9e−02	*9.0e−01*	**1.8e−02**	
DTLZ5	3.793e−01	1.497e−01	1.245e−01	**1.091e−01**	*1.104e−01*	+
	4.0e−02	1.9e−02	1.9e−02	**1.7e−02**	*2.0e−02*	
DTLZ6	8.642e−01	8.249e−01	**1.025e−01**	*1.502e−01*	2.308e−01	+
	3.0e−01	9.3e−02	**2.1e−02**	*4.3e−02*	6.3e−02	
DTLZ7	6.235e−01	5.437e−01	5.199e−01	*5.193e−01*	**5.187e−01**	+
	2.5e−02	1.3e−02	3.7e−03	*2.9e−02*	**1.3e−03**	
WFG1	7.182e−01	*6.514e−01*	1.148e+00	**5.811e−01**	6.662e−01	+
	5.4e−02	*4.8e−02*	1.2e−01	**9.4e−02**	5.8e−02	
WFG2	7.931e−01	7.529e−01	7.603e−01	*7.469e−01*	**7.462e−01**	+
	1.7e−02	1.3e−02	2.7e−03	*2.2e−03*	**4.3e−03**	
WFG3	6.125e−01	4.394e−01	*3.650e−01*	**3.638e−01**	3.726e−01	+
	3.6e−02	1.2e−02	*6.9e−03*	**6.3e−03**	*8.7e−03*	
WFG4	3.786e−01	2.720e−01	3.940e−01	*1.363e−01*	**1.356e−01**	+
	3.9e−02	2.5e−02	5.2e−02	*2.2e−02*	**2.1e−02**	
WFG5	4.126e−01	2.795e−01	1.360e−01	*1.323e−01*	**1.307e−01**	+
	5.1e−02	2.3e−02	2.0e−02	*2.2e−02*	**2.1e−02**	
WFG6	3.898e−01	2.488e−01	**1.187e−01**	*1.266e−01*	1.448e−01	+
	4.2e−02	3.1e−02	**1.9e−02**	*4.0e−02*	4.3e−02	
WFG7	3.794e−01	2.468e−01	1.287e−01	**1.074e−01**	*1.167e−01*	+
	4.6e−02	1.8e−02	1.7e−02	**1.8e−02**	*3.0e−02*	
WFG8	6.452e−01	6.165e−01	**5.418e−01**	*5.570e−01*	5.856e−01	+
	5.5e−02	8.1e−02	**3.6e−02**	*4.2e−02*	7.1e−02	
WFG9	3.956e−01	2.919e−01	2.031e−01	**1.436e−01**	*1.503e−01*	+
	4.1e−02	2.0e−02	2.0e−02	**1.7e−02**	*2.2e−02*	

TABLE 5.5 Median and Interquartile Range of the HV Quality Indicator, $\tilde{x}_{\text{IQR}}$

Problem	NSGAII	SPEA2	OMOPSO	MOCell	AbYSS	Stat. Conf.
ZDT1	6.593e−01	6.600e−01	*6.614e−01*	6.610e−01	**6.614e−01**	+
	4.4e−04	3.9e−04	*3.2e−04*	2.5e−04	**3.2e−04**	
ZDT2	3.261e−01	3.264e−01	*3.283e−01*	**3.284e−01**	3.282e−01	+
	4.3e−04	8.1e−04	*2.4e−04*	**4.3e−04**	2.8e−04	
ZDT3	5.148e−01	5.141e−01	5.147e−01	*5.152e−01*	**5.158e−01**	+
	2.3e−04	3.6e−04	8.8e−04	*3.1e−04*	**3.5e−03**	
ZDT4	*6.555e−01*	6.507e−01	0.000e+00	**6.586e−01**	6.553e−01	+
	4.5e−03	1.2e−02	0.0e+00	**3.0e−03**	6.0e−03	
ZDT6	3.880e−01	3.794e−01	**4.013e−01**	3.968e−01	*4.004e−01*	+
	2.3e−03	3.6e−03	**7.1e−05**	1.1e−03	*1.9e−04*	
DTLZ1	4.882e−01	*4.885e−01*	0.000e+00	**4.909e−01**	4.863e−01	+
	5.5e−03	*6.2e−03*	0.0e+00	**3.8e−03**	1.7e−02	
DTLZ2	2.108e−01	2.118e−01	2.119e−01	**2.124e−01**	*2.123e−01*	+
	3.1e−04	1.7e−04	2.8e−04	**4.5e−05**	*6.5e−05*	
DTLZ3	0.000e+00	0.000e+00	0.000e+00	0.000e+00	0.000e+00	+
	1.7e−01	0.0e+00	0.0e+00	0.0e+00	0.0e+00	
DTLZ4	2.091e−01	2.102e−01	2.098e−01	*2.107e−01*	**2.107e−01**	+
	2.1e−01	2.1e−01	4.0e−04	*2.1e−01*	**5.9e−05**	
DTLZ5	2.108e−01	2.118e−01	2.119e−01	**2.124e−01**	*2.123e−01*	+
	3.5e−04	1.7e−04	3.0e−04	**3.1e−05**	*6.8e−05*	
DTLZ6	1.748e−01	9.018e−03	**2.124e−01**	*1.614e−01*	1.109e−01	+
	3.6e−02	1.4e−02	**5.0e−05**	*4.2e−02*	4.1e−02	
DTLZ7	3.334e−01	3.336e−01	3.341e−01	*3.343e−01*	**3.344e−01**	+
	2.1e−04	2.2e−04	2.2e−04	*9.5e−05*	**7.8e−05**	
WFG1	**5.227e−01**	3.851e−01	1.603e−01	*4.955e−01*	2.270e−01	+
	1.3e−01	1.1e−01	9.0e−02	*1.7e−01*	1.3e−01	
WFG2	5.614e−01	5.615e−01	**5.642e−01**	*5.616e−01*	5.611e−01	+
	2.8e−03	2.8e−03	**6.8e−05**	*2.9e−04*	1.1e−03	
WFG3	4.411e−01	4.418e−01	**4.420e−01**	*4.420e−01*	4.417e−01	+
	3.2e−04	2.0e−04	**2.2e−05**	*1.9e−04*	5.9e−04	
WFG4	2.174e−01	2.181e−01	2.063e−01	*2.188e−01*	**2.191e−01**	+
	4.9e−04	3.0e−04	1.7e−03	*2.3e−04*	**2.0e−04**	
WFG5	1.948e−01	1.956e−01	**1.963e−01**	1.962e−01	*1.962e−01*	+
	3.6e−04	1.8e−04	**6.3e−05**	6.9e−05	*6.3e−05*	
WFG6	2.027e−01	2.040e−01	**2.102e−01**	1.955e−01	1.709e−01	+
	9.0e−03	8.6e−03	**1.1e−04**	3.4e−02	3.3e−02	
WFG7	2.089e−01	2.098e−01	2.103e−01	*2.105e−01*	**2.106e−01**	+
	3.3e−04	2.4e−04	1.0e−04	*1.3e−04*	**1.7e−04**	
WFG8	**1.473e−01**	1.472e−01	1.462e−01	*1.473e−01*	1.440e−01	+
	2.1e−03	2.2e−03	1.1e−03	*2.2e−03*	3.2e−03	
WFG9	2.374e−01	**2.388e−01**	2.369e−01	*2.387e−01*	2.377e−01	+
	1.7e−03	**2.3e−03**	5.8e−04	*2.6e−03*	3.6e−03	

was found (p-value > 0.05). Each data entry in the tables includes the median, $\tilde{x}$, and interquartile range, IQR, as measures of location (or central tendency) and statistical dispersion, because, on the one hand, some samples follow a Gaussian distribution whereas others do not, and on the other hand, mean and median are theoretically the same for Gaussian distributions. Indeed, whereas the median is a feasible measure of central tendency in both Gaussian and non-Gaussian distributions, using the mean makes sense only for Gaussian distributions. The same holds for the standard deviation and the interquartile range.

The best result for each problem is shown in boldface type. For the sake of better understanding, we have used italic to indicate the second-best result; in this way, we can visualize at a glance the best-performing techniques. Furthermore, as the differences between the best and second-best values are frequently very small, we try to avoid biasing our conclusions according to a unique best value. Next, we analyze in detail the results obtained.

5.5.1 Convergence

We include in Table 5.3 the results obtained when employing the IGD indicator. We observe that MOCell obtains the lowest (best) values in 10 of the 21 MOPs (the boldface values) and with statistical confidence in all cases (see the + signs in the last column of the table). This means that the resulting Pareto fronts from MOCell are closer to the optimal Pareto fronts than are those computed by the other techniques. OMOPSO achieves the lowest values in five MOPs, but it fails when solving the problems ZDT4, DTLZ1, and DTLZ3 (see the noticeable differences in the indicator values in those problems). The third algorithm in the ranking is AbYSS, and it is noticeable for the poor performance of the reference algorithms, NSGA-II and SPEA2, which get the best values in only 2 and 1 of the 21 MOPs considered, respectively. If we consider the second-best values (the italic values), our claims are reinforced, and MOCell is clearly the best algorithm according to the convergence of the Pareto fronts obtained.

5.5.2 Diversity

The values obtained after applying the Δ quality indicator are included in Table 5.4. The results indicate that MOCell gets the lowest values in 10 of the 21 MOPs, whereas AbYSS and OMOPSO produce the lowest values in 6 and 5 MOPs, respectively. NSGA-II and SPEA2 are the last algorithms in the ranking, and none of them produces the best results in any of the 21 problems considered. If we look at the second-lowest values, it is clear that MOCell is the most salient algorithm according to the diversity property in the MOPs solved: It provides the lowest or second-lowest Δ values in all the problems.

5.5.3 Convergence and Diversity: HV

The last quality indicator we considered is the hypervolume (HV), which gives us information about both convergence and diversity of the Pareto fronts. The values obtained are included in Table 5.5. Remember that the higher values of HV are desirable. At first glance, if we consider only the highest values, the ranking of the techniques would be, first, AbYSS and OMOPSO, and then MOCell, NSGA-II, and SPEA2. However, OMOPSO gets an HV value of 0 in three MOPs. This means that all the solutions obtained are outside the limits of the true Pareto front; when applying this indicator, these solutions are discarded, because otherwise the value obtained would be unreliable. We can observe that all of the metaheuristics get a value of 0 in problem DTLZ3. With these considerations in mind, the best technique would be AbYSS, followed by MOCell.

This conclusion could be a bit contradictory, because it is against the fact that MOCell was the more accurate algorithm according to the IGD and Δ indicators. However, closer observation shows that the differences in problems DTLZ4, WFG5, and WFG8, compared to the highest values, are negligible: the medians are the same, and they are worse only in the interquartile range. In fact, taking into account the second-highest values, we can conclude that MOCell is the most outstanding algorithm according to HV. Again, the poor behavior of NSGA-II and SPEA2 is confirmed by this indicator.

5.5.4 Discussion

Previous work has revealed that MOCell [14] and AbYSS [16] outperformed NSGA-II and SPEA2 in a similar benchmark of MOPs. Inclusion in the same study of these four techniques, plus OMOPSO, leads to the result that NSGA-II and SPEA2 appear to be the worst techniques. We note that all the results presented in the tables have statistical confidence.

Our study indicates that in the context of the problems, quality indicators, and parameter settings used, modern advanced metaheuristics can provide better search capabilities than those of the classical reference algorithms. In particular, MOCell is the most salient metaheuristic in this comparison. If we take into account that MOCell uses, as many metaheuristics do, the ranking scheme and the crowding distance indicator of NSGA-II (see ref. 15 for more details), the conclusion is that the cellular cGA model is particularly well suited to solving the MOPs considered in our study.

To finish this section, we include three graphics showing some Pareto fronts produced by the algorithms we have compared (we have chosen a random front out of the 100 independent runs). In each figure, the fronts are shifted along the Y-axis to provide a better way to see at a glance the varying behavior of the metaheuristics.

In Figure 5.4 we depict the fronts obtained when solving problem ZDT1, a convex MOP. The spread of the points in the figure corroborates the excellent values of the Δ indicator obtained by OMOPSO and MOCell (see Table 5.4). The poor performance of NSGA-II is also shown clearly.

We show next the fronts produced when solving the nonconvex MOP DTLZ2 in Figure 5.5. In this example, MOCell and AbYSS give the fronts that have the best diversity, as can be seen in Table 5.4. Although the convergence property

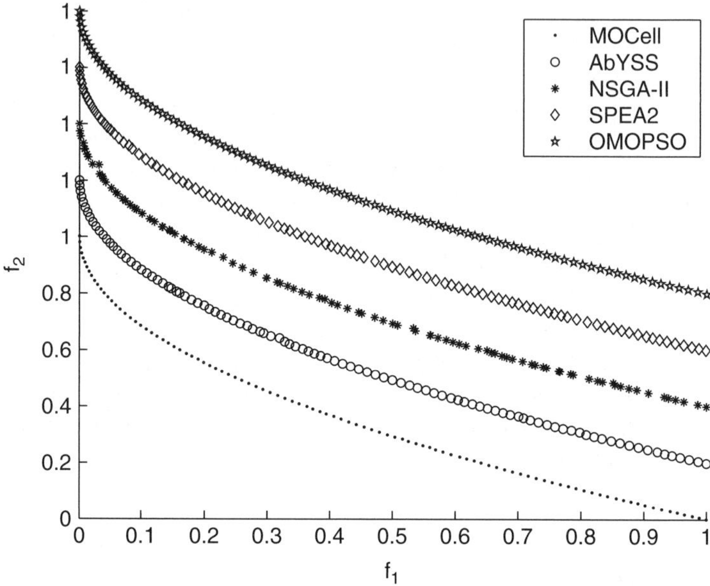

Figure 5.4 Pareto fronts obtained when solving problem ZDT1.

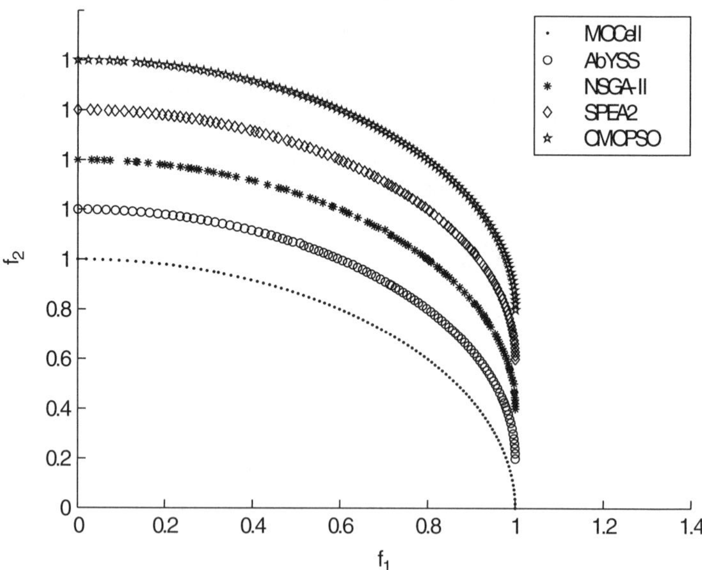

Figure 5.5 Pareto fronts obtained when solving problem DTLZ2.

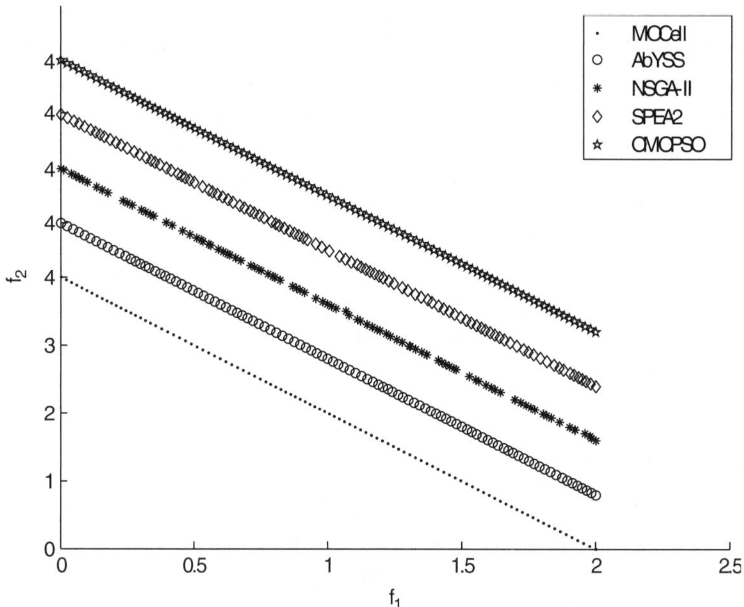

Figure 5.6 Pareto fronts obtained when solving problem WFG3.

cannot be observed in the figure, the IGD values of these algorithms in this problem (see Table 5.3) indicate that MOCell and AbYSS get the most accurate Pareto fronts.

The final example, shown in Figure 5.6, is the degenerated problem WFG3, which in its bi-objective formulation presents a linear Pareto front. This problem is another example of the typical behavior of the metaheuristics we have compared in many of the problems studied: MOCell, OMOPSO, and AbYSS produce fronts having the best diversity; SPEA2 and NSGA-II are a step beyond them.

5.6 CONCLUSIONS

In this chapter we have compared three modern multiobjective metaheuristics against NSGA-II and SPEA2, the reference algorithms in the field. To evaluate the algorithms, we have used a rigorous experimentation methodology. We have used three quality indicators, a set of MOPs having different properties, and a statistical analysis to ensure the confidence of the results obtained.

A general conclusion of our study is that advanced metaheuristics algorithms can outperform clearly both NSGA-II and SPEA2. In the context of this work, NSGA-II obtains the best result in only 2 problems out of the 20 that constitute the benchmark with regard to the IGD and HV quality indicators, and SPEA2 yields the best value in only one MOP using the same indicators.

The most outstanding algorithm in our experiments is MOCell. It provides the overall best results in the three quality indicators, and it is worth mentioning that it yields the best or second-best value in all the problems in terms of diversity. The scatter search algorithm, AbYSS, and the PSO technique, OMOPSO, are the two metaheuristics that follow MOCell according to our tests. However, we consider that the behavior of AbYSS is better than that of OMOPSO, because this is a less robust technique, as shown by the poor values in HV of OMOPSO in problems ZDT4, DTLZ1, and DTLZ3.

We plan to extend this study in the future by adding more recent metaheuristics. We have restricted our experiments to bi-objective problems, so a logical step is to include in further work at least the DTLZ and WFG problem families, using their three-objective formulation.

Acknowledgments

This work has been partially funded by the Ministry of Education and Science and FEDER under contract TIN2005-08818-C04-01 (the OPLINK project). Juan J. Durillo is supported by grant AP-2006-03349 from the Spanish government.

REFERENCES

1. C. Blum and A. Roli. Metaheuristics in combinatorial optimization: overview and conceptual comparison. *ACM Computing Surveys*, 35(3):268–308, Sept. 2003.
2. R. L. Carraway, T. L. Morin, and H. Moskowitz. Generalized dynamic programming for multicriteria optimization. *European Journal of Operational Research*, 44:95–104, 1990.
3. C. A. Coello, D. A. Van Veldhuizen, and G. B. Lamont. Genetic algorithms and evolutionary computation. In *Evolutionary Algorithms for Solving Multi-objective Problems*. Kluwer Academic, Norwell, MA, 2002.
4. K. Deb. *Multi-objective Optimization Using Evolutionary Algorithms*. Wiley, New York, 2001.
5. K. Deb, A. Pratap, S. Agarwal, and T. Meyarivan. A fast and elitist multiobjective genetic algorithm: NSGA-II. *IEEE Transactions on Evolutionary Computation*, 6(2):182–197, 2002.
6. K. Deb, L. Thiele, M. Laumanns, and E. Zitzler. Scalable test problems for evolutionary Multi-objective optimization. In A. Abraham, L. Jain, and R. Goldberg, eds., *Evolutionary Multiobjective Optimization: Theoretical Advances and Applications*. Springer-Verlag, New York, 2005, pp. 105–145.
7. J. Demšar. Statistical comparison of classifiers over multiple data sets. *Journal of Machine Learning Research*, 7:1–30, 2006.
8. J. J. Durillo, A. J. Nebro, F. Luna, B. Dorronsoro, and E. Alba. jMetal: a Java framework for developing multi-objective optimization metaheuristics. *Technical Report ITI- 2006-10*, Departamento de Lenguajes y Ciencias de la Computación, Universidad de Málaga, E.T.S.I. Informática, Campus de Teatinos, Spain, Dec. 2006.

9. F. W. Glover and G. A. Kochenberger. *Handbook of Metaheuristics* (International Series in Operations Research and Management Sciences). Kluwer Academic, Norwell, MA, 2003.
10. S. Huband, L. Barone, R. L. While, and P. Hingston. A scalable multi-objective test problem toolkit. In C. Coello, A. Hernández, and E. Zitler, eds., *Proceedings of the 3rd International Conference on Evolutionary Multicriterion Optimization* (*EMO'05*), vol. 3410 of *Lecture Notes in Computer Science*. Springer-Verlag, New York, 2005, pp. 280–295.
11. D. Jaeggi, G. Parks, T. Kipouros, and J. Clarkson. A multi-objective tabu search algorithm for constrained optimisation problems. In C. Coello, A. Hernández, and E. Zitler, eds., *Proceedings of the 3rd International Conference on Evolutionary Multicriterion Optimization* (*EMO'05*), vol. 3410 of *Lecture Notes in Computer Science*. Springer-Verlag, New York, 2005, pp. 490–504.
12. J. Knowles and D. Corne. The Pareto archived evolution strategy: a new baseline algorithm for multiobjective optimization. In *Proceedings of the 1999 Congress on Evolutionary Computation*. IEEE Press, Piscataway, NJ, 1999, pp. 9–105.
13. J. Knowles, L. Thiele, and E. Zitzler. A tutorial on the performance assessment of stochastic multiobjective optimizers. *Technical Report 214*. Computer Engineering and Networks Laboratory (TIK), ETH, Zurich, Switzerland, 2006.
14. A. J. Nebro, J. J. Durillo, F. Luna, B. Dorronsoro, and E. Alba. A cellular genetic algorithm for multiobjective optimization, In D. A. Pelta and N. Krasnogor, eds., *NICSO 2006*, pp. 25–36.
15. A. J. Nebro, J. J. Durillo, F. Luna, B. Dorronsoro, and E. Alba. Design issues in a multiobjective cellular genetic algorithm. In S. Obayashi, K. Deb, C. Poloni, T. Hiroyasu, and T. Murata, eds., *Proceedings of the 4th International Conference on Evolutionary MultiCriterion Optimization* (*EMO'07*), vol. 4403 of *Lecture Notes in Computer Science*. Springer-Verlag, New York, 2007, pp. 126–140.
16. A. J. Nebro, F. Luna, E. Alba, B. Dorronsoro, J. J. Durillo, and A. Beham. AbYSS: adapting scatter search to multiobjective optimization. Accepted for publication in *IEEE Transactions on Evolutionary Computation*, 12(4):439–457, 2008.
17. M. Reyes and C. A. Coello Coello. Improving pso-based multi-objective optimization using crowding, mutation and ϵ-dominance. In C. Coello, A. Hernández, and E. Zitler, eds., *Proceedings of the 3rd International Conference on Evolutionary MultiCriterion Optimization,* (*EMO'05*), vol. 3410 of *Lecture Notes in Computer Science*, Springer-Verlag, New York, 2005, pp. 509–519.
18. T. Sen, M. E. Raiszadeh, and P. Dileepan. A branch and bound approach to the bicriterion scheduling problem involving total flowtime and range of lateness. *Management Science*, 34(2):254–260, 1988.
19. D. J. Sheskin, *Handbook of Parametric and Nonparametric Statistical Procedures*. CRC Press, Boca Raton, FL, 2003.
20. B. S. Stewart and C. C. White. Multiobjective A*. *Journal of the ACM*, 38(4): 775–814, 1991.
21. D. A. Van Veldhuizen and G. B. Lamont. Multiobjective evolutionary algorithm research: a history and analysis. *Technical Report TR- 98-03*. Department of Electrical and Computer Engineering, Graduate School of Engineering, Air Force Institute of Technology, Wright-Patterson, AFB, OH, 1998.

22. M. Visée, J. Teghem, M. Pirlot, and E. L. Ulungu. Two-phases method and branch and bound procedures to solve knapsack problem. *Journal of Global Optimization*, 12:139–155, 1998.

23. A. Zhou, Y. Jin, Q. Zhang, B. Sendhoff, and E. Tsang, Combining model-based and genetics-based offspring generation for multi-objective optimization using a convergence criterion. In *Proceedings of the 2006 IEEE Congress on Evolutionary Computation*, 2006, pp. 3234–3241.

24. E. Zitzler, K. Deb, and L. Thiele, Comparison of multiobjective evolutionary algorithms: empirical results. *Evolutionary Computation*, 8(2):173–195, 2000.

25. E. Zitzler, M. Laumanns, and L. Thiele. SPEA2: Improving the strength Pareto evolutionary algorithm. *Technical Report 103*. Computer Engineering and Networks Laboratory (TIK), Swiss Federal Institute of Technology (ETH), Zurich, Switzerland, 2001.

26. E. Zitzler and L. Thiele, Multiobjective evolutionary algorithms: a comparative case study and the strength Pareto approach. *IEEE Transactions on Evolutionary Computation*, 3(4):257–271, 1999.

CHAPTER 6

Canonical Metaheuristics for Dynamic Optimization Problems

G. LEGUIZAMÓN, G. ORDÓÑEZ, and S. MOLINA

Universidad Nacional de San Luis, Argentina

E. ALBA

Universidad de Málaga, Spain

6.1 INTRODUCTION

Many real-world optimization problems have a nonstationary environment; that is, some of their components depend on time. These types of problems are called *dynamic optimization problems* (DOPs). Dynamism in real-world problems can be attributed to several factors: some are natural (e.g., weather conditions); others can be related to human behavior (e.g., variation in aptitude of different individuals, inefficiency, absence, and sickness); and others are business related (e.g., the addition of new orders and cancellation of old ones). Techniques that work for static problems may therefore not be effective for DOPs, which require algorithms that make use of old information to find new optima quickly. An important class of techniques for static problems are the metaheuristics (MHs), the most recent developments in approximate search methods for solving complex optimization problems [21].

In general, MHs guide a subordinate heuristic and provide general frameworks that allow us to create new hybrids by combining different concepts derived from classical heuristics, artificial intelligence, biological evolution, natural systems, and statistical mechanical techniques used to improve their performance. These families of approaches include, but are not limited to, evolutionary algorithms, particle swarm optimization, ant colony optimization, differential evolution, simulated annealing, tabu search, and greedy randomized adaptive search process. It is remarkable that MHs have been applied successfully to stationary

Optimization Techniques for Solving Complex Problems, Edited by Enrique Alba, Christian Blum, Pedro Isasi, Coromoto León, and Juan Antonio Gómez

optimization problems. However, there also exists increasing research activity on new developments of MHs applied to DOPs with encouraging results.

In this chapter we focus on describing the use of population-based MHs (henceforth, called simply MHs) from a unifying point of view. To do that, we first give a brief description and generic categorization of DOPs. The main section of the chapter is devoted to the application of canonical MHs to DOPs, including general suggestions concerning the way in which they can be implemented according to the characteristic of a particular DOP. This section includes an important number of references of specific proposals about MHs for DOPs. The two last sections are devoted, respectively, to considerations about benchmarks and to typical metrics used to assess the performance of MHs for DOPs. Finally, we draw some conclusions.

6.2 DYNAMIC OPTIMIZATION PROBLEMS

DOPs are optimization problems in which some of the components depend on time; that is, there exists at least one source of dynamism that affects the objective function (its domain, its landscape, or both). However, the mere existence of a time dimension in a problem does not mean that the problem is dynamic. Problems that can be solved ahead of time are not dynamic even though they might be time dependent. According to Abdunnaser [1], Bianchi [4], Branke [11], and Psaraftis [36], the following features can be found in most real-world dynamic problems: (1) the problem can change with time in such a way that future scenarios are not known completely, yet the problem is known completely up to the current moment without any ambiguity about past information; (2) a solution that is optimal or nearly optimal at a certain time may lose its quality in the future or may even become infeasible; (3) the goal of the optimization algorithm is to track the shifting optima through time as closely as possible; (4) solutions cannot be determined ahead of time but should be found in response to the incoming information; and (5) solving the problem entails setting up a strategy that specifies how the algorithm should react to environmental changes (e.g., to solve the problem from scratch or adapt some parameters of the algorithm at every change).

The goal in stationary optimization problems is to find a solution $s*$ that maximises the objective function O (note that maximizing a function O is the same as minimizing $-O$):

$$s^* = \arg\max_{s \in S} O(s) \tag{6.1}$$

where S is the set of feasible solutions and $O\colon S \to \mathbb{R}$ is the objective function. On the other hand, for DOPs the set of feasible solutions and/or the objective function depends on time; therefore, the goal at time t is to find a solution $s*$

that fulfills

$$s^* = \arg\max_{s \in S(t)} O(s, t) \tag{6.2}$$

where at time t_1, set $S(t)$ is defined for all $t \leq t_1$ and $O(s, t)$ is fully defined for $s \in S(t) \wedge t \leq t_1$; however, there exists uncertainty for $S(t)$ and/or $O(s, t)$ values for $t > t_1$. This uncertainty is required to consider a problem as dynamic. Nevertheless, instances of dynamic problems used in many articles to test algorithms are specified without this type of uncertainty, and changes are generated following a specific approach (see Section 6.3). Additionally, the particular algorithm used to solve DOPs is not aware of these changes and the way they are generated. Consequently, from the point of view of the algorithm, these are real DOP instances satisfying the definition above.

Regarding the above, the specification of a DOP instance should include (1) an initial state [conformed of the definition of $S(0)$ and a definition of $O(s, 0)$ for all $s \in S(t)$], (2) a list of sources of dynamism, and (3) the specifications of how each source of dynamism will affect the objective function and the set of feasible solutions. Depending on the sources of dynamism and their effects on the objective function, DOPs can be categorized in several ways by answering such questions as "what" (aspect of change), "when" (frequency of change), and "how" (severity of change, effect of the algorithm over the scenario, and presence of patterns). Each of these characteristics of DOPs is detailed in the following subsections.

6.2.1 Frequency of Change

The frequency of change denotes how often a scenario is modified (i.e., is the variation in the scenario continuous or discrete?). For example, when discrete changes occur, it is important to consider the mean time between consecutive changes. According to these criteria and taking into account their frequency, two broad classes of DOPs can be defined as follows:

1. *Discrete DOPs.* All sources of dynamism are discrete; consequently, for windowed periods of time, the instance seems to be a stationary problem. When at some time new events occur, they make the instance behave as another stationary instance of the same problem. In other words, DOPs where the frequency of change is discrete can be seen as successive and different instances of a stationary problem, called *stages*.

Example 6.1 (Discrete DOP) Let us assume an instance of a version of the dynamic traveling salesman problem (DTSP) where initially the salesman has to visit n cities; however, 10 minutes later, he is informed that he has to visit one more city; in this case the problem behaves as a stationary TSP with n cities for the first 10 minutes, and after that it behaves as a different instance of the stationary TSP, with $n+1$ cities.

2. *Continuous DOPs*. At least one source of dynamism is continuous (e.g., temperature, altitude, speed).

Example 6.2 (Continuous DOP) In the moving peaks problem [9,10] the objective function is defined as

$$O(s,t) = \max\left(B(s), \max_{i=1}^{m} P(s, h_i(t), w_i(t), p_i(t))\right) \tag{6.3}$$

where s is a vector of real numbers that represents the solution, $B(s)$ a time-invariant basis landscape, P a function defining the peak shape, m the number of peaks, and each peak has its own time-varying parameter: height (h_i), width (w_i), and location (p_i). In this way each time we reevaluate solution s the fitness could be different.

6.2.2 Severity of Change

The severity of change denotes how strong changes in the system are (i.e., is it just a slight change or a completely new situation?) In discrete DOPs, severity refers to the differences between two consecutive stages.

Example 6.3 (Two Types of Changes in DTSP) Let us suppose a DTSP problem with 1000 cities, but that after some time a new city should be incorporated in the tour. In this case, with a high probability, the good-quality solutions of the two stages will be similar; however, if after some time, 950 out of 1000 cities are replaced, this will generate a completely different scenario.

6.2.3 Aspect of Change

The aspect of change indicates what components of the scenario change: the optimization function or some constraint. There exist many categorizations of a DOP, depending on the aspect of change. Basically, a change could be in one of the following categories:

- *Variations in the values of the objective function* (i.e., only the objective values are affected). Thus, the goal is to find a solution $s*$ that fulfills the equation

$$s^* = \arg\max_{s \in S} O(s,t) \tag{6.4}$$

- *Variations in the set of feasible solutions* (i.e., only the set of problem constraints are affected). Thus, the goal is to find a solution $s*$ that fulfills the equation

$$s^* = \arg \max_{s \in S(t)} O(s) \tag{6.5}$$

In this case, a possible change in $S(t)$ will make feasible solutions infeasible, and vice versa.

- *Variations in both of the components above*, which is the most general case, and the goal is to find a solution $s*$ that fulfills Equation 6.2.

6.2.4 Interaction of the Algorithm with the Environment

Up to this point we have assumed that all the sources of dynamism are external, but for some problems the algorithm used to approach them is able to influence the scenario. In this way, DOPs can be classified [6] as (a) *control influence*: the algorithms used to tackle the problem have no control over the scenario (i.e., all sources of dynamism are external); and (b) *system influence*: the algorithms used to tackle the problem have some control over the scenario (i.e., the algorithms themselves incorporate an additional source of dynamism).

Example 6.4 (System Influence DOP) Let us assume that we have a production line with several machines, where new tasks arrive constantly to be processed by some of the machines. The algorithm chosen produces a valid schedule, but after some time a *machine idle event* is generated. The best solution known so far is used to select and assign the task to the idle machine. In this way the task assigned (selected by the algorithm) is no longer on the list of tasks to be scheduled (i.e., the algorithm has changed the scenario).

6.2.5 Presence of Patterns

Even if the sources of dynamism are external, the sources sometimes show patterns. Therefore, if the algorithm used to tackle the DOP is capable of detecting this situation, it can take some advantage to react properly to the new scenario [7,8,39]. Some characteristics of MHs make them capable of discovering some patterns about situations associated with the possible changes in the scenario and, additionally, keeping a set of solutions suitable for the next stage. In this way, when a change occurs, the MH could adapt faster to the newer situation. However, sometimes it is not possible to predict a change but it is still possible to predict when it will be generated. In these cases, a possible mechanism to handle the changes is to incorporate in the fitness function a measure of solution flexibility (see Section 6.3) to reduce the adapting cost to the new scenario [13]. Furthermore, a common pattern present in many DOPs, especially in theoretical problems, is the presence of cycles (an instance has cycles if the optimum returns to previous locations or achieves locations close to optimum).

Example 6.5 (Presence of Cycles) A classic DOP that fits in this category is a version of the dynamic knapsack problem [22], where the problem switches indefinitely between two stages.

6.3 CANONICAL MHs FOR DOPs

Algorithm 6.1 shows the skeleton of the most traditional population-based MHs for solving stationary problems as defined in Equation 6.2. However, its application to DOPs needs incorporation of the concept of time either by using an advanced library to handle real-time units or by using the number of evaluations [11], which is the more widely used method. When considering the number of evaluations, Algorithm 6.1 should be adapted to include dynamism in some way, as shown in Algorithm 6.2, which represents a general outline of an MH to handle the concept of time. However, the objective is not the same for stationary problems as for DOPs. Whereas in stationary problems the objective is to find a high-quality solution, in DOPs there exists an additional objective, which is to minimize the cost of adapting to the new scenario; that is, there are two opposing objectives: *solution quality* and *adaptation cost*. The first is similar to the fitness functions used with stationary problems and represents the quality of the solution

Algorithm 6.1 Static MH

```
Iteration = 0;
Metaheuristic-dependent steps I ();
while termination criteria not satisfied do
  Generate new solutions ();
  Metaheuristic-dependent steps II ();
  Iteration = Iteration + 1;
end while
```

Algorithm 6.2 DMH: Time Concept Incorporation

```
Iteration = 0;
Time = 0;
Metaheuristic-dependent steps I ();
Time = Time + Number of solutions generated;
while termination criteria not satisfied do
  Generate new solutions ();
  Time = Time + Number of solutions generated;
  Metaheuristic-dependent steps II ();
  Iteration = Iteration + 1;
end while
```

at a specific time; the second is focused on evaluating the adaptation capacity of the algorithm when the scenario has changed.

The first step is to determine the way in which the MHs will react when facing a new scenario. If considering discrete DOPs, the stages can be seen as independent stationary problems (see Section 6.2.1). In this case, a reinitialization step takes place when a particular event changes the scenario (see Algorithm 6.3). When these two stages are similar, it makes no sense to discard all the information obtained so far; therefore, an alternative to the reinitialization process is to reuse the information collected and adapt it to the new stage (see Algorithm 6.4). These two algorithms have been widely used in different studies and it

Algorithm 6.3 DMH: Reinitialization

```
Iteration = 0;
Time = 0;
Metaheuristic-dependent steps I ();
Time = Time+Number of solutions generated;
while termination criteria not satisfied do
  Generate new solutions ();
  Time = Time+Number of solutions generated;
  Metaheuristic-dependent steps II ();
  Iteration = Iteration+1;
  if exist input event then
     Process event to adapt problem ();
     Metaheuristic-dependent steps I ();
  end if
end while
```

Algorithm 6.4 DMH: Reuse Information

```
Iteration = 0;
Time = 0;
Metaheuristic-dependent steps I ();
Time = Time+Number of solutions generated;
while termination criteria not satisfied do
  Generate new solutions ();
  Time = Time+Number of solutions generated;
  Metaheuristic-dependent steps II ();
  Iteration = Iteration+1;
  if exist input event then
     Process event to adapt problem ();
     Adapt metaheuristic structures ();
  end if
end while
```

has been shown that Algorithm 6.4 performs significantly better than Algorithm 6.3 in most situations; however, some authors suggest that for problems with a high frequency of change, Algorithm 6.3 outperforms ref. 11. As a general rule, it can be said that for both low values of frequency and severity of change, Algorithm 6.4 outperforms Algorithm 6.3.

On the other hand, in continuous DOPs, each time a solution is evaluated the scenario could be different. Therefore, Algorithm 6.3 is not suitable for these types of problems, and similarly, Algorithm 6.4 cannot be applied directly. To do that, it is necessary to consider the aspect of change and the particular MH. To adapt Algorithm 6.4 for continuous DOPs, lines 10 to 13 should be moved into either "Generate new solutions" or "Metaheuristic-dependent steps II," according to the MH. A common approach when the severity of change is low is to ignore these changes for discrete periods of time and proceed as if solving a discrete DOP.

Although the second strategy (represented by Algorithm 6.4) is more advanced than the first (represented by Algorithm 6.3), the second has better convergence properties, which, indeed, are useful in stationary problems, but when solving DOPs, those properties can be a serious drawback, due to the lack of diversity. There are two principal approaches to augmenting population diversity: reactive and proactive [43]. The reactive approach consists of reacting to a change by triggering a mechanism to increase the diversity, whereas the proactive approach consists of maintaining population diversity through the entire run of the algorithm. In the following, some examples of these two approaches are listed.

- *Reactive approaches*: hypermutation [15], variable local search [40], shaking [18], dynamic updating of the MH parameters to intensify either exploration or exploitation [42,43], turbulent particle swarm optimization [26], pheromone conservation procedure [17].
- *Proactive approaches*: random immigrants [23], sharing/crowding [2,14], thermodynamical GA [31], sentinels [33], diversity as second objective [29], speciation-based PSO (SPSO) [27,28].
- *Reactive and/or proactive approaches* (*multipopulation*): maintain different subpopulations on different regions [11], self-organizing scouts [11,12] multinational EA [38], EDA multipopulation approach with random migration [30], multiswarm optimization in dynamic environments [5].

It is worth remarking that for the different MHs there exist different alternatives for implementing approaches to handling diversity. However, particular implementations should avoid affecting the solution quality (e.g., an excessive diversity will avoid a minimum level of convergence, also needed when facing DOPs).

Another important concept in MHs for DOPs is solution flexibility. A solution can be considered flexible if after a change in the scenario, its fitness is similar to that of the preceding stage. Clearly, flexibility is an important and desirable feature for the solutions which allow the MHs to lower the adaptation cost.

To generate solutions as flexible as possible, the following approaches can be considered:

- *Multiobjective*. The MH is designed to optimize two objectives, one represented by the original fitness function and the other by a metric value indicating solution flexibility [12].
- *Dynamic fitness*. When the time to the next change is known or can be estimated, the fitness function is redefined as

$$F'(x) = [1 - a(t)]F(x) + a(t)\ \mathrm{Flex}(x) \tag{6.6}$$

  where $F(x)$ is the old fitness function, Flex(x) measures the solution flexibility, and $a(t) \in [0, 1]$ is an increasing function except when changes occur. In this way, function a is designed such that (a) when the time for the next change is far away, it gives very low values (i.e., solutions with higher fitness values are preferred); and (b) when approaching the next change, it gives higher values in order to prefer more flexible solutions.

In addition, the literature provides plenty of specific techniques for specific classes of DOPs approached by means of MHs. For instance, for DOPs that show similar or identical cyclical stages, it could be useful to consider an approach that uses implicit or explicit memory to collect valuable information to use in future stages (Algorithm 6.5 shows the overall idea of an MH that uses explicit memory). On the other hand, algorithms proposed for system control DOPs have been developed [6]. Other authors have redefined a MH specifically designed to be used for DOPs (e.g, the population-based ACO) [24,25].

Algorithm 6.5 DMH: With Explicit Memory

```
Iteration = 0;
Time = 0;
Metaheuristic-dependent steps I ();
Time = Time+Number of solutions generated;
while termination criteria not satisfied do
  Generate new solutions ();
  Time = Time+Number of solutions generated;
  Metaheuristic-dependent steps II ();
  Iteration = Iteration+1;
  if exist input event then
     Save information for future use ()
     Process event to adapt problem ();
     Adapt Metaheuristic structures ();{Using the informa-
     tion saved in previous stages}
  end if
end while
```

6.4 BENCHMARKS

Benchmarks are useful in comparing optimization algorithms to highlight their strengths, expose their weaknesses, and to help in understanding how an algorithm behaves [1]. Therefore, selection of an adequate benchmark should take into account at least how to (1) specify and generate a single instance and (2) select an appropriate set of instances.

6.4.1 Instance Specification and Generation

As mentioned earlier, DOP instances are defined as (a) an initial state, (b) a list of sources of dynamism, and (c) the specification of how each source of dynamism will affect the objective function and/or the set of feasible solutions. However, for a particular benchmark the behavior of the sources of dynamism has to be simulated. To do that, two approaches can be considered: use of (1) a stochastic event generator or (2) a predetermined list of events. The instances modeled using a stochastic event generator generally have the advantage that they are parametric and allow us to have a wide range of controlled situations. On the other hand, instances that use a predetermined list of events allow us to repeat the experiments, which provides some advantages for the statistical test. If all the sources of dynamism are external (control influence DOPs), it is a good idea to use the stochastic event generator to create a list of events. In addition, it is worth remarking that the researcher could know beforehand about the behavior of the instances for all times t; in real-life situations, only past behavior is known. Therefore, algorithms used to tackle the problem should not take advantage of knowledge of future events.

6.4.2 Selection of an Adequate Set of Benchmarks

When defining a new benchmark we must take into account the objective of our study: (a) to determine how the dynamic problem components affect the performance of a particular MH, or (b) to determine the performance of a particular MH for a specific problem. Although these are problem-specific considerations, it is possible to give some hints as to how to create them.

For the first case it is useful to have a stochastic instance generator that allows us to specify as many characteristics of the instances as possible (i.e., frequency of change, severity of change, etc.). However, to analyze the impact of a specific characteristic on an MH, it is generally useful to choose one characteristic and fix the remaining ones. In this situation, a possible synergy among the characteristics must be taken into account (i.e., sometimes the effect of two or more characteristics cannot be analyzed independently). On the other hand, when comparing the performance of an MH against other algorithms or MHs, it is useful to consider existing benchmarks when available. As an example, Table 6.1 displays a set of well-known benchmarks for different DOPs. Only when no benchmark is available or when an available benchmark is not adequate for the problem

TABLE 6.1 Examples of Benchmarks

Problem	Reference
Dynamic bit matching	[35]
Moving parabola	[3]
Time-varying knapsack	[22]
Moving peaks	[34]
Dynamic job shop scheduling	[19]

considered is it necessary to generate a new benchmark considering at least the inclusion of instances: (a) with different characteristics to cover the maximum number of situations with a small set of instances (the concept of synergy is also very important here), (b) with extreme and common characteristics, and (c) that avoids being tied to a specific algorithm.

6.5 METRICS

To analyze the behavior of an MH applied to a specific DOP, it is important to remember that in this context, there exist two opposite objectives: (1) solution quality and (2) adaptation cost. In Sections 6.5.1 and 6.5.2 we describe some metrics for each of these objectives, but it is also important to track the behavior of the MHs through time. Clearly, for stationary problems the usual metrics give some isolated values, whereas for DOPs, they are usually curves, which are generally difficult to compare. For example, De Jong [16] decides that one dynamic algorithm is better than another if the respective values for the metrics are better at each unit of time; however, this in an unusual situation. On the other hand, a technique is proposed that summarizes the information by averaging: (a) each point of the curve [11], or (b) the points at which a change happened [37].

In this section we present the more widely used metrics for DOPs according to the current literature. Although these metrics are suitable for most situations, other approaches must be considered for special situations.

6.5.1 Measuring Solution Quality

To measure the quality of a solution at time t, it is generally useful to transform the objective function to values independent of range. This is more important in DOPs, where the range varies over time, making it difficult to analyze the time versus objective value curve.

Example 6.6 (Change in the Objective Values) Let us assume a DTSP instance where from $t = 0$ to $t = 10$ the salesman has to visit 100 cities and the objective function ranges between 30 and 40 units (the cost to complete the tour); but after 10 minutes (i.e., $t > 10$), 50 cities are removed. Accordingly,

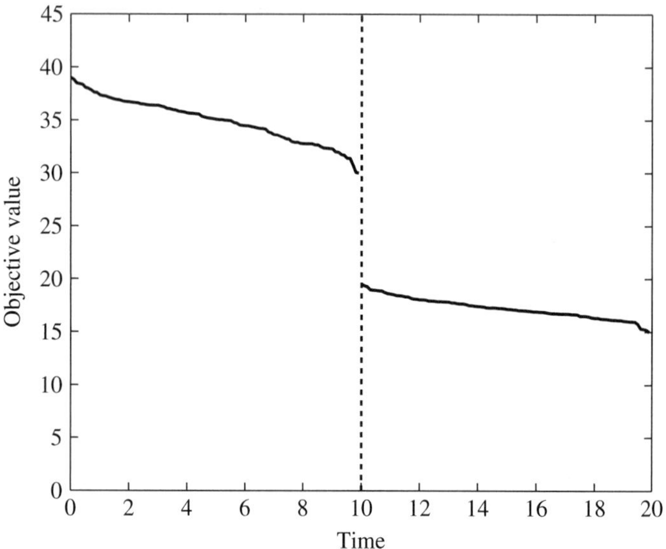

Figure 6.1 Hypothetical objective value obtained through the run for an instance of the DTSP problem.

the objective function now ranges between 15 and 20 units. Figure 6.1 shows the time versus objective value for this problem; notice that the change at time 10 could be misunderstood as a significant improvement in the quality of the solution.

To assess the quality of the solution independent of the range of the objective values, the following metrics can be considered:

- Error:

$$\text{error}(t) = \frac{o^{\text{optimum}}(t) - o^{\text{best}}(t)}{o^{\text{optimum}}(t)} \tag{6.7}$$

- Accuracy [31]:

$$\text{accuracy}(t) = \frac{o^{\text{best}}(t) - o^{\text{worst}}(t)}{o^{\text{optimum}}(t) - o^{\text{worst}}(t)} \tag{6.8}$$

where $o^{\text{optimum}}(t)$ is the objective value of the optimum solution at time t, $o^{\text{worst}}(t)$ is the objective value of the worst solution at time t, and $o^{\text{best}}(t)$ is the objective value of the best solution known at time t. The values of accuracy vary between 0 and 1 on the basis of accuracy $(t) = 1$ when $o^{\text{best}}(t) = o^{\text{optimum}}(t)$. Additional metrics can be found in the work of Morrison [32] and Weicker [41]. Unfortunately, use of these metrics is not a trivial task for some problems, since it is necessary to know the values for $o^{\text{optimum}}(t)$ and $o^{\text{worst}}(t)$. When these values are not available, a common approach is to use the upper and lower bounds given by the state of the art.

6.5.2 Measuring the Adaptation Cost

Another major goal in dynamic environments is mimimization of the adaptation cost when the scenario has changed (in this situation a decreasing quality of the current solutions is generally observed). To measure the adaptation cost, the following two approaches can be considered: (1) measuring the loss of quality and (2) measuring the time that the algorithm needs to generate solutions of quality similar to that found before the change took place. To do that, Feng et al. [20] propose two metrics:

1. Stability(t), a very important measure in DOPs, which is based on the accuracy values (see Equation 6.8) and is defined as

$$\text{stability}(t) = \max\{0, \text{accuracy}(t-1) - \text{accuracy}(t)\} \tag{6.9}$$

2. ϵ-reactivity(t), which denotes the ability of an adaptive algorithm to react quickly to changes and is defined as

$$\epsilon\text{-reactivity}(t) = \underset{<t' \leq \text{ maxIter}}{\overset{\min}{t' - t | t' \in \mathbb{N} \wedge t}} \times \left\{ t' = \text{maxIter} \vee \frac{\text{accuracy}(t')}{\text{accuracy}(t)} \geq 1 - \epsilon \right\} \tag{6.10}$$

As with other metrics that measure the solution quality, the metrics above can be represented by curves. However, the stability curve measures the reaction of an MH to changes, and the more common approach (in discrete DOPs) to show the respective values is to consider the average values of stability at the time when changes occur [37].

6.6 CONCLUSIONS

In this chapter we have presented from a unifying point of view a basic set of canonical population-based MHs for DOPs as well as a number of general considerations with respect to the use of adequate benchmarks for DOPs and some standard metrics that can be used to assess MH performance. It is important to note that for specific DOPs and MHs, there exist an important number of implementation details not covered here that should be considered; however, we have given some important suggestions to guide researchers in their work.

Acknowledgments

The first three authors acknowledge funding from the National University of San Luis (UNSL), Argentina and the National Agency for Promotion of Science and Technology, Argentina (ANPCYT). The second author also acknowledges the National Scientific and Technical Research Council (CONICET), Argentina. The fourth author acknowledges funding from the Spanish Ministry of Education and Science and FEDER under contract TIN2005-08818-C04-01 (the OPLINK project).

REFERENCES

1. Y. Abdunnaser. Adapting evolutionary approaches for optimization in dynamic environments. Ph.D. thesis, University of Waterloo, Waterloo, Ontario, Canada, 2006.
2. H. C. Andersen. An investigation into genetic algorithms, and the relationship between speciation and the tracking of optima in dynamic functions. Honours thesis, Queensland University of Technology, Brisbane, Australia, Nov. 1991.
3. P. J. Angeline. Tracking extrema in dynamic environments. In P. J. Angeline, R. G. Reynolds, J. R. McDonnell, and R. Eberhart, eds., *Proceedings of the 6th International Conference on Evolutionary Programming* (*EP'97*), Indianapolis, IN, vol. 1213 of *Lecture Notes in Computer Science*. Springer-Verlag, New York, 1997, pp. 335–345.
4. L. Bianchi. Notes on dynamic vehicle routing: the state of the art. *Technical Report IDSIA 05-01*. Istituto dalle Mòlle di Studi sull' Intelligènza Artificiale (IDSIA), Manno-Lugano, Switzerland, 2000.
5. T. Blackwell and J. Branke. Multi-swarm optimization in dynamic environments. In G. R. Raidl, S. Cagnoni, J. Branke, D. Corne, R. Drechsler, Y. Jin, C. G. Johnson, P. Machado, E. Marchiori, F. Rothlauf, G. D. Smith, and G. Squillero, eds., *Proceedings of Applications of Evolutionary Computing: EvoBIO, EvoCOMNET, EvoHOT, EvoIASP, EvoMUSART, and EvoSTOC* (*EvoWorkshops'04*), vol. 3005 of *Lecture Notes in Computer Science*. Springer-Verlag, New York, 2004, pp. 489–500.
6. P. A. N. Bosman. Learning, anticipation and time-deception in evolutionary online dynamic optimization. In S. Yang and J. Branke, eds., *Proceedings of the 7th Genetic and Evolutionary Computation Conference: Workshop on Evolutionary Algorithms for Dynamic Optimization Problems* (*GECCO:EvoDOP'05*). ACM Press, New York, 2005, pp. 39–47.
7. P. A. N. Bosman and H. La Poutrè. Computationally intelligent online dynamic vehicle routing by explicit load prediction in an evolutionary algorithm. In T. P. Runarsson, H.-G. Beyer, E. Burke, J. J. Merelo-Guervos, and X. Yao, eds., *Proceedings of the 9th International Conference on Parallel Problem Solving from Nature* (*PPSN IX*), Reykjavik, Iceland, vol. 4193 of *Lecture Notes in Computer Science*. Springer-Verlag, New York, 2006, pp. 312–321.
8. P. A. N. Bosman and H. La Poutrè. Learning and anticipation in online dynamic optimization with evolutionary algorithms: the stochastic case. In D. Thierens, ed., *Proceedings of the 9th Genetic and Evolutionary Computation Conference* (*GECCO'07*), London. ACM Press, New York, 2007, pp. 1165–1172.

9. J. Branke. Moving peaks benchmark. http://www.aifb.uni-karlsruhe.de-/~jbr/MovPeaks/movpeaks/.

10. J. Branke. Memory enhanced evolutionary algorithms for changing optimization problems. In P. J. Angeline, Z. Michalewicz, M. Schoenauer, X. Yao, and A. Zalzala, eds., *Proceedings of the 1999 IEEE Congress on Evolutionary Computation* (*CEC'99*), Mayflower Hotel, Washington, DC, vol. 3. IEEE Press, Piscataway, NJ, 1999, pp. 1875–1882.

11. J. Branke. Evolutionary Optimization in Dynamic Environments, vol. 3 of *On Genetic Algorithms and Evolutionary Computation.* Kluwer Academic, Norwell, MA, 2002.

12. J. Branke, T. Kaufler, C. Schmidt, and H. Schmeck. A multipopulation approach to dynamic optimization problems. In I. C. Parmee, ed., *Proceedings of the 4th International Conference on Adaptive Computing in Design and Manufacture* (*ACDM'00*), University of Plymouth, Devon, UK., Springer-Verlag, New York, 2000, pp. 299–308.

13. J. Branke and D. Mattfeld. Anticipation in dynamic optimization: the scheduling case. In M. Schoenauer, K. Deb, G. Rudolph, X. Yao, E. Lutton, J. J. Merelo, and H.-P. Schwefel, eds., *Proceedings of the 5th International Conference on Parallel Problem Solving from Nature* (*PPSN V*), Inria Hotel, Paris, vol. 1917 of *Lecture Notes in Computer Science*, Springer-Verlag, New York, 2000, pp. 253–262.

14. W. Cedeno and V. R. Vemuri. On the use of niching for dynamic landscapes. In *Proceedings of the International Conference on Evolutionary Computation* (*ICEC'97*), Indianapolis, IN. IEEE Press, Piscataway, NJ, 1997, pp. 361–366.

15. H. G. Cobb. An investigation into the use of hypermutation as an adaptive operator in genetic algorithms having continuous, time-dependent nonstationary environments. *Technical Report AIC-90-001*. Navy Center for Applied Research in Artificial Intelligence, Washington, DC, 1990.

16. K. A. De Jong. An analysis of the behavior of a class of genetic adaptive systems. Ph.D. thesis, University of Michigan, Ann Arbor, MI, 1975.

17. A. V. Donati, L. M. Gambardella, N. Casagrande, A. E. Rizzoli, and R. Montemanni. Time dependent vehicle routing problem with an ant colony system. *Technical Report IDSIA-02-03*. Istituto dalle Mòlle Di Studi Sull' Intelligènza Artificiale (IDSIA), Manno-Lugano, Switzerland, 2003.

18. C. J. Eyckelhof and M. Snoek. Ant systems for dynamic problems—the TSP case: ants caught in a traffic jam. In M. Dorigo, G. Di Caro, and M. Sampels, eds., *Proceedings of the 3rd International Workshop on Ant Algorithms* (*ANTS'02*), London. vol. 2463 of *Lecture Notes in Computer Science*. Springer-Verlag, New York, 2002, pp. 88–99.

19. H.-L. Fang, P. Ross, and D. Corne. A promising genetic algorithm approach to job-shop scheduling, re-scheduling, and open-shop scheduling problems. In S. Forrest, ed., *Proceedings of the 5th International Conference on Genetic Algorithms* (*ICGA'93*), San Mateo, CA. Morgan Kaufmann, San Francisco, CA, 1993, pp. 375–382.

20. W. Feng, T. Brune, L. Chan, M. Chowdhury, C. K. Kuek, and Y. Li. Benchmarks for testing evolutionary algorithms. *Technical Report CSC-97006*. Center for System and Control, University of Glasgow, Glasgow, UK, 1997.

21. F. Glover and G. A. Kochenberg, editors. *Handbook of Metaheuristics*. Kluwer Academic, London, 2003.

22. D. E. Goldberg and R. E. Smith. Nonstationary function optimization using genetic algorithms with dominance and diploidy. In J. J. Grefenstette, ed., *Proceedings of the 2nd International Conference on Genetic Algorithms* (*ICGA'87*), Pittsburgh, PA. Lawrence Erlbaum Associates Publishers, Hillsdale, NJ, 1987, pp. 59–68.

23. J. J. Grefenstette. Genetic algorithms for changing environments. In R. Männer and B. Manderickm, eds., *Proceedings of the 2nd International Conference on Parallel Problem Solving from Nature* (*PPSN II*), Brussels, Belgium. Elsevier, Amsterdam, 1992, pp. 137–144.

24. M. Guntsch and M. Middendorf. Applying population based ACO to dynamic optimization problems. In M. Dorigo, G. Di Caro, and M. Sampels, eds., *Proceedings of the 3rd International Workshop on Ant Algorithms* (*ANTS'02*), London, vol. 2463 of *Lecture Notes in Computer Science*. Springer-Verlag, New York, 2002, pp. 111–122.

25. M. Guntsch and M. Middendorf. A population based approach for ACO. In S. Cagnoni, J. Gottlieb, E. Hart, M. Middendorf, and G. R. Raidl, eds., *Proceedings of Applications of Evolutionary Computing: EvoCOP, EvoIASP, EvoSTIM/EvoPLAN* (*EvoWorkshops'02*), Kinsale, UK, vol. 2279 of *Lecture Notes in Computer Science*. Springer-Verlag, New York, 2002, pp. 71–80.

26. L. Hongbo and A. Ajith. Fuzzy adaptive turbulent particle swarm optimization. In N. Nedjah, L. M. Mourelle, A. Vellasco, M. M. B. R. and Abraham, and M. Köppen, eds., *Proceedings of 5th International Conference on Hybrid Intelligent Systems* (*HIS'05*), Río de Janeiro, Brazil. IEEE Press, Piscataway, NJ, 2005, pp. 445–450.

27. X. Li. Adaptively choosing neighbourhood bests using species in a particle swarm optimizer for multimodal function optimization. In R. Poli et al., eds., *Proceedings of the 6th Genetic and Evolutionary Computation Conference* (*GECCO'04*), Seattle, WA, vol. 3102 of *Lecture Notes in Computer Science*. Springer-Verlag, New York, 2004, pp. 105–116.

28. X. Li, J. Branke, and T. Blackwell. Particle swarm with speciation and adaptation in a dynamic environment. In M. Cattolico, ed., *Proceedings of the 8th Genetic and Evolutionary Computation Conference* (*GECCO'06*), Seattle, WA. ACM Press, New York, 2006, pp. 51–58.

29. L. T. Bui, H. A. Abbass, and J. Branke. Multiobjective optimization for dynamic environments. In *Proceedings of the 2005 IEEE Congress on Evolutionary Computation* (*CEC'05*), vol. 3, Edinburgh, UK. IEEE Press, Piscataway, NJ, 2005, pp. 2349–2356.

30. R. Mendes and A. Mohais. DynDE: a differential evolution for dynamic optimization problems. In *Proceedings of the 2005 IEEE Congress on Evolutionary Computation* (*CEC'05*), Edinburgh, UK. IEEE Press, Piscataway, NJ, 2005, pp. 2808–2815.

31. N. Mori, H. Kita, and Y. Nishikawa. Adaptation to a changing environment by means of the thermodynamical genetic algorithm. In H. Voigt, W. Ebeling, and I. Rechenberg, eds., *Proceedings of the 4th International Conference on Parallel Problem Solving from Nature* (*PPSN IV*), Berlin, vol. 1141 of *Lecture Notes in Computer Science*. Springer-Verlag, New York, 1996, pp. 513–522.

32. R. W. Morrison. Performance measurement in dynamic environments. In E. Cantú-Paz, J. A. Foster, K. Deb, L. Davis, R. Roy, U.-M. O'Reilly, H.-G. Beyer, R. K. Standish, G. Kendall, S. W. Wilson, M. Harman, J. Wegener, D. Dasgupta, M. A. Potter, A. C. Schultz, K. A. Dowsland, N. Jonoska, and J. F. Miller, eds., *Proceedings of the 5th Genetic and Evolutionary Computation Conference: Workshop on*

Evolutionary Algorithms for Dynamic Optimization Problems (*GECCO:EvoDOP'03*), Chigaco, vol. 2723 of *Lecture Notes in Computer Science*. Springer-Verlag, New York, 2003, pp. 99–102.

33. R. W. Morrison. *Designing Evolutionary Algorithms for Dynamic Environments*. Springer-Verlag, New York, 2004.
34. R. W. Morrison and K. A. De Jong. A test problem generator for non-stationary environments. In P. J. Angeline, Z. Michalewicz, M. Schoenauer, X. Yao, and A. Zalzala, eds., *Proceedings of the 1999 IEEE Congress on Evolutionary Computation* (*CEC'99*), Mayflower Hotel, Washington, DC, vol. 3. IEEE Press, Piscataway, NJ, 1999, pp. 2047–2053.
35. E. Pettit and K. M. Swigger. An analysis of genetic-based pattern tracking and cognitive-based component tracking models of adaptation. In M. Genesereth, ed., *Proceedings of the National Conference on Artificial Intelligence* (*AAAI'83*), Washington, DC. Morgan Kaufmann, San Francisco, CA, 1983, pp. 327–332.
36. H. N. Psaraftis. Dynamic vehicle routing: status and prospect. *Annals of Operations Research*, 61:143–164, 1995.
37. K. Trojanowski and Z. Michalewicz. Searching for optima in non-stationary environments. In P. J. Angeline, Z. Michalewicz, M. Schoenauer, X. Yao, and A. Zalzala, eds., *Proceedings of the 1999 IEEE Congress on Evolutionary Computation* (*CEC'99*), Mayflower Hotel, Washington, DC. IEEE Press, Piscataway, NJ, 1999, pp. 1843–1850.
38. R. K. Ursem. Multinational GA optimization techniques in dynamic environments. In D. Whitley, D. Goldberg, E. Cantú-Paz, L. Spector, I. Parmee, and H.-G. Beyer, eds., *Proceedings of the 2nd Genetic and Evolutionary Computation Conference* (*GECCO'00*), Las Vegas, NV. Morgan Kaufmann, San Francisco, CA, 2000, pp. 19–26.
39. J. I. van Hemert and J. A. La Poutrè. Dynamic routing problems with fruitful regions: models and evolutionary computation. In Y. Yao, E. Burke, J. A. Lozano, J. Smith, J. J. Merelo-Guervós, J. A. Bullinaria, J. Rowe, P. Tiňo, A. Kabán, and H.-P. Schwefel, eds., *Proceedings of the 8th International Conference on Parallel Problem Solving from Nature* (*PPSN VIII*), Birmingham, UK, vol. 3242 of *Lecture Notes in Computer Science*. Springer-Verlag, New York, 2004, pp. 692–701.
40. F. Vavak, K. Jukes, and T. C. Fogarty. Adaptive combustion balancing in multiple burner boiler using a genetic algorithm with variable range of local search. In T. Bäck, ed., *Proceedings of the 7th International Conference on Genetic Algorithms* (*ICGA'97*), Michigan State University, East Lansing, MI. Morgan Kaufmann, San Francisco, CA, 1997, pp. 719–726.
41. K. Weicker. Performance measures for dynamic environments. In J. J. Merelo-Guervós, P. Adamidis, H.-G. Beyer, J.-L. Fernández-Villacañas, and H.-P. Schwefel, eds., *Proceedings of the 7th International Conference on Parallel Problem Solving from Nature* (*PPSN VII*), Granada, Spain, vol. 2439 of *Lecture Notes in Computer Science*. Springer-Verlag, New York, 2002, pp. 64–76.
42. D. Zaharie. Control of population diversity and adaptation in differential evolution algorithms. In R. Matousek and P. Osmera, eds., *Proceedings of the 9th International Conference of Soft Computing* (*MENDEL'03*), Faculty of Mechanical Engineering, Brno, Czech Republic, June 2003, pp. 41–46.

43. D. Zaharie and F. Zamfirache. Diversity enhancing mechanisms for evolutionary optimization in static and dynamic environments. In *Proceedings of the 3rd Romanian–Hungarian Joint Symposium on Applied Computational Intelligence and Informatics (SACI'06)*, Timisoara, Romania, May 2006, pp. 460–471.

CHAPTER 7

Solving Constrained Optimization Problems with Hybrid Evolutionary Algorithms

C. COTTA and A. J. FERNÁNDEZ

Universidad de Málaga, Spain

7.1 INTRODUCTION

The foundations for *evolutionary algorithms* (EAs) were established at the end of the 1960s [1,2] and strengthened at the beginning of the 1970s [3,4]. EAs appeared as an alternative to exact or approximate optimization methods whose performance in many real problems was not acceptable. When applied to real problems, EAs provide a valuable relation between the quality of the solution and the efficiency in obtaining it; for this reason these techniques attracted the immediate attention of many researchers and became what they now represent: a cutting-edge approach to real-world optimization. Certainly, this has also been the case for related techniques such as *simulated annealing* (SA) [5] and *tabu search* (TS) [6]. The term *metaheuristics* has been coined to denote them.

The term *hybrid evolutionary algorithm* (HEA) or *hybrid metaheuristics* refers to the combination of an evolutionary technique (i.e., metaheuristics) and another (perhaps exact or approximate) technique for optimization. The aim is to combine the best of both worlds, with the objective of producing better results than can be obtained by any component working alone. HEAs have proved to be very successful in the optimization of many practical problems [7,8], and as a consequence, there currently is increasing interest in the optimization community for this type of technique.

One crucial point in the development of HEAs (and hybrid metaheuristics in general) is the need to exploit problem knowledge, as was demonstrated clearly in the formulation of the *no free lunch theorem* (NFL) by Wolpert and Macready

Optimization Techniques for Solving Complex Problems, Edited by Enrique Alba, Christian Blum, Pedro Isasi, Coromoto León, and Juan Antonio Gómez

[9] (a search algorithm that performs in strict accordance with the amount and quality of the problem knowledge incorporated). Quite interestingly, this line of thinking had already been advocated by several researchers in the late 1980s and early 1990s (e.g., Hart and Belew [10], Davis [11], and Moscato [12]). This is precisely the foundation of one of the most known instances of HEAs, the *memetic algorithms* (MAs), a term used first in the work of Moscato [13,14,15]. Basically, an MA is a search strategy in which a population of optimizing agents synergistically cooperate and compete [12]. These agents are concerned explicitly with using knowledge from the problem being solved, as suggested in both theory and practice [16]. The success of MAs is evident, and one of the consequences is that currently the term *memetic algorithm* is used as synonym of for *hybrid evolutionary algorithm*, although in essence MAs are a particular case of HEAs.

As already mentioned, HEAs were born to tackle many problems that are very difficult to solve using evolutionary techniques or classical approaches working alone. This is precisely the case for *constrained problems*. Generally speaking, a constrained problem consists of a set of constraints involving a number of variables restricted to having values in a set of (possibly different) finite domains. Basically, a constraint is a relation maintained between the entities (e.g., objects or variables) of a problem, and constraints are used to model the behavior of systems in the real world by capturing an idealized view of the interaction between the variables involved. Solving a constrained problem consists of finding a possible assignment (of values in the computation domains) for the constrained variables that satisfies all the constraints. This type of problem can be solved using different techniques, ranging from traditional to modern. For example, some approaches to solving a problem are in the area of operational research (OR), genetic algorithms, artificial intelligence (AI) techniques, rule-based computations, conventional programs, and constraint-based approaches. Usually, solving is understood as the task of searching for a single solution to a problem, although sometimes it is necessary to find the set of all solutions. Also, in certain cases, because of the cost of finding all solutions, the aim is simply to find the best solution or an approximate solution within fixed resource bounds (e.g., in a reasonable time). Such types of constrained problems are called *partially constrained problems* (PCPs). An example of a PCP is a *constrained combinatorial optimization problem* (CCOP), which assigns a cost to each solution and tries to find an optimal solution within a given time frame [17]. In this chapter we focus on CCOPs.

Not surprisingly, HEAs have been used extensively to solve these types of problems. This chapter represents our particular share in this sense. In this work we analyze the deployment of HEAs on this domain. Initially, in Section 7.2 we provide a generic definition for these types of problems and an overview of general design guidelines to tackle CCOPs. In Section 7.3 we discuss a number of interesting and not well known (at least, in the EA community) CCOPs that the authors have recently attacked via HEAs. In particular, for each problem we provide a formal definition and discuss some interesting proposals that have tackled the problem. The chapter ends with a summary of lessons learned and some current and emerging research trends in HEAs for managing CCOPs.

7.2 STRATEGIES FOR SOLVING CCOPs WITH HEAs

In general terms, an unconstrained COP is often defined as a tuple $\langle S, f\rangle$, where $S \triangleq D_1 \times D_2 \times \cdots \times D_N$ is the search space, and $f\colon S \longrightarrow \mathbb{Z}$ is the objective function. Each D_i is a discrete set containing the values that a certain variable x_i can take. Therefore, any solution $\vec{x} \in S$ is a list $\langle x_1, \ldots, x_n\rangle$, and we are interested in finding the solution $\vec{x}$ minimizing (w log) the objective function f.

A constrained COP arises when we are just interested in optimizing the objective function within a subset $S_{\text{val}} \subset S$. This subset S_{val} represents *valid* or *feasible solutions*, and it may even be the case that f is a partial function defined only on these feasible solutions. The incidence vector of S_{val} on S defines a Boolean function $\phi\colon S \longrightarrow \mathbb{B}$ [i.e., $\phi(\vec{x}) = 1$ iff $\vec{x}$ is a valid solution]. In practice, constrained COPs include a well-structured function ϕ such that it can be computed efficiently. Typically, this is achieved via the conjunction of several simpler Boolean functions $\phi_i\colon S \longrightarrow \mathbb{B}$ [i.e., $\phi(\vec{x}) = \prod_i \phi_i(\vec{x})$]. Each of these functions is a *constraint*.

As an example, consider the *vertex cover problem*: Given a graph $G(V, E)$, find a subset $V' \subseteq V$ of minimal size such that for any $(u, v) \in E$, it holds that $\{u, v\} \cap V' \neq \emptyset$. In this case, $S = \{0, 1\}^{|V|}$, the objective function (to be minimized) is $f(\vec{x}) = \sum_{i=1}^{|V|} x_i$, and there is a collection of binary constraints $\phi_{uv} = \min(1, x_u + x_v)$, $(u, v) \in E$.

The previous definition can easily be generalized to weighted CCOPs, where we do not simply have to know whether or not a solution satisfies a particular constraint, but we also have an indication *how far* that solution is from satisfying that constraint (in case it does not satisfy it at all). For this purpose, it is more convenient to denote each constraint as a function $\delta_i\colon S \longrightarrow \mathbb{N}$, where $\delta(\vec{x}) = 0$ indicates fulfillment of the corresponding constraint, and any strictly positive value indicates an increasingly higher degree of violation of that constraint. Note that $\phi_i(\vec{x}) = \max(0, 1 - \delta_i(\vec{x}))$. The formulation above of weighted CCOPs allows a rather straightforward approach to tackling such a problem with HEAs: namely, incorporating the constraint functions within the objective function. This can be done in different ways [18], but the simplest method is aggregating all constraint functions within a *penalty term* that is added to the objective function: for example,

$$f'(\vec{x}) = f(\vec{x}) + \sum_i \delta_i(\vec{x}) \tag{7.1}$$

Different variants can be defined here, such as raising each δ_i term to a certain power (thus avoiding linear compensation among constraints and/or biasing the search toward low violation degrees) or adding an offset value in case of infeasibility to ensure that any feasible solution is preferable to any nonfeasible solution.

This penalty approach can obviously be used only in those cases in which the objective function is defined on nonfeasible solutions. A typical example is the *multidimensional 0–1 knapsack* problem (MKP). This problem is defined via a row vector of profits $\vec{p}$, a column vector of capacity constraints $\vec{c}$, and a constraint matrix M. Solutions are binary column vectors $\vec{x}$, the objective function is $f(\vec{x}) = \vec{p} \cdot \vec{x}$, and the feasibility constraint is $M \cdot \vec{x} \leq \vec{c}$. Let $\vec{d} = \vec{c} - M \cdot \vec{x}$. Then $\delta_i = -\min(0, d_i)$.

This approach has the advantage of being very simple and allowing the use of rather standard EAs. Nevertheless, a HEA can provide a notably better solution if it is aware of these penalty terms and focuses on their optimization (see, e.g., the *maximum density still life problem* described in Section 7.3.2, or the *social golfer problem* described in Section 7.3.3). An alternative approach is trying to enforce the search being carried within the feasible region of the search space. This can be done in two ways:

1. By allowing the algorithm to traverse the infeasible region temporarily during the reproduction phase, but using a repairing procedure to turn nonfeasible solutions into feasible solutions before evaluation.
2. By restricting the search to the feasible region at all times. This can in turn be done in two ways:
   (a) By defining appropriate initialization, recombination, and mutation operators that take and produce feasible solutions.
   (b) By defining an unconstrained auxiliary search space S_{aux} and an adequate mapping *dec*: $S_{\text{aux}} \longrightarrow S_{\text{val}}$.

The repair approach is possibly the simplest option after the penalty approach, although it must be noted that a straightforward repair procedure is not always available. In any case it is interesting to note that in some sense (and depending on the particular procedure chosen for repairing), this stage can be regarded as a local search phase and some repair-based EAs can therefore qualify as memetic algorithms. An example of repairing can be found in the GA defined by Chu and Beasley for the MKP [19]. They define a heuristic order for traversing the variables and keep setting them to zero as long as the solution is nonfeasible (this procedure is complemented by a subsequent improvement phase in which variables are set to 1 in inverse order as long as the solution remains feasible). Another example of repairing can be found in the *protein structure prediction problem* [20], in which the search space is composed of all embeddings of a given string in a certain fixed lattice, and solutions are feasible only if they are self-avoiding. The repairing procedure is in this case more complex and requires the use of a backtracking algorithm to produce a feasible embedding. Nevertheless, even accounting for this additional cost, the HEA performs better than a simpler penalty-based approach.

Restricting the search to the feasible space via appropriate operators is in general much more complex, although it is the natural approach in certain problems, most notably in permutational problems. Feasible initialization and mutation are rather straightforward in this case, and there exists an extensive literature dealing with the definition of adequate operators for recombination, (see, e.g., ref. 21). Fixed-size subset problems are also dealt with easily using this method. As to the use of decoders, it is an arguably simpler and more popular approach. A common example, again on the MKP, is to define an auxiliary search space composed of all permutations of objects and to decode a particular permutation using a procedure that traverses the list of objects in the order indicated and includes objects in the solution if doing so does not result in a nonfeasible solution. Also for the MKP problem, a different approach was defined by Cotta and Troya [22] using lists of integers representing perturbations of the original instance data, and utilizing a greedy algorithm to obtain a solution from the instance modified. Note that contrary to some suggestions in the literature, it is not necessary (and often not even recommended) to have a decoder capable of producing any feasible solution or producing them with the same frequency: It is perfectly admissible to ignore suboptimal solutions and/or to bias the search to promising regions via overrepresentation of certain solutions (this is precisely the case with the previous example). Another example of a general approach can be found in a GRASP-like decoding procedure [23] in which solutions are encoded via a list of natural numbers that are used to control the level of greediness of a constructive heuristic. This approach has been used with success in the Golomb ruler *problem* (see Section 7.3.1).

A particular problem with this decoder approach is the fact that locality is easily lost; that is, a small change in the genotype can result in a large change in the phenotype [24,25]. This was observed, for example, in the indirect encoding of trees via Prüfer numbers [26].

7.3 STUDY CASES

In this section we review our recent work on solving CCOPs by applying hybrid collaborative techniques involving evolutionary techniques. In particular, we focus on four constrained problems that are not very well known in the evolutionary programming community and that we have tackled with a certain level of success in recent years.

7.3.1 Optimal Golomb Rulers

The concept of Golomb rulers was introduced by Babcock in 1953 [27], and described further by Bloom and Golomb [28]. *Golomb rulers* are a class of undirected graphs that, unlike most rulers, measure more discrete lengths than the number of marks they carry. The particularity of Golomb rulers is that on any given ruler, all differences between pairs of marks are unique, which makes them really interesting in many practical applications [29,30].

Traditionally, researchers are interested in discovering the *optimal golomb ruler* (OGR), that is, the shortest Golomb ruler for a number of marks. The task of finding optimal or nearly optimal Golomb rulers is computationally very difficult and results in an extremely challenging combinatorial problem. Proof of this is the fact that the search for an optimal 19-mark Golomb ruler took approximately 36,200 CPU hours on a Sun Sparc workstation using a very specialized algorithm [31]. Also, optimal solutions for 20 to 24 marks were obtained using massive parallelism projects, taking from several months up to four years (for the 24-mark instance) for each of those instances [30,32–34].

The OGR problem can be classified as a fixed-size subset selection problem, such as the p-median problem [35], although it exhibits some very distinctive features.

Formal Definition An *n-mark Golomb ruler* is an ordered set of n distinct nonnegative integers, called *marks*, $a_1 < \cdots < a_n$, such that all the differences $a_i - a_j$ $(i > j)$ are distinct. We thus have a number of constraints of the form $a_i - a_j \neq a_k - a_m$ $[i > j, k > m, (i, j) \neq (k, m)]$. Clearly, we may assume that $a_1 = 0$. By convention, a_n is the *length of the Golomb ruler*. A Golomb ruler with n marks is an optimal Golomb ruler if and only if:

- There exist no other n-mark Golomb rulers of shorter length.
- The ruler is canonically "smaller" with respect to equivalent rulers. This means that the first differing entry is less than the corresponding entry in the other ruler.

Figure 7.1 shows an OGR with four-marks. Observe that all distances between any two marks are different. Typically, Golomb rulers are represented by the values of the marks on the ruler [i.e., in a n-mark Golomb ruler, $a_i = x$ $(1 \leq i \leq n)$ means that x is the mark value in position i]. The sequence $(0, 1, 4, 6)$ would then represent the ruler in Figure 7.1. However, this representation turns out to be inappropriate for EAs (e.g., it is problematic with respect to developing good crossover operators [36]). An alternative representation consists of representing the Golomb ruler via the lengths of its segments, where the length of a segment of a ruler is defined as the distance between two consecutive marks. Therefore, a Golomb ruler can be represented by $n - 1$ marks, specifying the

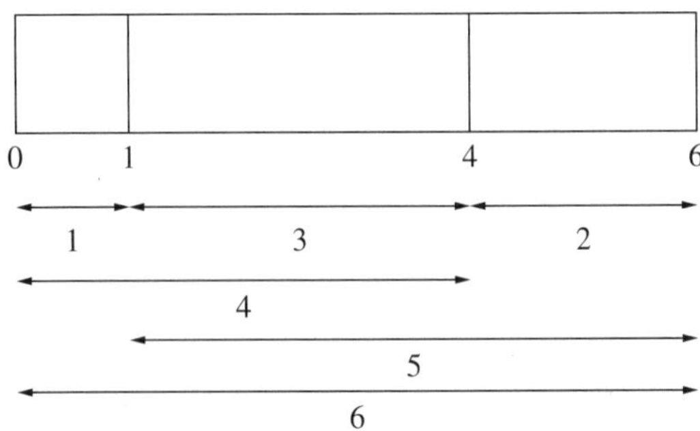

Figure 7.1 Golomb ruler with four marks.

lengths of the $n - 1$ segments that compose it. In the previous example, the sequence $(1, 3, 2)$ would encode the ruler depicted in Figure 7.1.

Solving OGRs In addition to related work discussed above, here we discuss a variety of techniques that have been used to find OGRs: for instance, systematic (exact) methods such as the method proposed by Shearer to compute OGRs up to 16 marks [37]. Basically, this method was based on the utilization of branch-and-bound algorithms combined with a depth-first search strategy (i.e., backtracking algorithms), making its use in experiments of upper bound sets equal to the length of the best known solution. Constraint programming (CP) techniques have also been used, although with limited success [38,39]. The main drawback of these complete methods is that they are costly computationally (i.e., time consuming). A very interesting hybridization of local search (LS) and CP to tackle the problem was presented by Prestwich [40]; up to size 13, the algorithm is run until the optimum is found, and for higher instances the quality of the solutions deteriorates.

Hybrid evolutionary algorithms were also applied to this problem. In this case two main approaches can be considered in tackling the OGR problem with EAs. The first is the *direct approach*, in which the EA conducts the search in the space $\mathcal{S}_G$ of all possible Golomb rulers. The second is the *indirect approach*, in which an auxiliary $\mathcal{S}_{\text{aux}}$ space is used by the EA. In the latter case, a decoder [41] must be utilized to perform the $\mathcal{S}_{\text{aux}} \longrightarrow \mathcal{S}_G$ mapping. Examples of the former (direct) approach are the works of Soliday et al. [36] and Feeney [29]. As to the latter (indirect) approach, we cite work by Pereira et al. [42] (based on the notion of random keys [43]); also, Cotta and Fernández [23] used a problem-aware procedure (inspired in GRASP [44]) to perform genotype-to-phenotype mapping with the aim of ensuring the generation of feasible solutions; this method was shown to outperform other approaches.

A HEA incorporating a tabu search algorithm for mutation was proposed by Dotú and Van Hentenryck [45]. The basic idea was to optimize the length of the rulers indirectly by solving a sequence of feasibility problems (starting from an upper bound l and producing a sequence of rulers of length $l_1 > l_2 > \cdots > l_i > \cdots$). This algorithm performed very efficiently and was able to find OGRs for up to 14 marks; in any case we note that this method requires an estimated initial upper bound, something that clearly favored its efficiency. At the same time, we conducted a theoretical analysis on the problem, trying to shed some light on the question of what makes a problem difficult for a certain search algorithm for the OGR problem. This study [46] consisted of an analysis of the fitness landscape of the problem. Our analysis indicated that the high irregularity of the neighborhood structure for the direct formulation introduces a drift force toward low-fitness regions of the search space. The indirect formulation that we considered earlier [23] does not have this drawback and hence would in principle be more amenable for conducting a local search. Then we presented an MA in which our indirect approach (i.e., a GRASP-based EA) was used in the phases of initialization and restarting of the population [47], whereas a direct approach in the form of a local improvement method based on the tabu search (TS) algorithm [45] was considered in the stages of recombination and local improvement. Experimental results showed that this algorithm could solve OGRs up to 15 marks and produced Golomb rulers for 16 marks that are very close to the optimal value (i.e., 1.1% distant), thus significantly improving the results reported in the EA literature.

Recently [48], we combined ideas from greedy randomized adaptive search procedures (GRASP) [49], scatter search (SS) [50,51], tabu search (TS) [52,53], clustering [54], and constraint programming (CP), and the resulting algorithm was able of solving the OGR problem for up to 16 marks, a notable improvement with regard to previous approaches reported in the literature in all the areas mentioned (i.e., EAs, LS, and CP). This algorithm yields a metaheuristic approach that is currently state of the art in relation to other metaheuristic approaches.

7.3.2 Maximum Density Still Life Problem

Conway's game of life [55] consists of an infinite checkerboard in which the only player places checkers on some of its squares. Each square has eight neighbors: the eight cells that share one or two corners with it. A cell is alive if there is a checker on it, and dead otherwise. The state of the board evolves iteratively according to three rules: (1) if a cell has exactly two living neighbors, its state remains the same in the next iteration; (2) if a cell has exactly three living neighbors, it is alive in the next iteration; and (3) if a cell has fewer than two or more than three living neighbors, it is dead in the next iteration.

One challenging constrained optimization problem based on the game of life is the *maximum density still life problem* (MDSLP). To introduce this problem, let us define a stable pattern (also called a *still life*) as a board configuration that does not change through time, and let the *density* of a region be its percentage of living cells. The MDSLP in an $n \times n$ grid consists of finding a still life of maximum density. This problem is very difficult to solve, and although to the best of our knowledge, it has not been proved to be NP-hard, no polynomial-time algorithm for it is known. The problem has a number of interesting applications [56–58], and a dedicated web page (http://www.ai.sri.com/~nysmith/life) maintains up-to-date results. Figure 7.2 shows some maximum density still lives for small values of n.

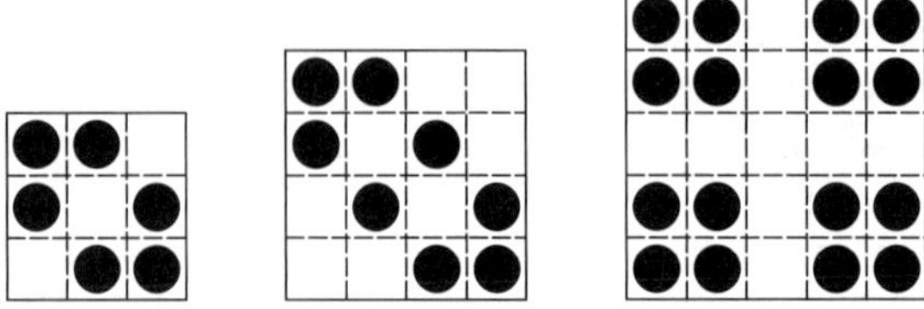

Figure 7.2 Maximum density still lives for $n \in \{3, 4, 5\}$.

Formal Definition The constraints and objectives of the MDSLP are formalized in this section, in which we follow a notation similar to that used by Larrosa et al. [59,60]. To state the problem formally, let r be an $n \times n$ binary matrix, such that $r_{ij} \in \{0, 1\}$, $1 \leq i, j \leq n$ [$r_{ij} = 0$ if cell (i, j) is dead, and 1 otherwise]. In addition, let $\mathcal{N}(r, i, j)$ be the set comprising the neighborhood of cell r_{ij}:

$$\mathcal{N}(r, i, j) = \{ r_{(i+x)(j+y)} \mid x, y \in \{-1, 0, 1\} \wedge x^2 + y^2 \neq 0 \wedge 1 \leq (i + x), (j + y) \leq \|r\| \} \tag{7.2}$$

where $\|r\|$ denotes the number of rows (or columns) of square matrix r, and let the number of living neighbors for cell r_{ij} be noted as $\eta(r, i, j)$:

$$\eta(r, i, j) = \sum_{c \in \mathcal{N}(r,i,j)} c \tag{7.3}$$

According to the rules of the game, let us also define the following predicate that checks whether cell r_{ij} is stable:

$$S(r, i, j) = \begin{cases} 2 \leq \eta(r, i, j) \leq 3 & r_{ij} = 1 \\ \eta(r, i, j) \neq 3 & r_{ij} = 0 \end{cases} \tag{7.4}$$

To check boundary conditions, we further denote by $\tilde{r}$ the $(n + 2) \times (n + 2)$ matrix obtained by embedding r in a frame of dead cells:

$$\tilde{r}_{ij} = \begin{cases} r_{(i-1)(j-1)} & 2 \leq i, j \leq n + 1 \\ 0 & \text{otherwise} \end{cases} \tag{7.5}$$

The maximum density still life problem for an $n \times n$ board, MDSLP(n), can now be stated as finding an $n \times n$ binary matrix r such that

$$\sum_{1 \leq i, j \leq n} (1 - r_{ij}) \text{ is minimal} \tag{7.6}$$

subject to

$$\bigwedge_{1 \leq i, j \leq n+2} S(\tilde{r}, i, j) \tag{7.7}$$

Solving MDSLPs The MDSLP has been tackled using different approaches. Bosch and Trick [61] compared different formulations for the MDSLP using integer programming (IP) and constraint programming (CP). Their best results were obtained with a hybrid algorithm that mixed the two approaches. They were able to solve the cases for $n = 14$ and $n = 15$ in about 6 and 8 days of CPU time, respectively. Smith [62] used a pure constraint programming approach to attack the problem and proposed a formulation of the problem as a constraint satisfaction problem with 0–1 variables and nonbinary constraints; only instances up to $n = 10$ could be solved. The best results for this problem, reported by Larrosa et al. [59,60], showed the usefulness of an exact technique based on variable elimination and commonly used in solving constraint satisfaction problems: *bucket elimination* (BE) [63]. Their basic approach could solve the problem for $n = 14$ in about 10^5 seconds. Further improvements pushed the solvability boundary forward to $n = 20$ in about twice as much time. Recently, Cheng and Yap [64,65] tackled the problem via the use of ad hoc global `case` constraints, but their results are far from those obtained previously by Larrosa et al.

Note that all the techniques discussed previously are exact approaches inherently limited for increasing problem sizes, whose capabilities as "anytime" algorithms are unclear. To avoid this limitation, we proposed the use of HEAs to tackle the problem. In particular, we considered an MA consisting of an EA endowed with tabu search, where BE is used as a mechanism for recombining solutions, providing the best possible child from the parental set [66]. Experimental tests indicated that the algorithm provided optimal or nearly optimal results at an acceptable computational cost. A subsequent paper [67] dealt with expanded multilevel models in which our previous exact/metaheuristic hybrid was further hybridized with a branch-and-bound derivative: beam search (BS). The experimental results show that our hybrid evolutionary proposals were a practical and efficient alternative to the exact techniques employed so far to obtain still life patterns.

We recently proposed [48] a new hybrid algorithm that uses the technique of minibuckets (MBs) [68] to further improve the lower bounds of the partial solutions that are considered in the BS part of the hybrid algorithm. This new algorithm is obtained from the hybridization, at different levels, of complete solving techniques (BE), incomplete deterministic methods (BS and MB), and stochastic algorithms (MAs). An experimental analysis showed that this new proposal consistently finds optimal solutions for MDSLP instances up to $n = 20$ in considerably less time than do all the previous approaches reported in the literature. Moreover, this HEA performed at the state-of-the-art level, providing solutions that are equal to or better than the best ones reported to date in the literature.

7.3.3 Social Golfer Problem

The *social golfer problem* (SGP), first posted at http://sci.op-research in May 1998, consists of scheduling $n = g \cdot s$ golfers into g groups of s players every week for w weeks so that no two golfers play in the same group more than once. The problem can be regarded as an optimization problem if for two given values of g and s, we ask for the maximum number of weeks w the golfers can play together. SGP is a combinatorial constrained problem that raises interesting issues in symmetry breaking (e.g., players can be permuted within groups, groups can be ordered arbitrarily within every week, and even the weeks themselves can be permuted). Note that symmetry is also present in both the OGR problem and the MDSLP. Notice, however, that problem symmetry is beyond the scope of this chapter and thus symmetry issues will not be discussed explicitly here.

Formal Definition As mentioned above, SGP consists of scheduling $n = g \cdot s$ golfers into g groups of s players every week for w weeks, so that no two golfers play in the same group more than once. An instance of the social golfer is thus specified by a triplet $\langle g, s, w \rangle$. A (potentially infeasible) solution for such an instance is given by a schedule $\sigma : \mathbb{N}_g \times \mathbb{N}_w \longrightarrow 2^{\mathbb{N}_n}$, where $\mathbb{N}_i = \{1, 2, \ldots, i\}$, and $|\sigma(i, j)| = s$ for all $i \in \mathbb{N}_g$, $j \in \mathbb{N}_w$, that is, a function that on input (i, j) returns the set of s players that constitute the ith group of the jth week. There are many possible ways of modeling for SGP, which is one of the reasons that it is so interesting. In a generalized way, this problem can be modeled as a constraint satisfaction problem (CSP) defined by the following constraints:

- A golfer plays exactly once a week:

$$\forall p \in \mathbb{N}_n : \forall j \in \mathbb{N}_w : \exists! i \in \mathbb{N}_g : p \in \sigma(i, j) \tag{7.8}$$

  This constraint can be also formalized by claiming that no two groups in the same week intersect:

$$\forall j \in \mathbb{N}_w : \forall i, i' \in \mathbb{N}_g, i \neq i' : \sigma(i, j) \cap \sigma(i', j) = \emptyset \tag{7.9}$$

- No two golfers play together more than once:

$$\forall j, j' \in \mathbb{N}_w : \forall i, i' \in \mathbb{N}_g, i \neq i' : |\sigma(i, j) \cap \sigma(i', j')| \leq 1 \tag{7.10}$$

  This constraint can also be formulated as a weighted constraint: Let $\#_\sigma(a, b)$ be the number of times that golfers a and b play together in schedule σ:

$$\#_\sigma(a, b) = \sum_{i \in \mathbb{N}_g} \sum_{j \in \mathbb{N}_w} [\{a, b\} \subseteq \sigma(i, j)] \tag{7.11}$$

  where $[\cdot]$ is the Iverson bracket: namely, [`true`]= 1 and [`false`]= 0. Then we can define the degree of violation of a constraint a-and-b-play-together-at-most-once as $\max(0, \#_\sigma(a, b) - 1)$.

As already noted, symmetries can appear in (and can be removed from) this problem in several forms; see refs. 69–71 for more details.

Solving the SGP The SGP was first attacked by CP techniques that addressed the SGP mainly by detecting and breaking symmetries (see, e.g., refs. 72–79). Due to the interesting properties of the problem, it also attracted attention in other optimization areas and has been tackled extensively using different techniques. Here we mention just some of the most recent advances in solving the SGP. For example, Harvey and Winterer [69] have proposed constructing solutions to the SGP by using sets of mutually orthogonal Latin squares. Also, Gent and Lynce [79] have introduced a satisfiability (SAT) encoding for the SGP. Barnier and Brisset [70] have presented a combination of techniques to find efficient solutions to a specific instance of SGP, *Kirkman's schoolgirl problem*. Global constraints for lexicographic ordering have been proposed by Frisch et al. [80], being used for breaking symmetries in the SGP. Also, a tabu-based local search algorithm for the SGP is described by Dotú and Van Hentenryck [81].

To the best of our knowledge, we presented the first attempt of tackling the SGP by evolutionary techniques [71]; it consisted of a memetic algorithm (MA) that is based on the hybridization of evolutionary programming and tabu search. The flexibility of MAs eased the handling of the problem symmetries. Our MA, based on selection, mutation, and local search, performed at a state-of-the-art level for this problem.

7.3.4 Consensus Tree Problem

The inference (or reconstruction) of phylogenetic trees is a problem from the bioinformatics domain that has direct implications in areas such as multiple sequence alignment [82], protein structure prediction [83], and molecular epidemiological studies of viruses [84], just to cite a few. This (optimization) problem seeks the best tree representing the evolutionary history of a collection of species, therefore providing a hierarchical representation of the degree of closeness among a set of organisms. This is typically done on the basis of molecular information (e.g., DNA sequences) from these species, and can be approached in a number of ways: maximum likelihood, parsimony, distance matrices, and so on [85]. A number of different high-quality trees (with quality measured in different ways) can then be found, each possibly telling something about the *true* solution. Furthermore, the fact that data come from biological experiments, which are not exact, makes nearly optimal solutions (even nearly optimal with respect to different criteria) be almost as relevant as the actual optimum. It is in this situation where the *consensus tree problem* (often called *supertree*) comes into play [86]. Essentially, a consensus method tries to summarize a collection of trees provided as input, returning a single tree [87]. This implies identifying common substructures in the input trees and representing these in the output tree.

Formal Definition Let T be a strictly binary rooted tree; a LISP-like notation will be used to denote the structure of the tree. Thus, (sLR) is a tree with root s and with L and R as subtrees, and $(\cdot)$ is an empty tree. The notation (a) is a shortcut for $(a(\cdot)(\cdot))$. Let $\mathcal{L}(T)$ be the set of leaves of T. Each edge e in T defines a bipartition $\pi_T(e) = \langle S_1, S_2 \rangle$, where S_1 are the leaves in $\mathcal{L}(T)$ that can be reached from the root passing through e, and S_2 are the remaining leaves. We define $\Pi(T) = \{\pi_T(e) \mid e \in T\}$.

The consensus tree problem consists of representing a collection of trees $\{T_1, \ldots, T_m\}$ as a single tree that should be optimal with respect to a certain model. This model can be approached in several ways [87]; for instance, the models described by Gusfield [88] (i.e., tree compatibility problem) and by Day [89] (i.e., strict consensus) focus on finding a tree such that $\Pi(T) = \cup_{i=1}^{m} \Pi(T_i)$ and $\Pi(T) = \cap_{i=1}^{m} \Pi(T_i)$, respectively; also the model presented by Barthélemy and McMorris [90] (i.e., the median tree) tries to minimize the sum of differences between T and the input trees [i.e., $\min \sum_{i=1}^{m} d(T, T_i)$].

As a consequence, the meaning of *global optimum* is different from that of typical CCOPs since it is very dependent on the model selected as well as other metrics, such as the way of evaluating the differences between the trees (i.e., the distance between trees). This is so because the distance $d(T, T')$ between trees is not standard, and different alternatives can be considered. Perhaps the most typical distance is defined as the number of noncommon bipartitions in $\Pi(T)$ and $\Pi(T')$: This is also termed the *partition metric*. In any case, alternative metrics can be considered. For example, one can cite edit distance [measured in terms of some edition operations, such as nearest-neighbor interchange (NNI), subtree prune and regraft (SPR), or tree bisection and reconnection (TBR), among others; see ref. 91] or the TreeRank measure [92]. All of these metrics have advantages and drawbacks, so it is not always easy to determine which is the best election.

The fact that the problem admits different formulations is an additional reason that it so interesting. For example, recently, the problem was formulated as a constraint satisfaction problem [93]: Gent et al. presented constraint encoding based on the observation that any rooted tree can be considered as being minimum ultrametric [94] when we label interior nodes with their depth in that tree. This guarantees that any path from the root to a leaf corresponds to a strictly increasing sequence. See refs. 93 and 95 for more details about this encoding.

Solving the CTP Regarding the inference of phylogenetic trees, the use of classical exact techniques can be considered generally inappropriate in this context. Indeed, the use of heuristic techniques in this domain seems much more adequate. These can range from simple constructive heuristics (e.g., greedy agglomerative techniques such as UPGMA [96]) to complex metaheuristics [97] (e.g., evolutionary algorithms [98,99] or local search [100]). At any rate, it is well known that any heuristic method is going to perform in strict accordance with the amount of problem knowledge it incorporates [9,16]. Gallardo et al. [101] explored this possibility precisely and presented a model for the integration of branch-and-bound techniques [102] and memetic algorithms [13–15]. This model resulted in a synergistic combination yielding better results (experimentally speaking) than those of each of its constituent techniques, as well as classical agglomerative algorithms and other efficient tree construction methods.

Regarding the consensus tree problem, very different methods have been used (e.g., polynomial time algorithms [103], constraint programming (CP) [93,95,104], and evolutionary algorithms (EAs) [99], among others). With the aim of improving the results obtained so far, we are now experimenting with hybrid methods that combine the best of the CP and EA proposals. More specifically, we are evaluating two algorithms: The first is an MA [15] in which we use a CP method for supertree construction based on the method published by Gent et al. [93] as a recombination operator. The results should soon be reported [105].

7.4 CONCLUSIONS

Constrained COPs are ubiquitous and are representative of a plethora of relevant real-world problems. As such, they are also typically difficult to solve and demand the use of flexible cutting-edge optimization technologies for achieving competitive results. As we have shown, the framework of evolutionary computation offers solid ground on which powerful optimizations algorithms for CCOPs can be built via hybridization.

HEAs provide different options for dealing with CCOPs, ranging from the inclusion of constraints as side objectives to the definition of ad hoc operators working within the feasible region, including the use of repairing mechanisms or complex genotype–phenotype mappings. Each of these approaches is suited to different (not necessarily disjoint) optimization scenarios. Hence, they can avail practitioners in different ways, providing them with alternative methods for solving the problem at hand.

It is difficult to provide general design guidelines, since there are many problem specificities to be accounted for. The availability of other heuristics (either classical or not) may suggest the usefulness of performing a direct search in the space of solutions, so that these heuristics can be exploited by the HEA. The particulars of the objective function or the sparseness of valid solutions may dictate whether or not a penalty-based approach is feasible (e.g., nonfeasible solutions could not be evaluated at all, the search algorithm could spend most of the time outside the feasible region). The availability of construction heuristics can in turn suggest the utilization of indirect approaches (e.g., via decoder functions). This approach is sometimes overestimated, in the sense that one has to be careful to provide the HEA with a search landscape that is easy to navigate. Otherwise, the benefits of the heuristic mapping can be counteracted by the erratic search dynamics.

From a global perspective, the record of success of HEAs on CCOPs suggests that these techniques will be used increasingly to solve problems in this domain. It is then expected that the corpus of theoretical knowledge on the deployment of HEAs on CCOPs will grow alongside applications of these techniques to new problems in the area.

Acknowledgments

The authors are partially supported by the Ministry of Science and Technology and FEDER under contract TIN2005-08818-C04-01 (the OPLINK project). The second author is also supported under contract TIN-2007-67134.

REFERENCES

1. L. J. Fogel, A. J. Owens, and M. J. Walsh. *Artificial Intelligence Through Simulated Evolution*. Wiley, New York, 1966.
2. H.-P. Schwefel. Kybernetische Evolution als Strategie der experimentellen Forschung in der Strömungstechnik. Diplomarbeit, Technische Universität Berlin, Hermann Föttinger–Institut für Strömungstechnik, März, Germany, 1965.
3. J. Holland. *Adaptation in Natural and Artificial Systems*. University of Michigan Press, Ann Arbor, MI, 1975.
4. I. Rechenberg. *Evolutionsstrategie: optimierung technischer Systeme nach Prinzipien der biologischen Evolution*. Frommann-Holzboog, Stuttgart, Germany, 1973.
5. S. Kirkpatrick, C. D. Gelatt, Jr., and M. P. Vecchi. Optimization by simulated annealing. *Science*, 220(4598):671–680, 1983.
6. F. Glover and M. Laguna. *Tabu Search*. Kluwer Academic, Norwell, MA, 1997.
7. C. Reeves. Hybrid genetic algorithm for bin-packing and related problems. *Annals of Operation Research*, 63:371–396, 1996.
8. A. P. French, A. C. Robinson, and J. M. Wilson. Using a hybrid genetic-algorithm/branch and bound approach to solve feasibility and optimization integer programming problems. *Journal of Heuristics*, 7(6):551–564, 2001.
9. D. H. Wolpert and W. G. Macready. No free lunch theorems for optimization. *IEEE Transactions on Evolutionary Computation*, 1(1):67–82, 1997.
10. W. E. Hart and R. K. Belew. Optimizing an arbitrary function is hard for the genetic algorithm. In R. K. Belew and L. B. Booker, eds., *Proceedings of the 4th International Conference on Genetic Algorithms*, San Mateo CA. Morgan Kaufmann, San Francisco, CA, 1991, pp. 190–195.
11. L. D. Davis. *Handbook of Genetic Algorithms*. Van Nostrand Reinhold Computer Library, New York, 1991.
12. P. Moscato. On evolution, search, optimization, genetic algorithms and martial arts: towards memetic algorithms. *Technical Report 826*. Caltech Concurrent Computation Program, California Institute of Technology, Pasadena, CA, 1989.
13. P. Moscato. Memetic algorithms: a short introduction. In D. Corne, M. Dorigo, and F. Glover, eds., *New Ideas in Optimization*, McGraw-Hill, Maidenhead, UK, 1999, pp. 219–234.
14. P. Moscato and C. Cotta. A gentle introduction to memetic algorithms. In F. Glover and G. Kochenberger, eds., *Handbook of Metaheuristics*, Kluwer Academic, Norwell, MA, 2003, pp. 105–144.
15. P. Moscato, C. Cotta, and A. S. Mendes. Memetic algorithms. In G. C. Onwubolu and B. V. Babu, eds., *New Optimization Techniques in Engineering*. Springer-Verlag, Berlin, 2004, pp. 53–85.

16. J. Culberson. On the futility of blind search: an algorithmic view of "no free lunch." *Evolutionary Computation*, 6(2):109–127, 1998.
17. E. C. Freuder and R. J. Wallace. Partial constraint satisfaction. *Artificial Intelligence*, 58:21–70, 1992.
18. A. E. Eiben and J. E. Smith. *Introduction to Evolutionary Computating*. Natural Computing Series. Springer-Verlag, New York, 2003.
19. P. C. Chu and J. E. Beasley. A genetic algorithm for the multidimensional knapsack problem. *Journal of Heuristics*, 4(1):63–86, 1998.
20. C. Cotta. Protein structure prediction using evolutionary algorithms hybridized with backtracking. In J. Mira and J.R. Álvarez, eds., *Artificial Neural Nets Problem Solving Methods*, vol. 2687 of *Lecture Notes in Computer Science*. Springer-Verlag, New York, 2003, pp. 321–328.
21. C. Cotta and J. M. Troya. Genetic forma recombination in permutation flowshop problems. *Evolutionary Computation*, 6(1):25–44, 1998.
22. C. Cotta and J. M. Troya. A hybrid genetic algorithm for the 0–1 multiple knapsack problem. In G. D. Smith, N. C. Steele, and R. F. Albrecht, eds., *Artificial Neural Nets and Genetic Algorithms 3*, Springer-Verlag, New York, 1998, pp. 251–255.
23. C. Cotta and A. J. Fernández. A hybrid GRASP: evolutionary algorithm approach to Golomb ruler search. In Xin Yao et al., eds., *Parallel Problem Solving from Nature VIII*, vol. 3242 of *Lecture Notes in Computer Science*. Springer-Verlag, New York, 2004, pp. 481–490.
24. E. Falkenauer. The lavish ordering genetic algorithm. In *Proceedings of the 2nd Metaheuristics International Conference* (*MIC'97*), Sophia-Antipolis, France, 1997, pp. 249–256.
25. J. Gottlieb and G. R. Raidl. The effects of locality on the dynamics of decoder-based evolutionary search. In L. D. Whitley et al., eds., *Proceedings of the 2000 Genetic and Evolutionary Computation Conference*, Las Vegas, NV. Morgan Kaufmann, San Francisco, CA, 2000, pp. 283–290.
26. J. Gottlieb, B. A. Julstrom, F. Rothlauf, and G. R. Raidl. Prüfer numbers: a poor representation of spanning trees for evolutionary search. In L. Spector et al., eds., *Proceedings of the 2001 Genetic and Evolutionary Computation Conference*, San Francisco, CA. Morgan Kaufmann, San Francisco, CA, 2001, pp. 343–350.
27. W. C. Babcock. Intermodulation interference in radio systems. *Bell Systems Technical Journal*, pp. 63–73, 1953.
28. G. S. Bloom and S. W. Golomb. Applications of numbered undirected graphs. *Proceedings of the IEEE*, 65(4):562–570, 1977.
29. B. Feeney. Determining optimum and near-optimum Golomb rulers using genetic algorithms. Master's thesis, Computer Science, University College Cork, UK, Oct. 2003.
30. W. T. Rankin. Optimal Golomb rulers: an exhaustive parallel search implementation. Master's thesis, Electrical Engineering Department, Duke University, Durham, NC, Dec. 1993.
31. A. Dollas, W. T. Rankin, and D. McCracken. A new algorithm for Golomb ruler derivation and proof of the 19 mark ruler. *IEEE Transactions on Information Theory*, 44:379–382, 1998.
32. M. Garry, D. Vanderschel, et al. In search of the optimal 20, 21 & 22 mark Golomb rulers. GVANT project. http://members.aol.com/golomb20/index.html, 1999.

33. J. B. Shearer. Golomb ruler table. Mathematics Department, IBM Research. http://www.research.ibm.com/people/s/shearer/grtab.html, 2001.
34. W. Schneider. Golomb rulers. MATHEWS: the archive of recreational mathematics. http://www.wschnei.de/number-theory/golomb-rulers.html, 2002.
35. P. Mirchandani and R. Francis. *Discrete Location Theory*. Wiley-Interscience, New York, 1990.
36. S. W. Soliday, A. Homaifar, and G. L. Lebby. Genetic algorithm approach to the search for Golomb rulers. In L. J. Eshelman, ed., *Proceedings of the 6th International Conference on Genetic Algorithms* (*ICGA'95*), Pittsburgh, PA. Morgan Kaufmann, San Francisco, CA, 1995, pp. 528–535.
37. J. B. Shearer. Some new optimum Golomb rulers. *IEEE Transactions on Information Theory*, 36:183–184, Jan. 1990.
38. B. M. Smith and T. Walsh. Modelling the Golomb ruler problem. Presented at the Workshop on Non-binary Constraints (IJCAI'99), Stockholm, Sweden, 1999.
39. P. Galinier, B. Jaumard, R. Morales, and G. Pesant. A constraint-based approach to the Golomb ruler problem. Presented at the 3rd International Workshop on Integration of AI and OR Techniques (CP-AI-OR'01), 2001.
40. S. Prestwich. Trading completeness for scalability: hybrid search for cliques and rulers. In *Proceedings of the 3rd International Workshop on the Integration of AI and OR Techniques in Constraint Programming for Combinatorial Optimization Problems* (*CPAIOR'01*), Ashford, UK, 2001, pp. 159–174.
41. S Koziel and Z. Michalewicz. A decoder-based evolutionary algorithm for constrained parameter optimization problems. In T. Bäeck, A. E. Eiben, M. Schoenauer, and H.-P. Schwefel, eds., *Parallel Problem Solving from Nature V*, vol. 1498 of *Lecture Notes in Computer Science*. Springer-Verlag, New York, 1998, pp. 231–240.
42. F. B. Pereira, J. Tavares, and E. Costa. Golomb rulers: The advantage of evolution. In F. Moura-Pires and S. Abreu, eds., *Progress in Artificial Intelligence, 11th Portuguese Conference on Artificial Intelligence*, vol. 2902 of *Lecture Notes in Computer Science*, Springer-Verlag, New York, 2003, pp. 29–42.
43. J. Bean. Genetic algorithms and random keys for sequencing and optimization. *ORSA Journal on Computing*, 6:154–160, 1994.
44. M. G. C. Resende and C. C. Ribeiro. Greedy randomized adaptive search procedures. In F. Glover and G. Kochenberger, eds., *Handbook of Metaheuristics*, Kluwer Academic, Norwell, MA, 2003, pp. 219–249.
45. I. Dotú and P. Van Hentenryck. A simple hybrid evolutionary algorithm for finding Golomb rulers. In D.W. Corne et al., eds., *Proceedings of the 2005 Congress on Evolutionary Computation* (*CEC'05*), Edinburgh, UK, vol. 3. IEEE Press, Piscataway, NJ, 2005, pp. 2018–2023.
46. C. Cotta and A. J. Fernández. Analyzing fitness landscapes for the optimal Golomb ruler problem. In J. Gottlieb and G. R. Raidl, eds., *Evolutionary Computation in Combinatorial Optimization*, vol. 3248 of *Lecture Notes in Computer Science*. Springer-Verlag, New York, 2005, pp. 68–79.
47. C. Cotta, I. Dotú, A. J. Fernández, and P. Van Hentenryck. A memetic approach to Golomb rulers. In T. P. Runarsson et al., eds., *Parallel Problem Solving from Nature IX*, vol. 4193 of *Lecture Notes in Computer Science*. Springer-Verlag, New York, 2006, pp. 252–261.

48. C. Cotta, I. Dotú, A. J. Fernández, and P. Van Hentenryck. Local search-based hybrid algorithms for finding Golomb rulers. *Constraints*, 12(3):263–291, 2007.

49. T. A. Feo and M. G. C. Resende. Greedy randomized adaptive search procedures. *Journal of Global Optimization*, 6:109–133, 1995.

50. F. Glover. A template for scatter search and path relinking. In Jin-Kao Hao et al., eds., *Selected Papers of the 3rd European Conference on Artificial Evolution* (*AE'97*), Nîmes, France, vol. 1363 of *Lecture Notes in Computer Science*. Springer-Verlag, New York, 1997, pp. 1–51.

51. M. Laguna and R. Martí. *Scatter Search: Methodology and Implementations in C*. Kluwer Academic, Norwell, MA, 2003.

52. F. Glover. Tabu search: part I. *ORSA Journal of Computing*, 1(3):190–206, 1989.

53. F. Glover. Tabu search: part II. *ORSA Journal of Computing*, 2(1):4–31, 1989.

54. A. K. Jain, N. M. Murty, and P. J. Flynn. Data clustering: a review. *ACM Computing Surveys*, 31(3):264–323, 1999.

55. M. Gardner. The fantastic combinations of John Conway's new solitaire game. *Scientific American*, 223:120–123, 1970.

56. M. Gardner. On cellular automata, self-reproduction, the garden of Eden and the game of "life." *Scientific American*, 224:112–117, 1971.

57. E. R. Berlekamp, J. H. Conway, and R. K. Guy. *Winning Ways for Your Mathematical Plays*, vol. 2 of *Games in Particular*. Academic Press, London, 1982.

58. M. Gardner. *Wheels, Life, and Other Mathematical Amusements*. W.H. Freeman, New York, 1983.

59. J. Larrosa and E. Morancho. Solving "still life" with soft constraints and bucket elimination. In Rossi [106], pp. 466–479.

60. J. Larrosa, E. Morancho, and D. Niso. On the practical use of variable elimination in constraint optimization problems: "still life" as a case study. *Journal of Artificial Intelligence Research*, 23:421–440, 2005.

61. R. Bosch and M. Trick. Constraint programming and hybrid formulations for three life designs. In *Proceedings of the 4th International Workshop on Integration of AI and OR Techniques in Constraint Programming for Combinatorial Optimization Problems* (*CP-AI-OR*), Le Croisic, France, 2002, pp. 77–91.

62. B. M. Smith. A dual graph translation of a problem in "life." In P. Van Hentenryck, ed., *Principles and Practice of Constraint Programming* (*CP'02*), Ithaca, NY, vol. 2470 of *Lecture Notes in Computer Science*. Springer-Verlag, New York, 2002, pp. 402–414.

63. R. Dechter. Bucket elimination: a unifying framework for reasoning. *Artificial Intelligence*, 113(1–2):41–85, 1999.

64. K. C. K. Cheng and R. H. C. Yap. Ad-hoc global constraints for life. In P. van Beek, ed., *Principles and Practice of Constraint Programming* (*CP'05*), Sitges, Spain, vol. 3709 of Lecture Notes in Computer Science. Springer-Verlag, New York, 2005, pp. 182–195.

65. K. C. K. Cheng and R. H. C. Yap. Applying ad-hoc global constraints with the case constraint to still-life. *Constraints*, 11:91–114, 2006.

66. J. E. Gallardo, C. Cotta, and A. J. Fernández. A memetic algorithm with bucket elimination for the still life problem. In J. Gottlieb and G. Raidl, eds., *Evolutionary Computation in Combinatorial Optimization*, Budapest, Hungary

vol. 3906 of *Lecture Notes in Computer Science*. Springer-Verlag, New York, 2006, pp. 73–85.

67. J. E. Gallardo, C. Cotta, and A. J. Fernández. A multi-level memetic/exact hybrid algorithm for the still life problem. In T. P. Runarsson et al., eds., *Parallel Problem Solving from Nature IX*, Reykjavik, Iceland, vol. 4193 of *Lecture Notes in Computer Science*. Springer-Verlag, New York, 2006, pp. 212–221.
68. R. Dechter. Mini-buckets: a general scheme for generating approximations in automated reasoning. In *Proceedings of the 15th International Joint Conference on Artificial Intelligence,* 1997, pp. 1297–1303.
69. W. Harvey and T. Winterer. Solving the MOLR and social golfers problems. In P. van Beek, ed., *Proceedings of the 11th International Conference on Principles and Practice of Constraint Programming*, Sitges, Spain, vol. 3709 of *Lecture Notes in Computer Science*. Springer-Verlag, New York, 2005, pp. 286–300.
70. N. Barnier and P. Brisset. Solving Kirkman's schoolgirl problem in a few seconds. *Constraints*, 10(1):7–21, 2005.
71. C. Cotta, I. Dotú, A. J. Fernández, and P. Van Hentenryck. Scheduling social golfers with memetic evolutionary programming. In F. Almeida et al., eds., *Hybrid Metaheuristics*, vol. 4030 of *Lecture Notes in Computer Science*. Springer-Verlag, New York, 2006, pp. 150–161.
72. T. Fahle, S. Schamberger, and M. Sellmann. Symmetry breaking. In T. Walsh, ed., *Proceedings of the 7th International Conference on Principles and Practice of Constraint Programming*, Paphos, Cyprus, vol. 2239 of *Lecture Notes in Computer Science*. Springer-Verlag, New York, 2001, pp. 93–107.
73. B. M. Smith. Reducing symmetry in a combinatorial design problem. In *Proceedings of the 3rd International Workshop on the Integration of AI and OR Techniques in Constraint Programming for Combinatorial Optimization Problems*, Ashford, UK, 2001, pp. 351–359.
74. M. Sellmann and W. Harvey. Heuristic constraint propagation. In P. Van Hentenryck, ed., *Proceedings of the 8th International Conference on Principles and Practice of Constraint Programming*, Ithaca, NY, vol. 2470 of *Lecture Notes in Computer Science*. Springer-Verlag, New York, 2002, pp. 738–743.
75. A. Ramani and I. L. Markov. Automatically exploiting symmetries in constraint programming. In B. Faltings, A. Petcu, F. Fages, and F. Rossi, eds., *Recent Advances in Constraints, Joint ERCIM/CoLogNet International Workshop on Constraint Solving and Constraint Logic Programming, (CSCLP'04)*, Lausanne, Switzerland, vol. 3419 of *Lecture Notes in Computer Science*, Revised Selected and Invited Papers. Springer-Verlag, New York, 2005, pp. 98–112.
76. S. D. Prestwich and A. Roli. Symmetry breaking and local search spaces. In R. Barták and M. Milano, eds., *Proceedings of the 2nd International Conference on the Integration of AI and OR Techniques in Constraint Programming for Combinatorial Optimization Problems*, Prague, Czech Republic, vol. 3524 of *Lecture Notes in Computer Science*. Springer-Verlag, New York, 2005, pp. 273–287.
77. T. Mancini and M. Cadoli. Detecting and breaking symmetries by reasoning on problem specifications. In J.-D. Zucker and L. Saitta, eds., *International Symposium on Abstraction, Reformulation and Approximation (SARA'05)*, Airth Castle, Scotland, UK, vol. 3607 of *Lecture Notes in Computer Science*. Springer-Verlag, New York, 2005, pp. 165–181.

78. M. Sellmann and P. Van Hentenryck. Structural symmetry breaking. In L. Pack Kaelbling and A. Saffiotti, eds., *Proceedings of the 19th International Joint Conference on Artificial Intelligence* (*IJCAI'05*). Professional Book Center, Edinburgh, UK, 2005, pp. 298–303.

79. I. P. Gent and I. Lynce. A SAT encoding for the social golfer problem. Presented at the IJCAI'05 Workshop on Modelling and Solving Problems with Constraints, Edinburgh, UK, July 2005.

80. A. M. Frisch, B. Hnich, Z. Kiziltan, I. Miguel, and T. Walsh. Global constraints for lexicographic orderings. In P. Van Hentenryck, ed., *Proceedings of the 8th International Conference on Principles and Practice of Constraint Programming*, Ithaca, NY, vol. 2470 of *Lecture Notes in Computer Science*. Springer-Verlag, New York, 2002, pp. 93–108.

81. I. Dotú and P. Van Hentenryck. Scheduling social golfers locally. In R. Barták and M. Milano, eds., *Proceedings of the International Conference on Integration of AI and OR Techniques in Constraint Programming for Combinatorial Optimization Problems 2005*, Prague, Czech Republic, vol. 3524 of *Lecture Notes in Computer Science*. Springer-Verlag, New York, 2005, pp. 155–167.

82. J. Hein. A new method that simultaneously aligns and reconstructs ancestral sequences for any number of homologous sequences, when the phylogeny is given. *Molecular Biology and Evolution*, 6:649–668, 1989.

83. B. Rost and C. Sander. Prediction of protein secondary structure at better than 70% accuracy. *Journal of Molecular Biology*, 232:584–599, 1993.

84. C.-K. Ong, S. Nee, A. Rambaut, H.-U. Bernard, and P. H. Harvey. Elucidating the population histories and transmission dynamics of papillomaviruses using phylogenetic trees. *Journal of Molecular Evolution*, 44:199–206, 1997.

85. J. Kim and T. Warnow. Tutorial on phylogenetic tree estimation. In T. Lengauer et al., eds., *Proceedings of the 7th International Conference on Intelligent Systems for Molecular Biology*, Heidelberg, Germany. American Association for Artificial Intelligence Press, Merlo Park, CA, 1999, pp. 118–129.

86. O. R. P. Bininda-Emonds, ed. *Phylogenetic Supertrees: Combining Information to Reveal the Tree of Life*. Computational Biology Series. Kluwer Academic, Boston, 2004.

87. D. Bryant. A classification of consensus methods for phylogenetics. In M. Janowitz et al., eds., *Bioconsensus*, DIMACS-AMS, 2003, pp. 163–184.

88. D. Gusfield. Efficient algorithms for inferring evolutionary trees. *Networks*, 21:19–28, 1991.

89. W. H. E. Day. Optimal algorithms for comparing trees with labeled leaves. *Journal of Classiffication*, 2:7–28, 1985.

90. J.-P. Barthélemy and F. R. McMorris. The median procedure for n-trees. *Journal of Classiffication*, 3:329–334, 1986.

91. B. L. Allen and M. Steel. Subtree transfer operations and their induced metrics on evolutionary trees. *Annals of Combinatorics*, 5:1–15, 2001.

92. J. T. L. Wang, H. Shan, D. Shasha, and W. H. Piel. Treerank: A similarity measure for nearest neighbor searching in phylogenetic databases. In *Proceedings of the 15th International Conference on Scientific and Statistical Database Management*, Cambridge MA. IEEE Press, Piscataway, NJ, 2003, pp. 171–180.

93. I. P. Gent, P. Prosser, B. M. Smith, and W. Wei. Supertree construction with constraint programming. In Rossi [106], pp. 837–841.

94. B. Y. Wu, K.-M. Chao, and C. Y. Tang. Approximation and exact algorithms for constructing minimum ultrametric trees from distance matrices. *Journal of Combinatorial Optimization*, 3(2):199–211, 1999.

95. P. Prosser. Supertree construction with constraint programming: recent progress and new challenges. In *Workshop on Constraint Based Methods for Bioinformatics, (WCB'06)*, 2006, pp. 75–82.

96. R. R. Sokal and C. D. Michener. A statistical method for evaluating systematic relationships. *University of Kansas Science Bulletin*, 38:1409–1438, 1958.

97. A. A. Andreatta and C. C. Ribeiro. Heuristics for the phylogeny problem. *Journal of Heuristics*, 8:429–447, 2002.

98. C. Cotta and P. Moscato. Inferring phylogenetic trees using evolutionary algorithms. In J. J. Merelo et al., eds., *Parallel Problem Solving from Nature VII*, volume 2439 of *Lecture Notes in Computer Science*. Springer-Verlag, New York, 2002, pp. 720–729.

99. C. Cotta. On the application of evolutionary algorithms to the consensus tree problem. In G.R. Raidl and J. Gottlieb, eds., *Evolutionary Computation in Combinatorial Optimization*, vol. 3448 of *Lecture Notes in Computer Science*. Springer-Verlag, New York, 2005, pp. 58–67.

100. D. Barker. LVB: parsimony and simulated annealing in the search for phylogenetic trees. *Bioinformatics*, 20:274–275, 2004.

101. J. E. Gallardo, C. Cotta, and A. J. Fernández. Reconstructing phylogenies with memetic algorithms and branch-and-bound. In S. Bandyopadhyay, U. Maulik, and J. Tsong-Li Wang, eds., *Analysis of Biological Data: A Soft Computing Approach*. World Scientific, Hackensack, NJ, 2007, pp. 59–84.

102. E. L. Lawler and D. E. Wood. Branch and bounds methods: a survey. *Operations Research*, 4(4):669–719, 1966.

103. P. Daniel and C. Semple. A class of general supertree methods for nested taxa. *SIAM Journal of Discrete Mathematics*, 19(2):463–480, 2005.

104. A. Ozäygen. Phylogenetic supertree construction using constraint programming. Master's thesis, Graduate School of Natural and Applied Sciences, Çankaya University, Ankara, Turkey, 2006.

105. C. Cotta, A. J. Fernández, and A. Gutiérrez. On the hybridization of complete and incomplete methods for the consensus tree problem. Manuscript in preparation, 2008.

106. F. Rossi, ed. *Proceedings of the 9th International Conference on Principles and Practice of Constraint Programming*, Kinsale, UK, vol. 2833 of *Lecture Notes in Computer Science*. Springer-Verlag, New York, 2003.

CHAPTER 8

Optimization of Time Series Using Parallel, Adaptive, and Neural Techniques

J. A. GÓMEZ, M. D. JARAIZ, M. A. VEGA, and J. M. SÁNCHEZ

Universidad de Extremadura, Spain

8.1 INTRODUCTION

In many science and engineering fields it is necessary to dispose of mathematical models to study the behavior of phenomena and systems whose mathematical description is not available a priori. One interesting type of such systems is the time series (TS). Time series are used to describe behavior in many fields: astrophysics, meteorology, economy, and so on. Although the behavior of any of these processes may be due to the influence of several causes, in many cases the lack of knowledge of all the circumstances makes it necessary to study the process considering only the TS evolution that represents it. Therefore, numerous methods of TS analysis and mathematical modeling have been developed. Thus, when dealing with TS, all that is available is a signal under observation, and the physical structure of the process is not known. This led us to employ planning system identification (SI) techniques to obtain the TS model. With this model a prediction can be made, but taking into account that the model precision depends on the values assigned to certain parameters.

The chapter is structured as follows. In Section 8.2 we present the necessary background on TS and SI. The problem is described in Section 8.3, where we have focused the effort of analysis on a type of TS: the sunspot (solar) series. In Section 8.4 we propose a parallel and adaptive heuristic to optimize the identification of TS through adjusting the SI main parameters with the aim of improving the precision of the parametric model of the series considered. This methodology could be very useful when the high-precision mathematical modeling of dynamic complex systems is required. After explaining the proposed heuristics and the

Optimization Techniques for Solving Complex Problems, Edited by Enrique Alba, Christian Blum, Pedro Isasi, Coromoto León, and Juan Antonio Gómez

tuning of their parameters, in Section 8.5 we show the results we have found for several series using different implementations and demonstrate how the precision improves.

8.2 TIME SERIES IDENTIFICATION

SI techniques [1,2] can be used to obtain the TS model. The model precision depends on the values assigned to certain parameters. In SI, a TS is considered as a sampled signal $y(k)$ with period T that is modeled with an ARMAX [3] polynomial description of dimension na (the model size), as we can see:

$$y(k) + a_1\, y(k_1) + \cdots + a_{na}\, y(k_{na}) = 0 \tag{8.1}$$

Basically, the identification consists of determining the ARMAX model parameters a_i (θ in matrix notation) from measured samples $y(k_i)$ [$\varphi(k)$ in matrix notation]. In this way it is possible to compute the estimated signal $y_e(k)$

$$y_e(k) = [-a_1\, y(k-1) - \cdots - a_{na}\, y(k-na)] = \varphi^T(k)\,\theta(k-1) \tag{8.2}$$

and compare it with the real signal $y(k)$, calculating the generated error:

$$\text{error}(k) = y(k) - y_e(k) \tag{8.3}$$

The recursive estimation updates the model (a_i) in each time step k, thus modeling the system. The greater number of sample data that are processed, the more precise the model because it has more system behavior history. We consider SI performed by the well-known recursive least squares (RLS) algorithm [3]. This algorithm is specified primarily by the forgetting factor constant λ and the samples observed, $y(k)$. There is no fixed value for λ; a value between 0.97 and 0.995 is often used [4]. From a set of initial conditions [$k = p$, $\theta(p) = 0$, $P(p) = 1000I$, where I is the identity matrix and p is the initial time greater than the model size na], RLS starts building $\varphi^T(k)$, doing a loop of operations defined in Equations 8.2 to 8.6.

$$K = \frac{P(k-1)\varphi(k)}{\lambda + \varphi^T(k)P(k-1)\varphi(k)} \tag{8.4}$$

$$P(k) = \frac{P(k-1) - K\varphi^T(k)P(k-1)}{\lambda} \tag{8.5}$$

$$\theta(k) = \theta(k-1) + K\ \text{error}(k) \tag{8.6}$$

The recursive identification is very useful when it is a matter of predicting the following behavior of the TS from the data observed up to the moment. For many purposes it is necessary to make this prediction, and for predicting it is necessary to obtain information about the system. This information, acquired by means of

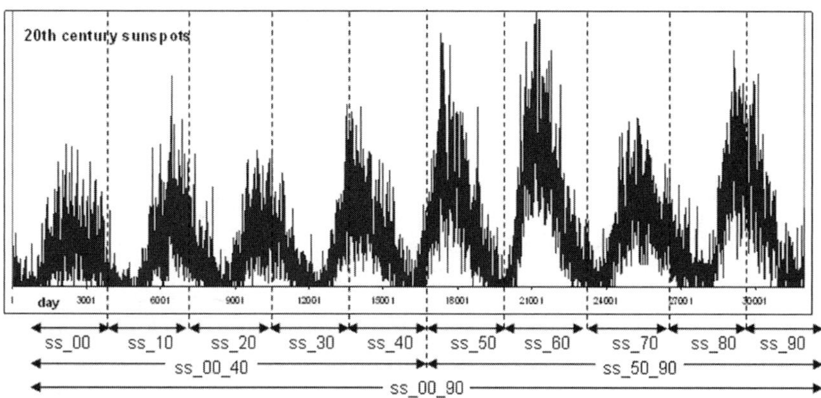

Figure 8.1 Sunspot series used as a benchmark to evaluate the algorithmic proposal.

the SI, consists in elaborating a mathematical parametric model to cover system behavior.

We use a TS set as benchmark to validate the algorithmic proposal that we present in this chapter. Some of these benchmarks have been collected from various processes [5] and others correspond to sunspot series acquired from measured observations of the sun activity [6,7]. For this type of TS we have used 13 series (Figure 8.1) showing daily sunspots: 10 series (*ss*_00, *ss*_10, *ss*_20, *ss*_30, *ss*_40, *ss*_50, *ss*_60, *ss*_70, *ss*_80, and *ss*_90) corresponding to sunspot measurements covering 10 years each (e.g., *ss*_20 compiles the sunspots from 1920 to 1930), two series (*ss*_00_40 and *ss*_50_90) covering 50 years each, and one series (*ss*_00_90) covering all measurements in the twentieth century.

8.3 OPTIMIZATION PROBLEM

When SI techniques are used, the model is generated a posteriori by means of the data measured. However, we are interested in system behavior prediction in running time, that is, while the system is working and the data are being observed. So it would be interesting to generate models in running time in such a way that a processor may simulate the next behavior of the system. At the same time, our first effort is to obtain high model precision. SI precision is due to several causes, mainly to the forgetting factor (Figure 8.2). Frequently, this value is critical for model precision. Other sources can also have some degree of influence (dimensions, initial values, the system, etc.), but they are considered as problem definitions, not parameters to be optimized. On the other hand, the precision problem may appear when a system model is generated in sample time for some dynamic systems: If the system response changes quickly, the sample frequency must be high to avoid the key data loss in the system behavior description. If the system is complex and its simulation from the model

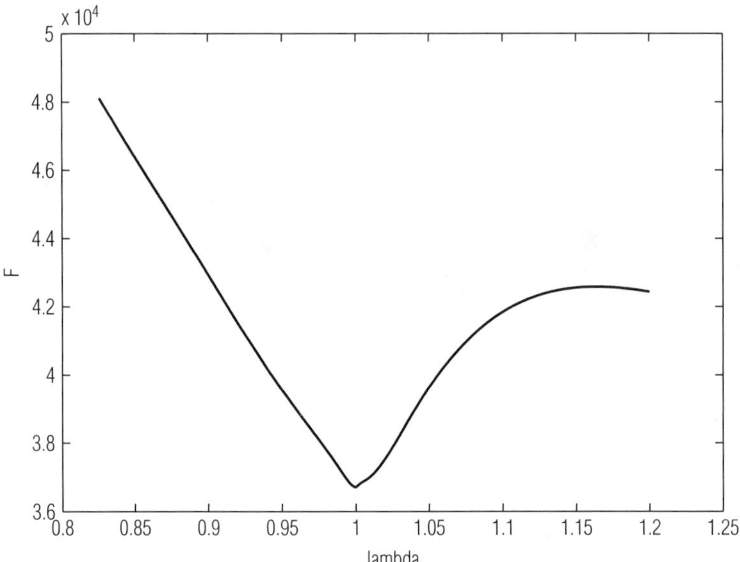

Figure 8.2 Error in identification for several λ values using RLS for *ss*_80 benchmark when $na = 5$. It can be seen that the minimum error is reached when $\lambda = 0.999429$.

to be found must be very trustworthy, the precision required must be very high, and this implies a great computational cost. Therefore, we must find a trade-off between high sample frequency and high precision in the algorithm computation. However, for the solar series, the interval of sampling is not important, the real goal being to obtain the maximum accuracy of estimation.

Summarizing, the optimization problem consists basically of finding the λ value such that the identification error is minimum. Thus, the fitness function F,

$$F(\lambda) = \sum_{k=k_0}^{k=k_0+\mathrm{SN}-1} |y_e(k) - y(k)| \tag{8.7}$$

is defined as the value to minimize to obtain the best precision, where SN is the number of samples. The recursive identification can be used to predict the behavior of the TS [Figure 8.3(a)] from the data observed up to the moment. It is possible to predict system behavior in $k+1$, $k+2$, ... by means of the model $\theta(k)$ identified in k time. As identification advances in time, the predictions improve, using more precise models. If *ks* is the time until the model is elaborated and for which we carry out the prediction, we can confirm that this prediction will have a larger error while we will be farther away from *ks* [Figure 8.3(b)]. The value predicted for $ks+1$ corresponds to the last value estimated until ks. When we have more data, the model starts to be reelaborated for computing the new estimated values [Figure 8.3(c)].

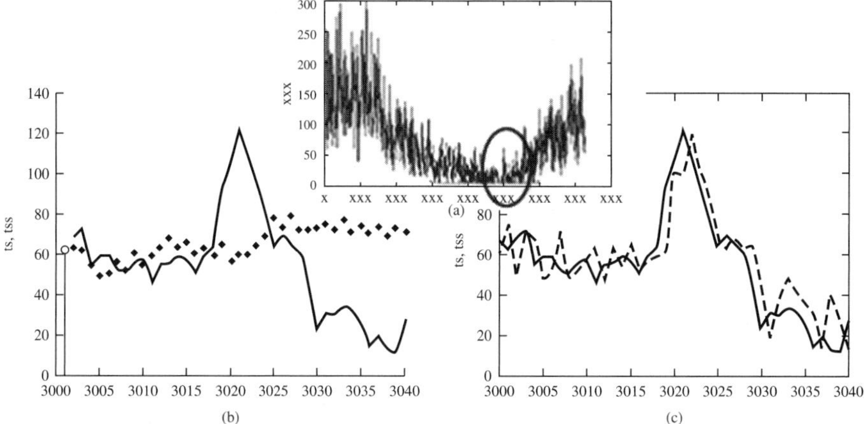

Figure 8.3 (a) The *ss_90* series. If $ks = 3000$, we can see in (b) the long-term prediction based on the model obtained up to ks ($na = 30$ and $\lambda = 0.98$). The prediction precision is reduced when we are far from ks. (c) Values predicted for the successive $ks + 1$ obtained from the updated models when the identification (and ks) advances.

The main parameters of the identification are na and λ. Both parameters have an influence on the precision of prediction results, as shown in Figures 8.4 and 8.5. As we can see, establishing an adequate value of na and λ may be critical in obtaining a high-quality prediction. Therefore, in the strategy to optimize the prediction we try, first, to establish a satisfactory dimension of the mathematical model (i.e., a model size giving us good results without too much computational cost), which will serve us later in finding the optimal λ value (the optimization problem). To do that, many experiments have been carried out. In these experiments the absolute difference between the real and predicted values has been used to quantify the prediction precision. Fitness in the prediction experiments is measured as

$$\text{DIFA}(X) = \sum_{i-1}^{i=X} |y_s(ks + i) - y(ks + i)| \tag{8.8}$$

We use our own terminology for the prediction error [DIFA(X), where X is an integer], allowing the reader to understand more easily the degree of accuracy that we are measuring. Figure 8.6 shows the results of many experiments, where different measurements of the prediction error [for short-term DIFA(1) and for long-term DIFA(50)] for 200 values of the model size are obtained. These experiments have been made for different ks values to establish reliable conclusions.

For the experiments made we conclude that the increase in model size does not imply an improvement in the prediction precision (however, a considerable increase in computational cost is achieved). This is easily verified using large

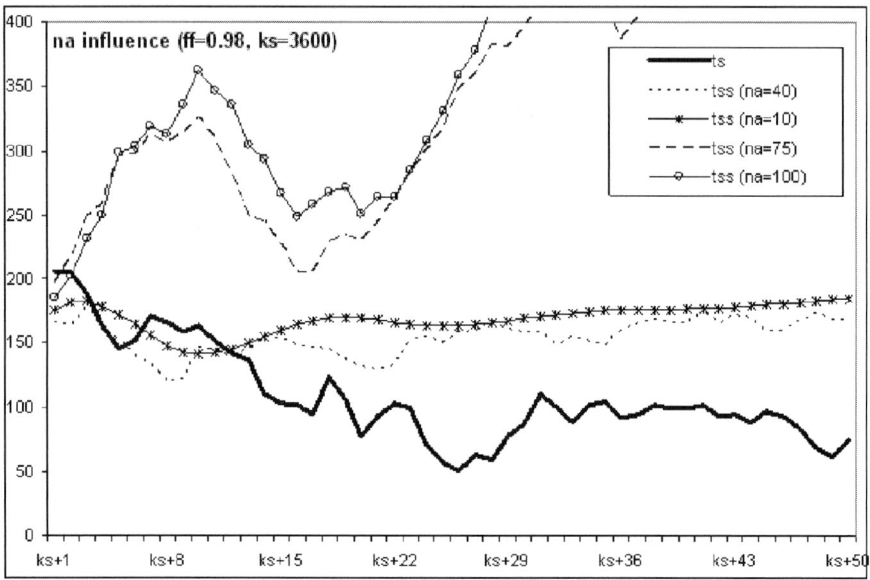

Figure 8.4 Influence of the model size *na* on short-term (*ks*) and long-term ($ks + 50$) prediction for the *ss*_90 series.

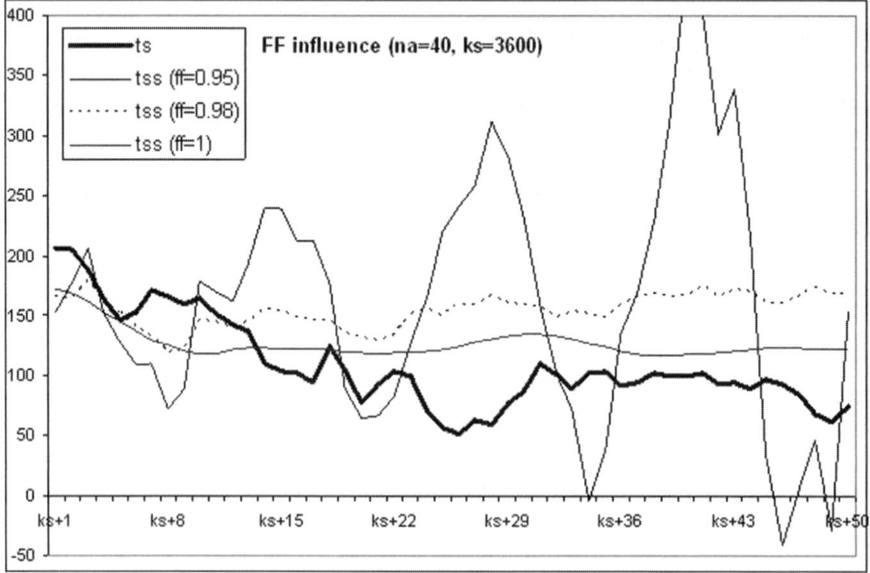

Figure 8.5 Influence of the forgetting factor λ on short-term (*ks*) and long-term ($ks + 50$) prediction for the *ss*_90 series.

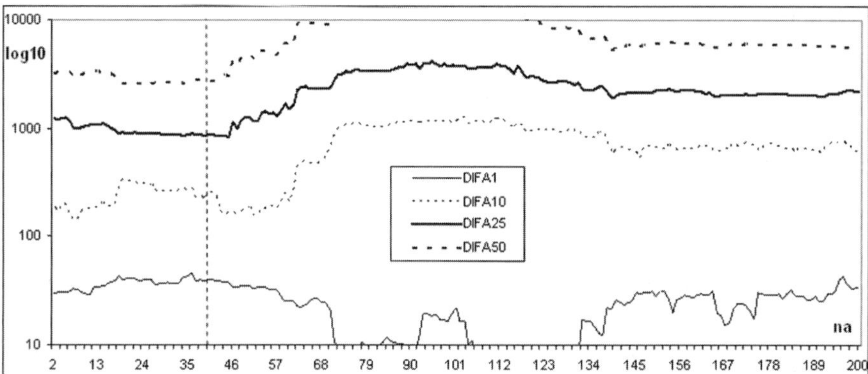

Figure 8.6 Precision of the prediction for the *ss_90* series from $ks = 3600$ using $\lambda = 0.98$. The measurements have been made for models with sizes between 2 and 200.

ks values (the models elaborated have more information related to the TS) and long-term predictions. From these results and as a trade-off between prediction and computational cost, we have chosen $na = 40$ to fix the model size for future experiments.

The forgetting factor is the key parameter for prediction optimization. Figure 8.7 shows, with more detail than Figure 8.3(b), an example of the λ influence on the short- and long-term predictions by means of the representation of the measurement of its precision. It is clear that for any λ value, short-term prediction is more precise than long-term prediction. However, we see that the value chosen for λ is critical to finding a good predictive model (from the four values chosen, $\lambda = 1$ produces the prediction with minimum error, even in the

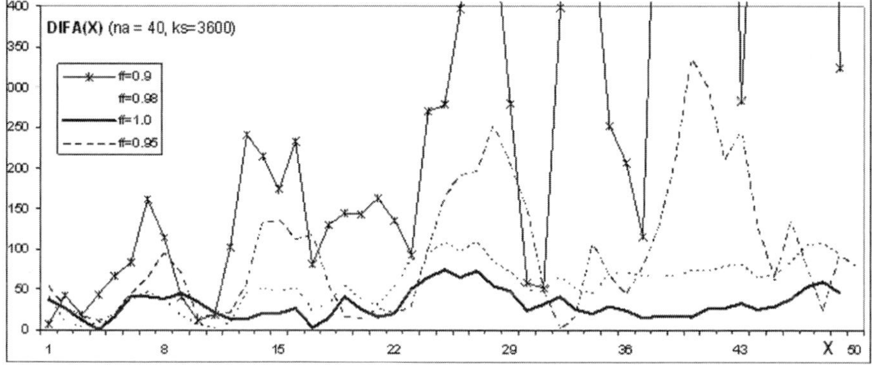

Figure 8.7 Influence of λ in the prediction precision for *ss_90* with $na = 40$ and $ks = 3600$.

long term). This analysis has been confirmed by carrying out a great number of experiments with the sunspot series, modifying the initial prediction time ks.

8.4 ALGORITHMIC PROPOSAL

To find the optimum value of λ, we propose a parallel adaptive algorithm that is partially inspired by the concept of artificial evolution [8,9] and the simulated annealing mechanism [10]. In our proposed PARLS (parallel adaptive recursive least squares) algorithm, the optimization parameter λ is evolved for predicting new situations during iterations of the algorithm. In other words, λ evolves, improving its fitness.

The evolution mechanism (Figure 8.8) is as follows. In its first iteration, the algorithm starts building a set of λ values (individuals in the initial population) covering the interval R (where the individuals are generated) uniformly from the middle value selected (λc). Thus, we assume that the population size is the length R, not the number of individuals in the population. A number of parallel processing units (PUN) equal to the number of individuals perform RLS identification with each λ in the population. Therefore, each parallel processing unit is an identification loop that considers a given number of samples (PHS) and an individual in the population as the forgetting factor. At the end of the iteration, the best individual is the one whose corresponding fitness has the minimum value of all those computed in the parallel processing units, and from it a new set of λ values (the new population) is generated and used in the next iteration to perform new identifications during the following PHS samples.

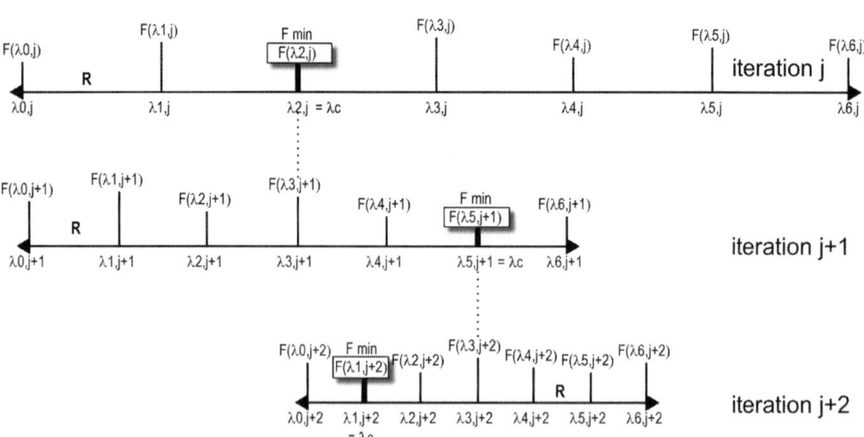

Figure 8.8 For an iteration of the algorithm, the individuals λ of the population are the inputs for RLS identifications performed in parallel processing units. At the end of iteration the best individual is the one whose fitness F is minimal, and from it a new population is generated covering a more reduced interval.

Fitness F is defined as the accumulated error of the samples in the iteration according to Equation 8.7:

$$F(\lambda_{PU_X}) = \sum_{k=k_0}^{k=k_0+PHS-1} |y_e(k) - y(k)| \tag{8.9}$$

Generation of the new population is made with a more reduced interval, dividing R by a reduction factor (RED). In this way the interval limits are moved so that the middle of the interval corresponds to the best found individual in the preceding iteration, such way that the population generated will be nearer the previous best individual found. The individuals of the new population always cover the new interval R uniformly. Finally, PARLS stops when the maximum number of iterations (PHN) is reached, according to the total number of samples (TSN) of the TS or when a given stop criterion is achieved. PARLS nomenclature is summarized in Table 8.1, and pseudocode of this algorithmic proposal is given in Algorithm 8.1

TABLE 8.1 PARLS Nomenclature and Tuned Values

Name	Description	Value
λ	Forgetting factor (individual)	
R	Generation interval (population size)	0.05
λc	Central in R (individual in the middle)	1
PUN	Number of parallel processing units	11
PHN	Maximum number of iterations of the algorithm	4
RED	Reduction factor of R	2
TSN	Total number of samples of the time series	[a]
PHS	Samples processed in an iteration of the algorithm	[a]

[a]Depends on the TS selected and the maximum number of iterations.

Algorithm 8.1 Parallel Adaptive Algorithm

```
Select initial individual in the middle of the population
Select initial population size
while stop criterion not reached do
  Generate the population
  while the PHS samples are being processed do
    Evaluate fitness in each parallel processing unit
  end while
  Determine the individual with minimum fitness
  New individual in the middle = best individual
  New population size = population size/RED
end while
```

The goal of the iterations of the algorithm is that the identifications performed by the processing units will converge to optimum λ values, so the final TS model will be of high precision. Therefore, PARLS could be considered as a population-based metaheuristic rather than a parallel metaheuristic, because each processing unit is able to operate in isolation, as well as the relevant problem itself, as only a single real-valued parameter (λ) is optimized.

Each processing unit performs the RLS algorithm as it is described in Equations 8.2 to 8.6. NNPARLS is the PARLS implementation, where the parallel processing units are built by means of neural networks [11]. With this purpose we have used the NNSYSID toolbox [12] of Matlab [13]. In this toolbox the multilayer perceptron network is used [14] because of its ability to model any case. For this neural net it is necessary to fix several parameters: general architecture (number of nodes), stop criterion (the training of the net is concluded when the stop criterion is lower than a determined value), maximum number of iterations (variable in our experiments), input to hidden layer and hidden to output layer weights, and others.

8.5 EXPERIMENTAL ANALYSIS

PARLS offers great variability for their parameters. All the parameters have been fully checked and tested in order to get a set of their best values, carrying out many experiments with a wide set of benchmarks. According to the results we have obtained, we can conclude that there are not common policies to tune the parameters in such a way that the best possible results will always be found. But the experiments indicate that there is a set of values of the main parameters for which the optimal individual found defines an error in the identification less than the one obtained with classical or randomly chosen forgetting factor values. We can establish fixed values for PARLS parameters (the right column in Table 8.1) to define a unique algorithm implementation easily applicable to any TS. This has the advantage of a quick application for a series without the previous task of tuning parameters.

A question of interest relates to the adequate model size. There is a large computational cost in time when na is high, as we can see in Figure 8.9. Therefore, the model size should be selected in relation to the computational cost required, and this size is considered as part of the definition of the problem, not as part of the optimization heuristic. We have computed several series with different values for na with processing units implemented with neural networks. Our conclusion is that an accurate tuning for general purposes could be from $na = 20$, although this value can be increased considerably for systems with wide sampling periods. Since a larger na value increases the model precision, for the sunspot series an adequate value of na can be 40, since the time required to compute is tolerable because the series has a long sampling interval (one day).

Another important question relates to how to establish the initial population by selecting the limits of the valid range. We have performed lots of experiments

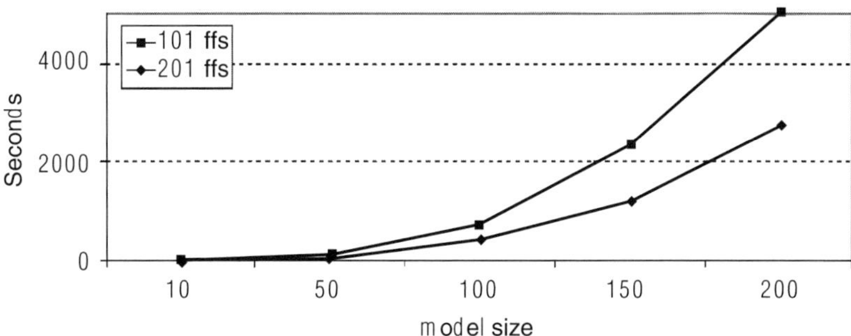

Figure 8.9 Increment of the computation time with model size for the *kobe* benchmark using a Pentium-41.7 GHz–based workstation, evaluating 101 or 201 values of the forgetting factor.

to determine the approximate optimal individual for several TSs using massive RLS computations. Some of these experiments are shown in Figure 8.10. We can see that there is a different best solution for each TS, but in all cases there is a smooth V-curve that is very useful in establishing the limits of the initial population. We have selected the tuned values $\lambda c = 1$ and $R = 0.05$, from which we can generate the initial population. Observing the results, it could be said that the optimum value is $\lambda = 1$, but it is dangerous to affirm this if we have not reduced the range more. Then, by reducing the interval of search we will be able to find a close but different optimum value than 1, as we can see in Figure 8.11.

In Figure 8.12 compare the results of an RLS search and the PARLS algorithm for several sunspot series. In all cases the same tuned parameters have been used ($na = 20$, $PUN = 11$, $\lambda c = 1$, $R = 0.05$, $RED = 2$, $PHN = 4$). The computational effort of 11 RLS identifications is almost equal to PARLS with 11 processing units, so the results can be compared to establish valid conclusions. With these tuned parameters, PARLS always yields better results than RLS. This fact contrasts with the results obtained for other types of TSs, for which PARLS yields better results in most, but for a few TSs the difference in the corresponding fitness oscillates between 2 and 5% in favor of RLS. Therefore, if for the series studied we said that PARLS improves or holds the results found by RLS using the same computational effort, for the sunspot solar series we can say that PARLS always gives us better results [15].

Due to the fact that PARLS has obtained good results, we believe that it can be useful for prediction purposes. In this way we have computed some experiments to predict the next values from the last value registered for sun activity in the second half of the twentieth century. The results are shown in Figure 8.13. Prediction precision is given by the measurements DIFA(1) (short term) and DIFA(50) (long term). These results are compared, for evaluation

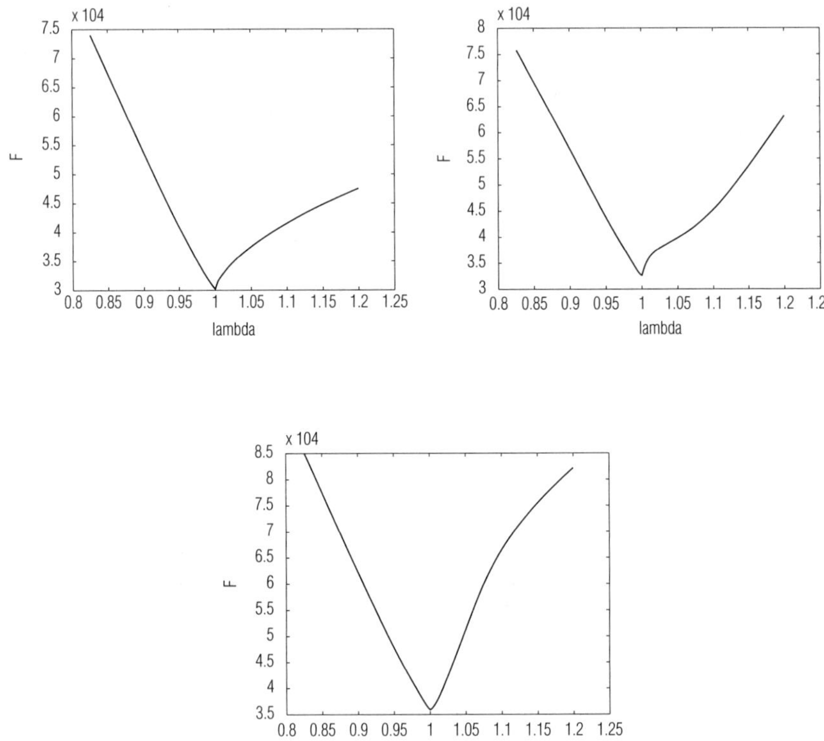

Figure 8.10 Fitness of the sunspot series *ss*_10, *ss*_20, and *ss*_40, calculated by RLS for 200 λ values in the same range ($\lambda c = 1$, $R = 0.4$) using the same model size ($na = 20$). We can see a smooth V-curve in all the cases.

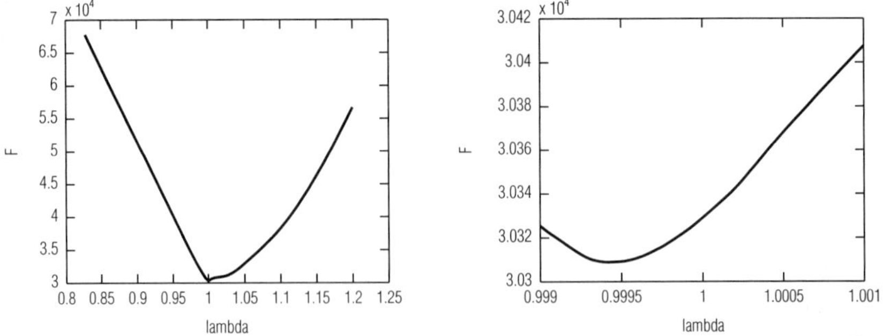

Figure 8.11 Fitness of the *ss*_90 series calculated by RLS for 100 λ values in two different ranges ($R = 0.4$ and $R = 0.002$), both centered in $\lambda c = 1$ and using the same model size ($na = 20$). It can be seen that the optimal λ value is not 1, and how it can be found by using a narrower range.

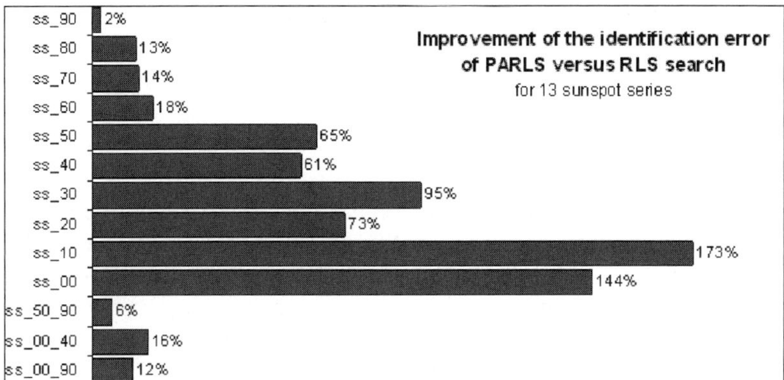

Figure 8.12 Review of some results. This figure shows a comparison of results between an RLS search (where each RLS execution searches the best fitness among 11 values of λ) and PARLS for several series with the same tuned parameters. The Minimum error was always found for the PARLS algorithm for all the series.

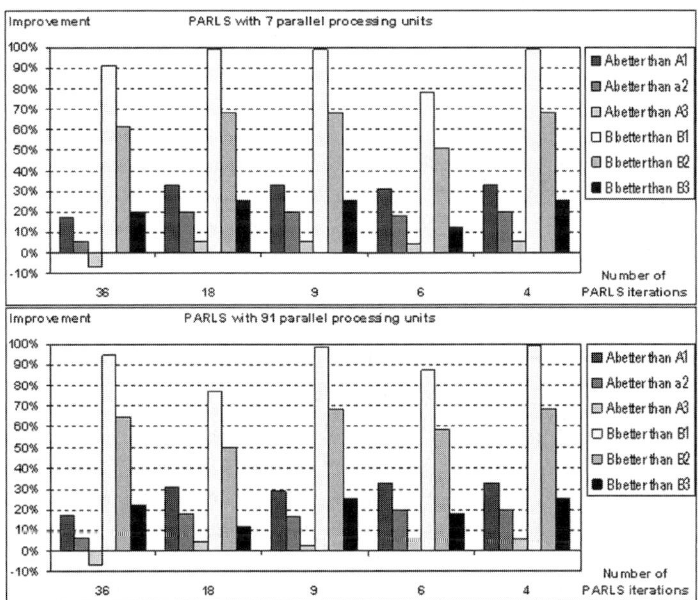

Figure 8.13 Sample of the precision of for short- and long-term prediction results compared with those obtained from three classical values of λ: 0.97, 0.98, and 0.995. A and B are DIFA1 and DIFA50 for λ_{opt}, respectively. $A1$ and $B1$ are DIFA1 and DIFA50 for $\lambda = 0.97$, $A2$ and $B2$ are DIFA1 and DIFA50 for $\lambda = 0.98$, and $A3$ and $B3$ are DIFA1 and DIFA50 for $\lambda = 0.995$. The settings for these experiment were: benchmark ss_50_90; $na = 40$; $ks = 18{,}210$; $TSN = 18{,}262$; $\lambda c = 1$; $R = 0.2$; $RED = 2$. PARLS uses the tuned values of its paremeters in Table 8.1, where the processing units were implemented with neural networks.

purposes, with those obtained using RLS identification with some values of λ inside the classic range [4]. It can be seen that PARLS yields better precision.

8.6 CONCLUSIONS

As a starting point, we can say that the parallel adaptive algorithm proposed offers a good performance, and this encourages us to follow this research. The results shown in Section 8.5 are part of a great number of experiments carried out using different TS and initial prediction times. In the great majority of the cases, PARLS offers a smaller error in the identification than if λ random values are used. Also, the neural network implementation of the parallel processing units has offered better results than in conventional programming [11]. However, we are trying to increase the prediction precision. Thus, our future working lines suggest using genetic algorithms or strategies of an analogous nature, on the one hand, to find the optimum set of values for the parameters of PARLS, and on the other hand, to find the optimum pair of values $\{na,\lambda\}$ (in other words, to consider the model size as an optimization parameter; in this case, multiobjective heuristics are required).

Acknowledgments

The authors are partially supported by the Spanish MEC and FEDER under contract TIC2002-04498-C05-01 (the TRACER project).

REFERENCES

1. T. Soderstrom. *System Identification*. Prentice-Hall, Englewood Cliffs, NJ, 1989.
2. J. Juang. *Applied System Identification*. Prentice Hall, Upper Saddle River, NJ, 1994.
3. L. Ljung. *System Identification*. Prentice Hall, Upper Saddle River, NJ, 1999.
4. L. Ljung. *System Identification Toolbox*. MathWorks, Natick, MA, 1991.
5. I. Markovsky, J. Willems, and B. Moor A database for identification of systems. In *Proceedings of the 17th International Symposium on Mathematical Theory of Networks and Systems*, 2005, pp. 2858–2869.
6. P. Vanlommel, P. Cugnon, R. Van Der Linden, D. Berghmans, F. Clette. The Sidc: World Data Center for the Sunspot Index. *Solar Physics* 224:113–120, 2004.
7. *Review of NOAA s National Geophysical Data Center*. National Academies Press, Washington, DC, 2003.
8. D. Goldberg. *Genetic Algorithms in Search, Optimization and Machine Learning*. Addison-Wesley, Reading, MA, 1989.
9. D. Fogel. *Evolutionary Computation: Toward a New Philosophy of Machine Intelligence*. IEEE Press, Piscataway, NJ, 2006.
10. S. Kirkpatrick, C. Gelatt, and M. Vecchi. Optimization by simulated annealing. *Science*, 220:671–680, 1983.

11. J. Gomez, J. Sanchez, and M. Vega. Using neural networks in a parallel adaptive algorithm for the system identification optimization. *Lecture Notes in Computer Science*, 2687:465–472, 2003.

12. M. Nørgaard. System identification and control with neural networks. Ph.D. thesis, Department of Automation, University of Denmark, 1996.

13. R. Pratap. *Getting Started with Matlab*. Oxford University Press, New York, 2005.

14. H. Demuth and M. Beale. *Neural Network Toolbox*. MathWorks, Natick, MA, 1993.

15. J. Gomez, M. Vega, and J. Sanchez. Parametric identification of solar series based on an adaptive parallel methodology. *Journal of Astrophysics and Astronomy*, 26:1–13, 2005.

CHAPTER 9

Using Reconfigurable Computing for the Optimization of Cryptographic Algorithms

J. M. GRANADO, M. A. VEGA, J. M. SÁNCHEZ, and J. A. GÓMEZ

Universidad de Extremadura, Spain

9.1 INTRODUCTION

Cryptography is a tool used since the early days of civilization: for example, to avoid the enemy reading messages from captured emissaries. Nowadays, the Internet has made it necessary to employ cryptography to transmit information through an insecure channel. One of the first cryptography machines was Enigma, which employs rotor machines to make the encryption, early in the twentieth century. Later, with the appearance of computers, the cryptographic algorithms became more complex. Nowadays, secure algorithms such as DES, Triple-DES, and IDEA exist. Finally, with the appearance of wireless networks, new specific cryptographic algorithms have been developed to be used in these networks, such as MARS, SERPENT, RC6, or Rijndael, which has turned into the new AES standard. Besides, due to the new fast network standards, these algorithms must be very fast, and a very interesting solution is to implement them using hardware such as field-programmable gate arrays (FPGAs) . At this point we can find works [2,11,28] in which FPGA-based implementations of IDEA, AES, and RC6 algorithms are described, respectively. We have implemented these algorithms using FPGAs, achieving very good results in all of them, as we will see in the next sections.

The chapter is structured as follows. In Section 9.2 we describe the three cryptographic algorithms implemented. Next we explain the solutions and component implementation within an FPGA. In Section 9.4 the results are shown. Finally, we describe the conclusions obtained.

Optimization Techniques for Solving Complex Problems, Edited by Enrique Alba, Christian Blum, Pedro Isasi, Coromoto León, and Juan Antonio Gómez

9.2 DESCRIPTION OF THE CRYPTOGRAPHIC ALGORITHMS

Next, we describe the IDEA, AES, and RC6 algorithms.

9.2.1 IDEA Algorithm

The international data encryption algorithm (IDEA) [22] is a 64-bit block cryptographic algorithm that uses a 128-bit key. This key is the same for both encryption and decryption (it is a symmetric algorithm, and it is used to generate fifty-two 16-bit subkeys). The algorithm consists of nine phases: eight identical phases [Figure 9.1(a)] and a final transformation phase [Figure 9.1(b)]. The encryption takes place when the 64-bit block is propagated through each of the first eight phases in a serial way, where the block, divided into four 16-bit subblocks, is modified using the six subkeys corresponding to each phase (elements Z^i_j of Figure 9.1: six subkeys per phase and four subkeys for the last phase). When the output of the eighth phase is obtained, the block goes through a last phase, the transformation one, which uses the last four subkeys.

The decryption follows an identical pattern but computing the sum or multiplication inverse of the subkey, depending on the case, and altering the order of use of the subkeys. As we can suppose, the major problem in the FPGA implementation of this algorithm lies in the multipliers, since apart from taking a great amount of computation and resources, they are executed four times in each phase. The improvement of this component is one of the most studied aspects of the literature. In our case we use KCM multipliers and partial and dynamic reconfiguration.

9.2.2 128-Bit AES Algorithm

The advanced encryption standard (AES) [9] is a symmetric block cipher algorithm that can process data blocks of 128 bits using cipher keys with lengths of 128, 192, and 256 bits. This algorithm is based on the Rijndael algorithm [7], but Rijndael can be specified with key and block sizes in any multiple of 32 bits, with a minimum of 128 bits and a maximum of 256 bits. In our case, we have used a 128-bit key.

The AES algorithm is divided into four phases, which are executed sequentially, forming rounds. The encryption is achieved by passing the plain text through an initial round, nine equal rounds and a final round. In all of the phases of each round, the algorithm operates on a 4×4 array of bytes (called the *state*). In Figure 9.2 we can see the structure of this algorithm. Reference 8 provides a complete mathematical explanation of the AES algorithm. In this chapter we explain only the MixColumns phase of the algorithm in more detail because it will be necessary to understand its implementation.

MixColumns Phase The MixColumns transformation operates on the state column by column, treating each column as a four-term polynomial. The columns

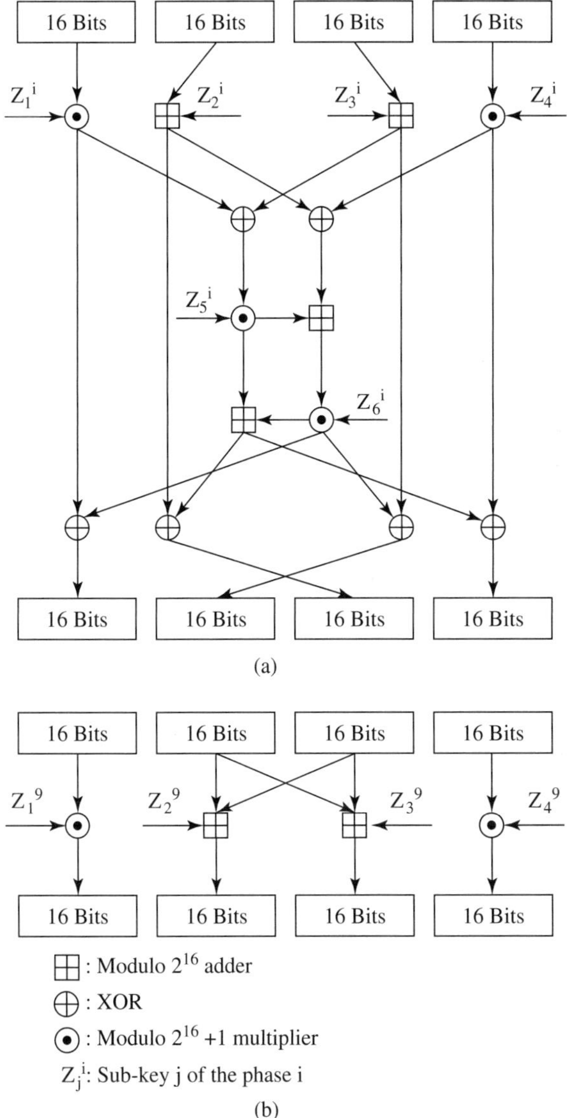

Figure 9.1 Description of the IDEA cryptographic algorithm phases: (a) normal phase; (b) transformation phase.

are considered as polynomials over GF(2^8) and multiplied modulo $x^4 + 1$ by a fixed polynomial $a(x)$, given by

$$a(x) = \{03\}x^3 + \{01\}x^2 + \{01\}x + \{02\} \tag{9.1}$$

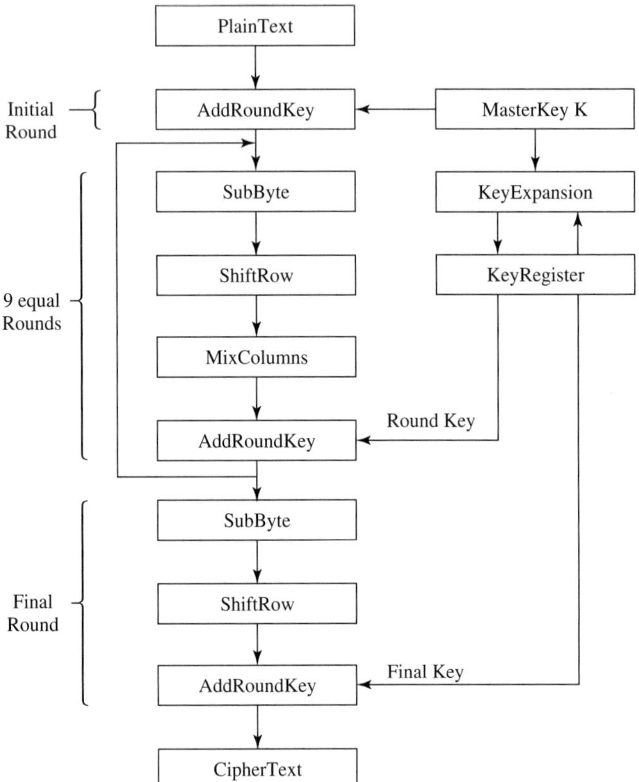

Figure 9.2 Description of the AES cryptographic algorithm.

This can be written as a matrix multiplication:

$$S'(x) = A(x) \otimes S(x):$$

$$\begin{bmatrix} S'_{0,c} \\ S'_{1,c} \\ S'_{2,c} \\ S'_{3,c} \end{bmatrix} = \begin{bmatrix} 02 & 03 & 01 & 01 \\ 01 & 02 & 03 & 01 \\ 01 & 01 & 02 & 03 \\ 03 & 01 & 01 & 02 \end{bmatrix} \begin{bmatrix} S_{0,c} \\ S_{1,c} \\ S_{2,c} \\ S_{3,c} \end{bmatrix} 0 \le c < 4 \tag{9.2}$$

As a result of this multiplication, the four bytes in a column are replaced as follows:

$$\begin{aligned} S'_{0,c} &= (\{02\} \bullet S_{0,c}) \oplus (\{03\} \bullet S_{1,c}) \oplus S_{2,c} \oplus S_{3,c} \\ S'_{1,c} &= S_{0,c} \oplus (\{02\} \bullet S_{1,c}) \oplus (\{03\} \bullet S_{2,c}) \oplus S_{3,c} \\ S'_{2,c} &= S_{0,c} \oplus S_{1,c} \oplus (\{02\} \bullet S_{2,c}) \oplus (\{03\} \bullet S_{3,c}) \\ S'_{3,c} &= (\{03\} \bullet S_{0,c}) \oplus S_{1,c} \oplus S_{2,c} \oplus (\{02\} \bullet S_{3,c}) \end{aligned} \tag{9.3}$$

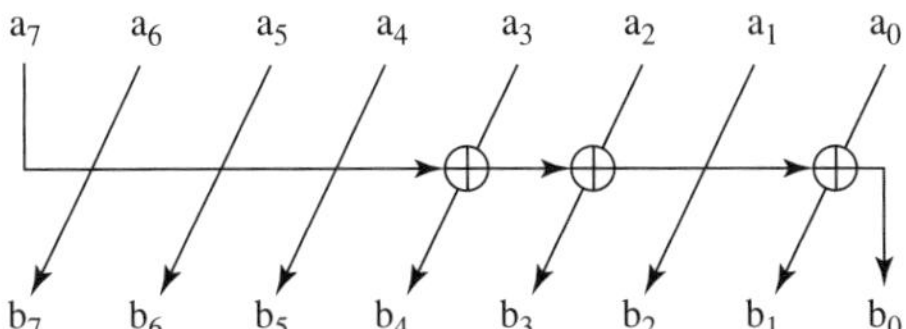

Figure 9.3 The *xtime* function.

where $\oplus$ represents the XOR operation and $\bullet$ is a multiplication modulo the irreducible polynomial $m(x) = x^8 + x^4 + x^3 + x + 1$. Figure 9.3 shows the implementation of the function B = xtime(A), which is used to make multiplications of a number by 2 modulo $m(x)$. So we will only have binary operations:

$$\{02\} \bullet S_{x,c} = \text{xtime}(S_{x,c}) \qquad (9.4)$$
$$\{03\} \bullet S_{x,c} = \text{xtime}(S_{x,c}) \oplus S_{x,c}$$

9.2.3 128-Bit RC6 Algorithm

RC6 [19] is a symmetric cryptographic algorithm and one of the five finalists of the advanced encryption standard (AES) competition [16]. It was also submitted to the New European Schemes for Signatures, Integrity and Encryption (NESSIE) [5] and the Cryptography Research and Evaluation Committee (CRYPTREC) [6] projects. It is a proprietary algorithm, patented by RSA Security [20]. This algorithm can employ different block and key sizes, depending on three parameters:

1. w: the word size in bits
2. r: the number of rounds made by the algorithm
3. b: the length of the encryption key in bytes

So the RC6 versions are denoted by RC6-w/r/b. In our case we use the version established by AES, that is, $w = 32$ (128-bit data), $r = 20$ (20 loop rounds), and $b = 16$ (128 bit key). So in the future we call RC6 to the RC6-32/20/16 version. In all versions of this algorithm, the following operations are used:

- $a + b$: integer addition modulo 2^w (2^{32} in our case).
- $a - b$: integer subtraction modulo 2^w (2^{32} in our case).
- $a \oplus b$: bitwise exclusive-or.
- $a \times b$: integer multiplication modulo 2^w (2^{32} in our case).
- $a <<< b$: rotate the word a to the left by the amount given by the least significant lg_2 w bits of b (the least significant 5 bits of b in our case).
- $a >>> b$: rotate the word a to the right by the amount given by the least significant lg_2 w bits of b (the least significant 5 bits of b in our case).

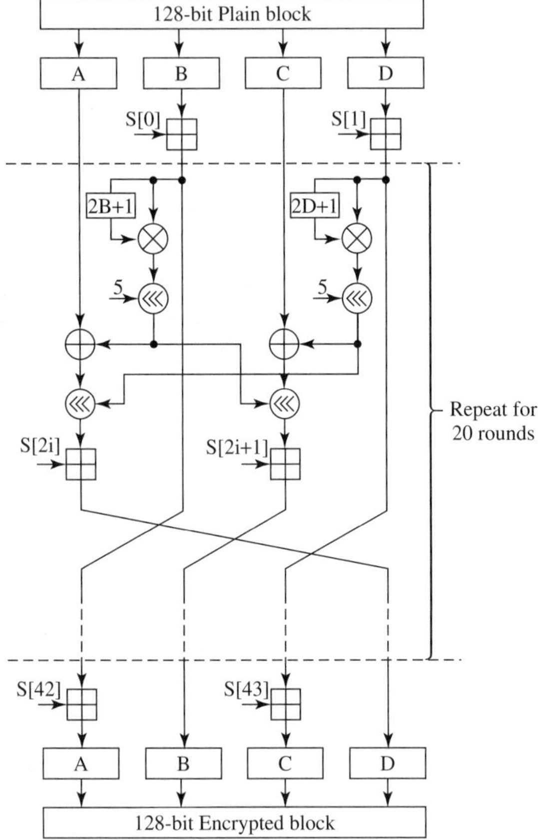

Figure 9.4 RC6 encryption algorithm.

RC6 works with four w-bit registers/words (A, B, C, and D), which contain the initial input plaintext as well as the output cipher text at the end of encryption. To encrypt a block, we will pass it through a loop of r iterations (in our case, 20) in which two subkeys are used per iteration. Besides, two subkeys operate before the loop and two after the loop. The algorithm can be seen in Figure 9.4. The decryption algorithm is practically equal to the encryption, but executing the operations in the inverse order. Besides, the additions are replaced by subtractions and the third and fourth left rotations of each round are replaced by right rotations.

9.3 IMPLEMENTATION PROPOSAL

To implement the cryptographic algorithms described previously, we combine three different techniques: partial and dynamic reconfiguration, pipelining, and in the IDEA case, parallelism. Let us see how we have implemented each algorithm.

In all cases we employ VHDL [18] to implement the reconfigurable elements, Handel-C [3] to implement nonreconfigurable elements, and JBits [24] to make the partial and dynamic reconfiguration. On the other hand, a Virtex-II 6000 FPGA [27] has been employed to implement all the algorithms.

9.3.1 Implementation of the IDEA

To implement the IDEA, we have used constant coefficient multipliers (KCMs) [25] and constant coefficient adders (KCAs) [12]. These elements operate a variable datum with a constant datum, and as we can see in Figure 9.1, all the multipliers and several adders of the IDEA algorithm operate with a constant. Thanks to these elements, we reach a high performance level since KCMs are much faster than the traditional multipliers (KCAs are only a little faster than normal adders, but we need them to do the key reconfiguration and reduce the clock cycle). We have employed an inner pipelining in the IDEA. This pipelining is done in the KCM and KCA elements since they are the slowest.

KCM and KCA Pipelining In the KCM as we can see in Figure 9.5 (a detailed explanation of KCM implementation is given in ref. 12), we have used a total of six stages and several pipelining registers (one 32-bit register, six 20-bit registers, eight 16-bit registers, two 4-bit registers and five 1-bit registers) to execute the KCM completely. On the other hand, we use only four stages to execute the KCA completely, and the number of pipelining registers is also lower (one 19-bit register, three 5-bit registers, and five 4-bit registers), as we can see in Figure 9.6 (ref. 12 includes a detailed explanation of the KCA implementation). As we said previously, this component is not much faster than a normal adder, but it allows us to increment the clock frequency (because it uses small adders) and to make the key reconfiguration.

Operation Pipelining Besides the inner pipelining, we use an operation pipelining. This pipelining is done among all simple operations (e.g., adders or XORs) of all the phases of the IDEA: In Figure 9.7 we can see how several pipelining registers are included to wait for the result of the KCMs because these elements have the majority of pipelining stages of the algorithm.

We can see that the pipelining done is a very fine grain (the total number of pipelining stages is 182), which greatly reduces the clock frequency (the clock frequency is 4.58 ns). This fact leads to a high latency (182 cycles), but if we are encrypting a great number of blocks, the benefit achieved by the high clock frequency clearly overcomes the disadvantage of the latency.

Dual Path Finally, we have implemented two separate and parallel data paths, a reason why we encrypt two blocks at the same time, that is, two blocks per cycle. Besides, these data paths allow us both to duplicate the performance of the encryption/decryption by means of reconfiguring the KCM and KCA with the same subkeys and to use one path to encryption and another to decryption.

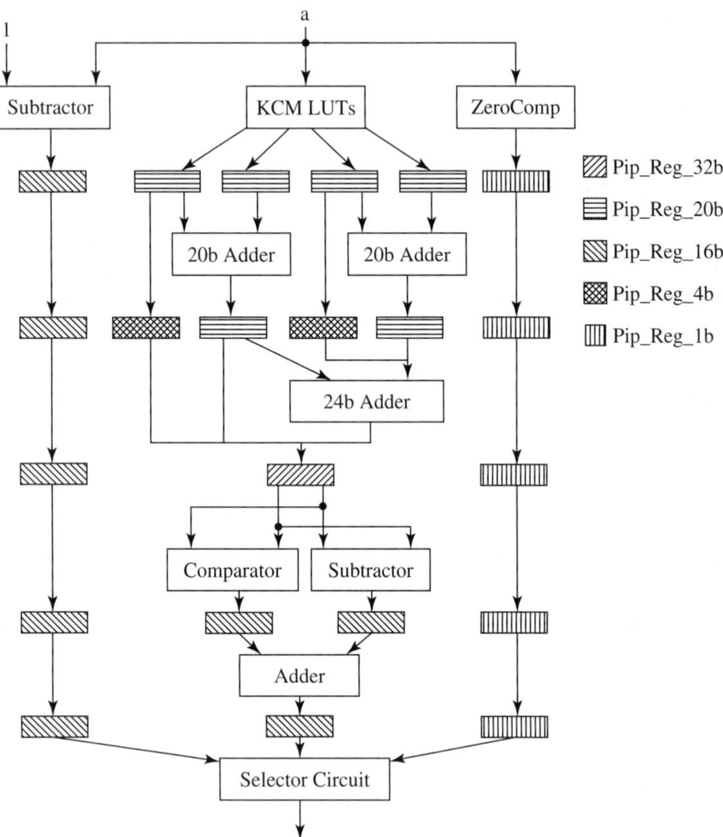

Figure 9.5 KCM pipelining.

This last option is best in case the FPGA is installed into a server that has the responsibility of encrypting the data coming from an extranet (e.g., the Internet) to, for example, a wireless insecure intranet, and vice versa.

9.3.2 Implementation of the AES Algorithm

Opposite to the IDEA, AES does not have complex operations, but on the other hand, it has a high memory cost. If we want to do the transformation of all bytes of the state in the SubByte phase in a parallel way, we must implement 16 SBox by phase. Besides, if we want to implement a pipelined version of the algorithm, we must implement 10 SubByte phases, each of which has 16 SBox tables. Furthermore, as we want to implement a pipelined version, we must define 11 4×4 1-byte subkeys, because we have 11 AddRoundKey phases. All these storing elements give us a total of 329,088 bits ($=$ 10 SubBytePhases $\times$ 16 SBoxTablesPerSubBytePhase $\times$ 256 ElementsPerSBox $\times$ 8 BitsPerElement $+$ 11

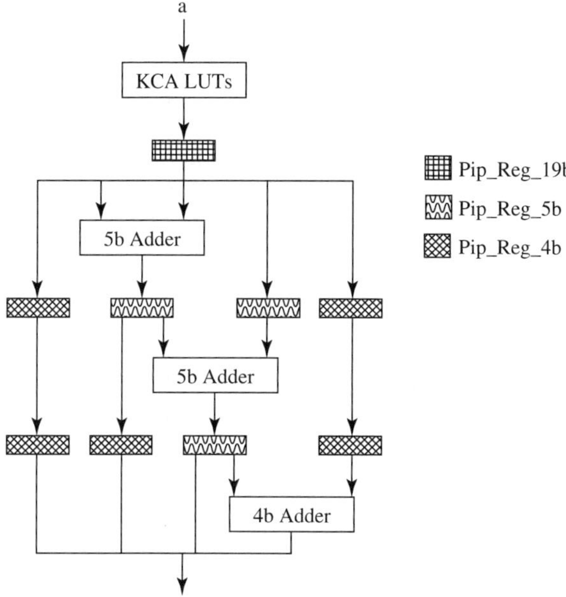

Figure 9.6 KCA pipelining.

AddRoundKeyPhases × 16 ElementsPerAddRoundKeyPhase × 8 BitsPerElement). On the other hand, we have only forty-one 128-bit pipelining registers (Figure 9.8), less than for the IDEA.

MixColumns Phase Implementation As we have said, the AES algorithm has no complex phases, and the implementation of all of them is very simple. However, the MixColumns phase implementation deserves special consideration. Formerly, we saw a brief mathematical description of this phase and now we will see how to implement it. To explain the implementation, we take the calculation of the element $S'_{0,0}$ of the $S'(x)$ matrix (Equation 9.2). The equation to solve this clement is

$$S'_{0,0} = (\{02\} \bullet S_{0,0}) \oplus (\{03\} \bullet S_{1,0}) \oplus S_{2,0} \oplus S_{3,0} \tag{9.5}$$

Let us remember that the • operation is done by means of the xtime function (Equation 9.2). So Equation 9.5 changes to

$$S'_{0,0} = \text{xtime}(S_{0,0})) \oplus (\text{xtime}(S_{1,0})) \oplus S_{1,0}) \oplus S_{2,0} \oplus S_{3,0} \tag{9.6}$$

Finally, taking into account the representation of the xtime function in Figure 9.3, we will use the final result described in Table 9.1 (the result of bit n will be the XOR among the four columns of row n).

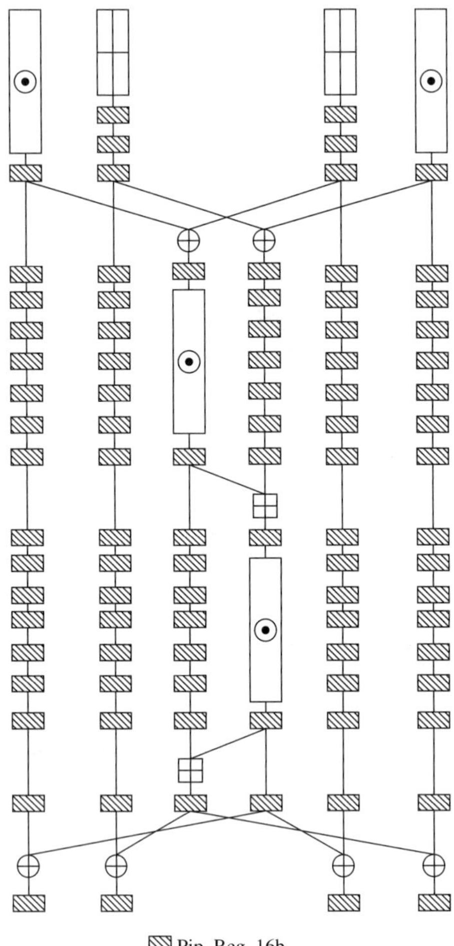

Figure 9.7 Operation pipelining of one phase of the IDEA.

KeyExpansion Phase Observing Figure 9.2, we can see that the KeyExpansion phase calculates the subkeys that will be used by the different AddRoundKey phases. However, we do not implement this method; instead, we use partial reconfiguration. The process is similar to the IDEA run-time reconfiguration, but in this case, we reconfigure the LUTs that will be used in the AddRoundKey phases (i.e., the LUTs that store the subkeys) instead of the components that use these LUTs (as happens in the IDEA).

AES Pipelining As we did in the IDEA, we use operation-level pipelining. However, as in the AES algorithm, all the operations of one phase (remember that in the AES algorithm, a round has several phases; see Figure 9.2) are done

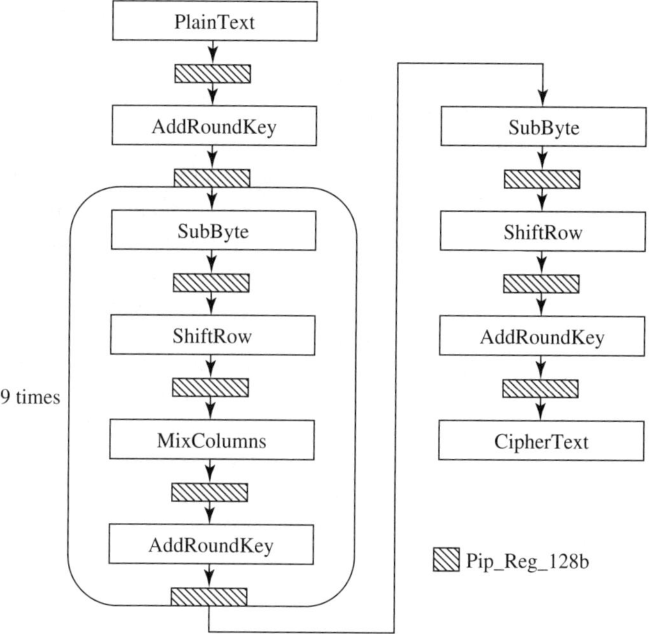

Figure 9.8 Phase-level pipelining of the AES algorithm.

TABLE 9.1 Calculation of the State's $S'_{0,0}$ Byte in the MixColumns Phase

$S'_{0,0}$ Bit	xtime($S_{0,0}$)	xtime($S_{1,0}$)$\oplus S_{1,0}$	$S_{2,0}$	$S_{3,0}$
7	$S_{0,0}[6]$	$S_{1,0}[7] \oplus S_{1,0}[6]$	$S_{2,0}[7]$	$S_{3,0}[7]$
6	$S_{0,0}[5]$	$S_{1,0}[6] \oplus S_{1,0}[5]$	$S_{2,0}[6]$	$S_{3,0}[6]$
5	$S_{0,0}[4]$	$S_{1,0}[5] \oplus S_{1,0}[4]$	$S_{2,0}[5]$	$S_{3,0}[5]$
4	$S_{0,0}[7] \oplus S_{0,0}[3]$	$S_{1,0}[7] \oplus S_{1,0}[4] \oplus S_{1,0}[3]$	$S_{2,0}[4]$	$S_{3,0}[4]$
3	$S_{0,0}[7] \oplus S_{0,0}[2]$	$S_{1,0}[7] \oplus S_{1,0}[3] \oplus S_{1,0}[2]$	$S_{2,0}[3]$	$S_{3,0}[3]$
2	$S_{0,0}[1]$	$S_{1,0}[2] \oplus S_{1,0}[1]$	$S_{2,0}[2]$	$S_{3,0}[2]$
1	$S_{0,0}[7] \oplus S_{0,0}[0]$	$S_{1,0}[7] \oplus S_{1,0}[1] \oplus S_{1,0}[0]$	$S_{2,0}[1]$	$S_{3,0}[1]$
0	$S_{0,0}[7]$	$S_{1,0}[7] \oplus S_{1,0}[0]$	$S_{2,0}[0]$	$S_{3,0}[0]$

at the same time, actually, it is phase-level pipelining. This pipelining can be seen in Figure 9.8.

9.3.3 Implementation of the RC6 Algorithm

In this case, the critical element is the multiplier, but opposite to the IDEA, both input data are variable. However, the subkeys are used in the adders, but in this case, 32-bit adders (in the IDEA we used 16-bit adders), which is why we use a variation of the IDEA KCAs. Let us see how we have implemented these elements.

32-Bit Adders To implement these elements, we have designed a 32-bit KCA based on the 16-bit IDEA KCA. So we are going to implement two slightly different 16-bit IDEA KCAs: The first, the least significant, will return 17 bits, that is, 16 bits for the addition of the 16 least significant bits and 1 bit for the output carry, and the second adder, the most significant, will return 16 bits, that is, the addition of the 16 bits most significant and the previous output carry. Implementation of these two elements can be seen in Figures 9.9 and 9.10.

In the figures the 4×5 LUTs return 4 bits related to the addition and one bit related to the carry. The four LSBs of each 4×5 LUT are used to generate a 16-bit operand, and the carry bits of the four LUTs plus the input carry in the most significant KCA are used to generate another 16-bit operand. Finally, these two operands are added, calculating the two 16-bit parts of the operation, and the complete 32-bit adder will be the result of joining the output data of the two KCAs. The resulting circuit can be seen in Figure 9.11. It is important

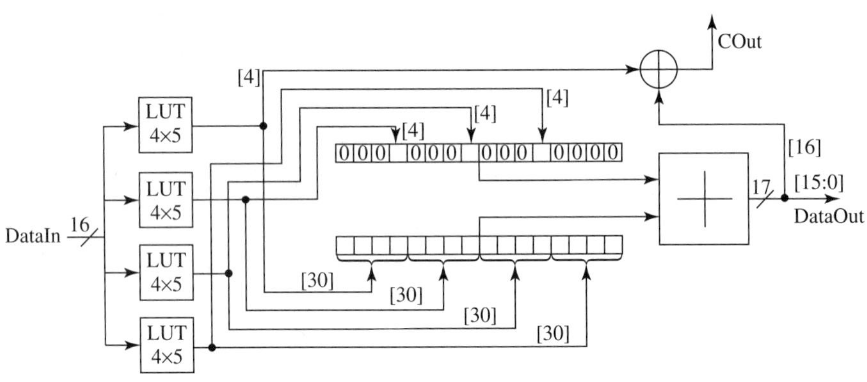

Figure 9.9 Least significant KCA with output carry.

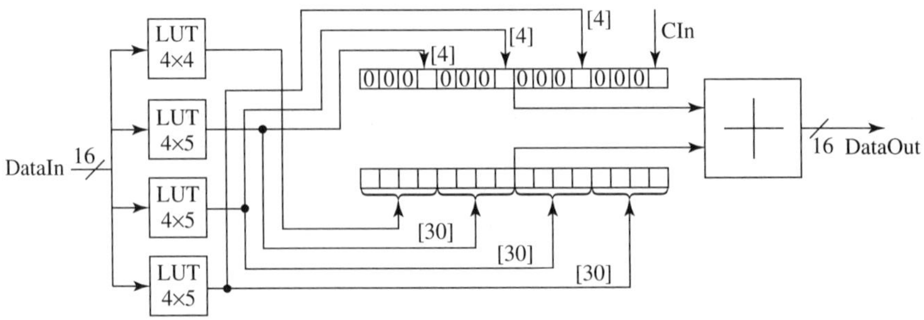

Figure 9.10 Most significant KCA with input carry.

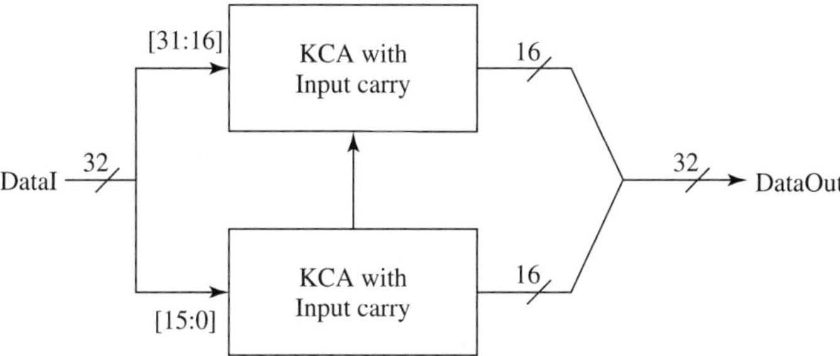

Figure 9.11 32-bit complete KCA.

to emphasize that the second adder uses a 4×4 LUT (the most significant one) because the adder is modulo 2^{32}, so the carry is discarded.

32-Bit Multipliers In this algorithm the multipliers are the critical elements, so it is important to implement them in an efficiently. To implement the multipliers we have used the MULT18x18s (18×18 bit multipliers) embedded in the Virtex-II 6000 FPGA [27].

Before beginning the multiplier design, we must take into account two facts: First, we have 18×18 multipliers, so we are going to divide the multiplication by parts using the divide-and-conquer method [17]; and second, the multiplication is modulo 2^{32}, so several operation bits will be discarded.

Now we can begin to design the circuit. As we have said, we have employed the divide-and-conquer method but have modified it because the multiplication is modulo 2^{32}. In the original divide-and-conquer algorithm

$$\begin{aligned} A^{32} \times B^{32} &= (A_{16}^H \times 2^{16} + A_{16}^L) \times (B_{16}^H \times 2^{16} + B_{16}^L) \\ &= A_{16}^H \times B_{16}^H \times 2^{32} + A_{16}^H \\ &\quad \times B_{16}^L \times 2^{16} + A_{16}^L \times B_{16}^H \times 2^{16} + A_{16}^L \times B_{16}^L \end{aligned} \tag{9.7}$$

the subscript indicates the size in bits of the operand and the superscript indicates either the high or low part of the operand. This equation shows the original divide-and-conquer method, but as we have multiplications modulo 2^{32}, Equation 9.7 is replaced by

$$\begin{aligned} (A_{16}^H \times B_{16}^H \times 2^{32} + A_{16}^H \times B_{16}^L \times 2^{16} + A_{16}^L \times B_{16}^H \times 2^{16} + A_{16}^L \times B_{16}^L)\%2^{32} \\ = A_{16}^H \times B_{16}^L \times 2^{16} + A_{16}^L \times B_{16}^H \times 2^{16} + A_{16}^L \times B_{16}^L \end{aligned} \tag{9.8}$$

Now, we need only add the multiplications, but not all the bits are used. The bits involved in the multiplication can be seen in Figure 9.12. So we have three

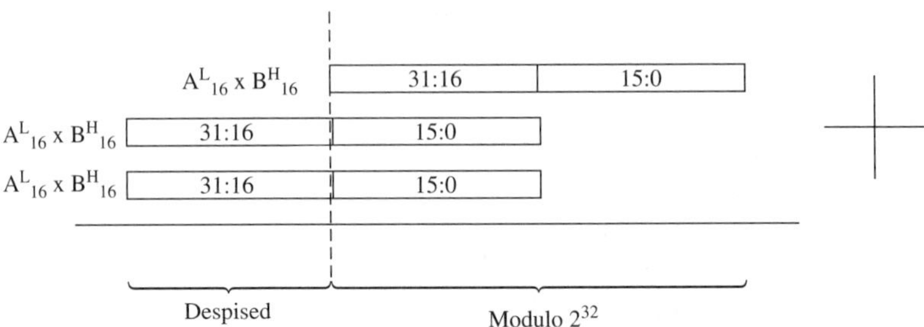

Figure 9.12 Addition modulo 2^{32} of the partial multiplications.

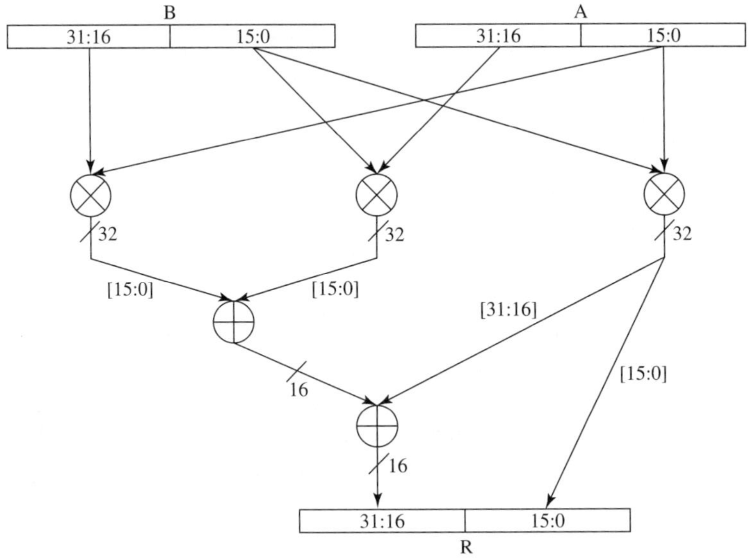

Figure 9.13 Complete multiplier modulo 2^{32}.

16-bit multipliers and two 16-bit adders instead of the four 16-bit multipliers and three 32-bit adders of the original divide-and-conquer method. Therefore, the final circuit of the multiplier modulo 2^{32} is shown in Figure 9.13.

RC6 Pipelining Until now we have explained how the operators are implemented. Now we are going to see how we have used pipelining to encrypt several parts of several blocks at the same time. We have implemented a middle-grain pipelining because the critical elements, that is, the MULT18x18s multipliers, expend more time than do the remaining elements, which is why we have preferred to balance the pipelining. Figure 9.14 shows the pipelining implementation. Also, the multipliers are actually pipelined, using two phases to complete the operation. So the final pipelined implementation has a total of 82 stages, where

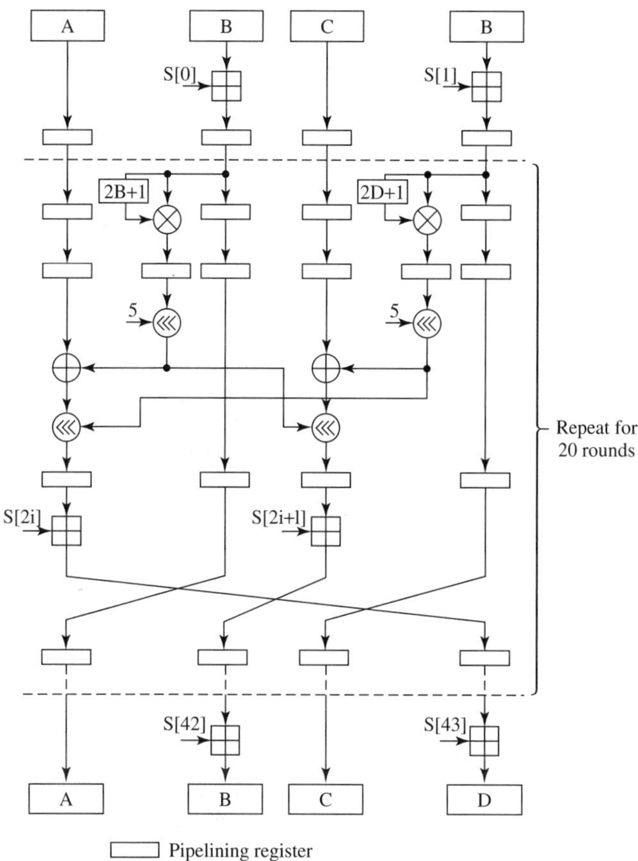

Figure 9.14 Final pipelined implementation for the RC6 algorithm.

each round of the main loop has four of these stages, and the total number of 32-bit pipelining registers is 484 (each multiplier has three inner registers).

9.4 EXPERMENTAL ANALYSIS

To obtain the following results, in all the cases we have used Xilinx ISE 8.1i and Celoxica DK4.

9.4.1 IDEA Results

After synthesis with Xilinx ISE, our implementation takes only 45% of the FPGA slices (exactly 15,016 slices) and has a clock of 4.58 ns. In conclusion, after the initial latency, it encrypts two 64-bit blocks every 4.58 ns. This gives us a total performance of 26.028 Gb/s, surpassing the best results achieved by other

TABLE 9.2 Comparative Results of the IDEA

Device	Performance (Gb/s)	Occupation (Slices)	Latency (cycles, ns)	Reference
XC2V6000-6	26.028	15,016	182,833	This work
XCV600-6	8.3	6,078	158,1205	[11]
XCV2000E-6	6.8	8,640	132,1250	[13]
XCV1000E-6	5.1	11,602	??,2134	[4]
XCV1000E-4	4.3	??	??	[1]
XCV1000E-6	2.3	11,204	??,7372	[15]
XCV800-6	1.28	6,312	25,1250	[10]

authors, as Table 9.2 shows. In addition, our results are better than the results of other authors in the latency. It is important to emphasize that although we have a high latency (182 cycles), which is bad for small data size encryption, the applications usually encrypt data of great size, particularly when are speaking of an encryption server. On the other hand, we have a double data path, and therefore our occupation and latency (in Table 9.2) encrypt two data blocks (other authors encrypt only one block).

9.4.2 AES Results

Now let's look at the AES results. Our implementation uses a total of 3720 slices, only 11% of the FPGA resources. This result is one of the best in that field, as we can see in Table 9.3 (it is important to emphasize that Table 9.3 does not include the slices or BRams which are used for the SBox storage because these data are very similar for all the implementations). On the other hand, thanks to the pipelined implementation, we can reach a clock cycle of 5.136 ns, that is, a performance of 23.211 Gb/s. This datum is not the best of Table 9.3 (it is the second best), but we improve the work [28] in the rest of columns of this table, so our result is comparable with the ref. [28] implementation and better than implementations by other authors. Finally, the latency column is also favorable to us, where we again have the best result.

9.4.3 RC6 Results

As we can see in Table 9.4, our implementation employs a total of 14,305 slices and reaches a clock of 7.926 ns, the best in the table. This clock allows us to achieve a performance of 15.040 Gb/s, surpassing the results reached by the other authors.

9.5 CONCLUSIONS

In this work we have presented the FPGA-based implementation of three important cryptographic algorithms (IDEA, AES, and RC6) using partial and dynamic

TABLE 9.3 Comparative Results of the AES Algorithm

Device	Performance (Gb/s)	Occupation (Slices)	Latency (Cycles, ns)	Reference
XC2V6000-6	23.211	3,720	41,210.5	This work
XC2VP70-7	29.8	7,761	60,253.8	[28]
XC2VP20-7	21.54	5,177	46,292.8	[14]
XCV2000E-8	20.3	5,810	??	[21]
XCV3200E-8	18.56	4,872	??	[23]
XCV812E-8	12.2	12,600	??,1,768.8	[9]
XCV3200E-8	11.77	2,784	??	[23]

TABLE 9.4 Comparative Results of the RC6 Algorithm

Device	Performance (Gb/s)	Occupation (Slices)	Clock (ns)	MULT-18 × 18s	Reference
XC2V6000-6	15.040	14,305	7.926	120	This work
XC2V3000-6	14.192	8,554	8.400	80	[2]
Four XCV1000-6	13.100	46,900	—	0	[9]
XCV1600E-8	9.031	14,110	13.200	0	[2]
0.5-μm CMOS	2.200	—	—	—	[26]

reconfiguration, pipelining, and in the IDEA case, parallelism. To implement these algorithms we have used a Virtex-II 6000 FPGA, VHDL and Handel-C as hardware description languages, and JBits to make the runtime reconfiguration. Taking all of these into account, we have reached very fast implementations of the three algorithms, so we can say that the techniques used in this work to implement cryptographic algorithms, that is, pipelining, parallelism, and partial and dynamic reconfiguration, are very good techniques for optimizing cryptographic algorithms.

Acknowledgments

The authors are partially supported by the Spanish MEC and FEDER under contract TIC2002-04498-C05-01 (the TRACER project).

REFERENCES

1. J. Beuchat, J. Haenni, C. Teuscher, F. Gómez, H. Restrepo, and E. Sánchez. Approches met érielles et logicielles de l'algorithme IDEA. *Technique et Science Informatiques*, 21(2):203–204, 2002.

2. J. Beuchat. *High Throughput Implementations of the RC6 Block Cipher Using Virtex-E and Virtex-II Devices*. Institut National de Recherche en Informatique et en Automatique, Orsay, France, 2002.
3. Celoxica. *Handel-C Language Reference Manual*, version 3.1. http://www.celoxica.com, 2005.
4. O. Cheung, K. Tsoi, P. Leong, and M. Leong. Tradeoffs in parallel and serial implementations of the international data encryption algorithm (IDEA). *Workshop on Cryptographic Hardware and Embedded Systems*, 2162:333–347, 2001.
5. Computer Security and Industrial Cryptography. https://www.cosic.esat.kuleuven.be/nessie, 2007.
6. Cryptography Research and Evaluation Committees. http://www.cryptrec.jp, 2007.
7. J. Daemen and V. Rijmen. The block cipher Rijndael. *Lecture Notes in Computer Science*, 1820:288–296, 2000.
8. *Federal Information Processing Standards Publication 197*. http://csrc.nist.gov/publications/fips/fips197/fips-197.pdf, 2001.
9. K. Gaj and P. Chodowiec. Fast implementation and fair comparison of the final candidates for advanced encryption standard using field programmable gate arrays. *Lecture Notes in Computer Science* 2020:84–99, 2001.
10. I. González. (2002) Codiseño en sistemas reconfigurables basado en Java. *Internal Technical Report*, Universidad Autonoma de Madrid, Spain.
11. I. González, S. Lopez-Buedo, F. J. Gómez, and J. Martínez. Using partial reconfiguration in cryptographic applications: an implementation of the IDEA algorithm. *In Proceedings of the 13th International Conference on Field Programmable Logic and Application*, pp. 194–203, 2003.
12. J. M. Granado-Criado, M. A. Vega-Rodríguez, J. M. Sánchez-Pérez, and J. A. Gómez-Pulido. A dynamically and partially reconfigurable implementation of the IDEA algorithm using FPGAs and Handel-C. *Journal of Universal Computer Science*, 13(3):407–418, 2007.
13. A. Hamalainen, M. Tomminska, and J. Skitta. 6.78 Gigabits per second implementation of the IDEA cryptographic algorithm. In *Proceedings of the 12th International Conference on Field Programmable Logic and Application*, pp. 760–769, 2002.
14. A. Hodjat and I. Verbauwhede. A 21.54 Gbits/s fully pipelined AES processor on FPGA. In *Proceedings of the 12th Annual IEEE Symposium on Field-Programmable Custom Computing Machines*, 2004.
15. M. P. Leong, O. Y. Cheung, K. H. Tsoi, and P. H. Leong A bit-serial implementation of the international data encryption algorithm (IDEA). In *Proceedings of the IEEE Symposium on Field-Programmable Custom Computing Machines*, pp. 122–131, 2000.
16. National Institute of Standards and Technology. http://csrc.nist.gov/archive/aes/index.html, 2007.
17. B. Parhami. *Computer Arithmetic*. Oxford University Press, New York, 2000.
18. V. A. Pedroni. *Circuit Design with VHDL*. MIT Press, Cambridge, MA, 2004.
19. R. Rivest, M. Robshaw, R. Sidney, and Y. Yin. The RC6 block cipher. http://www.rsa.com/rsalabs/node.asp?id=2512, 2007.
20. RSA security. http://www.rsa.com, 2007.

21. G. P. Saggese, A. Mazzeo, N. Mazzoca, and A. G. Strollo. An FPGA-based performance analysis of the unrolling, tiling, and pipelining of the AES algorithm. *Lecture Notes in Computer Science*, 2778:292–302, 2003.
22. B. Schneier. *Applied Cryptography*, 2nd ed. Wiley, New York, 1996.
23. F. X. Standaert, G. Rouvroy, J. Quisquater, and J. Legat. Efficient implementation of Rijndael encryption in reconfigurable hardware: improvements and design tradeoffs. *Lecture Notes in Computer Science*, 2779:334–350, 2003.
24. *JBits User Guide*. Sun Microsystems, 2004.
25. R. Vaidyanathan and J. L. Trahan. *Dynamic Reconfiguration: Architectures and Algorithms*. Kluwer Academic/Plenum Publishers, New York, 2003.
26. B. Weeks, M. Bean, T. Rozylowicz, and C. Ficke. (2000) Hardware performance simulations of round 2 advanced encryption standard algorithms. *Technical Report*. National Security Agency, Washington, DC
27. *Virtex-II Platform FPGAs: Complete Data Sheet*. Xilinx, 2005.
28. S. M. Yoo, D. Kotturi, D. W. Pan, and J. Blizzard. An AES crypto chip using a high-speed parallel pipelined architecture. *Microprocessors and Microsystems* 29(7):317–326, 2005.

CHAPTER 10

Genetic Algorithms, Parallelism, and Reconfigurable Hardware

J. M. SÁNCHEZ
Universidad de Extremadura, Spain

M. RUBIO
Centro Extremeño de Tecnologías Avanzadas, Spain

M. A. VEGA and J. A. GÓMEZ
Universidad de Extremadura, Spain

10.1 INTRODUCTION

The grid computing model offers a powerful and high-availability platform that allows efficient running of applications that are intensive in terms of calculations and data. It is a distributed model based on the coordinated use of several spread management services and computational resources. Its main characteristic is resource sharing; collaboration is the philosophy. New nodes can easily aggregate to an organization (named V.O., virtual organization, in grid terms), independent of its geographical situation. All resources shared within the V.O. will conform to a federated resource available for any member of it. That is, the EELA project [1] gathers power from sites placed in both Latin American and Europe, and any member of it can benefit from its aggregated resources. This gives the grid model the power to establish a resource with virtually unlimited computing power. The grid computing model is the best suitable for applications whose throughput can be increased by multiplying the executions per unit time (e.g., parametric jobs, which perform huge amounts of similar calculations just varying the input parameters). Grid applications are a type of parallelized application designed or modified for accessing the resources of the infrastructure (i.e., distribution along different nodes, access to the storage catalog, etc.).

Optimization Techniques for Solving Complex Problems, Edited by Enrique Alba, Christian Blum, Pedro Isasi, Coromoto León, and Juan Antonio Gómez

In this work we present the steps performed to enable an existing application to be run on the grid. The application solves the placement and routing problem of a logic circuit on an FPGA in the physical design stage. This is an NP problem that requires high computational resources, so the grid implementation is a good candidate for solving it. FPGAs are devices used to implement logic circuits. Representative FPGA manufacturers include Xilinx [2], Altera, and Actel. This work deals with the island model [3] FPGA architecture. An FPGA based on the island model consists of three main classes of elements distributed along the FPGA (see Figure 10.1). The elements are configurable logic blocks (CLBs), used for implementing the logic functions, input/output blocks (IOBs), which join the FPGA with external devices, and interconnection resources (IRs), which consist of interconnection wires (programmable nets) and interconnection switches, used to connect a CLB to another, or a CLB to an IOB.

The physical design stage is one of the most complex of the FPGA design cycle [4]. At this stage an optimal way of putting the CLB over the FPGA (placement) and interconnecting them (routing) must be found. A balance between the CLB area and the wire density will produce the most suitable solutions. To maintain this balance, placement and routing stages must be tackled simultaneously. If we attempt these stages sequentially, the CLB area may tend to be too small and then the interconnection scheme gets complex and confusing, because the wire density increases considerably. On the other hand, the use of a fully optimized routing scheme could result in high CLB area needs. This causes an NP problem with two simultaneous factors to take in account. For this reason we have used genetic algorithms (GAs) as the optimization tool. GAs offer a good resolution model for this type of problem (complex optimization problems [5,6]).

We use the grid to run a tool that solves the problem mentioned before. As the grid middleware, we have used gLite Middleware [7] and the GridWay metascheduler. To get the grid-enabled version of our current sequential tool

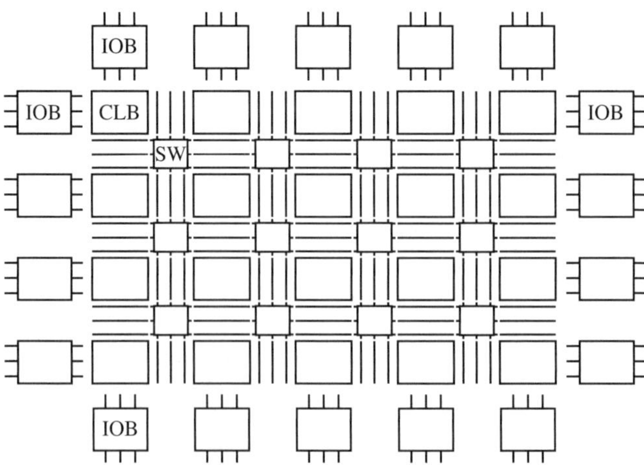

Figure 10.1 FPGA island architecture.

(more details of this tool are given in Section 10.3.2) we start identifying the part of the code to be parallelized (and run remotely). GAs are very suitable applications to be run over the grid, since they can easily be parallelized into several similar executions. If we consider each of these executions as an island of a DGA, we are adding to the parallel execution the advantages of distributing a GA (i.e., the chromosome migration). This parallelization could be considered as a process-level parallelization [8].

By taking the random ring distributed topology for the GA, several islands (populations) evolve in the working nodes (WNs) as grid jobs. A system for communicating the evolving islands (i.e., migration) has been implemented. The migration has a process that sends the islands to a local node and exchanges their optimal chromosomes. Once the new islands have been produced, they are sent again to continue the evolution. When the stop criterion is reached, the process finalizes and the best chromosomes of each island are taken as solutions of the FPGA placement and routing.

The results we have obtained by comparing both the performance and the quality results between sequential and distributed implementations demonstrate the benefits of using a grid, especially in terms of execution time and performance. So scientists have a larger platform to experiment with. Using a grid, they can launch more complex tasks and compared with sequential execution, increase considerably the number of experiments while expending less time and effort.

The chapter is organized as follows. In Section 10.2 we list some previous work developed in this domain. In Section 10.3 we explain the problem to solve. Sections 10.4 and 10.5 cover the algorithm proposals and the results obtained. Finally, conclusions are presented in Section 10.6.

10.2 STATE OF THE ART

We mention in this section previous work to run the placement and routing stages in FPGAs. Most of them are based on evolutionary algorithms and are designed to solve both placement and routing substages. Alexander et al. [9] distinguish between global and detailed routing. Their algorithm solves placement and routing substages with a recursive geometric strategy for simultaneous placement and global routing. Then they use a graph for the detailed routing phase. This technique heuristically minimizes the global wire length. In opposition, Fernandez et al. [10] use genetic programming (GP) instead of GAs. This work applies GP for an experimental study of a technique known as distributed parallel genetic programming (PGPD). This experimental study demonstrates the usefulness of the PGPD technique.

Gudise and Venayagamoorthy [11] use particle swarm optimization (PSO). PSO imitates behavior in which an element evolves by interacting with other elements in a population. Each particle (or individual) represents a possible circuit solution. By making interactions between the particles that represent a better placement scheme, the process for finding the optimal solution is carried out.

Altuna and Falcon [12] use a GA for the routing substage and also for pin assignation. It has been designed to act as an additional stage in the design cycle of the Xilinx toolset. So the output of one of this toolset stages acts as the input for this work. Then the output generated is integrated again in the next stage.

Finally, Rubio et al. [13] show an implementation for solving the placement and routing substages by using a distributed genetic algorithm (DGA). Several populations are defined in different nodes, performing communication for the migration process by using MPI libraries. In this work a results viewer was also developed.

10.3 FPGA PROBLEM DESCRIPTION AND SOLUTION

In this section we explain the problem of solving how GAs have been applied to this outcome and the way to enable application for the grid. Since the problem to solve is a key consideration in the FPGA design and architectures, the design cycle and the physical stage in that cycle are introduced. Then we explain how to apply certain GA libraries to solve the problem. This refers to the sequential tool version. Finally, the grid architecture, its management, and the middleware used are introduced.

10.3.1 FPGAs

FPGA are devices with programmable logical components. These components can range from simple elements such as OR and AND gates to more complicated logical constructions such as multiplexers and decoders. To enhance the functionality, the logical components are supported with memory elements. So an FPGA device can be reprogrammed. That means that the same piece of hardware can be used for many applications, such as signal processing, image analysis, computer vision, and cryptography. Since these applications are implemented and run directly on hardware, the efficiency and performance grow considerably. Another typical use for FPGA is ASIC prototyping. ASICs (application-specific integrated circuits) are nonreprogrammable circuits, which are used for concrete objectives such as the developing of a common chip or a circuit to be integrated into a device. In the ASIC design, an FPGA can be used to simulate the design cycle. If the design obtains successful results and all the design requirements are met, this design is validated and it is passed to the manufacturing process. But if the reports say that the design is not valid, the FPGA can be used to test a new ASIC design.

There are several architectures, depending on its physical topology and elements. One example is the hierarchical topology, in which the logical elements belong to different levels and a superior-level logical element can be decomposed on several lower-level logical elements. So there are two types of connections: local connections, connecting elements within the same level, and long connections, which communicate between logical elements. Figure 10.2(a)

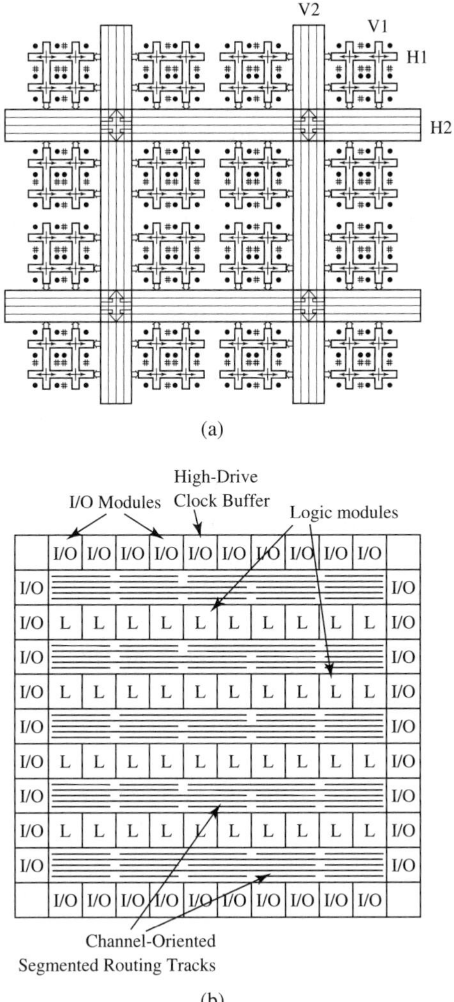

Figure 10.2 FPGA (a) hierarchical and (b) channel architecture.

shows an example of FPGA architecture with three nested levels of logical elements. Another common FPGA architecture is the channel architecture, on which both the logical elements and the interconnection resources (channels) are disposed in rows [Figure 10.2(b)]. Finally, another architecture (and the one in which our problem is based) is the island architecture (Figure 10.1), in which the logical elements are named CLBs (configurable logic blocks). The CLBs contain the combinational logic. They are composed of lookup tables (LUTs), memory elements, and function generators. The horizontal and vertical interconnection resources (IRs) are the connection lines. Since the IRs are based on fixed lines, the connection configuration (i.e., the routing) is performed by programming the SWs (switches) placed among the IRs. SWs are based on a static RAM.

When it is planned to implement a function of certain complexity into a FPGA, a fixed and well-defined procedure must be followed in order to save resources and efforts by avoiding errors caused by bad planning. This procedure is known as the FPGA design cycle [10]. This cycle is composed of seven stages:

1. *Specification*. In this stage the functions, requirements, and features of the final product must be defined. This stage reports a list defining terms such as size, speed, performance, functions, and architecture.
2. *Functional design*. We separate the entire system into functional units, each with a set of parameters and connection requirements that must be identified. This stage reposts a general scheme describing the unit relationship.
3. *Logical design*. In this stage, a HDL such as VHDL or Handel-C defines the system logic. The control flow, operation, word size, and so on, are defined. Then, logical optimization and minimization techniques are applied, to reduce the design as much as possible.
4. *Circuit design*. Parting from the logical design, which generates the minimum logical representation, a global circuit representation is made.
5. *Physical design*. A geometrical representation of the circuit is obtained. This stage identifies the physical placement of the functional units, elements, and interconnection resources. The physical design is a complex stage that has to be divided into substages. This is the stage in which this problem optimizes the various substages.
6. *Manufacturing*. The layout obtained from the preceding stage is ready to be manufactured. The silicon pieces are obtained.
7. *Encapsulation, tests, and debugging*. The chips obtained are encapsulated and verified to check its behavior. The initial specifications and requirements are also checked.

The physical design substages are partitioning, placement, and routing. The partitioning substage divides the logical elements into modules. Each module will be placed over a CLB. We consider only the automatic design of placement and routing substages, so we suppose that FPGA partitioning is well made using suitable tools.

The placement substage consists of deciding which CLBs of the FPGA must be chosen to implement the function for each module. The best way to implement this substage is to choose nearby CLBs. This will reduce the interconnection requirements, and the FPGA speed will increase [14] because the delays that slow down a FPGA are produced by connection delays instead of a badly implemented CLB logic. So this substage must find a proper CLB placement scheme. The objective of this substage is to minimize the distance of the CLBs in the scheme without affecting the interconnection density.

The routing subscheme connects the various CLBs. Both the SWs that direct the routing scheme and the physical lines that support the communication are identified. So the objective is to connect all the CLB modules over the fixed lines by following the logical and design requirements.

The placement and routing stages must be tackled simultaneously by improving both FPGA area and connection density. If we board these stages sequentially, it may occur that the CLB area tends to be too small and the interconnection scheme gets complex and confusing because the wire density grows up. On the other side, the use of a fully optimized routing scheme could derive on high CLB area needs.

The requirement for tackling placement and routing at the same time increases the problem complexity, because this constraint turns the problem to an NP problem in the sense of obtaining the best solution within a huge solution space. The problem is not linear now, and many interlaced factors must be taken in account. To solve NP problems in reasonable time, heuristic techniques are used. In our case we have used GAs as the heuristic search tool.

Two GA implementations have been implemented: sequential and distributed. Details about sequential development may be found in Section 10.3.2. The distributed implementation follows the random ring topology model, in which several populations or islands evolve independently and after a determined number of generations exchange their best chromosomes. After the exchange process, the evolution continues. In each migration of chromosomes, the destination island is chosen randomly. More details about the distributed version and its adaptation to grid computing may be found in Section 10.3.3.

10.3.2 Sequential Tool Implementation

The circuit structure corresponding to Boolean functions is given by a schematic file in electronic design interchange format (EDIF). EDIF format is designed to ease the exchange of designs and projects between CAD tools and designers and with the premise of achieving an easy parsing process using a CAD tool; its syntax is based on LISP syntax.

To read and process EDIF files, we have developed a parser to convert the .edf (EDIF extension) file to the internal representation used by the GA. To build a GA-based tool, we have used the DGA2K tool. DGA2K is a set of libraries for developing applications based on GAs. To obtain a GA-based application the user just needs to define the chromosome representation, the fitness function, and the genetic parameters, such as cross method, cross rate, mutation type, and number of generations.

The chromosome representation is based on a bit string whose length depends on the problem size (number of CLBs and nets, size of the FPGA, etc). The decoded chromosome consists of a set of coordinates (Xn,Yn) that represents each CLB position on the FPGA. Figure 10.3 shows a chromosome representation in coded and decoded ways containing only three CLBs.

During processing, nonvalid chromosomes can arise when certain operations are performed, such as population random generation, crossover, and mutation operations. For this reason we must be sure that any gene position is repeated,

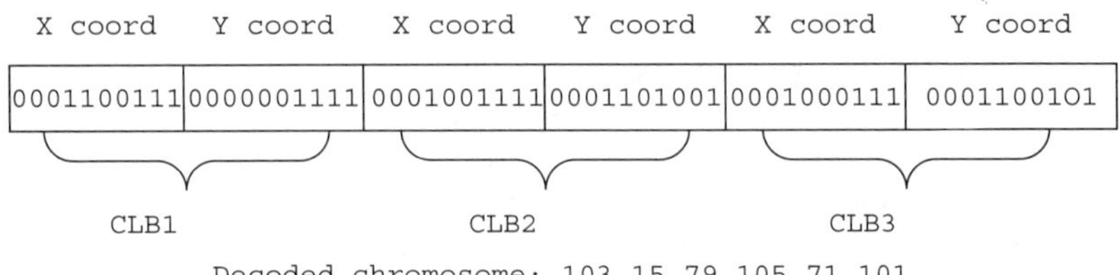

Figure 10.3 Chromosome representation.

or that it is not outside the bounds of the FPGA. For each nonvalid gene we change the position of its pseudorandom form, to guarantee the coherence of the chromosome.

The chromosome fitness is calculated by adding two terms, density and proximity. To calculate the proximity term, we consider the Euclidean distance in each coordinate pair, taken from the way they adjoin in the chromosomes and also in the first and last genes. For example, if a chromosome presents the structure $(x_0, y_0)(x_1, y_1)(x_2, y_2)$, its CLB proximity is calculated as follows:

$$p = \sqrt{(x_1 - x_0)^2 + (y_1 - y_0)^2} + \sqrt{(x_2 - x_1)^2 + (y_2 - y_1)^2} + \sqrt{(x_0 - x_2)^2 + (y_0 - y_2)^2} \quad (10.1)$$

Related to the need to tackle the placement and routing substages simultaneously in order to avoid solutions that present a confusing and muddy connection scheme, a main issue during the evolving process is to keep the density of the interconnection channels among the FPGA CLBs constant. For this reason we add the density term to the fitness function. To calculate this term, we act as follows: An internal connection is drawn up between each CLB source and destination pair. This connection joins both CLBs. Each time this connection crosses an SW we increase, by one unit, a weight assigned to it, in terms of penalty. In this way, a scheme with many highly weighted SWs (which indicates wire density in some area) produces a low-scored chromosome. So the fitness function is equivalent to the conjunction of both terms: CLB proximity and wire density. Most of the GA parameters for the tool are predefined; its values have been chosen according to the literature (see Table 10.1). nPointCrossover and mutationRate values have been modified with respect to the values recommended because we have found better values for them during the various experiments. This is due to the substantial gene length. Population size and generation number vary according to execution of the algorithm.

10.3.3 Grid Implementation and Architecture

We part from an application that runs sequentially on a single node. To port it to a grid environment, we must identify the portions of code that will be executed

TABLE 10.1 Genetic Algorithm Parameters

Parameter	Value	Explanation
CrossoverRate	1%	Probability of crossover operation each generation.
CrossoverMethod	nPointCrossover	Number of gene segmentation points to perform crossover operation.
nPointCrossover	12	Since the decoded gene has a big bit number (chromosome length: 260), 12 crossover points have been defined.
MutationRate	4/geneLength	Probability of mutation operation per gene in each generation.
IslandNumber	10	Applied in distributed execution; a number of islands evolving at the same time
MigrationRate	100	Applied in distributed execution; represents the number of generations between migrations.
Migrating Chromosome	15% of PopulationSize	Distributed execution; chromosomes exchanged in a migration.

remotely. Since we are using an application that is based on GAs, we can divide the execution into different populations or islands. We can break the code such that multiple islands can be run simultaneously. So we have obtained an execution scheme that is very well suited to run on a grid, since many independent jobs are defined (one of them per island), and they can be launched with different data inputs. With multiple islands we have used the random ring topology for distributing the GA. The overall DGA process is controlled by a shell script that generates the initial population, sends it to the islands, receives the intermediate results, performs migration between islands, and again sends the migrated population to the islands to be evolved. When a set of intermediate results is received from the islands and the migration process is performed, one stage is finished.

A tool (named inputMgm) for managing the populations has been developed. Depending on the current stage (either the initial or an intermediate stage), this tool creates the initial population or performs the migration among islands. To generate the initial population, a set of i files is created. The files contain the random population that will evolve on the islands.

The next step is to define the jobs (one per island) and send them to the WN. Since our grid infrastructure is based on GridWay [15], we generate JT files as job descriptor files, that is, configuration files on which the main job parameters are defined. The parameters we manage on the JT configuration are the executable file, the I/O files (specifying their remote location), and the arguments for the executable file.

The I/O filenames and the arguments passed to the executable file vary along the various DGA executions depending on the current experiment and on the current stage of this experiment, so different JT files have to be generated for each execution. The script requires the experiment number and the current stage. Then it automatically generates the JT files for the jobs to be sent with the parameters (executable arguments and I/O filenames). Finally, the script sends the jobs, and remote execution begins.

To perform the migration process, the remote jobs interrupt their execution every certain number of generations, returning an intermediate population file (IPF). An IPF represents the status of an island at a certain generation. When the IPFs of all islands are gathered on the local node, the inputMgm performs the migration by reading these files and then exchanging the chromosomes represented on them. Each IPF contains the various chromosomes decoded (the coordinates of each chromosome and its fitness function value). To translate the IPF information into the data structures managed by both the inputMgm and the islands, another parser has been developed using the Flex and Bison tools. This parser has been integrated as a C++ function into the inputMgm and into the executable file that runs in the islands.

The method of exchanging chromosomes between the islands follows the random ring topology model, which is explained as follows:

1. An integer array of size equal to the number of islands (i) is created. Each array element represents an island with its position set randomly. An example of a five-position random ring is shown in Table 10.2.
2. A migration from the island represented on position 1 to that represented on position 2 is performed (in the example, from 2 to 4).
3. To migrate means to exchange a fixed amount of chromosomes from one island to another. The chromosomes exchanged are those that present best fitness. The number of migrated chromosomes is 15% of the population size.

Figure 10.4 shows a scheme for the distributed island execution, giving the details of the data flow and the process location.

TABLE 10.2 Random Ring Example

Position	Island
1	2
2	4
3	5
4	3
5	1

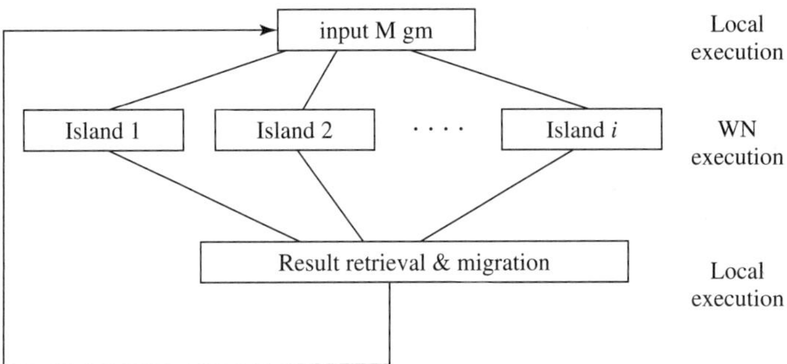

Figure 10.4 Distributed island execution.

The various evolution–migration stages described above are repeated 10 times. The output is a file representing the best chromosomes of each island. The grid implementation is based on gLite middleware. Over gLite infrastructure, the GridWay metascheduler has been installed, as we mentioned previously. We have used GridWay because it provides some interesting end-user facilities for grid operations such as submit, supervise, kill, or migrate jobs. Another reason for its use is the compatibility with the DRMAA API; use of this API is planned for future improvements of the distributed algorithm, as shown in Section 10.6. The GridWay commands used for the DGA execution are gwsubmit, gwps, and gwkill. Gwsubmit takes as parameter the JT files and sends the jobs, gwps checks the status of the remote job, and gwkill stops the remote execution of a job. GridWay performs the result retrieval automatically when the job finishes on the remote node.

10.4 ALGORITHMIC PROPOSAL

In this section we include pseudocode explaining sequential and distributed implementations. The building blocks for the sequential tool are explained, appearing as a black box into the distributed architecture explanation.

10.4.1 Sequential Tool Algorithm

The sequential tool follows the typical structure for a GA: parameter establishing, execution, and output results.

- *Parameter establishing* reads some configuration parameters from a file, such as the interconnection scheme, which contains the CLB definition and their interconnections. Those interconnection are links that join two CLBs, specifying both the source and destination CLBs.
- *GAExecution* represents the overall implemented GA (Algorithm 10.1).

Algorithm 10.1 GAExecution

```
for number of generations do
  Perform selection
  Perform nPointCrossover
  Perform mutation
  Evaluate population
end for
```

Algorithm 10.2 Selection

```
for numbers of chromosomes in population do
   Select N random chromosomes from population
   Select best chromosome from N
   Copy selected chromosome to new population
end for
```

Algorithm 10.3 nPointCrossover

```
Generate random number
if generated random number less than crossoverRate then
   Select two different chromosomes from population
   Locate nPoint positions from along the chromosome length
   Exchange the bits located between the positions
   Check chromosome validity
else
   Exit
end if
```

- The *selection* procedure is based on the tournament method (Algorithm 10.2).
- *nPointCrossover* selects two random chromosomes from the population then divides the chromosome into nPoints + 1 sections and exchanges the sections between them. This operation acts according to the mutation rate (Algorithm 10.3).
- The *mutation* operation selects a random chromosome from the population, then selects a random bit into the chromosomes and reverses it (Algorithm 10.4).
- The function that checks the chromosome validity prevents the creation of nonvalid chromosomes. A chromosome is not valid because one of its coordinates is outside the FPGA bounds or because there is another chromosome

Algorithm 10.4 Mutation

```
Choose random chromosome from population
Choose random bit from chromosome
Reverse chosen bit
Check chromosome validity
```

Algorithm 10.5 CheckChromosomeValidity

```
for coordinate in the decoded chromosome do
  if coordinate is outside FPGA bounds then
     Assign to the coordinate the remainder of the
     division between its value and the FPGA bound value
     repeat
       Compare coordinate selected with others if
       repetition found
       Generate small random number
       Alter coordinate with the number generated
     until no repetition found
  end if
end for
```

Algorithm 10.6 Evaluation

```
for chromosome in the population do
  Add to X the Euclidean distance between this
  chromosome position and the next chromosome position
end for
Order chromosome by its fitness value
```

with the same coordinate pairs. Since crossover and mutation alter the chromosome, its representation after performing these two operations could not be correct (Algorithm 10.5).

- The *evaluation* function calculates the fitness by adding density and proximity values, so these values have to be calculated into the function (Algorithm 10.6).
- Obtaining the *densityFactor* requires a longer procedure, so it has been implemented on a separated function. As we explained before, the SWs that are crossed by a connection have to be identified. To perform this step, a connection is drawn over an internal FPGA representation. In this representation the weights for each SW are stored in order to make a posterior calculation (Algorithm 10.7).

Algorithm 10.7 DensityFactor

```
for link defined on the interconnection scheme do
  Draw virtual connection
  for SW crossed by the connection do
    Increase SW weight
  end for
end for
for SW on the matrix do
  Add to Y the value associated with the SW
end for
return Y
```

Algorithm 10.8 SequentialToolOperation

```
Generate 10 initial random population files
Assign to current population files the random files
generated
repeat
  Create job definition file with the
  sequential GA and with the current population
  Submit job to grid
until 10 times
Perform migration
Assign to current population files  the result of the
previous function
```

- The sequential tool generates an output file with the CLB and interconnection positions corresponding to the best chromosomes obtained after the specified number of generations (Algorithm 10.8).
- The *performMigration* function (Algorithm 10.9) reads the population files from the islands. Then it creates an internal random ring for determining a migrating scheme randomly. In this migrating scheme all the islands appear, and each island acts as both a source for a migration and a destination for a different migration. The chromosomes are exchanged according to this scheme (refer to Section 10.3.3 for more details about the random ring). When migration is performed, the process generates one file for each island. These files are used as input for the next stage.

10.5 EXPERIMENTAL ANALYSIS

The experiments have been performed by running the GA under different parameters in both sequential and distributed models. The experiments are based on

Algorithm 10.9 PerformMigration

```
Read result from islands
Generate array with numbers from 1 to 10 spread in
random positions
for position K in the array do
  Exchange 15% of the chromosomes from the island referred
  to position K to the island referred to position K+1
  in the array
end for
Write output files with the islands exchanged
```

TABLE 10.3 Experiments Launched

Test	Characters	Generations	Test	Characters	Generations
1	100	500	7	100	1000
2	200	500	8	200	1000
3	300	500	9	300	1000
4	500	500	10	500	1000
5	750	500	11	750	1000
6	1000	500	12	1000	1000

the dimensions of the FPGA Xilinx 5 XC5VLX220 model, which consists of 160×108 CLBs [2]. The circuit that the FPGA optimizes is a 64-logical block function. This function has been taken from the ITC benchmarks [16]. Table 10.3 shows the genetic parameters for experiments launched.

To run both sequential and distributed experiments under the same circumstances, the sequential tasks have also been defined as grid jobs without dividing the population into islands or performing migration. All the experiments run on i686 machines with 4-Gbyte RAM, 3200-GHz dual-core processor, and Scientific Linux CERN v3. Both the time taken by the experiments and the fitness obtained by the GA are taken as the results. For the sequential model, 10 executions of all experiments have been run, so the fitness values we show below are equal to the average value of the best fitness from each execution for the various experiments. The execution time we show is also obtained by the average time taken by the 10 executions. The fitness is represented by a negative value.

The distributed execution is made under the same parameters (population size and generation number) as in case of sequential execution. For the distributed execution we have divided the population into 10 islands. The overall number of generations on each island has also been divided into 10 execution stages in order to perform migration at the end of each stage. So for each job in the sequential model we have defined 100 jobs in the distributed model (10 islands each evolving in 10 different stages). At the beginning of an experiment 10 initial random populations are generated, and then 10 first jobs are sent to the

WN. This is the first stage. When these first jobs have finished and their results are retrieved, the migration process is performed and the next package of jobs is defined; and then the second stage starts. This process is repeated during the 10 stages.

In the distributed case the fitness value is calculated as the average value of the fitness returned by each island in the tenth stage, when all islands have reached their maximum evolution. The time value is calculated as follows:

1. The total time is calculated for each independent island. This is made by adding the time spent by each of the 10 stages of the island.
2. The overall time value for an experiment is equal to the average of the total time spent by each independent island.
3. The communication times are included. This time includes the job definition, the time to send them to the WN, the time to wait for the jobs to get wrapped (to get associated to a queue in the grid), the time to wait for the job to finish its execution, the time to retrieve the job result, and finally, the time for chromosome migration. Independent of the length and size of the experiments, these communication times are similar, since the data movements have the same size. Also, both network and grid infrastructure were dedicated to these experiments alone, so no external perturbations appeared during the executions. For the experiments the average time spent for communication was 110 seconds.

In Table 10.4, time (measured in seconds) and fitness results are shown for the sequential and distributed executions. Figures 10.5 shows the fitness evolution comparing sequential and parallel executions. Two graphs are shown in the figure, one for experiments made with a 500-chromosome population size and another for experiments made with a 1000-chromosome population size. In

TABLE 10.4 Fitness Obtained and Time Taken for the Experiments

			Seq. Execution		Dist. Execution	
E	Char.	Gen.	Time (s)	Best Fitness	Time (s)	Best Fitness
1	100	500	5,887.14	−2,059.52	1,576.13	−2,123.45
2	200	500	10,676.14	−1,702.71	2,063.49	−1,686.14
3	300	500	15,329.86	−1,803.97	2,543.22	−1,596.36
4	500	500	25,133.00	−1,421.71	3,531.61	−1,347.15
5	750	500	38,595.57	−1,248.73	4,693.58	−1,395.29
6	1000	1000	52,133.00	−1,238.15	5,943.79	−1,387.76
7	100	1000	10,550.29	−1,620.76	2,064.69	−1,500.17
8	200	1000	20,242.57	−1,682.86	3,036.38	−1,335.22
9	300	1000	29,484.86	−1,644.60	4,013.49	−1,464.6
10	500	1000	50,485.29	−1,276.97	5,998.01	−1,091.3
11	750	1000	74,947.57	−1,270.04	8,383.11	−828.33
12	1000	1000	103,957.43	−1,168.17	10,889.37	−1,011.67

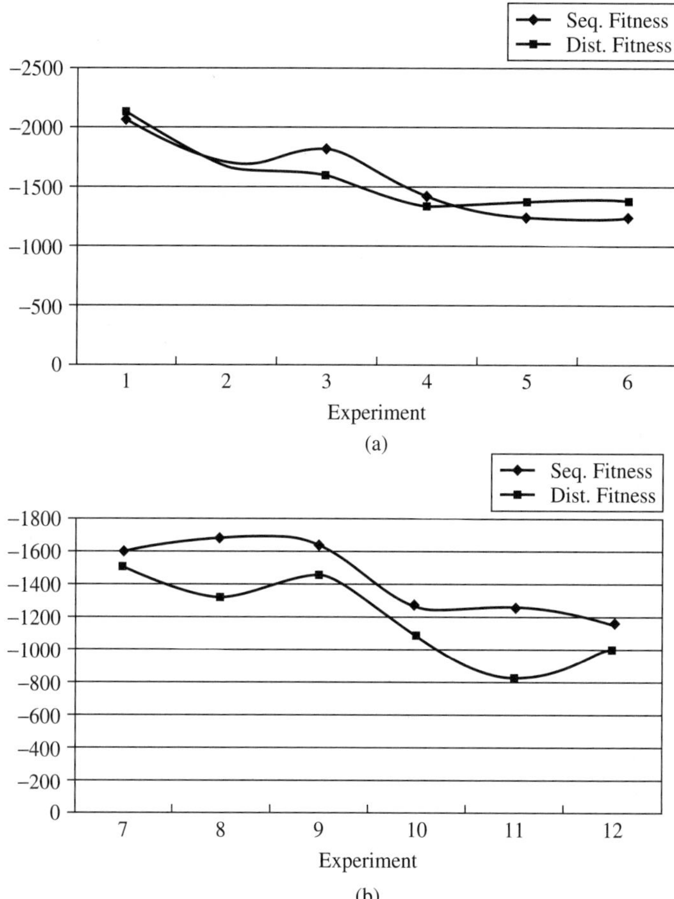

Figure 10.5 Fitness evolution for a population size of (a) 500 and (b) 1000 chromosomes.

Figure 10.5(a) the fitness evolution is similar in sequential and distributed execution. In Figure 10.5(b) we can see a bigger difference between the two execution models, especially in experiment 11, where the chromosomes obtained their best values. This experiment was configured with a population of 750 chromosomes. Until this population size, the fitness improves; if we increase the generation number, the fitness becomes worse. Populations larger than 750 chromosomes do not produce better results. The fitness improvement in the distributed execution is due to the diversification that migration between islands introduces. Better values are obtained with the distributed execution model than with the sequential model. Nevertheless, the really important improvement that grid-ported implementation introduces relates to execution times, as we can see from the comparisons shown in Figure 10.6(a) and (b).

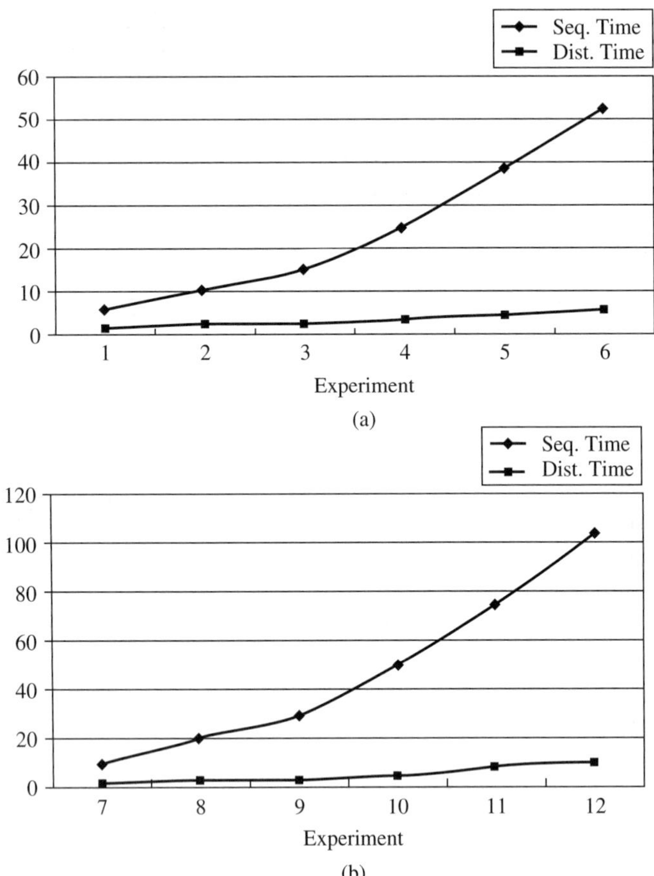

Figure 10.6 Time evolution for a population size of (a) 500 and (b) 1000 chromosomes.

Figure 10.6(a) and (b) show a great difference between the time taken by the sequential model and that taken by the distributed model. To increase the population size slightly affects the time taken by the distributed model, whereas the time spent in sequential generation grows considerably. The difference between the two lines' evolution becomes larger as the population size grows (Table 10.5 shows this tendency in percentage terms). The percentage values represent the time that distributed execution needs to run an experiment in comparison with the same experiment run on sequential execution.

As we said before, we have taken the communication times into account. If we considerate only the execution time (i.e., the time taken for the processes to run on the WN), all percentage values for the distributed execution would total 10% of the time taken by sequential execution. This 10% of time spent is directly proportional to the number of islands deployed (10).

TABLE 10.5 Distributed Time Compared to Sequential time

Exp.	%	Exp.	%	Exp.	%	Exp.	%
1	26.77	4	14.05	7	19.57	10	11.88
2	19.33	5	12.16	8	15.00	11	11.19
3	16.59	6	11.40	9	13.61	12	10.47

10.6 CONCLUSIONS

In this chapter we have presented the steps necessary to enable a GA-based application that solves a placement and routing problem to be run on the grid. We have developed an application in which we work with a circuit benchmark coming from ITC. To read and process the benchmarks, an EDIF parser has been developed. A genetic application has been developed using the DGA2K GAs libraries, customizing them to our problem. A distributed model was then developed. A piece of code to parallelize was identified as an island: Since the tool is based on GAs, we separate the code into different island execution. This distributed execution model is based on the random ring distributed GA topology, for which we define a set of WNs and use them to evolve the islands. A migration that exchanges the best-fitness chromosomes between islands is performed.

The grid implementation is based on gLite middleware. The GridWay meta-scheduler has been installed over gLite to obtain some end-user facilities for grid operations. These facilities have helped us in the task of defining massive jobs and the automation for retrieving results. Although the fitness improvement produced by the distributed model is not significant, the real advantage of this implementation takes place when considering the time taken for the experiments in comparison with the time that the sequential model spends. The new execution model offers a more powerful platform on which users can experiment with larger and more complex tasks, saving large amounts of both effort and time.

As future work, we plan to use the DRMAA API to manage the grid resources dynamically and to develop an automatic tool for generating and sending jobs. This tool will substitute for the scripts used, producing a more effective way to manage grid execution.

Acknowledgments

The authors are partially supported by the Spanish MEC and FEDER under contract TIC2002-04498-C05-01 (the TRACER project).

REFERENCES

1. I. Dutra I, J. Gutierrez J, V. Hernandez, B. Marechal, R. Mayo, and L. Nellen. e-Science applications in the EELA project. *Proceedings of IST-Africa* 2007:1–11, 2007.

2. C. Maxfield. *The Design Warrior Guide to FPGAs: Devices, Tools and Flows*. Elsevier, New York, 2004.
3. S. Brown, R. Francis, J. Rose, and Z. Vranesic. *Field-Programmable Gate Arrays*. Kluwer Academic, Norwell, MA, 1992.
4. M. Rubio, J. Sanchez, J. Gomez, and M. Vega. Placement and routing of Boolean functions in constrained FPGAs using a distributed genetic algorithm and local search. In *Proceedings of the 20th IEEE/ACM International Parallel and Distributed Processing Symposium*, pp. 1–10, 2006.
5. Z. Michalewicz. *Genetic Algorithms + Data Structures = Evolution Programs*. Springer-Verlag, Heidelberg, Germany, 1996.
6. D. Goldberg. *Genetic Algorithms in Search, Optimization and Machine Learning*. Addison-Wesley, Reading, MA, 1989.
7. S. Burke, S. Campana, A. Delgado-Peris, F. Donno, P. Mendez, R. Santinelli, and A. Sciaba. *gLite 3 User Guide*. Macquarie University, Sydney, Australia, 2007.
8. A. Sedighi. *The Grid from the Ground Up*. Springer-Verlag, New York, 2005.
9. M. Alexander, J. Cohoon, J. Ganley, and G. Robins. Placement and routing for performance-oriented FPGA layout. *VLSI Design: An International Journal of Custom-Chip Design, Simulation, and Testing*, 7(1), 1998.
10. F. Fernandez, J. Sanchez, and M. Tomassini. Placing and routing circuits on FPGAs by means of parallel and distributed genetic programming. *Proceedings of ICES*, 2001:204–215, 2001.
11. V. Gudise and G. Venayagamoorthy. FPGA placement and routing using particle swarm optimization. In *Proceedings of the IEEE 2004 Computer Society Annual Symposium*, 2004.
12. J. Altuna and W. Falcon. Using genetic algorithms for delay minimization in the FPGAs routing. In *Proceedings of the First International NAISO Congress on Neuro Fuzzy Technologies (NF'02)*, 2002, pp. 1–6.
13. M. Rubio, J. Sanchez, J. Gomez, and M. Vega. Genetic algorithms for solving the placement and routing problem of an FPGA with area constraints. *Proceedings of ISDA'04*:31–35, 2004.
14. J. Henning. SPEC CPU2000: measuring CPU performance in the new millennium. *IEEE Computer*, pp. 28–35, July 2000.
15. E. Huedo, R. Montero, and I. Llorente. The GridWay framework for adaptative scheduling execution on grids. *Scalable Computing: Practice and Experience*, 6(6):1–8, 2006.
16. S. Davidson. ITC99 benchmarks circuits, preliminary results. In *Proceedings of the ITC International Test Conference*, 99: 1125, 1999.

CHAPTER 11

Divide and Conquer: Advanced Techniques

C. LEÓN, G. MIRANDA, and C. RODRÍGUEZ

Universidad de La Laguna, Spain

11.1 INTRODUCTION

The divide-and-conquer (DnC) technique is a general method used to solve problems. Algorithms based on this technique divide the original problem in to smaller subproblems, solve them, and combine the subsolutions obtained to get a solution to the original problem. The scheme of this technique can be generalized to develop software tools that allow the analysis, design, and implementation of solvers for specific problems. Unfortunately, in many applications it is not possible to find a solution in a reasonable time. Nevertheless, it is possible to increase the size of the problems that can be approached by means of parallel techniques.

There are several implementations of general-purpose schemes using object-oriented paradigms [1–4]. Regarding divide and conquer, we can mention the following: Cilk [5] based on the C language and Satin [6] codified in Java. Satin [7] is a Java framework that allows programmers to easily parallelize applications based on the divide-and-conquer paradigm. The ultimate goal of Satin is to free programmers from the burden of modifying and hand-tuning applications to exploit parallelism. Satin is implemented on top of Ibis [8], a programming environment with the goal of providing an efficient Java-based platform for grid programming. Ibis consists of a highly efficient communication library and a variety of programming models, mostly for developing applications as a number of components exchanging messages through messaging protocols such as MPI [9]. The MALLBA [10,11] project is an effort to develop an integrated library of skeletons for combinatorial optimization, including exact, heuristic, and hybrid techniques. A skeleton [12] is a parameterized program

Optimization Techniques for Solving Complex Problems, Edited by Enrique Alba, Christian Blum, Pedro Isasi, Coromoto León, and Juan Antonio Gómez

schema with efficient implementations over diverse architectures. MALLBA allows the combined and nested use of all of its skeletons. Also, sequential and parallel execution environments are supported. This work focuses on the skeleton approach to parallel programming instead of the one based on programming language extensions. The skeleton approach can be viewed as an alternative to the current practice of low-level programming for parallel machines. Nevertheless, the formalism employed to define skeletons such as higher-order functions [13] do not comprise our objectives.

The MALLBA::DnC skeleton requires from the user the implementation of a C++ class Problem defining the problem data structures, a class `Solution` to represent the result, and a class `SubProblem` to specify subproblems. In some cases an additional `Auxiliar` class is needed to represent the subproblems in case they do not have exactly the same structure as the original problem.

In the `SubProblem` class the user has to provide the methods `easy()`, `solve()`, and `divide()`. The `easy()` function checks if a problem is simple enough to apply the simple resolution method `solve()`. The `divide()` function must implement an algorithm to divide the original problem into smaller problems with the same structure as that of the original problem.

The class `Solution` has to provide an algorithm to put together partial solutions in order to obtain the solution to a larger problem through the `combine()` function. MALLBA::DNC provides a sequential resolution pattern and a message-passing master–slave resolution pattern for distributed memory machines [14,15]. This work presents a new fully distributed parallel skeleton that provides the same user interface and consequently is compatible with codes already implemented. The new algorithm is a MPI asynchronous peer–processor implementation where all the processors are peers and behave in the same way (except during the initialization phase) and where decisions are taken based on local information.

The contents of the chapter are as follows. In Section 11.2 we describe the data structures used to represent the search space and the parallel MPI skeleton. To clarify the tool interface and flexibility, two problem implementation examples are introduced and discussed in Section 11.3. Conclusions and future prospectives are noted in Section 11.4.

11.2 ALGORITHM OF THE SKELETON

The algorithm has been implemented using object-oriented and message-passing techniques. Figure 11.1 graphically displays the data structure used to represent the tree space. It is a tree of subproblems (`sp`) where each node has a pointer to its father, the solution of the subproblem (`sol`), and the auxiliary variable (`aux`) for the exceptional cases when the subproblems do not have the same structure as the original problem. Additionally, a queue with the nodes pending to be explored is kept (`p` and `n`). The search tree is distributed among the processors. Registered in each node is the number of unsolved children and the number of children sent

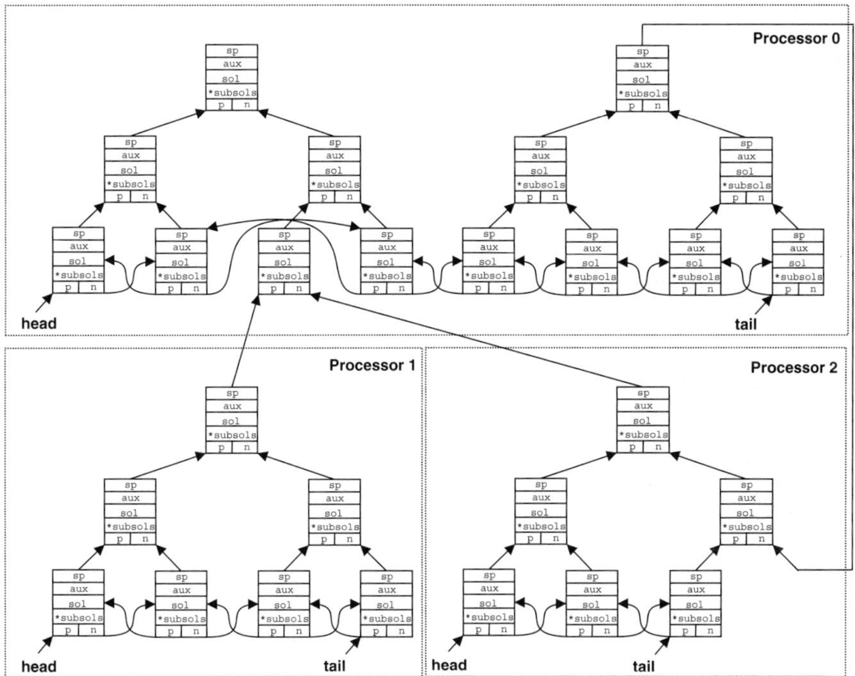

Figure 11.1 Data structure for the tree space.

to other processors (remote children). Also, an array of pointers to the solutions of the children nodes (`subsols`) is kept. When all the children nodes are solved, the solution of the actual node, `sol`, will be calculated combining the partial solutions stored in `subsols`. The solution is sent up to the father node in the tree and the partial solutions are disposed of. Since the father of a node can be located in another processor, the rank of the processor owning the father is stored on each node. Implementation of the algorithmic skeleton has been divided into two stages: initialization and resolution.

11.2.1 Initialization Stage

The data required to tackle resolution of a problem are placed into the processors. Initially, only the master processor (*master*) has the original problem and proceeds to distribute it to the remaining processors. Next, it creates the node representing the initial subproblem and inserts it into its queue. All the processors perform monitor and work distribution tasks. The state of the set of processors intervening in the resolution is kept by *all* the processors. At the beginning, all the processors except *master* are idle. Once this phase is finished, all the processors behave in the same way.

11.2.2 Resolution Stage

The goal of the resolution stage is to find a solution to the original problem. The majority of the skeleton tasks occur during this phase. The main aim of the design of the skeleton has been to achieve a *balanced distribution* of the workload. Furthermore, such an arrangement is performed in a *fully distributed* manner using only asynchronous communication and information that is local to each processor, avoiding in this way the use of a central control and synchronization barriers.

The control developed is based on a request–answer system. When a processor is idle, it requests work from the remaining processors. This request is answered by the remaining processors either with work or with a message indicating their inability to send work. Since the state of the processors has to be registered locally, each sends its answer not only to the processor requesting work but also to those remaining, informing them of the answer (whether work was or was not sent) and to which processor it was sent. This allows us to keep the state of *busy* or *idle* updated for each processor. According to this approach, the algorithm would finish when all the processors are marked idle. However, since request and answer messages may arrive in any arbitrary order, a problem arises: The answers may arrive before the requests or messages referring to different requests may be tangled. Additionally, it may also happen that a message is never received if the processors send a message corresponding to the resolution phase to a processor which has decided that the others are idle.

To cope with these problems, a *turn of requests* is associated with each processor, assigning a rank or order number to the requests in such a way that it can be determined which messages correspond to what request. They also keep a *counter of pending answers*. To complete the frame, additional constraints are imposed on the issue of messages:

- A processor that becomes idle performs a single request until it becomes busy again. This avoids an excess of request messages at the end of the resolution phase.
- A processor does not perform any request while the number of pending answers to its previous request is not zero. Therefore, work is not requested until checking that there is no risk to acknowledge work corresponding to a previous request.
- No processor sends messages not related to this protocol while it is idle.

Although the former restrictions solve the aforementioned protocol problems, a new problem has been introduced: If a processor makes a work request and none sends work as answer, that processor, since it is not allowed to initiate new petitions, will remain idle until the end of the stage (*starvation*). This has been solved by making the *busy* processors, when it is their turn to communicate, check if there are idle processors without pending answers to their last petition and, if there is one and they have enough work, to force the initiation of a new turn of request for that idle processor. Since the processor(s) initiating the new turns are working, such messages will be received before any subsequent messages for work requests produced by it (them), so there is no danger of a processor finishing the stage before those messages are received, and therefore they will not be lost.

The pseudocode in Algorithm 11.1 shows the five main tasks performed in this phase. The "conditional communication" part (lines 3 to 23) is in charge of the reception of all sort of messages and the work delivery. The communication is conditional since it is not made per iteration of the main loop but when the condition *time to communicate* takes a true value. The goal of this is to limit the

Algorithm 11.1 Resolution Phase Outline

```
 1: problemSolved ← false
 2: while (!problemSolved) do
 3:   {Conditional communication}
 4:   if (time to communicate) then
 5:       {Message reception}
 6:       while (pending packets) do
 7:         inputPacket ← packetComm.receive(SOLVING_TAG)
 8:         if (inputPacket.msgType == NOT_WORKING_MSG) then
 9:           ...
10:        else if (inputPacket.msgType ==
              CANT_SEND_WORK_MSG) then
11:           ...
12:        else if (inputPacket.msgType ==
              SEND_WORK_MSG_DONE) then
13:           ...
14:        else if (inputPacket.msgType == SEND_WORK_MSG)
              then
15:           ...
16:        else if (inputPacket.msgType == CHILD_SOL_MSG)
              then
17:           ...
18:        else
19:           ...
20:        end if
21:      end while
22:      {Avoid starvation when a request for work was
           neglected}
23:    end if
24:    {Computing subproblems ...}
25:    {Work request ...}
26:    {Ending the solution phase ...}
27: end while
```

time spent checking for messages, assuring a minimum amount of work between communications. The current implementation checks that the time that has passed is larger than a given threshold. The threshold value has been established as a small value, since for larger values the workload balance gets poorer and subsequent delay of the propagation of bounds leads to the undesired exploration of nonpromising nodes.

Inside the "message reception" loop (lines 5 to 21) the labels to handle the messages described in previous paragraphs are specified. Messages of the type CHILD_SOL_MSG are used to communicate the solution of a subproblem to the receiver processor.

The "computing subproblems" part (line 24) is where the generic divide-and-conquer algorithm is implemented. Each time the processor runs out of work, one and only one "work request" (line 25) is performed. This request carries the beginning of a new petition turn. To determine the "end of the resolution phase" (line 26) there is a check that no processor is working and that there are no pending answers. Also, it is required that all the nodes in the tree have been removed. In this case the problem is solved and the optimal solution has been found. The last condition is needed because the algorithm does not establish two separate phases for the division and combination processes, but both tasks are made coordinately. Therefore, if some local nodes have not been removed, there are one or more messages with a partial solution (CHILD_SOL_MSG) pending to be received, to combine them with the not-deleted local nodes.

Algorithm 11.2 shows the computing task in more depth. First, the queue is checked and a node is removed if it is not empty. If the subproblem is easy (lines 4 to 16), the user method `solve()` is invoked to obtain the partial solution. Afterward, the call to `combineNode()` will combine all the partial solutions that are available. The `combineNode()` method inserts the solution received as a parameter in the array of partial solutions `subsols` in the father node. If after adding this new partial solution the total number of subsolutions in the father node is completed, all of them are combined using the user method `combine()`. Then the father node is checked to determine if the combination process can be applied again. The combination process stops in any node not having all its partial solutions or when the root node is reached. This node is returned by the `combineNode()` method. If the node returned is the root node (it has no father) and the number of children pending to solutions is zero, the problem is solved (lines 8 to 11). If the node has no unsolved children but is an intermediate node, its father must be in another processor (lines 12 to 15). The solution is sent to this processor with a CHILD_SOL_MSG message, and it continues from this point with the combination process. If the subproblem removed from the queue is not simple (lines 16 to 21), it is divided into new subproblems using the `divide()` method provided by the user. The new subproblems are inserted at the end of the queue. To traverse the tree in depth, the insert and remove actions in the queue are always made at the end. This is a faster way to combine partial solutions and a more efficient use of memory.

Algorithm 11.2 Divide-and-Conquer Computing Tasks

```
if (!dcQueue.empty()) then
    node ← dcQueue.removeFromBack()
    sp ← node.getSubProblem()
    if sp.easy() then
        sp.solve(node.getSolution())
        node ← dcQueue.combineNode(pbm, node)
        if (node.getNumChildren() = = 0) then
            {solution of original}
            if (node.getFather() = = NULL) then
                sol ← node.getSolution();
            end if
        else
           {combination must continue in other processor}
           packetComm.send(node.getFatherProc(),
             SOLVING_TAG, ...)
        end if
    else
       sp.divide(pbm, subPbms[], node.getAuxiliar());
       for all i such that i < subPbms[].size do
          dcQueue.insertAtBack(dcNode.create(subpbms[i]),
            0, i, node)
       end for
    end if
end if
...
```

11.3 EXPERIMENTAL ANALYSIS

The experiments were performed instantiating the MALLBA::DnC skeleton for some classic divide-and-conquer algorithms: sorting (quicksort and mergesort), convex hull, matrix multiplication, fast Fourier transform, and huge integer product. Results were taken on a heterogeneous PC network, configured with four 800-MHz AMD Duron processors and seven AMD-K6 3D 500-MHz processors, each with 256 Mbytes of memory and a 32-Gbyte hard disk. The operating system installed was Debian Linux version 2.2.19 (herbert@gondolin), the C++ compiler used was GNU gcc version 2.95.4, and the message-passing library was *mpich* version 1.2.0.

Figure 11.2 shows the speedups obtained for the huge integer product implementation with a problem of size 4096. The parallel times were the average of five executions. The experiments labeled "ULL 500 MHz" and "ULL 800 MHz" were carried out on a homogeneous set of machines of sizes 7 and 4, respectively. ULL 800-500 MHz depicts the experiment on a heterogeneous set of machines

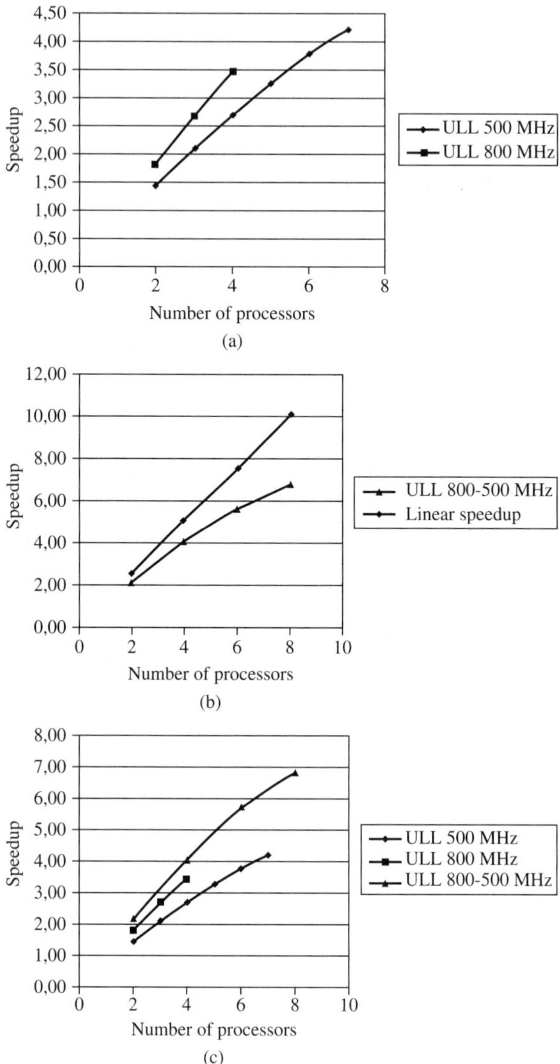

Figure 11.2 Times and speedup for the huge integer product, $n = 4096$: (a) homogeneous case; (b) heterogeneous case; (c) speedups.

where half the machines (four) were at 800 MHz and the other half were at 500 MHz. The sequential execution for the ULL 800–500 MHz experiment was performed on a 500-MHz processor. To interpretate the ULL 800–500 MHz line, take into account that the ratio between the sequential executions was 1.53. For eight processors the maximum speedup expected will be 10.12; that is, 1.53×4(fast processors) $+4$ (slow processors); see Figure 11.2(b). Comparing the three experiments depicted [Figure 11.2 (c)], we conclude that the algorithm does not experiment any loss of performance due to the fact of being executed on a heterogeneous network. Figure 11.3(a) represents the average number of nodes visited, for the experiment labeled "ULL 800–500 MHz." It is clear that an increase in the number of processors carries a reduction in the number of nodes visited. This shows the good behavior of the parallel algorithm.

A parameter to study is the load balance among the different processors intervening in the execution of the algorithm. Figure 11.3(b) shows the per processor average of the number of visited nodes for the five executions. Observe how the

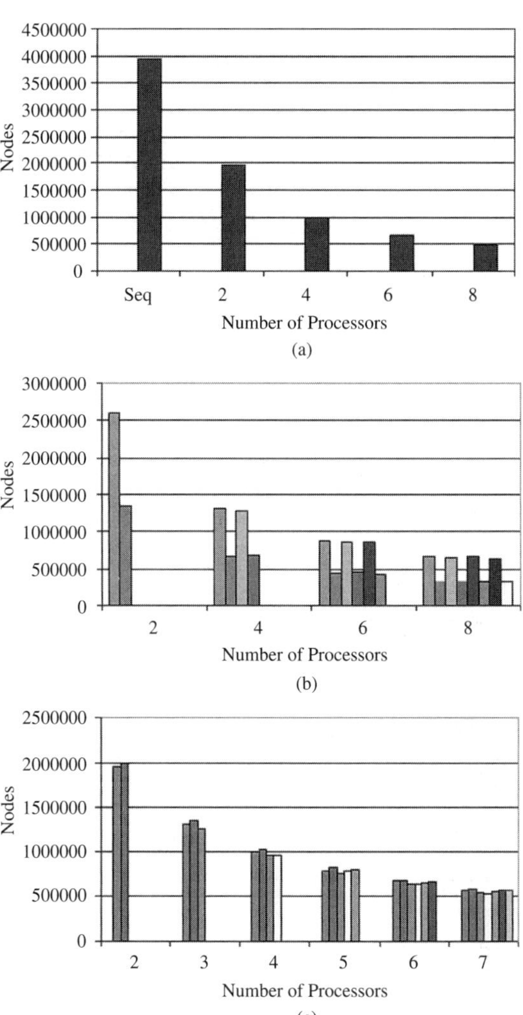

Figure 11.3 Nodes visited for the huge integer product: (a) average number of nodes visited (800–500 MHz); (b) heterogeneous case (800–500 MHz); (c) homogeneous case (500 MHz).

slow processors examine fewer nodes than the faster ones. It is interesting to compare these results with those appearing in Figure 11.3(c), corresponding to the homogeneous executions. Both pictures highlight the fairness of the workload distribution.

Similar results are obtained for implementation of the other algorithms. Figures 11.4 and 11.5 present the same study for Strassen's matrix product algorithm. Even though the grain is finer than in the previous example, the performance behavior is similar.

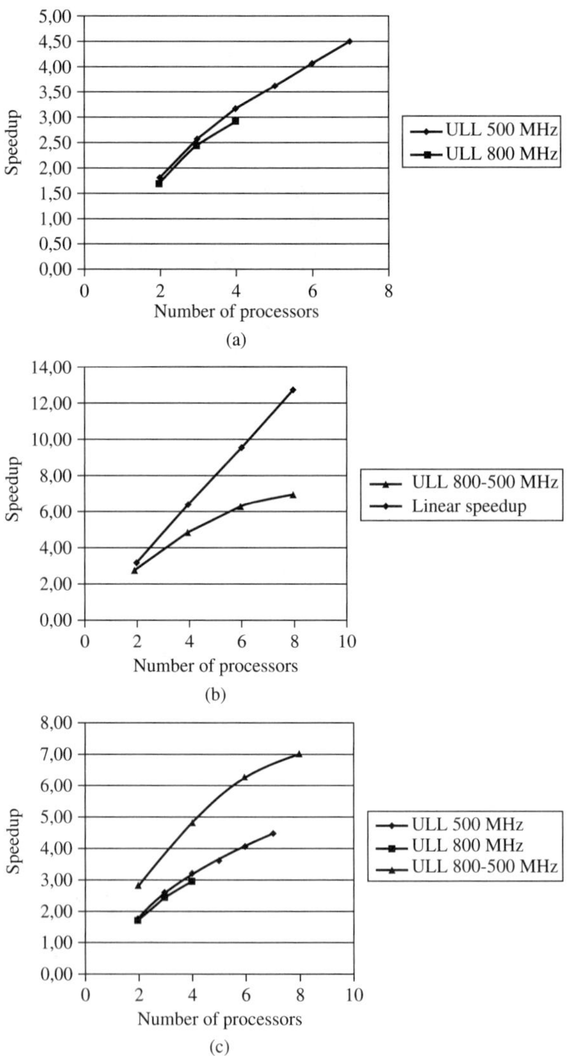

Figure 11.4 Times and speedup for Strassen's matrix multiplication ($n = 512$): (a) homogeneous case; (b) heterogeneous case; (c) speedups.

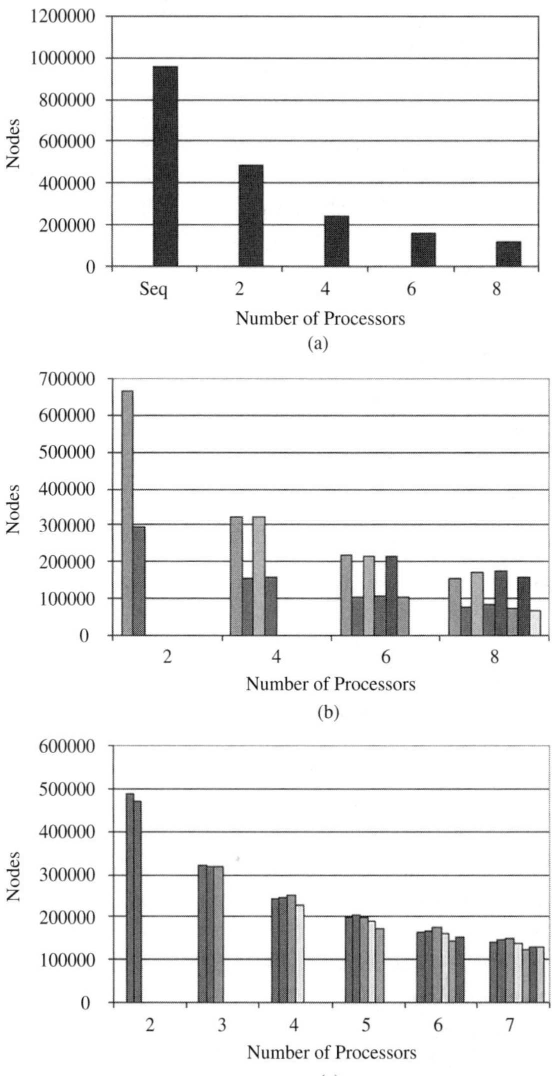

Figure 11.5 Nodes visited for Strassen's matrix multiplication: (a) average number of nodes visited (800–500 MHz); (b) heterogeneous case (800–500 MHz); (c) homogeneous case (500 MHz).

11.4 CONCLUSIONS

In this work chapter we describe a parallel implementation of a skeleton for the divide-and-conquer technique using the message-passing paradigm. The main contribution of the algorithm is the achievement of a balanced workload among the processors. Furthermore, such an arrangement is accomplished in a fully distributed manner, using only asynchronous communication and information local to each processor. To this extent, the use of barriers and a central control has been avoided. The results obtained show good behavior in the homogeneous and heterogeneous cases. Ongoing work focuses on lightening the replication of information relative to the state of a certain neighborhood. An OpenMP [16]-based resolution pattern for shared memory has also been developed. Due to the fine-grained parallelism scheme required for implementation, the results obtained are not scalable.

Acknowledgments

This work has been supported by the EC (FEDER) and by the Spanish Ministry of Education and Science under the Plan Nacional de I+D+i, contract TIN2005-08818-C04-04. The work of G. Miranda has been developed under grant FPU-AP2004-2290.

REFERENCES

1. J. Anvik and et al. Generating parallel programs from the wavefront design pattern. In *Proceedings of the 7th International Workshop on High-Level Parallel Programming Models and Supportive Environments* (*HIPS'02*), Fort Lauderdale, FL, 2002.
2. B. Le Cun, C. Roucairol, and TNN Team. BOB: a unified platform for implementing branch-and-bound like algorithms. *Technical Report 95/016*. PRiSM, Université de Versailles, France, 1995.
3. A. Grama and V. Kumar. State of the art in parallel search techniques for discrete optimization problems. *Knowledge and Data Engineering*, 11(1):28–35, 1999.
4. H. Kuchen. A skeleton library. In *Proceedings of Euro-Par*, vol. 2400 of *Lecture Notes in Computer Science.* Springer-Verlag, New York, 2002, pp. 620–629.
5. T. Kielmann, R. Nieuwpoort, and H. Bal. *Cilk-5.3 Reference Manual*. Supercomputing Technologies Group, Cambridge, MA, 2000.
6. T. Kielmann, R. Nieuwpoort, and H. Bal. Satin: efficient parallel divide-and-conquer in Java. In *Proceedings of Euro-Par*, vol. 1900 of *Lecture Notes in Computer Science.* Springer-Verlag, New York, 2000, pp. 690–699.
7. R. Nieuwpoort, J. Maassen, T. Kielmann, and H. Bal. Satin: simple and efficient Java-based grid programming. *Scalable Computing: Practice and Experience*, 6(3):19–32, 2005.
8. R. Nieuwpoort and et al. Ibis: a flexible and efficient Java-based grid programming environment. *Concurrency and Computation: Practice and Experience*, 17(7–8):1079–1107, 2005.
9. M. Snir, S. W. Otto, S. Huss-Lederman, D. W. Walker, and J. J. Dongarra. *MPI: The Complete Reference*. MIT Press, Cambridge, MA, 1996.
10. E. Alba and et al. MALLBA: A library of skeletons for combinatorial optimization. In *Proceedings of Euro-Par*, vol. 2400 of *Lecture Notes in Computer Science.* Springer-Verlag, New York, 2002, pp. 927–932.

11. E. Alba et al. Efficient parallel LAN/WAN algorithms for optimization: the MALLBA project. *Parallel Computing*, 32:415–440, 2006.

12. M. Cole. Algorithmic skeletons: a structured approach to the management of parallel computation. *Research Monographs in Parallel and Distributed Computing*. Pitman, London, 1989.

13. S. Gorlatch. Programming with divide-and-conquer skeletons: an application to FFT. *Journal of Supercomputing*, 12(1–2):85–97, 1998.

14. I. Dorta, C. León, and C. Rodríguez. A comparison between MPI and OpenMP branch-and-bound skeletons. In *Proceedings of the 8th International Workshop on High Level Parallel Programming Models and Supportive Environments*. IEEE Computer Society Press, Los Alamitos, CA, 2003, pp. 66–73.

15. I. Dorta, C. León, C. Rodríguez, and A. Rojas. Parallel skeletons for divide-and-conquer and branch-and-bound techniques. In *Proceedings of the 11th Euromicro Conference on Parallel, Distributed and Network Based Processing*, Geneva, 2003, pp. 292–298.

16. OpenMP Architecture Review Board. *OpenMP C and C++ Application Program Interface*, Version 1.0. http://www.openmp.org, 1998.

CHAPTER 12

Tools for Tree Searches: Branch-and-Bound and A* Algorithms

C. LEÓN, G. MIRANDA, and C. RODRÍGUEZ

Universidad de La Laguna, Spain

12.1 INTRODUCTION

Optimization problems appear in all aspects of our lives. In a combinatorial optimization problem, we have a collection of decisions to make, a set of rules defining how such decisions interact, and a way of comparing possible solutions quantitatively. Solutions are defined by a set of decisions. From all the feasible solutions, our goal is to select the best. For this purpose, many search algorithms have been proposed. These algorithms search among the set of possible solutions (*search space*), evaluating the candidates and choosing one as the final solution to the problem. Usually, the search space is explored following a tree data structure. The root node represents the initial state of the search, when no decision has yet been made. For every decision to be made, a node is expanded and a set of children nodes is created (one for each possible decision). We arrive at the leaf nodes when all the problem decisions have been taken. Such nodes represent a feasible solution. The tree structure can be generated explicitly or implicitly, but in any case, the tree can be explored in different orders: level by level (*breadth-first search*), reaching a leaf node first and backtracking (*depth-first search*), or at each step choosing the best node (*best-first search*). The first two proposals are brute-force searches, whereas the last is an informed search because it uses heuristic functions to apply knowledge about the problem to reduce the amount of time spent in the process.

Branch-and-bound (BnB) [1–4] is a general-purpose tree search algorithm to find an optimal solution to a given problem. The goal is to maximize (or minimize) a function $f(x)$, where x belongs to a given feasible domain. To apply this algorithmic technique, it is required to have functions to compute *lower* and

Optimization Techniques for Solving Complex Problems, Edited by Enrique Alba, Christian Blum, Pedro Isasi, Coromoto León, and Juan Antonio Gómez

upper bounds and methods to divide the domain and generate smaller subproblems (*branch*). The use of bounds for the function to be optimized combined with the value of the current best solution enables the algorithm to search parts of the solution space only implicitly. The technique starts by considering the original problem inside the full domain (called the *root problem*). The lower and upper bound procedures are applied to this problem. If both bounds produce the same value, an optimal solution is found and the procedure finishes. Otherwise, the feasible domain is divided into two or more regions. Each of these regions is a section of the original one and defines the new search space of the corresponding *subproblem*. These subproblems are assigned to the children of the root node. The algorithm is applied to the subproblems recursively, generating a tree of subproblems. When an optimal solution to a subproblem is found, it produces a feasible solution to the original problem, so it can be used to prune the remaining tree. If the lower bound of the node is larger than the best feasible solution known, no global optimal solution exists inside the subtree associated with such a node, so the node can be discarded from the subsequent search. The search through the nodes continues until all of them have been solved or pruned.

A* [5] is a best-first tree search algorithm that finds the least-cost path from a given initial node to one goal node (of one or more possible goals). It uses a distance-plus-cost heuristic function $f(x)$ to determine the order in which the search visits nodes in the tree. This heuristic is a sum of two functions: the path-cost function $g(x)$ (cost from the starting node to the current one) and an admissible *heuristic estimate* of the distance to the goal, denoted $h(x)$. A* searches incrementally all routes leading from the starting point until it finds the shortest path to a goal. Like all informed search algorithms, it searches first the routes that appear to be most likely to lead toward the goal. The difference from a greedy best-first search is that it also takes into account the distance already traveled. Tree searches such as depth-first search and breadth-first search are special cases of the A* algorithm.

Any heuristic exploration uses some information or specific knowledge about the problem to guide the search into the state space. This type of information is usually determined by a *heuristic evaluation function, $h'(n)$*. This function represents the estimated cost to get to the objective node from node n. The function $h(n)$ represents the real cost to get to the solution node. As other informed searches, A* algorithms and BnB techniques make use of an *estimated total cost function, $f'(n)$*, for each node n. This function is determined by

$$f'(n) = g(n) + h'(n) \tag{12.1}$$

where, $g(n)$ represents the accumulated cost from the root node to node n and $h'(n)$ represents the estimated cost from node n to the objective node (i.e., the estimated remaining cost). An A* search uses the value of function 12.1 to continue the tree exploration, following, at each step, the branch with better success expectations. However, in a BnB search this information is needed to discard branches that will not lead to a better solution. The general process in which these and others searches are based is quite similar [6,7]. For each problem, the evaluation function to apply must be defined. Depending on the function defined, the optimal solution will or will not be found, and it will be reached after exploring more or fewer nodes. Also, the type of search strategy to be realized is determined by the way in which nodes are selected and discarded at each moment.

This work presents an algorithmic skeleton that implements these type of searches in a general way. The user decides exactly which search strategy is going to be used by determining some specific configuration parameters. The problem to be solved must also be specified to obtain automatically the sequential and parallel solution patterns provided by the tool. The skeleton has been tested with a wide range of academic problems, including the knapsack problem, the traveling salesman problem, and the minimum cost path problem. However, to analyze the behavior of the skeleton in detail, we chose to implement and test two complex problems: the two-dimensional cutting stock problem (2DCSP) and the error-correcting-code problem (ECC). The rest of the chapter is structured as follows. In Section 12.2 the state of the art for sequential and parallel tree search algorithms and tools is presented. The interface and internal operation of our algorithmic skeleton proposal are described in Section 12.3. To clarify the tool interface and flexibility, two problem implementation examples are introduced in Section 12.4. In Section 12.5 we present the computational results for the implemented problems. Finally, conclusions and some directions for future work are given.

12.2 BACKGROUND

The similarities among A* algorithms, BnB techniques, and other tree searches are clearly noticeable [6,7]. In all cases we need to define a method for representing the decision done at each step of the search. This can be seen as a division or branching of the current problem in subproblems. It is also necessary to define a way of deciding which state or node is going to be analyzed next. This will set the selection order among nodes. Finally, a (heuristic) function is needed to measure the quality of the nodes. The quality-indicator function and the type of subdivision to apply must be designed by attending the special properties of the problem. However, the general operation of these types of searches can be represented independent of the problem. Considering this common operation, many libraries and tools have been proposed [8–14]. Many of these tools have been implemented as algorithmic skeletons. An *algorithmic skeleton* must be understood as a set of procedures that comprise the structure to use in the development of programs for the solution of a given problem using a particular algorithmic technique. They provide an important advantage by comparison with the implementation of an algorithm from scratch, not only in terms of code reuse but also in methodology and concept clarity. Several parallel techniques have been applied to most search algorithms to improve their performance [15–17]. Many tools incorporate parallel schemes in their solvers to provide more efficient solutions to the user. The great benefit of such improvements is that the user obtains more efficient solvers without any requirement to know how to do parallel programming.

12.3 ALGORITHMIC SKELETON FOR TREE SEARCHES

In general, the software that supplies skeletons presents declarations of empty classes. The user must fill these empty classes to adapt the given scheme for solving a particular problem. Once the user has represented the problem, the tool provides sequential and/or parallel solvers without any additional input. The algorithmic skeleton proposed here allows the user to apply general tree search algorithms for a given problem. The skeleton [18] is included in the MALLBA [13] library and has been implemented in C++. It extends initial proposals for BnB techniques [19] and A* searches [20], allowing us to better adapt the tool to obtain more flexible searches. Figure 12.1 shows the two different parts of the proposed skeleton: One part implements the *solution patterns* provided by the implementation, whereas the second part must be completed by the user with the particular characteristics of the *problem to be solved*. This last part is the one to be used by the solution patterns to carry out the specified search properly.

The user adjustment required consists of two steps. First, the problem has to be represented through data structures. Then, certain functionalities representing the

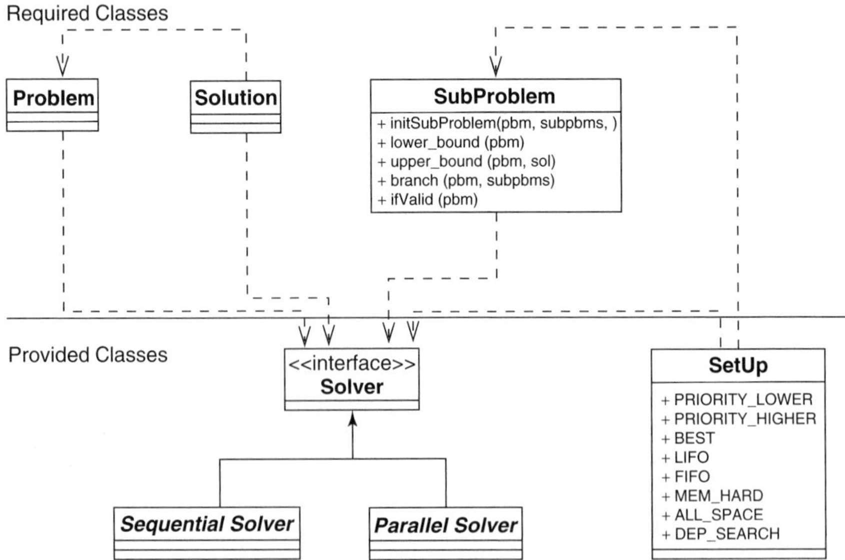

Figure 12.1 Skeleton architecture.

problem's properties must be implemented for each class. These functionalities will be invoked from the particular solution pattern so that when the application has been completed, the expected functionalities applied to the particular problem will be obtained. The skeleton requires the user to specify the following classes:

- `Problem`: stores the characteristics of the problem to be solved.
- `Solution`: defines how to represent the solutions.
- `SubProblem`: represents a node in the tree or search space.

The `SubProblem` class defines the search for a particular problem and must contain a `Solution`-type field in which to store the (partial) solution. The necessary methods to define this class are:

- `initSubProblem(pbm, subpbms)` creates the initial subproblem or subproblems from the original problem.
- `lower_bound(pbm)` calculates the subproblem's accumulated cost $g(n)$.
- `upper_bound(pbm, sol)` calculates the subproblem's estimated total cost $f'(n)$.
- `branch(pbm, subpbms)` generates a set of new subproblems from the current subproblem.
- `branch(pbm, sp, subpbms)` generates a set of new subproblems obtained from the combination of the current subproblem with a given subproblem `sp`.
- `similarTo(sp)` decides if the actual and given subproblems are similar or not.
- `betterThan(sp)` decides which of the actual or given subproblems is better.
- `ifValid(pbm)` decides if the actual subproblem has a probability of success. It is used in the parallel solver to discard subproblems that are not generated in the sequential solution.

The user can modify certain features of the search by using the configuration class `Setup` provided and depending on the definition given for some methods of the `SubProblem` class. The classes provided are:

- `Setup`. This class is used to configure all the search parameters and skeleton properties. The user can specify if an auxiliary list of expanded nodes is needed, the type of selection of nodes (exploration order), the type of subproblem generation (*dependent* or *independent*), if it is necessary to analyze similar subproblems, and whether to search the entire space or stop when the first solution is found. The meaning of each of these configuration parameters is explained in the following sections.
- `Solver`. Implements the strategy to follow: BnB, A*, and so on. One sequential solver and a shared-memory parallel solver are provided.

The skeleton makes use of two linked lists to explore the search space: `open` and `best`. The `open` list contains all the subproblems generated whose expansions are pending, while the subproblems analyzed are inserted in the `best` list. This list allows for the analysis of similar subproblems. Algorithm 12.1 shows the pseudocode of the sequential solver. First, the following node in the `open` list is removed. The criterion for the order in node selection is defined by the user through the `Setup` class. If the selected node is not a solution, the node is inserted in the `best` list. Insertions in `best` can check for node dominance and similarities in order to avoid exploring similar subproblems (subproblems that represent an equivalent solution to the problem). Then the node is branched and the newly generated subproblems that still have possibilities for improving the current solution are inserted into `open`. If a solution is found, the current best solution can be updated and the nonpromising nodes can be removed from `open`. Depending on the type of exploration specified by the user, the search can conclude when the first solution is found or can continue searching for more solutions so as eventually to keep the best.

The skeleton also provides the possibility of specifying *dependence* between subproblems. In this case the method used is `branch(pbm, sp, subpbms)`. It has to be implemented by the user and it generates new subproblems from the current one and a given subproblem `sp`. This method is necessary so as to

Algorithm 12.1 Sequential Solver Pseudocode

```
open ← createInitialSpbms(pbm)
best ← ∅
bestSolution ← ∅
while((!open.empty()) AND (!checkStopCondition())) do
  spbm ← open.remove()
  if((!spbm.isSolution()) AND (spbm.isPromising())) then
     best.insert(spbm)
     newSpbms ← branch(pbm)
     for all newSpbm in newSpbms do
        newSpbm.assignQualityIndicator()
        if newSpbm.isPromissing() then
           open.insert(newSpbm)
        end if
     end for
  else
     if node.isSolution() then
         bestSolution ← betterSolution(bestSolution, node)
         open.clearAllNonPromisingNodes()
     end if
  end if
end while
```

branch a subproblem that has to be combined with all the subproblems analyzed previously, that is, with all the subproblems in the `best` list. Instead of doing only a `branch` (as in the independent case shown in Algorithm 12.1), the method is called once with each of the elements in the `best` list and the current subproblem.

12.3.1 Shared-Memory Parallel Solver

In the literature, many message-passing proposals for different types of tree searches can be found, but no schemes based on shared memory [21]. This solver uses a shared-memory scheme to store the subproblem lists (`open` and `best`) used during the search process. Both lists have the same functionality as in the sequential skeleton. The difference is that the lists are now stored in shared memory. This makes it possible for several threads to work simultaneously on the generation of subproblems from different nodes. One of the main problems is maintaining the `open` list sorted. The order in the list is necessary when, for example, the user decides to do a best-first search. That is why standard mechanisms for the work distribution could not be used. The technique applied is based on a *master–slave model*. Before all the threads begin working together, the master generates the initial subproblems and inserts them into the `open` list. Until the master signals the end of the search, each slave branches subproblems from the unexplored nodes of the `open` list. Each slave explores the `open` list to find a node not assigned to any thread and with work to do. When such a node is found, the slave marks it as assigned, generates all its subproblems, indicating that the work is done, and finally leaves it unassigned. Meanwhile, the master extracts the first subproblem in `open`, checks for the stop conditions and similarities in `best`, and inserts in `open` the promising subproblems already generated.

When different slaves find an unassigned node and both decide to branch it at the same time or when the master checks a node's assignability just as a slave is going to mark it, some conflicts may appear. To avoid these types of conflicts and to ensure the integrity of the data, the skeleton adds some extra synchronization.

12.4 EXPERIMENTATION METHODOLOGY

To demonstrate the flexibility of the tool, two different problems were implemented and tested. It is important to note that the skeleton may or may not give the optimal solution to the problem and that it does a certain type of search, depending on the configuration specified by the user. So it is necessary to tune the tool setup properly for each problem. In this case, both problems required the skeleton to find an exact solution, but the search spaces specified for each were very different.

12.4.1 Error-Correcting-Code Problem

In a binary-code message transmission, interference can appear. Such interference may result in receiving a message that is different from the one sent originally. If the error is detected, one possible solution is to request the sender to retransmit the complete data block. However, there are many applications where data retransmission is not possible or is not convenient in terms of efficiency. In these cases, the message must be corrected by the receiver. For these particular situations, *error-correcting codes* [22] can be used.

In designing an error-correcting code, the objectives are:

- To minimize n (find the codewords of minimum length in order to decrease the time invested in message transmission)
- To maximize $d_{\min}$ (the Hamming distance between the codewords must be maximized to guarantee a high level of correction at the receiver). If the codewords are very different from each other, the likelihood of having a number of errors, so that a word can be transformed into a different one, is low)
- To maximize M (maximize the number of codewords in the code)

These objectives, however, are incompatible. So we try to optimize one of the parameters (n, M, or $d_{\min}$), giving a specific fixed value for the other two parameters. The most usual approach for the problem is to maximize $d_{\min}$ given n and M. The problem we want to solve is posed as follows: Starting with fixed values for the parameters M and n, we need to get (from all the possible codes that can be generated) the code with the maximum minimum distance. The total number of possible codes with M codewords of n bits is equal to 2^n M. Depending on the problem parameters (M and n), this value could be very high. In such cases, the approach to the problem could be almost nonfeasible in terms of computational resources. The execution time grows exponentially with an increase in either parameter M or n.

The approach to getting a code of M words of n bits with the maximum minimum distance is based on the *subset building algorithm*. This exact algorithm allows all possible codes with M words of length n to be generated. From all the possible codes, the one with the best minimum distance will be chosen as the final solution. Note that we can obtain several codes with the same best distance.

The algorithm used [23] is similar to an exhaustive tree search strategy. For this reason we also introduce the principles of a BnB technique to avoid exploring branches that will never result in an optimal solution.

Branching a node consists of building all the possible new codeword sets by adding a new codeword to the current codeword set. When the algorithm gets a solution code with a certain minimum distance, it will be set as the current best solution. In the future, branches with a current minimum distance lower or equal to the current best distance will be bound. This can be obtained by implementing the `lower_bound` and `upper_bound` methods properly. Finally, the code expected must be able to correct a certain number of errors (determined by the user or, by default, one). All the branches representing codes that break this constraint will be pruned.

In this case, the skeleton has been configured to execute a general and simple branch-and-bound search [23]. It is not necessary to keep the nodes in the `open` list ordered (the type of insertions in `open` is fixed to LIFO), nor is it essential to use the `best` list (the BEST parameter is not set to on). The subproblem branches are of the *independent* type, and the search will not stop with the first solution. It will continue exploring and will eventually keep the best of all the solutions found.

12.4.2 Two-Dimensional Cutting Stock Problem

The constrained 2DCSP targets the cutting of a large rectangle S of dimensions $L \times W$ in a set of smaller rectangles using orthogonal guillotine cuts: Any cut must run from one side of the rectangle to the other end and be parallel to one of the other two edges. The rectangles produced must belong to one of a given set of rectangle types $\mathcal{D} = \{T_1, \ldots, T_n\}$, where the ith type, T_i, has dimensions $l_i \times w_i$. Associated with each type T_i there is a profit c_i and a demand constraint b_i. The goal is to find a feasible cutting pattern with x_i pieces of type T_i maximizing the total profit:

$$\sum_{i=1}^{n} c_i x_i \text{ subject to } x_i \leq b_i \text{ and } x_i \in \mathcal{N} \qquad (12.2)$$

Wang [24] was the first to make the observation that all guillotine cutting patterns can be obtained by means of horizontal and vertical builds of meta-rectangles. Her idea was exploited by Viswanathan and Bagchi [25] to propose a best-first search A* algorithm (VB). The VB algorithm uses two lists and, at each step, the best build of pieces (or metarectangles) is combined with the best metarectangles already found to produce horizontal and vertical builds. All the approaches to parallelize VB make a special effort to manage the highly irregular computational structure of the algorithm. Attempts to deal with its intrinsically sequential nature inevitably appears either transformed on an excessively fine granularity or any other source of inefficiency [26,27].

The Viswanathan and Bagchi algorithm follows a scheme similar to an A* search, so that implementation with the skeleton is direct. At each moment, the best current build considered must be selected and placed at the left bottom corner of the surface S. Then it must be combined horizontally and vertically with the best builds selected previously (dependent-type branches). That is, the insertion in the `open` list must be done by order (depending on the value of the *estimated total profit*). This can be indicated by fixing the option PRIORITY_HIGHER in the features of the `open` list. The algorithm does *dependent* branches, so it is necessary to use the `best` list. The solver finishes the search when the first solution is found. If the `upper_bound` method has been defined properly, the first solution will be the best solution to the problem. Besides, in this particular case, the skeleton has been tuned to avoid analyzing subproblems similar and worse than others explored previously.

12.5 EXPERIMENTAL RESULTS

For the computational study of the ECC, a set of problem instances with different values of n and M were selected to test the behavior of the solvers provided. For all these instances, the minimum number of errors to correct was set to 1. For the 2DCSP, we have selected some instances from those available in refs. 28 and 29. The experiments were carried out on an Origin 3800 with 160-MIPS R14000 (600-MHz) processors. The compiler used was the MIPSpro v.7.41.

The execution times for the ECC problem obtained with the sequential solver and the one-thread parallel solver are shown in Table 12.1. The comparison of such times reveals the overhead introduced by the shared memory scheme. Due to the problem specification, all nodes generated need to be computed (i.e., the entire search space is explored to ensure achievement of the best solution). So the number of nodes computed for the sequential and parallel executions is always the same. The parallel results obtained with the shared-memory solver are presented in Table 12.2. Parallel times show that some improvements were improved, although the problem behavior does not seem very stable. The total number of nodes computed does not decrease in the parallel executions, so the speedup achieved is obtained by means of the parallel work among threads.

The sequential results obtained for the 2DCSP are presented in Table 12.3. Due to the implementation done, the number of nodes generated is different

TABLE 12.1 ECC Sequential Executions

	Sequential Solver		Parallel Solver	
Problem	Computed	Time (s)	Computed	Time (s)
$n = 8,\ M = 18$	100	2332.46	100	2682.62
$n = 9,\ M = 17$	221	223.79	221	244.27
$n = 10,\ M = 12$	469	24.61	469	25.60
$n = 13,\ M = 5$	4022	4238.54	4022	4253.52

TABLE 12.2 ECC Parallel Executions

Problem	2 Threads	4 Threads Time (s)	8 Threads	16 Threads
$n = 8,\ M = 18$	5149.48	9403.33	8605.96	733.83
$n = 9,\ M = 17$	244.32	743.55	13.75	13.65
$n = 10,\ M = 12$	25.73	23.25	23.94	33.80
$n = 13,\ M = 5$	4279.37	1717.34	954.08	627.31

TABLE 12.3 2DCSP Sequential Executions

	Sequential Solver			Parallel Solver		
Problem	Computed	Generated	Time (s)	Computed	Generated	Time (s)
1_	3,502	10,079	3.30	3,449	10,085	5.55
A4	864	38,497	50.53	841	39,504	65.78
A5	1,674	13,387	6.88	1,453	12,779	9.85

from the number of nodes computed because the search finishes when the first solution arrives. That leaves nodes in the `open` list that have been created but will not be explored. Sequential times for the most common test instances in the literature [28] are almost negligible, so some larger instances [29] were selected for the parallel study. Results obtained with the parallel solver are shown in Table 12.4. In the parallel executions, the total number of nodes computed may vary since the search order is altered. When searching in parallel, the global bounds can easily be improved, yielding a decreasing number of nodes computed (e.g., problem instance *cat3_2*). Although in cases where the bounds are highly accurate, the parallel search may produce some nodes that are not generated in the sequential search (e.g., problem instance *cat1_1*). In general, the parallel results for the 2DCSP get worse, with more threads participating in the solution. This usually occurs when trying to parallelize fine-grained problems. To provide an idea of the load balancing during the parallel executions, Figure 12.2 is shown. Computation of the nodes is distributed among the master and slave threads. The master leads the exploration order, while the slave threads generate nodes and their corresponding insertions into the local lists. For the ECC problem, the number of nodes computed by each master–slave thread is shown. In this case, all the nodes generated are inserted and computed during the search process. For the 2DCSP, the number of nodes inserted are shown. For this problem, the insertion of a node represents the heavier workload done by each thread (i.e., checking for similarities and duplicated nodes).

TABLE 12.4 2DCSP Parallel Executions

	1 Thread		2 Threads	
Problem	Computed	Time (s)	Computed	Time (s)
cat1_1	42,000	112.09	50,000	73.49
cat3_2	10,000	32.08	11,000	5.59
	4 Threads		**8 Threads**	
Problem	Computed	Time (s)	Computed	Time (s)
cat1_1	56,000	73.96	57,000	68.99
cat3_2	3,000	0.77	5,000	1.01

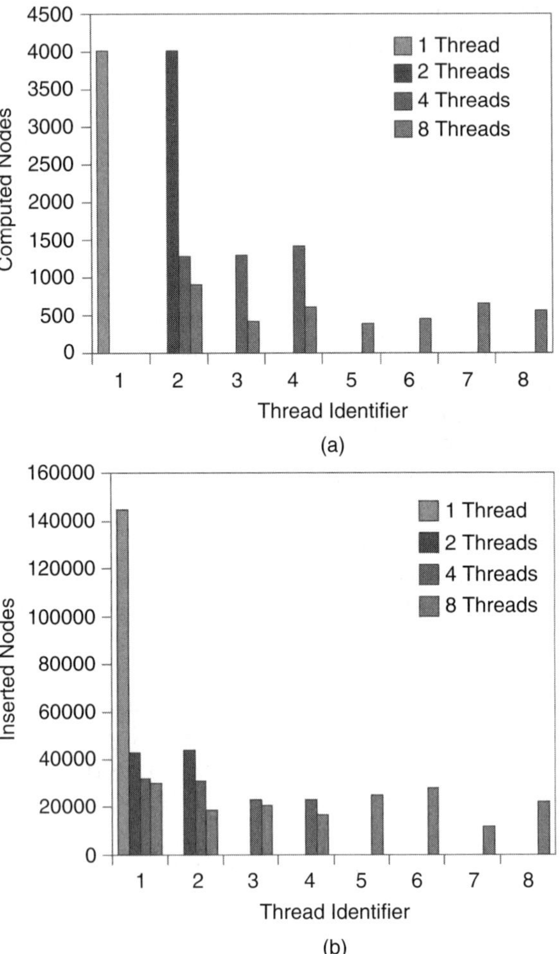

Figure 12.2 Load balancing: (a) ECC ($n = 13$, $M = 5$); (b) 2DCSP (*cat3_2*).

In general, a fair distribution of nodes is difficult to obtain since it is not only needed to distribute the subproblems fairly to be generated but also those to be inserted. Before doing a certain combination, we are unable to know if it will be valid or not (to be inserted). Anyway, even when a fair balance of the inserted nodes is achieved, it does not mean that the work involved is distributed homogeneously. Due to the internal operation of the parallel solver, it usually ensures a fair distribution of the workload. Although the "number of computed or inserted nodes" is not properly distributed among the threads, the work involved is always balanced. The load balancing is better achieved for fine-grained problems: Every time a thread finishes a node branching, it asks for the next one to be explored.

12.6 CONCLUSIONS

Considering the similarities among many different types of tree searches, an algorithmic skeleton was developed to simplify implementation of these general tree searches. The skeleton provides a sequential and a parallel solver as well as several setup options that make it possible to configure flexible searches. The main contribution is the parallel solver: It was implemented using the shared-memory paradigm. We presented computational results obtained for the implementation of two different complex problems: ECC and 2DCSP.

As we have shown, the synchronization needed to allow the list to reside in shared memory introduces significant overhead to the parallel version. The results seem to be improved when the skeleton executes a search with *independent* generation of subproblems (i.e., the ECC problem). When it is not necessary to maintain the list ordered and the problem has coarse grain, the threads spend less time looking for work and doing the corresponding synchronization through the list. That makes it possible to obtain lower overhead and better speedup. In such cases, some time improvements are obtained, thanks to the parallel analysis of nodes. The improvements introduced by the parallel solver in the case of the 2DCSP (fine-grained and *dependent* generation of subproblems) are obtained because the number of nodes generated decreases when more threads collaborate in the problem solution. That is because the update of the best current bound is done simultaneously by all the threads, allowing subproblems to be discarded that in the sequential case have to be inserted and explored.

In any type of tree search, sublinear and superlinear speedups can be obtained since the search exploration order may change in the parallel executions [30–32]. Parallel speedups depend strongly on the particular problem. That is a result of changing the search space exploration order when more than one thread is collaborating in the solution. Anyway, even in cases where the parallel execution generates many more nodes than does sequential execution, the parallel scheme is able to improve sequential times. Synchronization between threads when accessing the global list is a critical point, so the behavior and scalability of the parallel scheme is not completely uniform. As shown, the times (for both problems) in different executions are not very stable. In any case, the parallel results depend strongly on the platform characteristics (compiler, processor structure, exclusive use, etc.).

The main advantage of this tool is that the user can tune all the search parameters to specify many different types of tree searches. In addition, the user gets a sequential and a parallel solver with only simple specification of the problem. Another advantage of the parallel scheme is that the work distribution between the slaves is balanced. The parallel solver works fine for coarse-grained and independent generation-type problems, but for fine-grained problems with dependent generation of subproblems, the synchronization required for sharing the algorithm structures introduces an important overhead to the scheme. For this reason, a more coarse-grained parallelization is being considered to partially avoid such overhead.

Acknowledgments

This work has been supported by the EC (FEDER) and by the Spanish Ministry of Education and Science under the Plan Nacional de I+D+i, contract TIN2005-08818-C04-04. The work of G. Miranda has been developed under grant FPU-AP2004-2290.

REFERENCES

1. A. H. Land and A. G. Doig. An automatic method of solving discrete programming problems. *Econometrica*, 28(3):497–520, 1960.
2. E. Lawler and D. Wood. Branch-and-bound methods: a survey. *Operations Research*, 14:699–719, 1966.
3. L. G. Mitten. Branch-and-bound methods: general formulation and properties. *Operations Research*, 18(1):24–34, 1970.
4. T. Ibaraki. Theoretical comparisons of search strategies in branch-and-bound algorithms. *International Journal of Computer and Informatics Sciences*, 5(4):315–344, 1976.
5. P. E. Hart, N. J. Nilsson, and B. Raphael. A formal basis for the heuristic determination of minimum cost paths. *IEEE Transactions on Systems Science and Cybernetics*, SSC-4(2):100–107, 1968.
6. T. Ibaraki. Enumerative approaches to combinatorial optimization (Part I). *Annals of Operations Research*, 10(1–4):1–342, 1987.
7. D. S. Nau, V. Kumar, and L. Kanal. General branch and bound, and its relation to A* and AO*. *Articial Intelligence*, 23(1):29–58, 1984.
8. B. Le Cun, C. Roucairol, and TNN Team. BOB: a unified platform for implementing branch-and-bound like algorithms. *Technical Report 95/016. Laboratoire* PRiSM, Université de Versailles, France, 1995.
9. Y. Shinano, M. Higaki, and R. Hirabayashi. A generalized utility for parallel branch and bound algorithms. In *Proceedings of the 7th IEEE Symposium on Parallel and Distributed Processing*, 1995, pp. 392–401.
10. S. Tschöke and T. Polzer. Portable parallel branch-and-bound library. *PPBB-Lib User Manual, Library Version 2.0.* Department of Computer Science, University of Paderborn, Germany, 1996.
11. R. C. Correa and A. Ferreira. Parallel best-first branch-and-bound in discrete optimization: a framework. *Solving Combinatorial Optimization Problems in Parallel: Methods and Techniques*, vol. 1054 of *Lecture Notes in Computer Science*, Springer-Verlag, New York, 1996, pp. 171–200.
12. PICO: an object-oriented framework for parallel branch and bound. In *Proceedings of the Workshop on Inherently Parallel Algorithms in Optimization and Feasibility and their Applications*. Elsevier Scientific, New York, 2000, pp. 219–265.
13. E. Alba et al. MALLBA: A library of skeletons for combinatorial optimization. In *Proceedings of Euro-Par*, vol. 2400 of *Lecture Notes in Computer Science.* Springer-Verlag, Berlin, 2002, pp. 927–932.

14. T. K. Ralphs, L. Ladányi, and M. J. Saltzman. A library hierarchy for implementing scalable parallel search algorithms. *Journal of Supercomputing*, 28(2):215–234, 2004.

15. A. Grama and V. Kumar. State of the art in parallel search techniques for discrete optimization problems. *Knowledge and Data Engineering*, 11(1):28–35, 1999.

16. V. D. Cung, S. Dowaji, B. Le Cun, T. Mautor, and C. Roucairol. Parallel and distributed branch-and-bound/A* algorithms. *Technical Report 94/31*. Laboratoire PRiSM, Université de Versailles, France, 1994.

17. V. D. Cung and B. Le Cun. An efficient implementation of parallel A*. In *Proceedings of the Canada–France Conference on Parallel and Distributed Computing*, 1994, pp. 153–168.

18. G. Miranda and C. León. OpenMP skeletons for tree searches. In *Proceedings of the 14th Euromicro Conference on Parallel, Distributed and Network-based Processing*. IEEE Computer Society Press, Los Alamitos, CA, 2006, pp. 423–430.

19. J. R. González, C. León, and C. Rodríguez. An asynchronous branch and bound skeleton for heterogeneous clusters. In *Proceedings of the 11th European PVM/MPI Users' Group Meeting*, vol. 3241 of *Lecture Notes in Computer Science*. Springer-Verlag, New York, 2004, pp. 191–198.

20. G. Miranda and C. León. An OpenMP skeleton for the A* heuristic search. In *Proceedings of High Performance Computing and Communications*, vol. 3726 of *Lecture Notes in Computer Science*. Springer-Verlag, New York, 2005, pp. 717–722.

21. I. Dorta, C. León, and C. Rodríguez. A comparison between MPI and OpenMP branch-and-bound skeletons. In *Proceedings of the 8th International Workshop on High Level Parallel Programming Models and Supportive Enviroments*. IEEE Computer Society Press, Los Alamitos, CA, 2003, pp. 66–73.

22. R. Hill. *A First Course in Coding Theory*. Oxford Applied Mathematics and Computing Science Series. Oxford University Press, New York, 1986.

23. C. León, S. Martín, G. Miranda, C. Rodríguez, and J. Rodríguez. Parallelizations of the error correcting code problem. In *Proceedings of the 6th International Conference on Large-Scale Scientific Computations*, vol. 4818 of *Lecture Notes in Computer Science*. Springer-Verlag, New York, 2007, pp. 683–690.

24. P. Y. Wang. Two algorithms for constrained two-dimensional cutting stock problems. *Operations Research*, 31(3):573–586, 1983.

25. K. V. Viswanathan and A. Bagchi. Best-first search methods for constrained two-dimensional cutting stock problems. *Operations Research*, 41(4):768–776, 1993.

26. G. Miranda and C. León. OpenMP parallelizations of Viswanathan and Bagchi's algorithm for the two-dimensional cutting stock problem. In *Parallel Computing 2005*, NIC Series, vol. 33, 2005, pp. 285–292.

27. L. García, C. León, G. Miranda, and C. Rodríguez. Two-dimensional cutting stock problem: shared memory parallelizations. In *Proceedings of the 5th International Symposium on Parallel Computing in Electrical Engineering*. IEEE Computer Society Press, Los Alamitos, CA, 2006, pp. 438–443.

28. M. Hifi. An improvement of Viswanathan and Bagchi's exact algorithm for constrained two-dimensional cutting stock. *Computer Operations Research*, 24(8):727–736, 1997.

29. DEIS Operations Research Group. Library of instances: two-constraint bin packing problem. http://www.or.deis.unibo.it/research_pages/ORinstances/2CBP.html.

30. B. Mans and C. Roucairol. Performances of parallel branch and bound algorithms with best-first search. *Discrete Applied Mathematics and Combinatorial Operations Research and Computer Science*, 66(1):57–76, 1996.

31. T. H. Lai and S. Sahni. Anomalies in parallel branch-and-bound algorithms. *Communications of the ACM*, 27(6):594–602, 1984.

32. A. Bruin, G. A. P. Kindervater, and H. W. J. M. Trienekens. Asynchronous parallel branch and bound and anomalies. In *Proceedings of the Workshop on Parallel Algorithms for Irregularly Structured Problem*, 1995, pp. 363–377.

CHAPTER 13

Tools for Tree Searches: Dynamic Programming

C. LEÓN, G. MIRANDA, and C. RODRÍGUEZ
Universidad de La Laguna, Spain

13.1 INTRODUCTION

The technique known as dynamic programming (DP) is analogous to divide and conquer. In fact, it can be seen as a reformulation of the divide-and-conquer (DnC) technique. Consequently, it aims at the same class of problems. Dynamic programming usually takes one of two approaches:

1. *Top-down approach*. The problem is broken into subproblems, and these subproblems are solved, but (and this is how it differs from divide and conquer) a memory cache (usually, a multidimensional data structure) is used to remember the mapping between subproblems and solutions. Before expanding any subproblem, the algorithm checks to see if the subproblem is in such a cache. Instead of repeating the exploration of the subtree, the algorithm returns the stored solution if the problem has appeared in the past.

2. *Bottom-up approach*. Subproblems are solved in order of complexity. The solutions are stored (in a multidimensional data structure which is the equivalent of the one used in the top-down approach). The solutions of the simpler problems are used to build the solution of the more complex problems.

From this description follows the main advantages and disadvantages of dynamic programming: There is an obvious gain when the divide-and-conquer search tree is exponential, and subproblems with the same characterization appear again and again. DP fails when there are no repetitions or even though there are, the range of subproblems is too large to fit in computer memory. In

Optimization Techniques for Solving Complex Problems, Edited by Enrique Alba, Christian Blum, Pedro Isasi, Coromoto León, and Juan Antonio Gómez

such cases the memory cache multidimensional data structure grows beyond the limits of the largest supercomputers.

13.2 TOP-DOWN APPROACH

Dynamic programming is strongly related to a general-purpose technique used to optimize execution time, known as memoization. When a memoized function with a specific set of arguments is called for the first time, it stores the result in a lookup table. From that moment on, the result remembered will be returned rather than recalculated. Only pure functions can be memoized. A function is *pure* if it is referentially transparent: The function always evaluates the same result value given the same arguments. The result cannot depend on any global or external information, nor can it depend on any external input from I/O devices.

The term *memoization* derives from the Latin word *memorandum* (to be remembered) and carries the meaning of turning a function $f\colon X_1 \times X_2 \rightarrow Y$ into something to be remembered. At the limit, and although it can be an unreachable goal, memoization is the progressive substitution of f by an associative representation of the multidimensional table:

$$\{(x_1, x_2, y) \in X_1 \times X_2 \times Y \text{ such that } y = f(x_1, x_2)\}$$

So instead of computing $f(x_1, x_2)$, which will usually take time, we will simply access the associated table entry $f[x_1][x_2]$. Memoization trades execution time in exchange for memory cost; memoized functions become optimized for speed in exchange for a higher use of computer memory space.

Some programming languages (e.g., Scheme, Common Lisp, or Perl) can automatically memoize the result of a function call with a particular set of arguments. This mechanism is sometimes referred to as *call by need*. Maple has automatic memoization built in.

The following Perl code snippet (to learn more about Perl and memoization in Perl see refs. 1 and 2), which implements a general-purpose memoizer, illustrates the technique (you can find the code examples in ref. 3):

```
sub memoize {
  my $func = shift;

  my %cache;
  my $stub = sub {
    my $key = join ',', @_;
    $cache{$key} = $func->(@_) unless exists $cache{$key};
    return $cache{$key};
  };
  return $stub;
}
```

The code receives in `$func` the only argument that is a reference to a function (line 2). The call to the `shift` operator returns the first item in the list of arguments (denoted by `@_`). A reference to a memoized version of the function is created in `$stub` (lines 5 to 9). The new function uses a hash `%cache` to store the pairs evaluated. The unique key for each combination of arguments is computed at line 6. The function arguments in `@_` are coerced to strings and concatenated with commas via the `join` operator in such a way that different combinations of arguments will be mapped onto different strings. The original function is called only if it is the first time for such a combination. In such a case, the result of the call `$func->(@_)` is stored in the cache (line 7). The entry will then be used by subsequent calls. Finally, `memoize` returns the reference to the memoized version of `$func` (line 10).

To see `memoize` working, let us apply it to a 0–1 knapsack problem with capacity C and N objects. The weights will be denoted w_k and the profits p_k. If we denote by $f_{k,c}$ the optimal value for the subproblem considering objects $\{0, \dots, k\}$ and capacity c, the following recursive equation holds:

$$f_{k,c} = \max\{f_{k-1,c}\} \cup \{f_{k-1,c-w_k} + p_k \text{ if } c > w_k\} \tag{13.1}$$

The following Perl subroutine uses Equation 13.1 to solve the problem:

```
sub knap {
  my ($k, $c) = @_;
  my @s;

  return 0 if $k < 0;
  $s[0] = knap($k-1, $c);
  if ($w[$k] <= $c) {
    $s[1] = knap($k-1, $c-$w[$k])+$p[$k];
  }
  return max(@s);
}
```

The following program benchmarks 1000 executions of the divide-and-conquer version versus the top-down memoized dynamic programming version (the function referenced by `$mknap`, line 7):

```
$ cat -n memoizeknap.pl
 1  #!/usr/bin/perl -I../lib -w
 2  use strict;
 3  use Benchmark;
 4  use Algorithm::MemoizeKnap;
 5  use List::Util qw(max sum);
 6
 7  my $mknap = memoize(\&knap);
```

```

my $N = shift || 12;
my $w = [ map { 3 + int(rand($N-3)) } 1..$N ];
my $p = [ map { $_ } @$w ];
my $C = int(sum(@$w)/3);

set($N, $C, $w, $p);

my ($r, $m);
timethese(1000, {
  recursive => sub { $r = knap($N-1, $C) },
  memoized  => sub { $m = $mknap->($N-1, $C) },
  }
);

print "recursive = $r, memoized = $m\n";
```

The results are overwhelming: The CPU time for the DP version is negligible while the divide-and-conquer approach takes almost 4 seconds:

```
$ ./memoizeknap.pl
Benchmark: timing 1000 iterations of memoized, recursive...
  memoized:  0 wallclock secs (0.00 usr + 0.00 sys = 0.00 CPU)
 recursive:  4 wallclock secs (3.75 usr + 0.00 sys = 3.75 CPU)
recursive = 23, memoized = 23
```

The reason for such success is this: While the divide-and-conquer approach explores a search tree with a potential number of 2^N nodes, the memoized (dynamic programming) version never explores more than $N \times C$ nodes. Since $N \times C \ll 2^N$, *it follows that the same attribute values*, `($k, $c)`, *appear again and again during exploration of the nodes of the DnC search tree*.

The contents of this section can be summarized in an equation and a statement:

- *Top-down dynamic programming* = *divide and conquer* + *memoization*
- *Apply dynamic programming when the attributes of the nodes in the divide-and-conquer search tree have a limited variation range, which is much smaller than the size of the search tree. Such size variation must fit into the computer memory.*

13.3 BOTTOM-UP APPROACH

The bottom-up approach reproduces a traversing of the divide-and-conquer search tree from the leaves to the root. Subproblems are *sorted by complexity* in such a way that when a subproblem is processed, all the subproblems from which it depends have already being solved. The solutions of such subproblems are then *combined* to find the solution of the current subproblem. The new solution is stored in the *dynamic programming table* to be used later during resolution of the subproblems that depend on the current subproblem. Each subproblem is considered only once. A bottom-up DP algorithm is determined by these methods:

1. `init_table`: initializes the DP table with the values for nondependant states.
2. `next`: an iterator. Traverses the directed acyclic graph of the states (also called *iteration space*) in an order that is compatible with the dependence graph (i.e., a topological sort). Expressed in divide-and-conquer terms: The children are visited before their father.
3. `combine`: equivalent of the homonymus divide-and-conquer method. Therefore, it returns the solution for the current state/subproblem in terms of its children.
4. `table(problem, sol)`: updates the entry of the dynamic programming table for subproblem `problem` with the solution specified in `sol`.

The following Perl code (see ref. 3) shows a generic bottom-up dynamic programming solver (method `run`):

```
sub run {
  my $dp = shift;

  $dp->init_table;
  while (my $s = $dp->next) {
    $dp->table($s, $dp->combine($s));
  }
}
```

The following code solves the 0–1 knapsack problem by instantiating all the attributes enumerated in the former list. The `combine` method simply mimics the dynamic programming equation:

$$f_{k,c} = \max\{f_{k-1,c}, f_{k-1,c-w_k} + p_k \text{ if } c > w_k\}$$

```
 sub combine {
  my $dp = shift;
  my $state = shift;

   return max($dp->children($state));
}
```

The auxiliar method `children` returns the children of the current state:

```
sub children {
  my $dp = shift;
  my ($k, $c) = @{shift()};

  my $wk = $dp->{w}[$k];
  my $pk = $dp->{p}[$k];
  my @s;
  $s[0] = $dp->table([$k-1, $c]);
  $s[1] = $pk+$dp->table([$k-1, $c-$wk]) if $c >= $wk;

  return @s;
}
```

The method `init_table` populates the first row corresponding to the optimal values $f_{0,c} = p_0$ if $c > w_0$ and 0 otherwise:

```
sub init_table {
  my $dp = shift;
  my @table;

  my $C = $dp->{C};
  my $w0 = $dp->{w}[0];
  my $p0 = $dp->{p}[0];
  for(my $c = 0; $c <= $C; $c++) {
    $table[0][$c] = ($w0 <= $c)? $p0 : 0;
  }
  $dp->{table} = \@table;
}
```

The iterator `next` traverses the iteration space in an order that is compatible with the set of dependencies: index `[$k, $c]` must precede indices `[$k+1, $c]` and `[$k+1, $c+$w[$k]]`:

```
{
 my ($k, $c) = (1, 0);

 sub next {
   my $dp = shift;

   my ($C, $N) = ($dp->{C}, $dp->{N});

   return [$k, $c++] if $c <= $C;
   $k++;
```

```
    return [$k, $c = 0] if ($k < $N);
    return undef;
  }
}
```

Once the DP bottom-up algorithm has computed the dynamic table, the method `solution` traverses the table backward to rebuild the solution:

```
subsolution {
  my $dp = shift;

  my @sol;

  my ($C, $N) = ($dp->{C}, $dp->{N});
  my $optimal = $dp->table([$N-1, $C]);

  # Compute solution backward
  for(my $k = $N-1, my $c = $C; $k >= 0; $k--) {
    my @s = $dp->children([$k, $c]);
    if (max(@s) == $s[0]) {
      $sol[$k] = 0;
    }
    else {
      $sol[$k] = 1;
      $c -= $dp->{w}[$k];
    }
  }
  $dp->{optimal} = $optimal;
  $dp->{solution} = \@sol; # reference to @sol
  return ($optimal, @sol)
}
```

13.4 AUTOMATA THEORY AND DYNAMIC PROGRAMMING

In DP terminology the divide-and-conquer *subproblems* become *states*. We liberally mix both terms in this chapter. A bit more formally, we establish in the set of nodes of the divide-and-conquer search tree the equivalence relation which says that two nodes are equivalent if they share the same attributes[1]:

$$\text{node}_1 \equiv \text{node}_2 \iff \text{node}_1\{\text{attributes}\} = \text{node}_2\{\text{attributes}\} \tag{13.2}$$

The dynamic programming states are actually the *equivalence classes* resulting from this equivalence relation. That is, *states* are the *attributes* stored in the

[1]The word *attributes* refers here to the information saved in the nodes. In the former 0–1 knapsack problem example, the attributes were the pair `(k, c)` holding the object and the capacity.

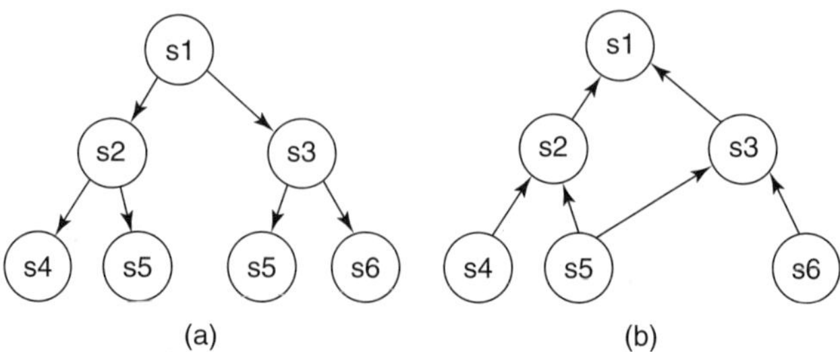

Figure 13.1 Divide-and-conquer nodes with the same attribute are merged: (a) the search tree becomes a DAG (labels inside the nodes stand for attributes); (b) in bottom-up DP the traversing is reversed.

nodes. As emphasized in the Section 13.3, the most relevant difference between DP and DnC is that nodes with the same attributes are considered only once. A consequence of establishing this equivalence between nodes is that the search space is no longer a tree but a directed acyclic graph (Figure 13.1).

The shape of the DAG in Figure 13.1 suggests a finite automaton. Edges correspond to transitions. An initial state q_0 can be added that represents the absence of decisions. In many cases the problem to solve using DP is an optimization problem.[2]

$$\min\ (\max) f(x) \text{ subject to } x \in S \tag{13.3}$$

in which the set of constraints S is modeled as a language; that is, $S \subset \Sigma^*$, where Σ is a finite alphabet of symbols called *decisions*. The elements of Σ^* are called *policies*. A policy $x \in \Sigma^*$ is said to be *feasible* if $x \in S$.

A large subset of the set of dynamic programming algorithms for optimization problems can be modeled by superimposing a cost structure to a deterministic finite automaton (DFA).

Definition 13.1 An *automaton with costs* is a triple $\Pi = (M, \mu, \xi_0)$ where:

1. $M = (Q, \Sigma, q_0, \delta, F)$ is a finite automaton. Q denotes the finite set of states, Σ is the finite set of decisions, δ: $Q \times \Sigma \to Q$ is the transition function, $q_0 \in Q$ is the initial state, and $F \subset Q$ is the set of final states.
2. μ: $R \times Q \times \Sigma \to R$ is the *cost function*, where R denotes the set of real numbers. The function μ is nondecreasing in its first argument:

$$\mu(\xi, q, a) \geq \xi \text{ and } \xi_1 \geq \xi_2 \implies \mu(\xi_1, q, a) \geq \mu(\xi_2, q, a)$$

3. $\xi_0 \in R$ is the setup cost of the initial state q_0.

[2]But not always; DP is not limited to optimization problems and has almost the same scope as DnC. As an example, the early parsing algorithm uses dynamic programming to parse any context-free language in $O(n^3)$ time [4]. Hardly parsing can be considered an optimization problem.

The transition function $\delta\colon Q \times \Sigma \rightarrow Q$ can be extended to the set of policies $\delta\colon Q \times \Sigma^* \rightarrow Q$ following these two formulas:

1. $\delta(q, \epsilon) = q$ for all q.
2. $\delta(q, xa) = \delta(\delta(q, x), a)$ for all q and for all a.

The notation $\delta(x) = \delta(q_0, x)$ is also common. Observe that with this formulation, the children of a given node q in the DnC search tree are $\delta^{-1}(q) = \{q' \in Q / \delta(q', a) = q,\ \exists a \in \Sigma\}$.

The cost function μ can also be extended to policies using these two inductive equations:

1. $\mu(\xi, q, \epsilon) = \xi$ for $\xi \in R$ and $q \in Q$.
2. $\mu(\xi, q, xa) = \mu(\mu(\xi, q, x), \delta(q, x), a)$ for $\xi \in R$, $q \in Q$, $x \in \Sigma^*$ and $a \in \Sigma$.

The notation $\mu(x) = \mu(\xi_0, q_0, x)$ is also used.

Definition 13.2 A finite *automaton with costs* $\Pi = (M, \mu, \xi_0)$ represents the optimization problem

$$\min\ f(x) \text{ subject to } x \in S$$

if and only if it satisfies:

1. The language accepted by the automaton M is the set of feasible solutions S, that is,

$$L(M) = \{x \in \Sigma^* / \delta(x) \in F\} = S$$

2. For any feasible policy $x \in S$, $f(x) = \mu(x)$ holds.

Definition 13.3 For any state q we define $G(q)$ as the *optimal value of any policy* leading from the start state q_0 to q [i.e., $G(q) = \min\{\mu(x)/x \in \Sigma^*$ and $\delta(x) = q\}$ assuming that q is reachable; otherwise, $G(q) = \infty$].

With these definitions it is now possible to establish the dynamic programming equations:

$$G(q) = \min\{\mu(G(q'), q', a)/q' \in Q, a \in \Sigma, \delta(q', a) = q\} \text{ and } G(q_0) = \xi_0 \tag{13.4}$$

Bellman's Optimality Principle The principle of optimality introduced by R. Bellman can now be summarized by saying that if $x \in S$ is an *optimal policy* between states q_1 and q_2,

$$\delta(q_1, x) = q_2 \text{ and } \mu(q_1, x) \leq \mu(q_1, u)\ \forall u \in \Sigma^* / \delta(q_1, u) = q_2$$

then any substring/subpolicy y of $x = zyw$ is also optimal between states $q_3 = \delta(q_1, z)$ and $q_4 = \delta(q_1, zy)$.

$$\mu(q_3, y) \leq \mu(q_3, u)\ \forall u \in \Sigma^*/\delta(q_3, u) = q_4$$

Not any DP optimization algorithm can be modeled through a finite automaton with costs. An example is the multiplication parenthesization problem (MPP):

Definition 13.4 In the MPP we have n matrices M_i, $i = 1, \ldots, n$. Matrix M_i has dimensions $c_{i-1} \times c_i$. The number of operations when doing the product of the n matrices,

$$P = M_1 \times M_2 \times \cdots \times M_n$$

depends on the way they are associated. For example, let $c_0 = 10, c_1 = 20, c_2 = 50, c_3 = 1$, and $c_4 = 100$. Then evaluation of $P = M_1 \times (M_2 \times (M_3 \times M_4))$ requires the order of

$$50 \times 1 \times 100 + 20 \times 50 \times 100 + 10 \times 20 \times 100 = 125{,}000$$

multiplications. Instead, the evaluation of $P = (M_1 \times (M_2 \times M_3)) \times M_4$ requires only 2200 multiplications. The problem is to find an association that minimizes the number of total operations.

Let $G(i, j)$ be the minimum cost for the multiplication of $M_i \times \cdots \times M_j$; then we have the following DP equations:

$$G(i, j) = \min\{G(i, k) + G(k + 1, j) + c_i \times c_k \times c_j / i \leq k < j\} \qquad (13.5)$$

To make a decision here is to decide the value for k: where to partition the interval $[i, j]$. But once you have decided about k, you have to make two decisions: one about interval $[i, k]$ and the other about $[k + 1, j]$. Observe that instead of a string as in the automaton with costs model, *policies here take the shape of a tree*. Thus, the policy for solution $P = (M_1 \times (M_2 \times M_3)) \times M_4$ will be described by the tree term $\times(\times(1, \times(2, 3)), 4)$. The solution $P = M_1 \times (M_2 \times (M_3 \times M_4))$ corresponds to the tree term $\times(1, \times(2, \times(3, 4)))$. DP algorithms like this are usually referred to as *polyadic problems*.

The language S here is a language whose phrases are trees instead of strings. The formal definition of a tree language is as follows:

Definition 13.5 An *alphabet with arity* (also called a *ranked alphabet*) is a pair (Σ, ρ) where Σ is a finite set of *operators* (also called constructors in functional languages) and $\rho : \Sigma \to (N_0)$ is the arity function. The set of items with arity k are denoted by Σ_k: $\Sigma_k = \{a \in \Sigma / \rho(a) = k\}$.

Definition 13.6 The *homogeneous tree language* $T(\Sigma)$ over a ranked alphabet (Σ, ρ) is defined by the following inductive rules:

- Any operator with arity *0* is in $T(\Sigma)$: $\Sigma_0 \subset T(\Sigma)$.
- If $f \in \Sigma_k$ and $t_1, \ldots, t_k \in T(\Sigma)$, then $f(t_1, \ldots, t_k)$ is in $T(\Sigma)$.

The elements of $T(\Sigma)$ are called *terms*.

Example 13.1 The trees to code the policies for the MPP are defined by the alphabet $\Sigma = \{\times, 1, 2, \ldots, n\}$, where $\rho(\times) = 2$ and $\rho(i) = 0$ for all $i \in \{1, \ldots, n\}$. Obviously, elements such as $\times(\times(1, \times(2, 3)), 4)$ belong to $T(\Sigma)$.

The former automata-based model for monadic dynamic programming algorithms must be extended to include these polyadic dynamic programming algorithms. But since the domain S of solutions of a polyadic problem is a tree language, we have to extend the concept of finite automaton to tree languages. This leads us to the concept of tree automaton (see Gonzalez Morales et al. [5]):

Definition 13.7 A *bottom-up deterministic tree finite automaton* (DTFA) is a 4-tuple $M = (Q, \Sigma, \delta, F)$, where:

1. Q is a finite set of states.
2. $F \subset Q$ is the set of *final* states.
3. $\Sigma = (\Sigma, \rho)$ is the input alphabet (with arity function ρ).
4. $\delta : \bigcup_{j \geq 0} \Sigma_j \times Q^j \to Q$ is the transition function.[3]

Example 13.2 See an example of DTFA that accepts the tree language of the solutions of the MPP:

1. $Q = \{q_{i,j} \mid i \leq j\}$.
2. $F = \{q_{1,n}\}$.
3. $\Sigma = \{\times, 1, 2, \ldots, n\}$ where $\rho(\times) = 2$ and $\rho(i) = 0$ for all $i \in \{1, \ldots, n\}$.
4. $\delta(\times, (q_{i,k}, q_{k+1,j})) = q_{i,j}$ and $\delta(i) = q_{i,i}$ for all $i \in \{1, \ldots, n\}$.

Given a DTFA $M = (Q, \Sigma, \delta, F)$, the partial function δ can be extended to $\hat{\delta}$: $T(\Sigma) \to Q$ by defining

$$\hat{\delta}(t) = \delta(a, \hat{\delta}(t_1), \ldots, \hat{\delta}(t_k)) \text{ if } t = a(t_1, \ldots, t_k)$$

$$\hat{\delta}(d) = \delta(d) \text{ if } d \in \Sigma_0$$

[3]Notice that j can be zero, and therefore transitions of the form $\delta(b) = q$, where $b \in \Sigma_0$, are allowed.

Definition 13.8 We say that $t \in T(\Sigma)$ is accepted by M if and only if $\hat{\delta}(t) \in F$. Thus, the tree language accepted by M is defined as

$$L(M) = \{t \in T(\Sigma) / \hat{\delta}(t) \in F\} \tag{13.6}$$

From now on we will not differentiate between δ and $\hat{\delta}$.

The idea is that $\delta(a, q_1, \ldots, q_j) = q$ means that of can substitute for the $T(\Sigma \cup Q)$ tree $a(q_1, \ldots, q_j)$. The tree $\times(\times(1, \times(2, 3)), 4)$ is thus accepted by the automaton defined in Example 13.4 (for $n = 4$) because $\delta(\times(\times(1, \times(2, 3)), 4)) = q_{1,4}$, as it proves the following successive applications of δ in a bottom-up traversing of the tree[4]:

$$\times(\times(1, \times(2, 3)), 4) \stackrel{*}{\Longrightarrow} \times(\times(q_{1,1}, \times(q_{2,2}, q_{3,3})), q_{4,4})$$
$$\Longrightarrow \times(\times(q_{1,1}, q_{2,3}), q_{4,4}) \Longrightarrow \times(q_{1,3}, q_{4,4}) \Longrightarrow q_{1,4} \in F$$

Definition 13.9 A *tree automaton with costs* is a pair $(M, \mu))$ where

1. $M = (Q, \Sigma, \delta, F)$ is a DTFA and
2. μ is a cost function

$$\mu\colon \bigcup_{j \geq 0} \mathrm{R}^j \times \Sigma_j \times Q^j \rightarrow \mathrm{R}$$
$$\mu(\xi_1, \ldots, \xi_j, a, q_1, \ldots, q_j) = \xi \in \mathrm{R}$$

That is, ξ is the cost of the transition $\delta(a, q_1, \ldots, q_j) = q$ when states $(q_1, \ldots, q_j)$ were reached at costs $(\xi_1, \ldots, \xi_j)$.

Observe that setting $j = 0$ in the former definition of μ, we have a vector $\vec{\xi}_0$ of initial values associated with the start rules $\delta(d)$ for symbols $d \in \Sigma_0$ [i.e., $\mu(d) = \xi_{0,d}$].

Example 13.3 For the MPP problem, μ can be defined as

$$\mu(0, q_{i,i}, i) = 0 \text{ for all } i \in \{1, \ldots n\}$$
$$\mu(\xi_1, \xi_2, q_{i,k}, q_{k+1,j}, \times) = \xi_1 + \xi_2 + c_{i-1} \times c_k \times c_j \in \mathrm{R}$$

[4]The notation $t \Longrightarrow t'$ for trees $t, t' \in T(\Sigma \cup Q)$, means that a subtree $a(q_1, \ldots, q_k) \in Q$ inside t has been substituted for by $q = \delta(a, q_1, \ldots, q_k)$. The notation $t \stackrel{*}{\Longrightarrow} t'$ is used to indicate that several one-step substitutions $t \Longrightarrow t_1 \Longrightarrow t_2 \Longrightarrow \cdots \Longrightarrow t'$ have been applied.

The cost function μ can be extended from the transitions Σ to a function $\hat{\mu}$ operating on the policies $T(\Sigma)$:

$$\hat{\mu}(a(t_1, \ldots, t_j)) = \mu(\hat{\mu}(t_1), \ldots, \hat{\mu}(t_j), a, \delta(t_1), \ldots, \delta(t_j))$$
$$\hat{\mu}(d) = \mu(d) \text{ for all } d \in \Sigma_0$$

As usual, we will not differentiate between μ and $\hat{\mu}$.

Definition 13.10 The *tree automaton with costs* $\Pi = (M, \mu)$ represents the optimization problem (f, S):

$$\min \ f(x) \text{ subject to } x \in S$$

if and only if it satisfies

$$S = L(M) \text{ and } f(t) = \mu(t) \text{ for all } t \in S$$

Definition 13.11 The optimal value $G(q)$ for a state q is defined as the optimal value of any tree policy $t \in T(\Sigma)$ such that $\delta(t) = q$:

$$G(q) = \min\{\mu(t)/\exists t \in T(\Sigma) \text{ such that } \delta(t) = q\} \tag{13.7}$$

With this definition it is now possible to generalize the dynamic programming equations established in Equation 13.4:

$$G(q) = \min\{\mu(G(q_1), \ldots, G(q_j), a, q_1, \ldots, q_j)/\delta(a, q_1, \ldots, q_j) = q\} \tag{13.8}$$

When we instantiate the former equation for the MPP problem we get Equation 13.5. It is now possible to revisit Bellman's optimality principle, stating that *any subtree of an optimal tree policy is optimal*.

Let us illustrate the set of concepts introduced up to here with a second example: the unrestricted two-dimensional cutting stock problem:

Definition 13.12 The *unrestricted two dimensional cutting stock problem* (U2DCSP) targets the cutting of a large rectangle S of dimensions $L \times W$ in a set of smaller rectangles using orthogonal guillotine cuts: Any cut must run from one side of the rectangle to the other end and be parallel to one of the other two edges. The rectangles produced must belong to one of a given set of rectangle types $\mathcal{D} = \{T_1, \ldots, T_n\}$, where the ith type T_i has dimensions $l_i \times w_i$. Associated with each type T_i there is a profit c_i. No constraints are set on the number of copies of T_i used. The goal is to find a feasible cutting pattern with x_i pieces of type T_i maximizing the total profit:

$$g(x) = \sum_{i=1}^{n} c_i x_i \qquad \text{where } x_i \in N$$

Observe that solutions to the problem can be specified by binary trees. Each node of the tree says whether the cut was vertical (i.e., horizontal composition of the solutions in the two subtrees that we denote by $-$) or horizontal (i.e., vertical composition of the two existent solutions that will be denoted by $|$):

$$\Sigma = \{T_1, \ldots, T_n\} \cup \{-, |\} \text{ with } \rho(T_i) = 1 \ \forall\, i \in \{1, \ldots, n\} \text{ and } \rho(|) = \rho(-) = 2$$

Figure 13.2 presents the solutions $\alpha = -(T_1, T_2)$, $\beta = |(t_1, T_2)$, $\gamma = -(T_2, T_3)$, $\delta = |(T_2, T_3)$, $-(\alpha, \beta) = -(-(T_1, T_2), |(t_1, T_2))$, $|(\alpha, \beta) = |(-(T_1, T_2), |(t_1, T_2))$, and so on, for some initials rectangles T_1, T_2, and T_3. The labels use postfix notation. Postfix notation provides a more compact representation of the tree. Instead of writing $-(\alpha, \beta)$, we write the operand and then the operator: $\alpha, \beta-$. The tree-feasible solutions will be also called *metarectangles*.

The set of states is $Q = \{q_{x,y}/0 \leq x \leq L \text{ and } 0 \leq y \leq W\}$. The idea is that the state $q_{x,y}$ represents a metarectangle of length x and width y. The transition function δ is given by

$$\begin{aligned} \delta(-, q_{x,y}, q_{z,y}) &= q_{x+z,y} \text{ if } x + z \leq L \\ \delta(|, q_{x,u}, q_{x,v}) &= q_{x,u+v} \text{ if } u + v \leq W \\ \delta(T_i) &= q_{l_i,w_i} \text{ for all } i \in \{1, \ldots, n\} \end{aligned}$$

The accepting state is $F = \{q_{L,W}\}$. Finally, the cost function is given by

$$\begin{aligned} \mu(\xi_0, \xi_1, -, q_{x,y}, q_{z,y}) &= \mu(\xi_0, \xi_1, -, q_{x,u}, q_{x,v}) = \xi_0 + \xi_1 \\ \mu(T_i) &= p_i \end{aligned}$$

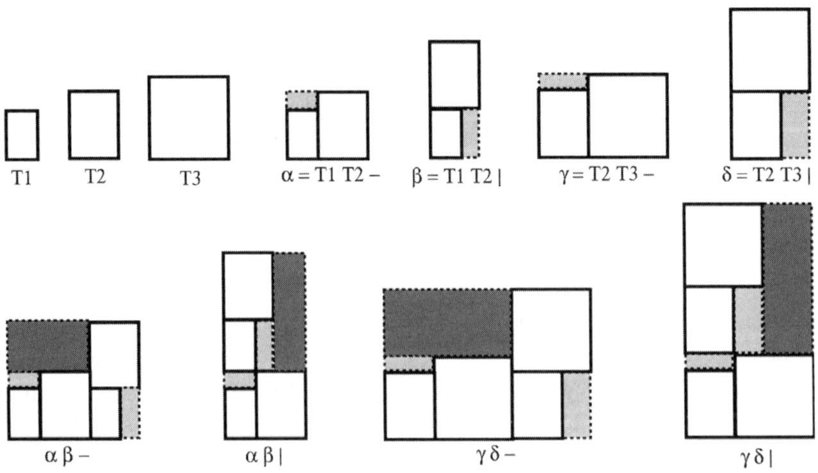

Figure 13.2 Examples of vertical and horizontal metarectangles. Instead of terms, postfix notation is used: $\alpha\ \beta\ |$ and $\alpha\ \beta\ -$ denote the vertical and horizontal constructions of rectangles α and β. Shaded areas represent waste.

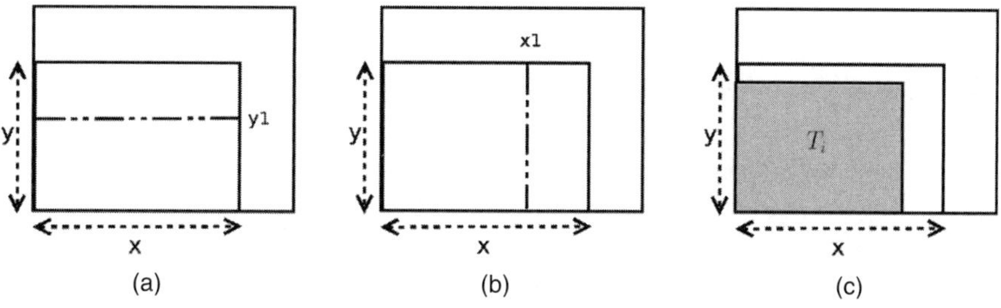

Figure 13.3 Calculation of $G(q_{x,y})$: (a) $S_{x,y,|}$; (b) $S_{x,y,-i}$; (c) $S_{x,y,i}$.

When we instantiate Equation 13.8 for this DTFA with costs, we get

$$G(q_{x,y}) = \max S_{x,y,-} \cup S_{x,y,|} \cup \bigcup_{i=1,n} S_{x,y,i} \tag{13.9}$$

where

$$S_{x,y,|} = \{G(q_{x,v}) + G(q_{x,y-v})/0 \leq v \leq y\}$$
$$S_{x,y,-} = \{G(q_{u,y}) + G(q_{x-u,y})/0 \leq u \leq x\}$$
$$S_{x,y,i} = p_i \text{ if } l_i \leq x \text{ and } w_i \leq y \text{ 0 otherwise}$$

Figure 13.3 illustrates the meaning of the three parts of the equation.

13.5 PARALLEL ALGORITHMS

We have again and again successfully exploited the same strategy to get an efficient parallelization of dynamic programming algorithms [6–14]. The DP DAG is partitioned in stages P_k, with $k \in \{1, \ldots, n\}$, that preserve the complexity order of the DP DAG (Algorithm 13.1). That is: problems/states in stage P_k depend only on subproblems/states on previous stages:

$$\forall\, q \in P_k \Longrightarrow \delta^{-1}(q) \subset \bigcup_{j \leq k} P_j$$

In not a few cases these stages P_k correspond to the levels of the divide-and-conquer search tree shrunk via the equivalence relation defined by Equation 13.2.

On a first high-level description the parallel algorithm uses n processors structured according to a pipeline topology. The processor k executes the algorithm (or an optimized version of it, if it applies) in Figure 13.1. The monadic formalism is used to keep the presentation simple.

Since the number of stages n is usually larger than the number of available processors p, a second, lower-level phase mapping the n processes onto the p

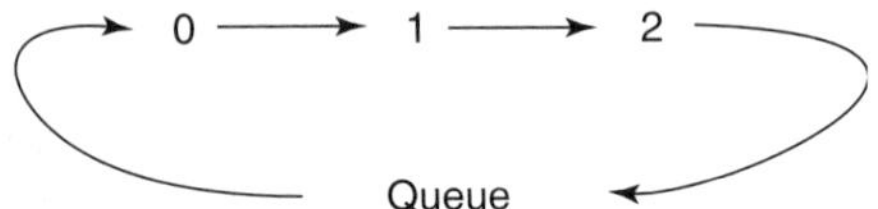

Figure 13.4 Ring topology ($p = 3$). The queue process decouples the last and first processors.

TABLE 13.1 Mapping 12 Processes onto a Three-Processor Ring: $k = 2$

Processor	Process
0	(P_0, P_1)
1	(P_2, P_3)
2	(P_4, P_5)
0	(P_6, P_7)
1	(P_8, P_9)
2	(P_{10}, P_{11})

Algorithm 13.1 Pipeline Algorithm for DP: Stage k

```
while (running) do
  receive from left (q', G(q'))
  send to right (q', G(q')) if ∃j/δ(q, Σ) ∩ P_j ≠ ∅
  for all q ∈ δ(q') ∩ P_k do
     update combination for (q, G(q))
     if received_all(δ^{-1}(q)) then
         send (q, G(q)) to the right neighbor
         update combinations for q'' ∈ P_k ∩ δ(q)
     end if
  end for
end while
```

processors is required. On homogeneous networks the strategy consists of structuring the p processors on a ring and using a block-cyclic mapping (with block size b). To avoid potential deadlocks on synchronous buffer-limited systems, the last processor is usually connected to the first via a queue process (Figure 13.4). Table 13.1 illustrates the mapping of 12 processes onto a three-processor ring ($b = 2$). There are two *bands*. During the first band, processor 0 computes the DP stages P_0 and P_1. The results of this band produced by processor 2 (i.e., P_5) are enqueued for its use by processor 0 during the process of the second band. During such a second band, processor 0 computes P_6 and P_7.

The approach presented leads to efficient parallel programs even if the value of b is not tuned. For example, to cite a typical gain for medium and large grain problems, speedups of 8.29 for 16 processors have been reported [5] for the MPP problem on a Cray T3E for an instance of $n = 1500$ matrices and $b = 1$.

Finding the optimal value of b for specific DP problems has been the object of study of a number of papers [811–14]. In some cases it is possible to find an analytical formula for the complexity of the parallel algorithm f(input sizes, n, b, p) and from there derive a formula for b that leads to the minimum time [11,12]. The problem to find the optimal mapping gets more complex when the network is heterogeneous: In such a case, different block sizes b_i for $i = 0, \ldots, p$ are considered [13,14].

13.6 DYNAMIC PROGRAMMING HEURISTICS

Dynamic programming has been extensively used in the land of exact algorithms. Although the search tree is reduced through the equivalence relation formulated in Equation 13.2, the graph is still traversed exhaustively, which usually takes a long time. Even worse, the states are usually elements of a multidimensional space. In the knapsack example, such a space is $[0, N-1] \times [0, C]$, in the MPP, is $[1, N] \times [1, N]$, and in the *unrestricted two-dimensional cutting stock problem*, is $[0, L] \times [0, W]$. More often than not, such a space is prohibitively large, including many dimensions. This is known as the "curse of dimensionality."[5] In the example we consider in this section, the constrained two-dimensional cutting stock problem (C2DCSP or simply 2DCSP), such a space is $[0, L] \times [0, W] \times [0, b_1] \times [0, b_2] \times \cdots \times [0, b_N]$, where b_i and N are integers and L and W are the length and width of a rectangular piece of material.

It is in general impossible to keep such huge DP tables in memory for reasonable values of the input parameters. It is, however, feasible, by declining the requirement of exactness, to project the large dimension state space $Q = Q_1 \times \cdots \times Q_n$ onto a smaller one, $Q' = Q_1 \times \cdots \times Q_k$ (with k much smaller than n), producing a projected acyclic graph that can be solved using DP. The solution obtained is an upper bound[6] of the optimal solution. In a variant of this strategy, during the combination of the projected DP a heuristic is called that extends the projected solution to a solution for the original DP. In this way we have a heuristic lower bound of the original problem. The two techniques are shown in the following two sections.

13.6.1 Constrained Two-Dimensional Cutting Stock Problem

The constrained two-dimensional cutting stock problem (2DCSP) is an extension of the unconstrained version presented in earlier sections. It targets the cutting

[5] A term coined by Richard Bellman, founder of the dynamic programming technique.

[6] Assuming that the problem is a maximization problem.

of a large rectangle S of dimensions $L \times W$ in a set of smaller rectangles using orthogonal guillotine cuts: Any cut must run from one side of the rectangle to the other end and be parallel to one of the other two edges. The rectangles produced must belong to one of a given set of rectangle types $\mathcal{D} = \{T_1, \ldots, T_n\}$, where the ith type T_i has dimensions $l_i \times w_i$. Associated with each type T_i there is a profit c_i and a demand constraint b_i. The goal is to find a feasible cutting pattern with x_i pieces of type T_i maximizing the total profit:

$$g(x) = \sum_{i=1}^{n} c_i x_i \text{ subject to } x_i \leq b_i \text{ and } x_i \in N$$

Wang [15] was the first to make the observation that all guillotine cutting patterns can be obtained by means of horizontal and vertical builds of metarectangles, as illustrated in Figure 13.2. Her idea has been exploited by several authors [16–19, 21,22] to design best-first search A˙ algorithms (see Chapter 12). The algorithm uses the Gilmore and Gomory DP solution presented earlier to build an upper bound.

13.6.2 Dynamic Programming Heuristic for the 2DCSP

The heuristic mimics the Gilmore and Gomory dynamic programming algorithm [23] but for substituting unbounded ones. feasible suboptimal vertical and horizontal combinations. Let $R = (r_i)_{i=1\ldots n}$ and $S = (s_i)_{i=1\ldots n}$ be sets of feasible solutions using $r_i \leq b_i$ and $s_i \leq b_i$ rectangles of type T_i. The cross product $R \otimes S$ of R and S is defined as the set of feasible solutions built from R and S without violating the bounding requirements [i.e., $R \otimes S$ uses $(\min\{r_i + s_i, b_i\})_{i=1\ldots n}$ rectangles of type T_i]. The lower bound is given by the value $H(L, W)$, computed as:

$$H(x, y) = \max \begin{cases} \max\left\{g(S(x, y_1) \otimes S(x, y - y_1)) & \text{such that } 0 < y_1 \leq \left\lfloor \dfrac{y}{2} \right\rfloor\right\} \\ \max\left\{g(S(x_1, y) \otimes S(x - x_1, y)) & \text{such that } 0 < x_1 \leq \left\lfloor \dfrac{x}{2} \right\rfloor\right\} \\ \max\{c_i & \text{such that } l_i \leq x \text{ and } w_i \leq y\} \end{cases}$$

being $S(x, y) = S(\alpha, \beta) \otimes S(\gamma, \delta)$ one of the cross sets [either a vertical construction $S(x, y_0) \otimes S(x, y - y_0)$ or a horizontal building $S(x_0, y) \otimes S(x - x_0, y)$] where the former maximum is achieved [i.e., $H(x, y) = g(S(\alpha, \beta) \otimes S(\gamma, \delta))$].

13.6.3 New DP Upper Bound for the 2DCSP

The upper bound improves existing upper bounds. It is trivial to prove that it is lower than the upper bounds proposed in refs. 16–19 and 21. The calculus of the new upper bound is made in three steps:

1. During the first step, the following bounded knapsack problem is solved using dynamic programming [18,19]:

$$V(\alpha) = \begin{cases} \max \sum_{i=1}^{n} c_i x_i \\ \text{subject to} \qquad \sum_{i=1}^{n} (l_i w_i) x_i \leq \alpha \\ \text{and} \qquad x_i \leq \min \left\{ b_i, \left\lfloor \frac{L}{l_i} \right\rfloor \times \left\lfloor \frac{W}{w_i} \right\rfloor \right\}, x_i \in \mathrm{N} \end{cases}$$

for all $0 \leq \alpha \leq L \times W$.

2. Then, $F_V(x, y)$ is computed for each rectangle using the equations

$$\overline{F}(x, y) = \max \begin{cases} \max \left\{ F_V(x, y_1) + F_V(x, y - y_1) \quad \text{such that } 0 < y_1 \leq \left\lfloor \frac{y}{2} \right\rfloor \right\} \\ \max \left\{ F_V(x_1, y) + F_V(x - x_1, y) \quad \text{such that } 0 < x_1 \leq \left\lfloor \frac{x}{2} \right\rfloor \right\} \\ \max \{ c_i \quad \text{such that } l_i \leq x \text{ and } w_i \leq y \} \end{cases}$$

where

$$F_V(x, y) = \min\{\overline{F}(x, y), V(x \times y)\}$$

3. Finally, substituting the bound of Gilmore and Gomory [23] by F_V in Viswanathan and Bagchi's upper bound [16], the proposed upper bound is obtained:

$$U_V(x, y) = \max \begin{cases} \max\{U_V(x + u, y) + F_V(u, y) \quad \text{such that } 0 < u \leq L - x\} \\ \max\{U_V(x, y + v) + F_V(x, v) \quad \text{such that } 0 < v \leq W - y\} \end{cases}$$

13.6.4 Experimental Analysis

This section is devoted to a set of experiments that prove the quality of the two DP heuristics presented earlier. The two heuristics are embedded inside the sequential and parallel algorithms presented in Chapter 12. The instances used in ref. 16–19 and 21 are solved by the sequential algorithm in a negligible time. For that reason, the computational study presented has been performed on some selected instances from the ones available in ref. 24. Tests have been run on a cluster of eight HP nodes, each consisting of two Intel Xeons at 3.20 GHz. The compiler used was gcc 3.3.

TABLE 13.2 Lower- and Upper-Bound Results

		Lower Bound		Upper Bound					
				V			U_V		
Problem	Solution Value	Value	Time(s)	Init.	Search	Nodes	Init.	Search	Nodes
25_03	21,693	21,662	0.44	0.031	2,835.1	179,360	0.031	2,308.8	157,277
25_05	21,693	21,662	0.44	0.031	2,892.2	183,890	0.030	2,304.8	160,932
25_06	21,915	21,915	0.45	0.032	35.6	13,713	0.033	20.8	10,310
25_08	21,915	21,915	0.45	0.032	205.6	33,727	0.028	129.0	25,764
25_09	21,672	21,548	0.50	0.031	37.3	17,074	0.030	25.5	13,882
25_10	21,915	21,915	0.51	0.032	1,353.9	86,920	0.033	1,107.2	73,039
50_01	22,154	22,092	0.73	0.106	2,132.2	126,854	0.045	1,551.2	102,662
50_03	22,102	22,089	0.79	0.043	4,583.4	189,277	0.045	3,046.6	148,964
50_05	22,102	22,089	0.78	0.045	4,637.7	189,920	0.045	3,027.8	149,449
50_09	22,088	22,088	0.80	0.046	234.4	38,777	0.043	155.4	29,124
100_08	22,443	22,443	1.22	0.077	110.2	25,691	0.076	92.9	22,644
100_09	22,397	22,377	1.28	0.076	75.6	20,086	0.076	61.8	17,708

Table 13.2 presents the results for the sequential runs. The first column shows the exact solution value for each problem instance. The next two columns show the solution value given by the initial lower bound and the time invested in its calculation (all times are in seconds). Note that the designed lower bound highly approximates the final exact solution value. In fact, the exact solution is reached directly in many cases. The last column compares two different upper bounds: the one proposed in ref. 17 and the new upper bound. For each upper bound, the time needed for its initialization, the time invested in finding the exact solution and the number of nodes computed, are presented. Nodes computed are nodes that have been transferred from `OPEN` to `CLIST` and combined with all previous `CLIST` elements. The new upper bound highly improves the previous bound: the number of nodes computed decreases, yielding a decrease in the execution time.

13.7 CONCLUSIONS

A dynamic programming algorithm is obtained simply by adding memoization to a divide-and-conquer algorithm. Automatas and tree automatas with costs provide the tools to model this technique mathematically, allowing the statement of formal properties and general sequential and parallel formulations of the strategy.

The most efficient parallelizations of the technique use a virtual pipeline that is later mapped onto a ring. Optimal mappings have been found for a variety of DP algorithms. Although dynamic programming is classified as an *exact technique*, it can be the source of inspiration for heuristics: both lower and upper bounds of the original problem.

Acknowledgments

Our thanks to Francisco Almeida, Daniel González, and Luz Marina Moreno; it has been a pleasure to work with them. Their ideas on parallel dynamic programming have been influential in the writing of this chapter.

This work has been supported by the EC (FEDER) and by the Spanish Ministry of Education and Science under the Plan Nacional de I + D + i, contract TIN2005-08818-C04-04. The work of G. Miranda has been developed under grant FPU-AP2004-2290.

REFERENCES

1. L. Wall, T. Christiansen, and R. L. Schwartz. *Programming Perl*. O'Reilly, Sebastopol, CA, 1996.
2. M. J. Dominus. *Higher-Order Perl*. Morgan Kaufmann, San Francisco, CA, 2005.
3. C. Rodriguez Leon. Code examples for the chapter "Tools for Tree Searches: Dynamic Programming." http://nereida.deioc.ull.es/Algorithm-DynamicProgramming-1.02.tar.gz.
4. J. Earley. An efficient context-free parsing algorithm. *Communications of the Association for Computing Machinery*, 13(2):94–102, 1970.
5. D. Gonzalez Morales, et al. Parallel dynamic programming and automata theory. *Parallel Computing*, 26:113–134, 2000.
6. D. Gonzalez Morales, et al. Integral knapsack problems: parallel algorithms and their implementations on distributed systems. In *Proceedings of the International Conference on Supercomputing*, 1995, pp. 218–226.
7. D. Gonzalez Morales, et al. A skeleton for parallel dynamic programming. In *Proceedings of Euro-Par*, vol. 1685 of *Lecture Notes in Computer Science*. Springer-Verlag, New York, 1999, pp. 877–887.
8. D. González, F. Almeida, L. M. Moreno, and C. Rodríguez. Pipeline algorithms on MPI: optimal mapping of the path planing problem. In *Proceedings of PVM/MPI*, vol. 1900 of *Lecture Notes in Computer Science*. Springer-Verlag, New York, 2000, pp. 104–112.
9. L. M. Moreno, F. Almeida, D. González, and C. Rodríguez. The tuning problem on pipelines. In *Proceedings of Euro-Par*, vol. 1685 of *Lecture Notes in Computer Science*. Springer-Verlag, New York, 2001, pp. 117–121.
10. L. M. Moreno, F. Almeida, D. González, and C. Rodríguez. Adaptive execution of pipelines. In *Proceedings of PVM/MPI*, vol. 2131 of *Lecture Notes in Computer Science*. Springer-Verlag, New York, 2001, pp. 217–224.
11. F. Almeida, et al. Optimal tiling for the RNA base pairing problem. In *Proceedings of SPAA*, 2002, pp. 173–182.
12. D. González, F. Almeida, L. M. Moreno, and C. Rodríguez. Towards the automatic optimal mapping of pipeline algorithms. *Parallel Computing*, 29(2):241–254, 2003.
13. F. Almeida, D. González, L. M. Moreno, and C. Rodríguez. Pipelines on heterogeneous systems: models and tools. *Concurrency—Practice and Experience*, 17(9):1173–1195, 2005.

14. F. Almeida, D. González, and L. M. Moreno. The master–slave paradigm on heterogeneous systems: a dynamic programming approach for the optimal mapping. *Journal of Systems Architecture*, 52(2):105–116, 2006.
15. P. Y. Wang. Two algorithms for constrained two-dimensional cutting stock problems. *Operations Research*, 31(3):573–586, 1983.
16. K. V. Viswanathan, and A. Bagchi. Best-first search methods for constrained two-dimensional cutting stock problems. *Operations Research*, 41(4):768–776, 1993.
17. M. Hifi. An improvement of Viswanathan and Bagchi's exact algorithm for constrained two-dimensional cutting stock. *Computer Operations Research*, 24(8): 727–736, 1997.
18. S. Tschoke, and N. Holthöfer. A new parallel approach to the constrained two-dimensional cutting stock problem. In A. Ferreira and J. Rolim, eds., *Parallel Algorithms for Irregularly Structured Problems*. Springer-Verlag, New York, 1995. http://www.uni-paderborn.de/fachbereich/AG/monien/index.html.
19. V. D. Cung, M. Hifi, and B. Le-Cun. Constrained two-dimensional cutting stock problems: a best-first branch-and-bound algorithm. *Technical Report 97/020*. Laboratoire PRiSM–CNRS URA 1525, Université de Versailles, France, Nov. 1997.
20. G. Miranda and C. León. OpenMP parallelizations of Viswanathan and Bagchi's algorithm for the two-dimensional cutting stock problem. *Parallel Computing 2005*, NIC Series, vol. 33, 2005, pp. 285–292.
21. L. García, C. León, G. Miranda, and C. Rodríguez. A parallel algorithm for the two-dimensional cutting stock problem. In *Proceedings of Euro-Par*, vol. 4128 of *Lecture Notes in Computer Science*. Springer-Verlag, New York, 2006, pp. 821–830.
22. L. García, C. León, G. Miranda, and C. Rodríguez. two-dimensional cutting stock problem: shared memory parallelizations. In *Proceedings of the 5th International Symposium on Parallel Computing in Electrical Engineering*. IEEE Computer Society Press, Los Alamitos, CA, 2006, pp. 438–443.
23. P. C. Gilmore and R. E. Gomory. The theory and computation of knapsack functions. *Operations Research*, 14:1045–1074, Mar. 1966.
24. DEIS Operations Research Group. Library of instances: two-constraint bin packing problem. http://www.or.deis.unibo.it/research_pages/ORinstances/2CBP.html.

PART II

APPLICATIONS

CHAPTER 14

Automatic Search of Behavior Strategies in Auctions

D. QUINTANA
Universidad Carlos III de Madrid, Spain

A. MOCHÓN
Universidad Nacional de Educación a Distancia, Spain

14.1 INTRODUCTION

Auctions have become a key mechanism in allocating items at different prices in many markets. Due to their special features and different formats, they are a useful tool for getting supply and demand to equilibrium. Some relevant markets where auctions are currently used are sales of personal communication services (PCSs), licenses, emission permits, electricity, and transportation services. It is very difficult to list the numerous papers on auctions but some examples are: Ref. 1 collects together in two volumes most of the relevant papers in the economics literature of auctions up to the year 2000; ref. 2 gives an account of developments in the field in the 40 years since Vickrey's pioneering paper; ref. 3 surveys the basic theory of how auctions work and describes the world-record-setting 3G mobile-phone license auctions; and ref. 4 gives an up-to-date treatment of both traditional theories of optimal auctions and newer theories of multiunit auctions and package auctions.

The study of auctions has traditionally been carried out by analytical means. However, this approach usually requires simplifying assumptions to reduce the complexity involved. This limitation is being tackled with the introduction of agent-based computational economics. This approach is based on simulations of auctions by intelligent artificial agents. The behavior of these agents is often adaptive and relies on different decision algorithms derived from the field of artificial intelligence, such as genetic algorithms (GAs), artificial neural networks, or evolutionary strategies.

Optimization Techniques for Solving Complex Problems, Edited by Enrique Alba, Christian Blum, Pedro Isasi, Coromoto León, and Juan Antonio Gómez

This chapter deals with the search for bidding strategies in a specific multiunit auction: the Ausubel auction. For this purpose, we develop a GA that tries to maximize the bidders' payoff. GAs are a robust method of adaptive searching, learning, and optimization in complex problem domains [5]. We start validating the model by testing it under circumstances for which there is an analytical solution, and then we focus our attention on the outcomes obtained when we face a new setting: the presence of synergies. The remainder of the chapter is structured as follows. A revision of previous work done in auctions with computational techniques is given in Section 14.2. In Section 14.3 we present the theoretical framework of the Ausubel auction. The description of the GA and the bidding functions tested are described in Section 14.4. In Section 14.5 we evaluate the experimental results. Finally, we summarize the main conclusions and outline potential futures work.

14.2 EVOLUTIONARY TECHNIQUES IN AUCTIONS

The use of computational simulation to analyze auctions is not new. There are several researchers who rely on computational experiments to study the complex and stochastic processes inherent in auction learning models [6]. The most popular approach examines the interaction of artificial adaptive agents (AAAs), developing their behavior with adaptive learning algorithms such as GAs, genetic programming (GP), or evolutionary strategies (ES).

A contentious issue in the design of the experimental framework concerns encoding the possible strategies that bidders can follow. Table 14.1 presents some of the most relevant pieces of work on the subject. Some authors use agents that are already implemented, such as agents with zero intelligent plus learning heuristics (ZIP). Cliff and Preist [7–9] tested this technique for the continuous double auction. Moreover, Cliff [10] proposed an automated online auction that allows a wide range of possible market mechanisms. Cliff models the market by means of a set of ZIP traders. These markets were allowed to evolve using a GA, in order to minimize the deviation of transaction prices from the theoretical equilibrium prices. Other authors prefer to develop agents with their own particular bidding protocol. Andreoni and Miller [6] developed a GA to capture the bidding patterns for first- and second-price auction formats with affiliated private values, independent private values, and common values. The algorithm searched over two-parameter linear functions in each auction. Dawid [11] modeled a two-population GA to study the learning behavior of buyers and sellers in a sealed-bid double auction with private information. Each individual in the population was encoded with a string, and a fitness value is calculated measuring the success of the individual by the average payoff earned in the last period (a group of m iterations).

Byde [12] also explored the possibility of using GAs in a sealed-bid auction where payments were determined by linear combinations of first- and

TABLE 14.1 Summary of Work with Auctions and EC

Title	Ref.	Year	Auction Format	Experimental Approach	Populations	Size	Generations
Auctions with artificial adaptive agents	[6]	1995	First-price, second-price	Agents with GA	1	40	1,000
A learning approach to auctions	[19]	1998	First-price	Game theory techniques			
Adaptive agents in persistent shout double auction	[8]	1998	Continuous double auction	Agents with zero intelligent plus (ZIP) learning heuristics			
Evolving parameter sets for adaptive trading agents in continuous double-auction markets	[7]	1998	Continuous double auction	Agents (ZIP) improved with GA	1	30	200
On the convergence of genetic learning in a double-auction market	[11]	1999	Sealed-bid double auction	Agents with GA	2	100	2,500
An adaptive agent bidding strategy based on stochastic modeling	[20]	1999	Continuous double auction	Agents with stochastic modeling and reinforcement learning			
Commodity trading using an agent-based lterated double auction	[9]	1999	Continuous double auction	Agents with zero intelligent plus (ZIP) learning heuristics			
Algorithms design for agents that participate in multiple simultaneous auctions	[21]	2000	Simultaneous English auctions	Agents with formal negotiation model			

(*continued overleaf*)

TABLE 14.1 ***(Continued)***

Title	Ref.	Year	Auction Format	Experimental Approach	Populations	Size	Generations
Experimental analysis of the efficiency of uniform-price versus discriminatory auctions in the England and Wales electricity market.	[22]	2001	Uniform and discriminatory price (multiunit)	Agents with reinforcement learning	1	20	
An optimal dynamic programming model for algorithm design in simultaneous auctions	[23]	2001	Simultaneous English auctions	Agents with dynamic programming, and nondynamic programming			
Evolving marketplace designs by artificial agents	[24]	2002	English and Dutch	Agents with zero intelligent plus (ZIP) learning heuristics and GA for parameter optimization			
Decision procedures for multiple auctions	[25]	2002	English, Dutch, first-price and second-price	Agents with GA	1	50	2,000
Coevolution of auction mechanisms and trading strategies: towards a novel approach to microeconomic design	[26]	2002	Continuous double auction	Agents with GP	1	60	10,000

A peer-to-peer agent auction	[27]	2002	Continuous double auction	Agents P2P	1	2.500–160.000	100
Developing a bidding agent for multiple heterogeneous auctions	[14]	2003	English, Dutch, and Vickrey	Agents with GA	1	50	2,000
Explorations in evolutionary design of online auction market mechanisms	[10]	2003	Continuous double auction	GA	1	30	500–1000
Exploring bidding strategies for market-based scheduling	[28]	2003	Simultaneous ascending auctions	Agents with different techniques			
Applying evolutionary game theory to auction mechanism design	[12]	2003	First-price, second-price	Agents with GA	1	6	500
Market-based recommendation: agents that compete for consumer attention	[29]	2004	Extended Vickrey	Agents with neural networks and ES	1	25	
Analysis of Ausubel auctions by means of evolutionary computation	[30]	2005	Ausubel	GA	1	30	1,000

second-price auctions. The strategy followed by an agent was defined by a piecewise linear function from input signal to bid. All agents followed optimal strategies that were determined by a GA. Anthony and Jennings [13,14] suggested an agent-based approach to bidding strategy evolution for specific environments using GAs. These authors tested the English, Dutch, and Vickrey auction. The behavior of the agents was subjected to a set of polynomial functions that represented bidding constraints. These functions were denoted as tactics and their combination was referred to as a *strategy*. The strategy requires the allocation by the agent of the appropriate weight to each tactic. These weights plus the set of parameters required to define the constraint functions mentioned conforms the array of floating-point values that define the individuals in the population to be evolved. Finally, Saez et al. [15] explored the bidding behavior and the effects on revenue and efficiency when one bidder evolves according to a GA. The auction format selected was the auction developed by Ausubel [16,17]. Saez et al. tested the algorithm for several rationing rules, and the results suggested that the selection of a rationing rule can be a key point in the final outcome of this auction mechanism.

14.3 THEORETICAL FRAMEWORK: THE AUSUBEL AUCTION

The research presented in this chapter is focused on the Ausubel auction (see [16,17]), which is defined as follows:

> In this auction format, the price increases continuously and for each price, p^l, each bidder i simultaneously indicates the quantity $q_i(p^l)$ he desires (demands are nonincremental in price). Bidders choose at what price to drop out of the bidding, with dropping out being irrevocable. When the price p^* is reached, such that aggregate demand no longer exceeds supply, the auction is over. When the auction is over each bidder i is assigned the quantity $q_i(p^*)$ and is charged the standing prices at which the respective objects were "clinched": With m object for sale at a price p^l, bidder i clinches an object when the aggregate demand of all other bidders drops, at least, from m to $m - 1$ but bidder i still demands two units or more, that is, when the demand of all bidders except him is smaller than the total supply but the total demand exceeds the total supply. In this situation bidder i is guaranteed at least one object no matter how the auction proceeds.

The analysis of auctions using adaptive agents can be performed using different approaches. The one that we are going to choose is focused on bidders and their behavior. We simulate Ausubel auctions where individual agents try to identify, in an autonomous way, strategies that result in higher profits for them. Therefore, we are facing an optimization problem. Given a set of agents (bidders) with specific personal values, each tries to find the bidding strategy that maximizes the payoff. The payoff is the difference between the value derived from the lots that they win and the price they paid for them.

Once we have set this starting point, we make a methodological proposal that will be built around the idea that agents, chasing individual targets, might obtain extraordinary profits in a dynamic environment with certain strategies. To understand the different strategies a bidder can have in an auction, it is useful to comprehend first the concept of personal value and marginal value. The *personal value* is the personal or estimated worth for a commodity that a person has. The *marginal value* is the increase in total personal value associated with an increment of a unit in consumption. Bidders can exhibit:

- *Diminishing marginal values* (substitutes in consumptions). This means that when a bidder increases his consumption by 1 unit, his total personal value increases but at a decreasing rate this is, he has diminishing marginal values. For example, consuming 1 unit can yield a personal value of 10 ($v_{i,1} = 10$), and consuming 2 units can yield a total personal value of 15, so the second unit has a marginal value of 5 ($v_{i,2} = 5$).
- *Synergies* (complementarities in consumptions). We are in the presence of synergies when the value of consuming multiple objects exceeds the sum of the objects' values separately. For example, consuming 1 unit can yield a personal value of 10, and consuming 2 units together can yield a total personal value of 25.

According to their values, bidders can follow three main strategies: underbid, overbid, and bid sincerely. Underbidding means to submit bids for lower prices than the personal values. Overbids are bids higher than the personal values. Finally, if bidders bid sincerely, they bid according to their true personal value.

To define the auction outcome, it is necessary first to make a basic specification of the model from the formulation in [16]. In each auction the seller offers M number of indivisible units of a homogeneous good to n number of bidders. Each bidder i obtains a marginal value of $v_{i,k}$ for the kth unit of the good, for $k = 1, \ldots, M$. Thus, if bidder i gets q_i^* units for a total payment of P_i^*, he obtains a payoff of

$$\sum_{k=1}^{q_i^*} V_{i,k} - P_i^* \text{ for } i = 1, \ldots, n \text{ and } q_i = 1, \ldots, M \tag{14.1}$$

For any round l, the aggregate demand by all bidders is $Q^l = \sum_{i=1}^{N} q_i^l$. The cumulative vector of quantities clinched C_i^l by bidder i at prices up to p^l is defined by

$$C_i^l = \max\left\{0, M - \sum_{j \neq i} q_j^l\right\} \text{ for } l = 1, \ldots, L \tag{14.2}$$

Given the individual quantities clinched at price p^l, set $c_i^0 = C_i^0$, defined as

$$c_i^l = C_i^l - C_i^{l-1} \tag{14.3}$$

The auction outcome associated with any final history $l = L$ is defined by

$$\text{Allocation:} \quad q_i^* = C_i^l \text{ for } l = 1, \ldots, L \tag{14.4}$$

$$\text{Payment:} \quad P_i^* = \sum_{l=0}^{L} p^l c_i^l \text{ for } l = 1, \ldots, L \tag{14.5}$$

In the same way, the seller's revenue per auction is defined as the total payments by all bidders:

$$R^* = \sum_{i=1}^{N} P_i^* \tag{14.6}$$

Table 14.2 includes an example of the Ausubel auction where participants bid sincerely and exhibit diminishing marginal values. The outcome of this example is reported in Table 14.3.

Ausubel demonstrated that in this auction format, when bidders have weakly diminishing marginal values (substitutes in consumptions), sincere bidding (SB) by every bidder is an *ex post* perfect equilibrium yielding to an efficient outcome.

TABLE 14.2 Auction Process in Which All Bidders Bid Sincerely and Exhibit Diminishing Marginal Values[a]

$v_{1,k}$: 124, 93, 47, 24, 13, 9, 5, 3, 2, 1, 1, 1, 1, 1, 1
$v_{2,k}$: 100, 72, 60, 44, 22, 18, 11, 6, 3, 2, 2, 2, 2, 1, 1
$v_{3,k}$: 198, 99, 73, 46, 23, 12, 7, 4, 3, 2, 1, 1, 1, 1, 1
$v_{4,k}$: 100, 58, 41, 39, 24, 16, 8, 7, 4, 2, 1, 1, 1, 1, 1

P^l	Q^l	q_1^l	C_1^l	q_2^l	C_2^l	q_3^l	C_3^l	q_4^l	C_4^l
10	24	5	0	7	0	6	0	6	0
11	23	5	0	6	0	6	0	6	0
12	22	5	0	6	0	5	0	6	0
13	21	4	0	6	0	5	0	6	0
16	20	4	0	6	1	5	0	5	0
18	19	4	0	5	1	5	1	5	1
22	18	4	1	4	1	5	2	5	2
23	17	4	2	4	2	4	2	5	3
24	15	3	3	4	4	4	4	4	4

[a]Four bidders and 15 lots; seller's revenue = 331.

TABLE 14.3 Auction Outcome When All Bidders Bid Sincerely

Bidder	Price Paid for the Units Clinched				Final Allocation	Final Payment	Payoff
	1st Unit	2nd Unit	3rd Unit	4th Unit			
1	22	23	24	–	$q_1^* = 3$	$P_1^* = 69$	195
2	16	23	24	24	$q_2^* = 4$	$P_2^* = 87$	189
3	18	22	24	24	$q_3^* = 4$	$P_3^* = 88$	328
4	18	22	23	24	$q_4^* = 4$	$P_4^* = 87$	151

Nevertheless, there are several markets where synergies or complementarities (when the value of multiple objects exceeds the sum of the objects' values separately) can be found. Probably, the best known are the personal communication services license sales, done in most countries (United States, UK, Germany, etc.). In this sector the agents might be interested in acquiring two or more geographically neighboring licenses (local synergies), or multiple licenses, in order to get economies of scale (global synergies) [18]. In this circumstance (i.e., in the presence of synergies), the theoretical equilibrium for the Ausubel auction is yet to be found.

As we mentioned earlier, in the experiments done, we consider two possible scenarios: The first one is a basic framework with decreasing marginal value (substitutes in consumption). The second deals with the presence of synergies (complementarities in consumption). Both can be studied using the codification of strategies presented in the following section.

14.4 ALGORITHMIC PROPOSAL

We present an artificial market where agents operate following bidding strategies that evolve with time. The algorithm that drives this evolution is a GA. This algorithm assesses the fitness of each strategy through the profit that the bidder obtains in a simulated auction. In every case, strategies are fixed for each auction but vary within successive auctions.

The experimental setting consists of 15 lots and four bidders that are risk neutral and have no limits on the amount they can spend (all variables are considered discrete). All bidders have independent private values that are observed privately by the respective bidders, making this an incomplete information game. Bidders have no bid information regarding the development of the auction process, as they only know whether the auction is still open (there is no dropout information). At the heart of this approach lie bidding functions that are determined by two main elements: personal values and potential strategies. We begin with the former.

In the basic experimental setting, all bidders have decreasing marginal values. This means that every bidder has a set of values organized from higher to lower that specify the marginal value from the consumption of each additional unit.

Since there is an analytical solution regarding the existence equilibrium, we tested this environment. To do so, we force each bidder to define his or her values for at least as many items as the total supply. Bidders' values are drawn independently and identically distributed from a uniform distribution with support [0, 200], with new random draws for each additional unit. For experiments that involve synergies, every bidder has a set of values organized from lower to higher that specify the marginal value for the consumption of each additional unit. Bidders' values are also drawn independently and are identically distributed from a uniform distribution with support [0, 10,000], with new random draws for each additional unit.

For both experimental settings, in each auction bidders use a strategy codified in the chromosome of the individual that represents that bidder. The definition of the GA strategy suggested requires the identification of actions. Each action is defined in terms of deviations (over- and underbidding) from the SB strategy. The quantity demanded according to the SB strategy of bidder i in round l is represented by $q^l_{\mathrm{SB}i}$, the units demanded if bidders bid according to their true values. The bidders have 13 possible actions to consider, which are represented in Table 14.4. All these strategies have an upper bound that is the lowest of either the number of units being auctioned or, alternatively, the units demanded in the previous round (as demand is required to be nonincreasing). The lower bound is the number of units that the participant has already clinched. It is possible that some strategies lead to noninteger numbers. In these circumstances the GA rounds up. The objective function that the GA tries to maximize is the payoff of each bidder. For this purpose, encoding the bidding strategies is a direct process. The 13 possible actions can be encoded with 4 bits (including the SB strategy, $q^l_{\mathrm{SB}i}$).

TABLE 14.4 Each Bidder Considers 13 Possible Actions or Bidding Strategies

Strategy		Action To Be Taken
0	$q^l_{\mathrm{SB}i}/4$	Underbidding
1	$q^l_{\mathrm{SB}i}/3$	Underbidding
2	$q^l_{\mathrm{SB}i}/2$	Underbidding
3	$q^l_{\mathrm{SB}i}/1.75$	Underbidding
4	$q^l_{\mathrm{SB}i}/1.5$	Underbidding
5	$q^l_{\mathrm{SB}i}/1.25$	Underbidding
6	$q^l_{\mathrm{SB}i}\times$	Bidding sincerely
7	$q^l_{\mathrm{SB}i} \times 1.25$	Overbidding
8	$q^l_{\mathrm{SB}i} \times 1.5$	Overbidding
9	$q^l_{\mathrm{SB}i} \times 1.75$	Overbidding
10	$q^l_{\mathrm{SB}i} \times 2$	Overbidding
11	$q^l_{\mathrm{SB}i} \times 3$	Overbidding
12	$q^l_{\mathrm{SB}i} \times 4$	Overbidding

14.5 EXPERIMENTAL ANALYSIS

In this section we analyze the experimental results in the presence of decreasing marginal utilities and synergies, with 15 items to be auctioned and four bidders. We study revenues and efficiency as well as the bidders' behavior and payoff. As we already mentioned, revenues are measured as the monetary units that the seller gets after each auction (see Equation 14.6). Furthermore, an auction is considered to be efficient when it allocates the object to the bidder who values it the most *ex post* the auction. Therefore, for these experiments, efficiency is defined as the sum of the personal values of the 15 units sold in an auction as a percentage of the sum of the 15 highest personal values in that auction (full efficiency = 100%). Finally, the bidders' payoff (see Equation 14.1) and bidding strategy are analyzed in each environment.

The assessment of the GA is made by running 10,000 auctions per experiment. The marginal values are initialized randomly for each experiment, but they are constant for all the GA executions, this is, they have the same marginal values for the 10,000 auctions. For the GA we used the standard implementation of the SimpleGA developed by GALib. The parameters used for the experiments are reported in Table 14.5. We use a vector of 4 bits for encoding each bidder (13 strategies). As we are working with four bidders, our population size is 4.

The first step in analyzing the auction outcome with the GA and decreasing marginal values is to study the bidders' behavior. Bidders try all possible strategies, but finally, bidders tend to bid sincerely (76.07% bidders follow this strategy). In terms of revenues and efficiency, if we compare the results obtained by the GA and the results predicted (i.e., those yielding an outcome based on all bidders bidding sincerely), we observe that the actual average revenue with the GA is higher than that predicted. However, this difference represents only 12%. The GA aims to improve the bidders' payoff. Therefore, all bidding strategies are tested. 72.84% of the auctions yield to the same revenue that would be obtained if all bidders did SB. The average efficiency level (98.8%) is also very close to the one predicted (100%), and 76.13% of the auctions yield to full efficiency. This difference is because the GA tries all possible bidding strategies and because of the evolution and mutation effects. At the beginning, all the strategies are settled randomly, and most of them correspond to bad strategies. Moreover, new individuals are always created with the mutation operator. However, bidders finally tend to bid sincerely.

TABLE 14.5 Parameters Used for the Experiments

GA Operator	Type	Parameter Tested
Selection	Roulette	
Crossover	Single point	100%
Mutation	Flip bit	$[1 \cdots 0.05]$

The bidders' payoff is another convenient way to study the participants' behavior. Table 14.6 includes the percentage of iterations that the bidders' strategy yield to the same payoff as bidding sincerely. The results reveal that most bidders using the GA strategies are getting the same payoff as those using the SB strategy. The GA has been tested in 50 experiments, and in all of them bidders tend to SB. This finding corroborates the theory established by Ausubel [17]: that SB is a weakly dominant strategy in this auction, with no bid information and weakly decreasing marginal values.

We repeat the same tests in an environment that involves synergies (when the value of multiple objects exceeds the sum of the objects' values separately). The effect over revenues and efficiency that the strategy proposed by the GA yields is summarized in Table 14.7. The average revenues obtained with the GA are very close to those that would be obtained if all bidders were to bid sincerely. The difference is only 2%. Of the 10,000 iterations, 75.12%

TABLE 14.6 Bidders' Payoff with Decreasing Marginal Values[a]

Bidder	Times (%) Payoff GA = Payoff SB
1	73.69
2	73.60
3	72.88
4	74.00

[a]Number of iterations in which each bidder earns the same payoff as if all bidders do SB.

TABLE 14.7 Revenue and Efficiency Analysis in the Presence of Synergies[a]

Revenue		Efficiency	
Revenue with sincere bidding (units)	76,660	Full efficiency (%)	100
Actual average revenue with the GA (units)	74,752	Actual average efficiency with the GA (%)	92.9
Standard error of the mean	83.15	Standard error of the mean	0.002
Difference	−1,907	—	
% Difference	2	% Difference	7
% of iterations that the GA yields to the same revenue as sincere bidding	75.12	% of iterations that the GA yields to the same revenue as sincere bidding	86.19

[a]Revenues are measured as the monetary units that the seller gets after each auction. Efficiency is defined as the sum of the values of the 15 units sold in an auction as a percentage of the sum of the 15 highest values in that auction. The average of both values is compared with the results if all bidders do SB.

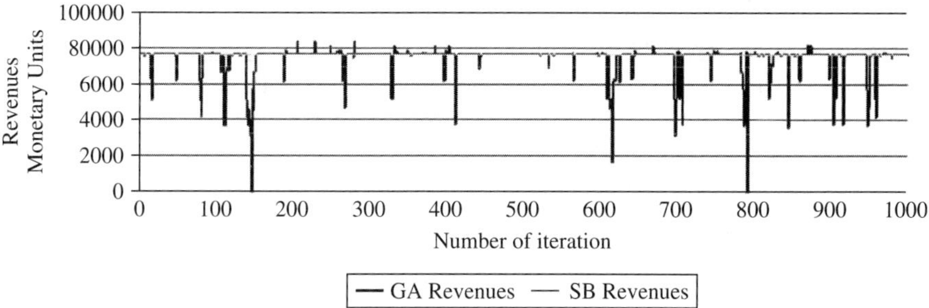

Figure 14.1 GA and SB revenue per iteration in the presence of synergies (last 1000 auctions).

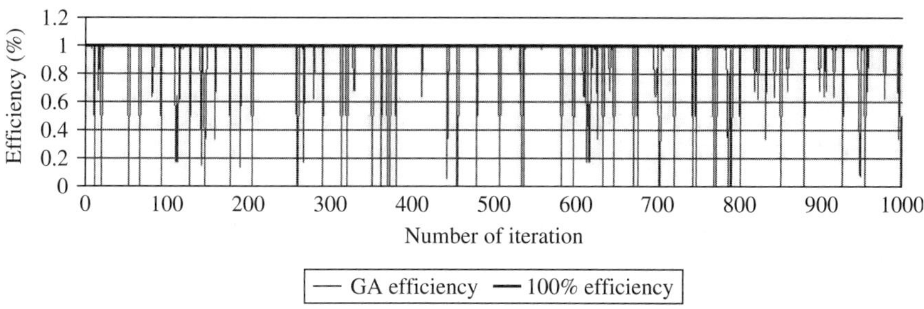

Figure 14.2 Efficiency per auction in the presence of synergies for the last 1000 auctions.

yield to the same revenues as if all bidders do SB. As bidders are primarily bidding sincerely, the average efficiency level is also very close to full efficiency (92.9%). Furthermore, 86.19% of the iterations gave an efficiency level of 100%.

Revenues per auction for the last 1000 iterations are represented in Figure 14.1. This figure just corroborates the intuition drawn from Table 14.7: SB is the most frequent strategy. The efficiency analysis for the last 1000 iterations is represented in Figure 14.2. Although the GA yields to some efficiency levels below 100%, those strategies are not stable. Nevertheless, the strategy that bidders tend to use yields full efficiency. When bidders bid sincerely, those participants with the highest valuations are awarded the items. However, it is also possible for this strategy to earn negative payoff in the presence of synergies. If one bidder does not clinch all the units demanded at the standing price, his or her valuation can be lower than the total payment. In this circumstance he or she will have a negative payoff (exposure problem). Even so, the GA best strategy without dropout information is to bid sincerely. Bidders have no information during the auction to respond to their rival's bids. As Table 14.8 shows, bidders usually earn the same payoff as if they all bid sincerely.

TABLE 14.8 Bidders' Payoff in the Presence of Synergies[a]

Bidder	Times(%) Payoff GA = Payoff SB
1	94.91
2	84.18
3	95.41
4	75.07

[a]Number of iterations during which each bidder earns the same payoff as if all bidders do SB.

The GA has been tested in 50 experiments within this environment. In all experiments conducted, bidders finally tend to bid sincerely. As there is no theory model developed for this assumption, the results predicted were uncertain. Similarly, as bidders tend to bid sincerely, efficient allocation of the items remains.

14.6 CONCLUSIONS

In this chapter we have dealt with the subject of bidding strategy optimization by means of evolutionary computation. This work is focused on a specific dynamic ascending multiunit auction which is referred to as the Ausubel auction. Given the complexity of studying this framework using traditional tools, we present an approach based on agent-based economic modeling.

We have developed a GA that can be employed successfully to evolve bidding strategies for this auction format. The GA aims to maximize each bidder's fitness, measured in terms of the the actual payoff. For this purpose, the algorithm generates 13 different bidding strategies or actions defined in terms of deviations from the SB strategy: overbidding, underbidding, or bidding sincerely. The experiments simulates auctions of 15 lots among four bidders. Each experiment was run for 10,000 iterations.

The results that we have obtained using decreasing marginal utilities and no bid information are consistent with the conclusions drawn by Ausubel [16,17] that established that SB is a weakly dominant strategy and an *ex post* perfect equilibrium. In all 50 computational experiments performed in this study, bidders tended to bid sincerely. The main contribution of this work is a study of this auction in the presence of synergies, as no theoretical model has yet been developed with this assumption. In the 50 experiments conducted, the algorithm evolves SB as the most frequent strategy. There are several directions in which the work presented could be extended. First, the information rule established can have important implications. Therefore, it could be interesting to analyze the outcome of the auction assuming that there is dropout information during the auctions that is, bidders have full bid information. Another alternative could be to let bidders have memories of their past experiences. Another possibility lying ahead is study of the final lot allocation when the bidders' values are interdependent (i.e., assuming common values). Finally, assessment of the sensitivity of the results to a wider range of combinations of number of bidders and items to be auctioned would be very interesting.

Acknowledgments

The authors are partially supported by the Spanish MEC and FEDER under contract TIN2005-08818-C04-02 (the OPLINK project).

REFERENCES

1. P. Klemperer. *The Economic Theory of Auctions*, vols. I and II. Edward Elgar, Cheltenham, UK, 2000.
2. V. Krishna. *Auction Theory*. Academic Press, San Diego, CA, 2002.
3. P. Klemperer. *Auctions: Theory and Practice*. Princeton University Press, Princeton, NJ, 2003.
4. P. Milgrom. *Putting Auction Theory to Work*. Cambridge University Press, New York, 2004.
5. J. H. Holland. *Adaptation in Natural and Artificial Systems*. University of Michigan Press, Ann Arbor, MI, 1975.
6. J. Andreoni and J. H. Miller. Auctions with artificial adaptive agents. *Games and Economic Behaviour*, 10:39–64, 1995.
7. D. Cliff. Evolving parameter sets for adaptive trading agents in continuous double-auction markets. In *Proceedings of the Agents'98 Workshop on Artificial Societies and Computational Markets*, 1998, pp. 38–47.
8. C. Preist and M. Van Tol. Adaptive agents in a persistent shout double auction. In *Proceedings of the 1st International Conference on the Internet, Computing and Economics*. ACM Press, New York, 1998, pp. 11–18.
9. C. Preist. Commodity trading using an agent-based iterated double auction. In *Proceedings of the 3rd Annual Conference on Autonomous Agents*, 1999, pp. 131–138.
10. D. Cliff. Exploration in evolutionary design of online auction market mechanisms. *Electronic Commerce Research and Applications*, 2(2):162–175, 2003.
11. H. Dawid. On the convergence of genetic learning in a double auction market. *Journal of Economic Dynamics and Control*, 23:1545–1567, 1999.
12. A. Byde. Applying evolutionary game theory to auction mechanism design. In *Proceedings of the ACM Conference on Electronic Commerce*, 2003, pp. 192–193.
13. P. Anthony and N. R. Jennings. Evolving bidding strategies for multiple auctions. In *Proceedings of the 15th European Conference on Artificial Intelligence*, 2002, pp. 178–182.
14. P. Anthony and N. R. Jennings. Developing a bidding agent for multiple heterogeneous auctions. *ACM Transactions on Internet Technology*, 3(3):185–217, 2003.
15. Y. Saez, D. Quintana, P. Isasi, and A. Mochon. Effects of a rationing rule on the Ausubel auction: a genetic algorithm implementation. *Computational Intelligence*, 23(2):221–235, 2007.

16. L. M. Ausubel. An efficient ascending-bid auction for multiple objects. *Working Paper 97-06*. University of Maryland, College Park, MD, 1997.
17. L. M. Ausubel. An efficient ascending-bid auction for multiple objects. *American Economic Review*, 94:1452–1475, 2004.
18. L. M. Ausubel, P. Cramton, P. McAfee, and J. McMillan. Synergies in wireless telephony: evidence from the broadband PCS auctions. *Journal of Economics and Management Strategy*, 6(3):497–527, 1997.
19. S. Hon-Snir, D. Monderer, and A. Sera. A learning approach to auctions. *Journal of Economic Theory*, 82 (1): 65–88, 1998.
20. S. Park, E. H. Durfee, and W. P. Birmingham. An adaptive agent bidding strategy based on autonomous agents. In *Proceedings of the 3rd Annual Conference on Autonomous Agents*, ACM Press, New York, 1999, pp. 147–153.
21. C. Preist. Algorithm design for agents which participate in multiple simultaneous auctions. Hewlett Packard Laboratories Technical Report HPL-2000-88, 2000.
22. J. Bower and D. Bunn. Experimental analysis of the efficiency of uniform-price versus discriminatory auctions in the England and Wales electricity market. *Journal of Economic Dynamics and Control*, 25: 561–592, 2001.
23. A. Byde. An optimal dynamic programming model for algorithm design in simultaneous auctions. Hewlett Packard Laboratories Technical Report HPL-2001-67, 2001.
24. Z. Qin. Evolving marketplace designs by artificial agents. M.Sc. Thesis, Computer Science Department, Bristol University, 2002.
25. A. Byde, C. Preist, and N. R. Jenning. Decision procedures for multiple auctions. In *Proceedings of the 1st International Joint Conference on Autonomous Agents and Multiagent Systems: Part 2*, ACM Press, New York, 2002, pp. 613–620.
26. S. Phelps, S. Parsons, P. McBurney, and E. Sklar. Co-evolution of auction mechanisms and trading strategies: towards a novel approach to microeconomic design. In *Proceedings of GECCO 2002*, 2002, pp. 65–72.
27. E. Ogston and S. Vassiliadis. A peer-to-peer agent auction. In *Proceedings of the 1st International Joint Conference on Autonomous Agents and Multiagent Systems: Part 1*, ACM Press, New York, 2002, pp. 151–159.
28. M. P. Wellman, J. K. Mackie-Mason, D. M. Reeves, and S. Swaminathan. Exploring bidding strategies for market-based scheduling. In *Proceedings of the 4th ACM Conference on Electronic Commerce*, ACM Press, New York, 2003, pp. 115–124.
29. S. M. Bohte, E. H. Gerding, and J. A. La Poutre. Market-based recommendation: agents that compete for consumer attention. *ACM Transactions on Internet Technologies*, 4:420–448, 2004.
30. A. Mochon, D. Quintana, Y. Saez, and P. Isasi. Analysis of Ausubel auctions by means of evolutionary computation. In *Proceedings of the IEEE Congress on Evolutionary Computation 2005*: Vol 3, 2005, pp. 2645–2652.

CHAPTER 15

Evolving Rules for Local Time Series Prediction

C. LUQUE, J. M. VALLS, and P. ISASI

Universidad Carlos III de Madrid, Spain

15.1 INTRODUCTION

A time series (TS) is a sequence of values of a variable obtained at successive, in most cases uniformly spaced, intervals of time. The goal is to predict future values of the variable y_i using D past values. In other words, the set $\{y_1, \ldots, y_D\}$ is used to predict $y_{D+\tau}$, where τ is a nonnegative integer, which receives the name of the prediction horizon. In TS related to real phenomena, a good model needs to detect which elements in the data set can generate knowledge leading to refusing those that are noise. In this work a new model has been developed, based on evolutionary algorithms to search for rules to detect local behavior in a TS. That model allows us to improve the prediction level in these areas.

Previous studies have used linear stochastic models, mainly because they are simple models and their computational burden is low. Autoregressive moving average (ARMA) models using data as to air pressure and water level at Venice have been used to forecast the water level at the Venice lagoon [1]. Following this domain, Zaldívar et al. [2] used a TS analysis based on nonlinear dynamic systems theory, and multilayer neural network models can be found. This strategy is applied to the time sequence of water-level data recorded in the Venice lagoon during the years 1980–1994. Galván et al. [3] used multilayer perceptrons trained using selective learning strategies to predict the Venice lagoon TS. Valls et al. [4] used radial basis neural networks trained with a lazy learning approach applied to the same TS and the well-known Mackey–Glass TS. Following Packard's work to predict dynamic systems [5–7], and using evolutionary algorithms [8–12] to generate prediction rules for a TS, some advanced techniques have been used in this work to obtain better results.

Optimization Techniques for Solving Complex Problems, Edited by Enrique Alba, Christian Blum, Pedro Isasi, Coromoto León, and Juan Antonio Gómez

15.2 EVOLUTIONARY ALGORITHMS FOR GENERATING PREDICTION RULES

In some domains, especially in the case of natural phenomena, the use of machine learning techniques is faced with certain problems. Usually, machine learning techniques, especially evolutionary algorithms, base their learning process on a set of examples. If these examples are distributed fundamentally throughout certain values, the learning process will focus on this range, considering the remaining values as noise. This fact is positive for some domains, but it becomes a disadvantage in others. For example, in the case of stock market or tide prediction, to mention two very different domains, most existing measures are over average values. On a few occasions, great increases or decreases take place. Nevertheless, those situations are indeed the situations that have more importance from the point of view of the prediction task. Our approach is based on finding rules that represent both the usual and atypical behaviors of the TS, to enable us to predict future values of the series. These rules will be obtained using an evolutionary algorithm.

To avoid the generalization problem, a Michigan approach has been implemented [13] in the evolutionary algorithm, using a steady-state strategy. In the Michigan approach, the solution to the problem is the total population instead of the individual that best fits. This allows us to evolve rules for common behaviors of the problem but also allows atypical behaviors. Doing that, we pay attention to those unusual behaviors which in other cases would be considered as noise.

Due to the fact that each rule is evaluated using only examples associated with it, it is only locally or partially applicable. This local characteristic allows the system to use specific rules for particular situations. On the other hand, this method does not assure that the system will make a prediction for all the TS. A balance must be found between system performance and the percentage of prediction.

15.3 EXPERIMENTAL METHODOLOGY

The approach suggested in this chapter is based on the generation of rules for making predictions. The steps, divided into encoding, initialization, evolution, and prediction, are explained in detail in this section.

15.3.1 Encoding

The first step consists of fixing a value for the constant D that represents the number of consecutive time instants used to make the prediction. For example, if $D = 5$, a rule could be a condition such as "if the value of the variable at time unit 1 is smaller than 100 and bigger than 50, at time unit 2 is smaller than 90 and bigger than 40, at time unit 3 is smaller than 5 and bigger than -10, at time unit 5 is smaller than 100 and bigger than 1, the measure at time unit $5 + \tau$ will be 33 with an expected error of 5." This rule could be expressed as

$$\begin{gathered}\text{IF}\,(50 < y_1 < 100)\ \text{AND}\ (40 < y_2 < 90)\\ \text{AND}\,(-10 < y_3 < 5)\ \text{AND}\ (1 < y_5 < 100)\\ \text{THEN prediction} = 33 \pm 5\end{gathered}$$

In Figure 15.1 a graphical representation of a rule is shown. For a rule (R), a conditional part (C_R) and a predicting part (P_R) are defined. The conditional part consists of a set of pairs of intervals $C_R = \{I_1^R, I_2^R, \ldots, I_i^R, \ldots, I_D^R\}$, where each I_i^R is an interval $I_i^R = \{LL_i^R, UL_i^R\}$, LL_i^R being the lower limit and UL_i^R the upper limit for the ith input data. The predicting part is composed of two values: the prediction value and the expected error, $P_R = \{p_R, e_R\}$. The conditional part could include void values ($*$), which means that the value for this interval is irrelevant; that is, we don't care which value it has. The previous rule is encoded as (50, 100, 40, 90, – 10, 5, $*$, $*$, 1, 100, 33, 5).

In this way, all the individuals in our evolutionary strategy can be encoded as vectors of real numbers of size $2 \times D + 2$, D being the number of time units used for prediction. Genetic operators can be used to generate new rules from an initial population. To do that we need to define a crossover process. We select two rules in the population, and they produce a new offspring rule. This offspring will inherit his genes from his parents, each gene being an interval I_j. For each $i < D$ the offspring can inherit two genes (one from each parent) with the same probability. This type of crossover is known as *uniform crossover*. This

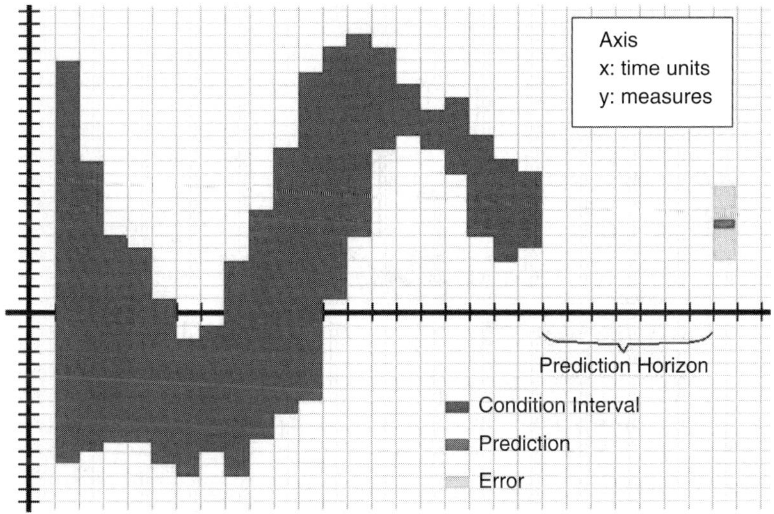

Figure 15.1 Graphical representation of a rule.

offspring will not inherit the values for "prediction" and "error," as we can see in the following example:

Parent A:	(50, 100, 40, 90, −10, 5, ∗, ∗, 1, 100, 33, 5)
Parent B:	**(60, 90, 10, 20, 15, 30, 40, 45,** ∗, ∗, **60, 8**)
Offspring:	(50, 100, **10, 20**, −10, 5, **40, 45**, ∗, ∗, *p*, *e*)

An example of this process is shown graphically in Figure 15.2. Once generated, an offspring may suffer an interval mutation. This event takes place with a 10% probability for each new individual. Should the individual be selected, one of its intervals will be altered.

Mutations over the interval (*LL*, *UL*) are shown in Table 15.1. All these mutations have the same probability of being selected. $R(x, y)$ means a random value between x and y with $x < y$, and W means the 10% of the width of the interval defined by (*LL*, *UL*). In other words, $W = 0.1(UL - LL)$. Obviously, the transformation of (*LL*,*UL*) into (LL',UL') must fit the condition $LL' < UL'$. Figure 15.3 summarizes these alternatives.

For each rule, several values are needed for the prediction made for that rule, and its expected accuracy or precision. As already mentioned, these values are not inherited from parents but must also be computed. With R representing an individual, the process to obtain the prediction and error values for R is the following:

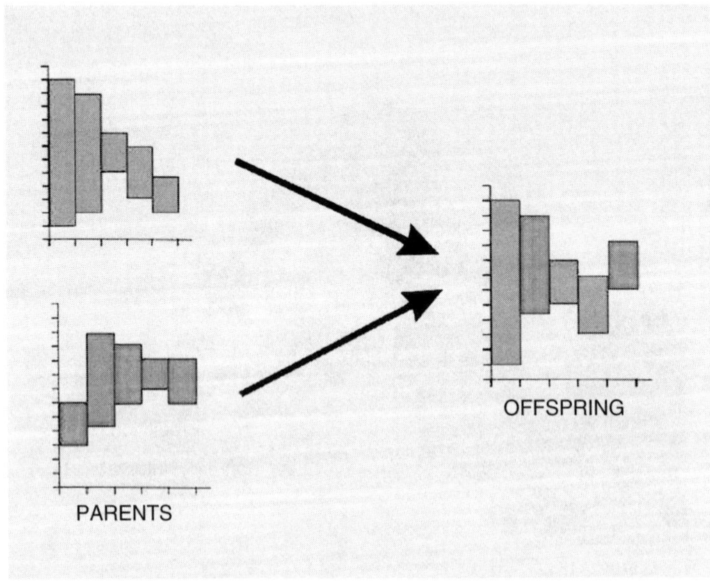

Figure 15.2 Crossover.

TABLE 15.1 Mutations

Name	Transformation
New random value	$(LL, UL) \rightarrow (R(0,1), R(0,1))$
Null condition	$(LL, UL) \rightarrow (*, *)$
Enlarge	$(LL, UL) \rightarrow (LL - R(0,1)W, UL + R(0,1)W)$
Shrink	$(LL, UL) \rightarrow (LL + R(0,1)W, UL - R(0,1)W)$
Move up	$(LL, UL) \rightarrow (LL + cW, UL + cW)\ c = R(0,1)$
Move down	$(LL, UL) \rightarrow (LL - cW, UL - cW)\ c = R(0,1)$

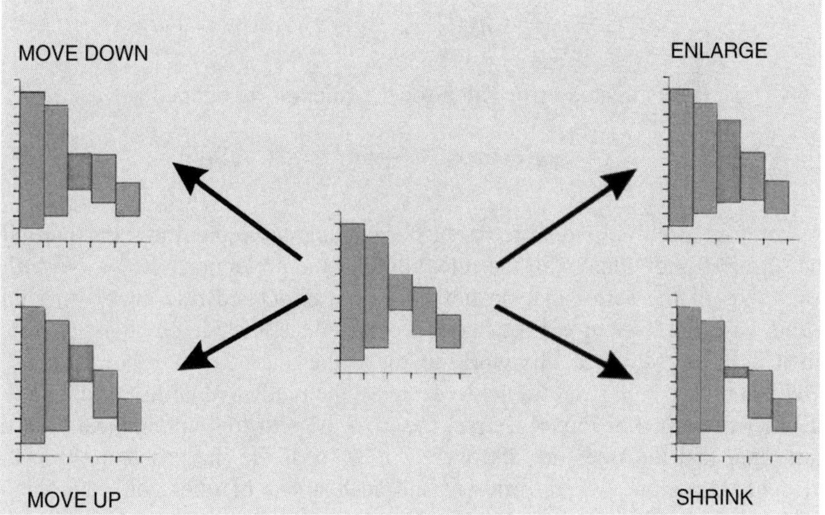

Figure 15.3 Mutation.

1. We must calculate the set of points of the TS $S = \{x_1, x_2, \ldots, x_k, \ldots, x_m\}$ such that it fits the conditional part of the rule R. This subset will be called $C_R(S)$ and is defined as

$$C_R(S) = \{\vec{X}_i | \vec{X}_i \text{ fits } C_R\} \tag{15.1}$$

where $\vec{X}_i = (x_i, x_{i+1}, \ldots, x_{i+D-1})$ and the vector $\vec{X}_i$ fits C_R if all its values fall inside the intervals of the rule:

$$LL_1^R \leq x_i \leq UL_1^R, LL_2^R \leq x_{i+1} \leq UL_2^R, \ldots, LL_D^R \leq x_{i+D-1} \leq UL_D^R \tag{15.2}$$

2. Once $C_R(S)$ is calculated, the next step consists of determining the output for the prediction horizon τ for each point. This value is $v_i = x_{i+D-1+\tau}$.

3. This new value is added to the vector $\vec{X}_i$ as a new component, so we have $C'_R(S) = \{\vec{X}'_i | \vec{X}_i \text{fits} C_R\}$, where $\vec{X}'_i$ is defined as

$$\vec{X}'_i = (x_i, x_{i+1}, \ldots, x_{i+(D-1)}, v_i) = (\vec{X}_i, v_i) \qquad (15.3)$$

4. The prediction p_R for the rule R is calculated by means of a linear regression with all the vectors in the set $C'_R(S)$. To do that it is necessary to calculate the coefficients of the regression $\vec{A} = (a_0, a_1, \ldots, a_D)$, which define the hyperplane that better approximates the set of points $C'_R(S)$. Let $\tilde{v}_i$ be the estimated value obtained by regression at the point $\vec{X}_i$, which follows

$$\tilde{v}_i = a_0 x_i + a_1 x_{i+1} + \cdots + a_{D-1} x_{i+D-1} + a_D \qquad (15.4)$$

5. Thus, the estimated error value for the rule, e_R, is defined as

$$e_R = \max_i \{|v_i - \tilde{v}_i| / \vec{X}_i \in C_R(S)\} \qquad (15.5)$$

Therefore, each individual represents a rule able to predict the series partially. The set of all individuals (all the rules) defines the prediction system. Nevertheless, zones of the series that do not have any associated rule could be found. In this case the system cannot make a decision in this region. It is desirable, and it is an objective of this work, to make the unpredicted zone as small as possible, always being careful not to decrease the predictive ability of the system for other regions. For this objective, the work tries to find automatically, at the same time that the rules are discovered, how to divide the problem space into independent regions, where different and ad hoc sets of rules could be applied without being affected by distortions that general rules have in some specific zones of the problem space.

This could be summarized in searching sets of rules that allow equilibrium between generalization and accuracy: rules that could achieve the best predictive ability in as many regions as possible but being as general as possible at the same time. Our system must therefore look for individuals that on the training set predict the maximum number of points with the minimum error. Following this idea, the *fitness* function for an individual R is defined in this way:

```
IF ((NR>1) AND (eR<EMAX)) THEN
        fitness = (NR*EMAX) - eR
ELSE
    fitness = f_min
```

where N_R is the number of points of the training set that fit the condition C_R [i.e., $N_R = \text{cardinal}(C_R(S))$]. EMAX is a parameter of the algorithm that punishes those with a maximum absolute error greater than EMAX. $f_{\min}$ is a minimum value assigned to those whose rule is not fitted at any point in the training set.

This means that if the number of examples accomplished by a rule is low, or the error in that set is big, the fitness will be the worst value, being the relative error as fitness otherwise.

The goal of this fitness function is to establish a balance between individuals whose prediction error is the lowest possible, and at the same time, the number of points of the series in which it makes a prediction [the number of points of $C_R(S)$] is the highest possible, that is, to reach the maximum possible generalization level for each rule.

15.3.2 Initialization

The method designed tends to maintain the diversity of the individuals in the population as well as its capacity to predict different zones of the problem space. But the diversity must exist previously. To do that, a specific procedure of population initialization has been devised. The main idea of this procedure is to make a uniform distribution throughout the range of possible output data. For example, in the case of Venice lagoon tide prediction, the output ranges from −50 to 150 cm. If the population has 100 individuals, the algorithm will create 100 intervals of 2 cm width, and in this way all the possible values of the output are included. The initialization procedure includes a rule for each interval mentioned previously, so intervals will have a default rule at the beginning of the process. The initialization procedure for an interval I has the following steps:

1. Select all the training patterns $\vec{X}_i$ whose output belongs to I.
2. Determine a maximum and a minimum value for each input variable of all the pattern selected in step 1. For example, if the patterns selected are (1,2,3), (2,2,2), (5,1,5), and (4,5,6), the maximum values should be 5, 5, and 6 for the first, second, and third input variables. So the minimum values should be 1, 1, and 2.
3. Those maximum and minimum values define the values assigned to the rule for each input variable. This rule's prediction is set as the mean of the output value of the patterns selected in step 1.

This procedure will produce very general rules, so they will cover all the prediction space. From this set of rules, thc evolutionary algorithm will evolve new, more specific rules, improving both the general predictive ability of the entire system and the prediction level in each specific zone of the space defined by each rule.

15.3.3 Evolution of Rules

Basically, the method described in this chapter uses a Michigan approach [6] with a steady state strategy. That means that at each generation two individuals are selected proportionally to the fitness function. We choose a tournament selection of size 3 as the selection operator. Those parents produce only one offspring by crossover. Then the algorithm replaces the individual nearest the offspring in phenotypic distance (i.e. it looks for the individual in the population that make predictions on similar zones in the prediction space). The offspring replaces the individual selected if and only if its fitness is higher; else the population does not change. This replacement method is used primarily in crowding methods [14]. Those methods try to maintain a diverse population to find several solutions to the problem. In the case of study in this chapter, this approach is justified by the fact that we are looking for several solutions to cover the space of prediction as widely as possible so that rules generated could predict the highest number of situations. Moreover, the diversity of the solutions allows the generation of rules for specific highly special situations. The pseudocode of the entire system is shown in Algorithm 15.1.

15.3.4 Prediction

The stochastic method defined previously obtains different solutions in different executions. After each execution the solutions obtained at the end of the process are added to those obtained in previous executions. The number of executions is determined by the percentage of the search space covered by the rules. The set of all the rules obtained in the various executions is the final solution of the system. Once the solution is obtained, it is used to produce outputs to unknown input patterns. This is done following the next steps:

- For each input pattern, we look for the rules that this pattern fits.
- Each rule produces an output for this pattern.
- The final system output (i.e., the prediction of the system) is the mean of the outputs for each pattern.

Those steps are summarized in Figure 15.4. In this case, when an unknown pattern is received, it may fit some of the rules in the system. In the example, rules 1, 2, 3, 5, 6, 8, 9, and 10 are fitted. Each fitted rule makes a prediction, so we have eight different predictions. To obtain the final prediction of the system for that pattern, the mean of all predictions is chosen as the definitive prediction for that pattern. If the pattern doesn't fit any rule, the system produces no prediction.

15.4 EXPERIMENTS

The method has been applied to three different domains: an artificial domain widely used in the bibliography (Mackey–Glass series) and two TSs corresponding to natural phenomena: the water level in Venice lagoon and the sunspot TS.

15.4.1 Venice Lagoon TS

One real-world TS represents the behavior of the water level at the Venice lagoon. Unusual high tides result from a combination of chaotic climatic elements with the more normal, periodic tidal systems associated with a particular area. The

Algorithm 15.1 Algorithm of Local Rules

```
Pseudocode of the local rules algorithm
R = Θ
while (Prediction level of R < θ) do
  P = Initial population made of uniformly
     distributed random rules
  Generate the prediction for each p_i ∈ P by linear
      regression
  Store the prediction zone z_pi as mean of every prediction
      made by individual p_i
  Evaluate each p_i ∈ P. Being e_pi the evaluation of
      individual i
  generation = 0; stopCondition = FALSE;
  while (generation < maxGenerations) AND (NOT stopCondition)
      do
    Select two individuals (p_1 and p_2) by tournament
        selection
    Generate an offspring (o) by crossing over the
        individuals selected
    o = Mutate (o,probability)
    Generate the prediction of o by linear regression
    z_o = Prediction zone of o
    e_o = Evaluation of o
    Find the q ∈ P individual whose prediction zone z_q is the
        closest to z_o
    if (e_q < e_o) then
       Replace q by o
    generation = generation + 1;
  end while
  R = R + P
end while
```

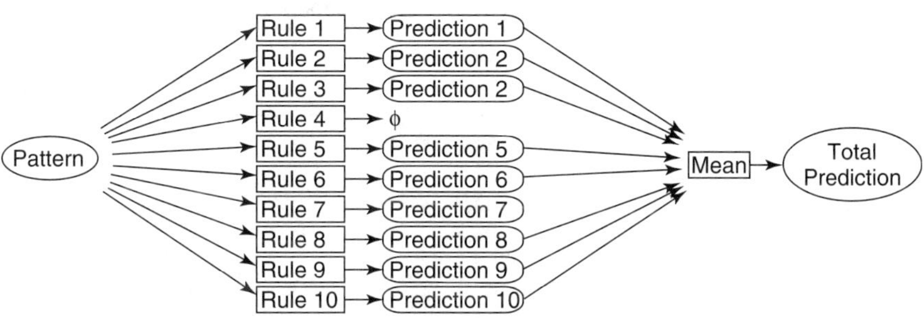

Figure 15.4 Schema of the prediction.

prediction of high tides has always been a subject of intense interest, not only from a human point of view but also from an economic one, and the water level of the Venice lagoon is a clear example of these events [1,15]. The most famous example of flooding in the Venice lagoon occurred in November 1966, when the water rose nearly 2 meters above the normal level. This phenomenon, known as *high water*, has spurred many efforts to develop systems to predict the sea level in Venice, mainly for the prediction of the high-water phenomenon [16]. Different approaches have been developed for the purpose of predicting the behavior of sea level at the lagoon [16,17]. Multilayer feedforward neural networks used to predict the water level [2] have demonstrated some advantages over linear and traditional models. Standard methods produce very good mean predictions, but they are not as successful for those unusual values. For the experiments explained above, the following values for the variables τ and D have been used: $D = 24$ and $\tau = 1, 4, 12, 24, 28, 48, 72, 96$. That means that the measures of the 24 consecutive hours of the water level have been used to predict the water level, measured in centimeters: 1, 4, 12, ... hours later.

The results from the experiments using the rule system (RS) are shown in Table 15.2, and the results obtained by other well-known machine learning algorithms (MLAs), in Table 15.3. The WEKA tool was used to perform a comparison with the MLA. Those experiments use a training set of patterns created by the first 5000 measures and a validation set of 2000 patterns. The populations of 100 persons evolved within 75,000 generations. The rules used as input 24 consecutive hours of the water level. The value "percentage of prediction" is the percentage of points in the validation set such that there is a prediction for it by at least one rule. No rule made a prediction for the remainder of the validation set. Columns SMO-Reg, IBK, LWL, Kstar, M5Rules, M5p, RBNN (radial basis neural networks), Perceptron NN, and Conjunctive Rules show the error obtained by those MLAs. The error measure used in those experiments is the root-mean-squared error (RMSE), normalized by the root of the variance of the output (NRMSE). Those error measures are defined as

TABLE 15.2 NRMSE and Percentage of Prediction Obtained by the RS for the Venice Lagoon TS

Pred. Hor.	Percent Pred.	NRMSE
1	98.5	0.0720
4	99.8	0.2536
12	95.2	0.3269
24	98.8	0.3547
28	98.2	0.4486
48	97.6	0.5254
72	100	0.6286
96	100	0.7057

TABLE 15.3 NRMSE Obtained by Other MLAs for the Venice Lagoon TS

Pred. Hor.	SMO-Reg	IBK	LWL	Kstar	M5-Rules	M5p	Percep. NN	RB NN	Conj. Rules
1	0.0724	0.3772	0.6046	0.4937	0.0760	0.0760	0.7698	0.0770	0.6391
4	0.2591	0.4878	0.6036	0.5688	0.3174	0.3174	0.7712	0.3493	0.6634
12	0.3244	0.4948	0.6048	0.6041	0.3464	0.3464	0.7132	0.7038	0.6926
24	0.3432	0.5131	0.6147	0.6786	0.3638	0.3638	0.7476	0.4679	0.7237
28	0.4471	0.6376	0.6824	0.7890	0.5018	0.5018	0.7964	0.5487	0.7943
48	0.5138	0.6331	0.6842	0.8098	0.5309	0.5309	0.7764	0.6955	0.7970
72	0.6202	0.7221	0.7702	0.8864	0.6556	0.6548	0.8178	0.8842	0.8420
96	0.7011	0.7863	0.8441	0.9336	0.7159	0.7208	0.8583	0.8537	0.9164

$$\text{RMSE} = \sqrt{\frac{\sum_{i=1}^{n} e^2}{n}} \tag{15.6}$$

$$\text{NRMSE} = \frac{\text{RMSE}}{\sqrt{\text{Var}}} \tag{15.7}$$

For those experiments, the objective was to maximize the percentage of validation set data predicted, avoiding a high mean error. It is interesting to observe that when the prediction horizon increases, the percentage of prediction does not diminish proportionally. Thus, the system seems to be stable to variations in the prediction horizon. This property seems very interesting, because it shows that the rules are adapted to the special and local characteristics of the series.

Studying Tables 15.2 and 15.3, it is important to remark that the prediction accuracy of the rule system outperforms most of the algorithms, being the percentage of prediction data very close to 100%. The SMO-Reg algorithm is the only one that improves most of the rule system results in most of the horizons for that problem. M5Rules and M5P algorithms' results are pretty close and slightly worse than the RS results. Real tide values and the prediction obtained by the RS with horizon 1 for a case of unusual high tide are compared in Figure 15.5. It can be seen how good the predicted value is, even for unusual behaviors.

15.4.2 Mackey–Glass TS

The Mackey–Glass TS is an artificial series widely used in the domain of TS forecasting [18–20] because it has specially interesting characteristics. It is a chaotic series that needs to be defined in great detail. It is defined by the differential equation

$$\frac{ds(t)}{dt} = -bs(t) + a\frac{s(t-\lambda)}{1 + s(t-\lambda)^{10}} \tag{15.8}$$

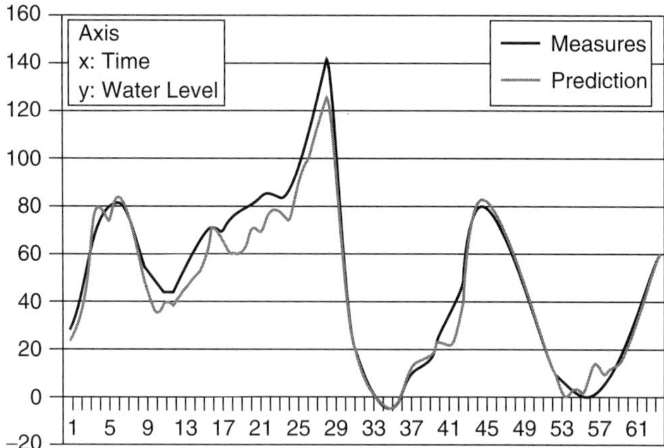

Figure 15.5 Prediction for an unusual tide with horizon 1.

As in refs. 18–20, the values $a = 0.2$, $b = 0.1$, and $\lambda = 17$ were used to generate the TS. Using Equation 15.8, 30,000 values of the TS are generated for each horizon. The initial 4000 samples are discarded to avoid the initialization transients. With the remaining samples, the training set used was composed of the points corresponding to the time interval [5000, 25,000]. The test set was composed of the samples [4000, 5000]. All data points are normalized in the interval [0, 1].

The results for the rule system algorithm are given in Table 15.4. For the same prediction horizon, the results of the algorithms IBK, LWL, KStar, M5Rules, M5p, Perceptron Neural Networks, Radial Basis Neural Networks, and Conjunctive Rule implemented in WEKA can be seen in Table 15.5. The error used for comparison is NRMSE (normalized root-mean-squared error). In both cases, the rule system improves over the result of most of the other algorithms (except IBK and KStar). This suggests that we have a better level of prediction for the difficult regions of the TS. The percentage of prediction for the test set (more than 75%) induces us to think that the discarded elements were certainly inductive of a high level of errors, since their discard allows better results than those obtained in the literature.

TABLE 15.4 NRMSE and Percentage of Prediction Obtained by the RS for the Mackey–Glass TS

Pred. Hor.	Percent Pred.	NRMSE
50	79.3	0.021
85	77.5	0.047

TABLE 15.5 NRMSE Obtained by Other MLA for the Mackey–Glass TS

Pred. Hor.	SMO-Reg	IBK	LWL	Kstar	M5-Rules	M5p	Percep. NN	RB NN	Conj. Rules
50	0.730	0.020	0.700	0.017	0.065	0.050	0.702	0.111	0.719
85	0.750	0.035	0.783	0.028	0.107	0.089	0.763	0.432	0.785

15.4.3 Sunspot TS

This TS contains the average number of sunspots per month measured from January 1749 to March 1977. These data are available at http://sidc.oma.be ("RWC Belgium World Data Center for the Sunspot"). That chaotic TS has local behaviors, noise, and even unpredictable zones using the archived knowledge. In Table 15.6 we can see the error and percentage of prediction obtained by the RS for different prediction horizons. Table 15.7 shows the results obtained for the same horizons by other well-known MLAs: SMO-Reg, IBK, LWL, KStar, M5Rules, M5p, and Conjunctive Rules are implemented WEKA. The results for Multilayer Feedforward NN and Recurrent NN have been obtained from Galván and Isasi [21]. The error measure used in both tables is defined as

$$e = \frac{1}{2(N+\tau)} \sum_{i=0}^{N} [x(i) - \tilde{x}(i)]^2 \tag{15.9}$$

In all cases the experiments were done using the same data set: from January 1749 to December 1919 for training, and from January 1929 to March 1977 for validation, standardized in the [0, 1] interval; in all cases, 24 inputs were used. In all cases the algorithm explained in this chapter (RS) improves over the results obtained by the other MLA, with a percentage of prediction very close to 100%. That fact suggests that we may obtain an even lower error if we prefer to sacrifice this high percentage of prediction. A deeper study of the results confirms the ability of this system to recognize, in a local way, the peculiarities of the series, as in the previous domains.

TABLE 15.6 NRMSE and Percentage of Prediction Obtained by the RS for the Sunspot TS

Pred. Horiz.	Percent Pred.	RS
1	100	0.00228
4	97.6	0.00351
8	95.2	0.00377
12	100	0.00642
18	99.8	0.01021

TABLE 15.7 NRMSE Obtained by Other MLA for the Sunspot TS

Pred. Horiz.	SMO-Reg	IBK	LWL	Kstar	M5 Rules	M5p	Conj. Rules	Feedforward NN	Recurr. NN
1	0.00233	0.00452	0.00756	0.00504	0.00230	0.00230	0.01178	0.00511	0.00511
4	0.00401	0.00565	0.01007	0.00658	0.00398	0.00485	0.01445	0.00965	0.00838
8	0.00522	0.00659	0.01238	0.00871	0.00532	0.00574	0.01600	0.01177	0.00781
12	0.00700	0.0079	0.01453	0.01043	0.00645	0.00665	0.01701	0.01587	0.01080
18	0.01138	0.01045	0.01906	0.01320	0.00746	0.00812	0.02057	0.02570	0.01464

15.5 CONCLUSIONS

In this chapter we present a new method based on prediction rules for TS forecasting, although it can be generalized for any problem that requires a learning process based on examples. One of the problems in the TS field is the generalization ability of artificial intelligence learning systems. On the one hand, general systems produce very good predictions over all the standard behaviors of the TS, but those predictions usually fail over extreme behaviors. For some domains this fact is critical, because those extreme behaviors are the most interesting.

To solve this problem, a rule-based system has been designed using a Michigan approach, tournament selections, and replacing new individuals by a steady-state strategy. This method includes a specific initialization procedure (Section 15.3.2) and a process designed to maintain the diversity of the solutions. This method presents the characteristic of not being able to predict the entire TS, but on the other hand, it has better accuracy, primarily for unusual behavior. The algorithm can also be tuned to attain a higher prediction percentage at the cost of worse prediction results.

Results show that for special situations, primarily for unusual behaviors (high tides, function peaks, etc.), the system is able to obtain better results than do the previous works, although the mean quality of the predictions over the entire series is not significantly better. Therefore, if it is possible, the system can find good rules for unusual situations, but it cannot find better rules for standard behaviors of the TS than can the previous works, where standard behaviors indicates behaviors that are repeated more often along the TS.

Another interesting characteristic of the system is its ability to find regions in the series whose behavior cannot be generalized. When the series contains regions with special particularities, the system cannot only localize them but can build rules for a best prediction of them. The method proposed has been devised to solve the TS problem, but it can also be applied to other machine learning domains.

Acknowledgments

This work has been financed by the Spanish-funded research MCyT project TRACER (TIC2002-04498-C05-04M) and by the Ministry of Science and Technology and FEDER contract TIN2005-08818-C04-01 (the OPLINK project).

REFERENCES

1. E. Moretti and A. Tomasin. Un contributo matematico all'elaborazione previsionale dei dati di marea a Venecia. *Boll. Ocean. Teor. Appl.* 2:45–61, 1984.
2. J. M. Zaldívar, E. Gutiérrez, I. M. Galván, and F. Strozzi. Forecasting high waters at Venice lagoon using chaotic time series analysis and nonlinear neural networks. *Journal of Hydroinformatics*, 2:61–84, 2000.
3. I. M. Galván, P. Isasi, R. Aler, and J. M. Valls. A selective learning method to improve the generalization of multilayer feedforward neural networks. *International Journal of Neural Systems*, 11:167–157, 2001.
4. J. M. Valls, I. M. Galván, and P. Isasi. Lazy learning in radial basis neural networks: a way of achieving more accurate models. *Neural Processing Letters*, 20:105–124, 2004.
5. T. P. Meyer and N. H. Packard. Local forecasting of high-dimensional chaotic dynamics. In M. Casdagli and S. Eubank, eds., *Nonlinear Modeling and Forecasting*. Addison-Wesley, Reading, MA, 1990, pp. 249–263.
6. M. Mitchell. *An Introduction to Genetic Algorithms*. MIT Press, Cambridge, MA, 1996, pp. 55–65.
7. N. H. Packard. A genetic learning algorithm for the analysis of complex data. *Complex Systems*, 4:543–572, 1990.
8. D. B. Fogel. An introduction to simulated evolutionary optimization. *IEEE Transactions on Neural Networks*, 5(1):3–14, 1994.
9. T. Bäck and H. P. Schwefel. Evolutionary algorithms: some very old strategies for optimization and adaptation. In *Proceedings of the 2nd International Workshop Software Engineering, Artificial Intelligence, and Expert Systems for High Energy and Nuclear Physics*, 1992, pp. 247–254.
10. J. H. Holland. *Adaptation in Natural and Artificial Systems*. University of Michigan Press, Ann Arbor, MI, 1975.
11. K. A. De Jong, W. M. Spears, and F. D. Gordon. Using genetic algorithms for concept learning. *Machine Learning*, 13:198–228, 1993.
12. C. Z Janikow. A knowledge intensive genetic algorithm for supervised learning. *Machine Learning*, 13:189–228, 1993.
13. L. B. Booker, D. E. Goldberg, and J. H. Holland. Classifier systems and genetic algorithms. *Artificial Intelligence*, 40:235–282, 1989.
14. K. A. De Jong. Analysis of the behavior of a class of genetic adaptive systems. Ph.D thesis, University of Michigan, 1975.
15. A. Michelato, R. Mosetti, and D. Viezzoli. Statistical forecasting of strong surges and application to the lagoon of Venice. *Boll. Ocean. Teor. Appl.*, 1:67–76, 1983.
16. A. Tomasin. A computer simulation of the Adriatic Sea for the study of its dynamics and for the forecasting of floods in the town of Venice. *Computer Physics Communications*, 5–51, 1973.
17. G. Vittori. On the chaotic features of tide elevation in the lagoon Venice. In *Proceedings of the 23rd International Conference on Coastal Engineering (ICCE'92)*, 1992, pp. 4–9.
18. M. C. Mackey and L. Glass. Oscillation and chaos in physiological control systems. *Science*, 197:287–289, 1997.

19. J. Platt. A resource-allocating network for function interpolation. *Neural Computation*, 3:213–225, 1991.
20. L. Yingwei, N. Sundararajan, and P. Saratchandran. A sequential learning scheme for function approximation using minimal radial basis function neural networks. *Neural Computation*, 9:461–478, 1997.
21. I. M. Galván and P. Isasi. Multi-step learning rule for recurrent neural models: an application to time series forecasting. *Neural Processing Letters*, 13:115–133, 2001.

CHAPTER 16

Metaheuristics in Bioinformatics: DNA Sequencing and Reconstruction

C. COTTA, A. J. FERNÁNDEZ, J. E. GALLARDO, G. LUQUE, and E. ALBA

Universidad de Málaga, Spain

16.1 INTRODUCTION

In recent decades, advances in the fields of molecular biology and genomic technologies have led to a very important growth in the biological information generated by the scientific community. The needs of biologists to utilize, interpret, and analyze that large amount of data have increased the importance of bioinformatics [1]. This area is an interdisciplinary field involving biology, computer science, mathematics, and statistics for achieving faster and better methods in those tasks.

Most bioinformatic tasks are formulated as difficult combinatorial problems. Thus, in most cases it is not feasible to solve large instances using exact techniques such as branch and bound. As a consequence, the use of metaheuristics and other approximate techniques is mandatory. In short, a metaheuristic [2,3] can be defined as a top-level general strategy that guides other heuristics to search for good solutions. Up to now there has been no commonly accepted definition for the term *metaheuristic*. It is just in the last few years that some researchers in the field have proposed a definition. Some fundamental characteristics:

- The goal is efficient exploration of the search space to find (nearly) optimal solutions.
- Metaheuristic algorithms are usually nondeterministic.
- They may incorporate mechanisms to avoid getting trapped in confined areas of the search space.
- The basic concepts of metaheuristics permit an abstract-level description.
- Metaheuristics are not problem specific.

Optimization Techniques for Solving Complex Problems, Edited by Enrique Alba, Christian Blum, Pedro Isasi, Coromoto León, and Juan Antonio Gómez

- Metaheuristics usually allow an easy parallel implementation.
- Metaheuristics must make use of domain-specific knowledge in the form of heuristics that are controlled by the upper-level strategy.

The main advantages of using metaheuristics to solve bioinformatics tasks are the following:

- Problems of bioinformatics seldom require us to find an optimal solution. In fact, they require robust, nearly optimal solutions in a short time.
- Data obtained from laboratories inherently involve errors. Due to their nondeterministic process, metaheuristics are more tolerant in these cases than deterministic processes.
- Several tasks in bioinformatics involve the optimization of different objectives, thereby making the application of (population-based) metaheuristics more natural and appropriate.

In this chapter we first present a brief survey of metaheuristic techniques and the principal bioinformatic tasks. Later we describe in more detail two important problems in the area of sequence analysis: the DNA fragment assembly and the shortest common supersequence problem. We use them to exemplify how metaheuristics can be used to solve difficult bioinformatic tasks.

16.2 METAHEURISTICS AND BIOINFORMATICS

In this section we present some background information about metaheuristics and problems of bioinformatics.

16.2.1 Metaheuristics

As we said before, a metaheuristic [2,3] can be defined as a top-level general strategy that guides other heuristics to search for good solutions. There are different ways to classify and describe metaheuristic algorithms. One of them classifies them depending on the number of solutions: population based (a set of solutions) and trajectory based (work with a single solution). The latter starts with a single initial solution. At each step of the search the current solution is replaced by another (often the best) solution found in its neighborhood. Very often, such a metaheuristic allows us to find a local optimal solution and so are called *exploitation-oriented methods*. On the other hand, the former make use of a population of solutions. The initial population is enhanced through a natural evolution process. At each generation of the process, the entire population or a part of the population is replaced by newly generated individuals (often, the best ones). Population-based methods are often called *exploration-oriented methods*. In the next paragraph we discuss the features of the most important metaheuristics.

Trajectory-Based Metaheuristics

1. *Simulated annealing* (SA) [4] is a stochastic search method in which at each step, the current solution is replaced by another one selected randomly from the neighborhood. SA uses a control parameter, temperature, to determine the probability of accepting nonimproving solutions. The objective is to escape from local optima and so to delay the convergence. The temperature is gradually decreased according to a cooling schedule such that few nonimproving solutions are accepted at the end of the search.

2. *Tabu search* (TS) [5] manages a memory of solutions or moves used recently, called the *tabu list*. When a local optimum is reached, the search carries on by selecting a candidate worse than the current solution. To avoid the selection of the previous solution again, and so to avoid cycles, TS discards the neighboring candidates that have been used previously.

3. The basic idea behind *variable neighborhood search* (VNS) [6] is to explore successively a set of predefined neighborhoods to provide a better solution. It uses the descent method to get the local minimum. Then it explores either at random or systematically the set of neighborhoods. At each step, an initial solution is shaken from the current neighborhood. The current solution is replaced by a new one if and only if a better solution has been found. The exploration is thus restarted from that solution in the first neighborhood. If no better solution is found, the algorithm moves to the next neighborhood, generates a new solution randomly, and attempts to improve it.

Population-Based Metaheuristics

1. *Evolutionary algorithms* (broadly called EAs) [7] are stochastic search techniques that have been applied successfully in many real and complex applications (epistatic, multimodal, multiobjective, and highly constrained problems). Their success in solving difficult optimization tasks has promoted research in the field known as *evolutionary computing* (EC) [3]. An EA is an iterative technique that applies stochastic operators to a pool of individuals (the population). Every individual in the population is the encoded version of a tentative solution. Initially, this population is generated randomly. An evaluation function associates a fitness value to every individual, indicating its suitability to the problem. There exist several well-accepted subclasses of EAs depending on representation of the individuals or how each step of the algorithm is performed. The main subclasses of EAs are the genetic algorithm (GA), evolutionary programming (EP), the evolution strategy (ES), and some others not shown here.

2. *Ant colony optimization* (ACO) [8] have been inspired by colonies of real ants, which deposit a chemical substance (called a *pheromone*) on the ground. This substance influences the choices they make: The larger the amount of pheromone on a particular path; the larger the probability that an ant selects the path. In this type of algorithm, artificial ants are stochastic construction procedures that probabilistically build a solution by iteratively adding problem

components to partial solutions by taking into account (1) heuristic information from the problem instances being solved, if available, and (2) pheromone trails that change dynamically at runtime to reflect the search experience acquired.

3. *Scatter search* (SS) [9] is a population-based metaheuristic that combines solutions selected from a reference set to build others. The method starts by generating an initial population of disperse and good solutions. The reference set is then constructed by selecting good representative solutions from the population. The solutions selected are combined to provide starting solutions to an improvement procedure. According to the result of such a procedure, the reference set and even the population of solutions can be updated. The process is iterated until a stopping criterion is satisfied. The SS approach involves different procedures, allowing us to generate the initial population, to build and update the reference set, to combine the solutions of such a set, to improve the solutions constructed, and so on.

Both approaches—trajectory-based and population-based—can also be combined to yield more powerful optimization techniques. This is particularly the case for memetic algorithms (MAs) [10], which blend ideas of different metaheuristics within the framework of population-based techniques. This can be done in a variety of ways, but most common approaches rely on the embedding of a trajectory-based technique within an EA-like algorithm (see refs. 11 and 12). It is also worth mentioning those metaheuristics included in the *swarm intelligence paradigm*, such as particle swarm optimization (PSO). These techniques regard optimization as an emergent phenomenon from the interaction of simple search agents.

16.2.2 Bioinformatic Tasks

In this subsection we describe the main bioinformatic tasks, giving in Table 16.1 the metaheuristic approaches applied to solve them. Based on the availability of the date and goals, we can classify the problems of bioinformatics as follows:

Alignment and Comparison of Genome and Proteome Sequences From the biological point of view, sequence comparison is motivated by the fact that all living organism are related by evolution. This implies that the genes

TABLE 16.1 Main Bioinformatic Tasks and Some Representative Metaheuristics Applied to Them

Bioinformatic Task	Metaheuristics
Sequence comparison and alignment	EA [13,14] MA [15] ACS [16] PSO [17]
DNA fragment assembly	EA [18,19] SA [20] ACS [21]
Gene finding and identification	EA [22,23]
Gene expression profiling	EA [24] MA [25,26] PSO [27]
Structure prediction	EA [28,29] MA [30] SA [31] EDA [32]
Phylogenetic trees	EA [33,34] SS [35] MA [36]

of species that are closer to each other should exhibit similarities at the DNA level. In biology, the sequences to be compared are either nucleotides (DNA, RNA) or amino acids (proteins). In the case of nucleotides, one usually aligns identical nucleotide symbols. When dealing with amino acids, the alignment of two amino acids occurs if they are identical or if one can be derived from the other by substitutions that are likely to occur in nature. A comparison of sequences comprises pairwise and simultaneous multiple sequence comparisons (and alignments). Therefore, algorithms for these problems should allow the deletion, insertion, and replacement of symbols (nucleotides or amino acids), and they should be capable of comparing a large number of long sequences. An interesting problem related to both sequence alignment and microarray production (see below) is described in Section 16.4.

DNA Fragment Assembly The fragment assembly problem consists of building the DNA sequence from several hundreds (or even, thousands) of fragments obtained by biologists in the laboratory. This is an important task in any genome project since the rest of the phases depend on the accuracy of the results of this stage. This problem is described in Section 16.3.

Gene Finding and Identification It is frequently the case in bioinformatics that one wishes to delimit parts of sequences that have a biological meaning. Typical examples are determining the locations of promoters, exons, and introns in RNA. In particular, automatic identification of the genes from the large DNA sequences is an important problem. Recognition of regulatory regions in DNA fragments has become particularly popular because of the increasing number of completely sequenced genomes and mass application of DNA chips.

Gene Expression Profiling This is the process for determining when and where particular genes are expressed. Furthermore, the expression of one gene is often regulated by the expression of another gene. A detailed analysis of all this information will provide an understanding of the internetworking of different genes and their functional roles. Microarray technology is used for that purpose.

Microarray technology allows expression levels of thousands of genes to be measured at the same time. This allows the simultaneous study of tens of thousands of different DNA nucleotide sequences on a single microscopic glass slide. Many important biological results can be obtained by selecting, assembling, analyzing, and interpreting microarray data correctly. Clustering is the most common task and allows us to identify groups of genes that share similar expressions and maybe, similar functions.

Structure Prediction Determining the structure of proteins is very important since there exists a strong relation between structure and function. This is one of the most challenging tasks in bioinformatics. There are four main levels of protein structure:

1. The primary structure is its linear sequence of amino acids.

2. The secondary structure is the folding of the primary structure via hydrogen bonds.
3. The tertiary structure refers to the three-dimensional structure of the protein and is generated by packing the secondary structural elements. Generally, the protein function depends on its tertiary structure.
4. The quaternary structure describes the formation of protein complexes composed of more than one chain of amino acids.

Also, the protein docking problem is related to the structure of the protein. This problem, to determine the interaction with other proteins, plays a key role in understanding the protein function.

Phylogenetic Analysis All species undergo a slow transformation process called evolution. Phylogenetic trees are labeled binary trees where leaves represent current species and inner nodes represent hypothesized ancestors. Phylogenetic analysis is used to study evolutionary relationships.

16.3 DNA FRAGMENT ASSEMBLY PROBLEM

In this section we study the behavior of several metaheuristics for the DNA fragment assembly problem. DNA fragment assembly is a problem solved in the early phases of the genome project and is thus very important, since the other steps depend on its accuracy. This is an NP-hard combinatorial optimization problem that is growing in importance and complexity as more research centers become involved in sequencing new genomes.

In the next subsection we present background information about the DNA fragment assembly problem. Later, the details of our approaches are presented and also how to design and implement these methods for the DNA fragment assembly problem. We finish this section by analyzing the results of our experiments.

16.3.1 Description of the Problem

To determine the function of specific genes, scientists have learned to read the sequence of nucleotides comprising a DNA sequence in a process called DNA sequencing. To do that, multiple exact copies of the original DNA sequence are made. Each copy is then cut into short fragments at random positions. These are the first three steps depicted in Figure 16.1, and they take place in the laboratory. After the fragment set is obtained, a traditional assemble approach is followed in this order: overlap, layout, and consensus. To ensure that enough fragments overlap, the reading of fragments continues until a coverage is satisfied. These steps are the last three in Figure 16.1. In what follows, we give a brief description of each of the three phases: overlap, layout, and consensus.

Overlap Phase Finding the overlapping fragments. This phase consists of finding the best or longest match between the suffix of one sequence and the

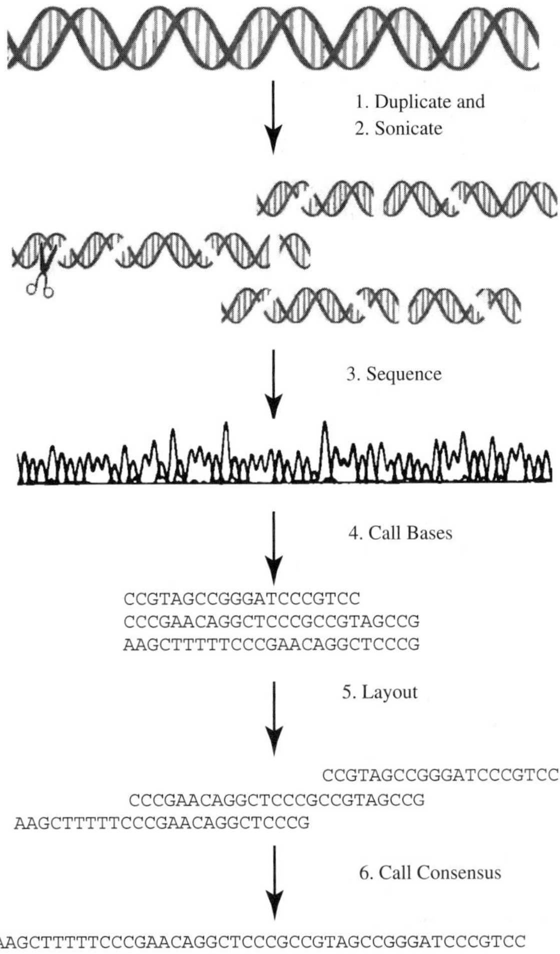

Figure 16.1 Graphical representation of DNA sequencing and assembly.

prefix of another. In this step we compare all possible pairs of fragments to determine their similarity. Usually, a dynamic programming algorithm applied to semiglobal alignment is used in this step. The intuition behind finding the pairwise overlap is that fragments with a high overlap are very likely to be next to each other in the target sequence.

Layout Phase Finding the order of fragments based on the computed similarity score. This is the most difficult step because it is difficult to tell the true overlap, due to the following challenges:

1. *Unknown orientation*. After the original sequence is cut into many fragments, the orientation is lost. One does not know which strand should be selected.

If one fragment does not have any overlap with another, it is still possible that its reverse complement might have such an overlap.

2. *Base call errors*. There are three types of base call errors: substitution, insertion, and deletion. They occur due to experimental errors in the electrophoresis procedure (the method used in laboratories to read DNA sequences). Errors affect the detection of fragment overlaps. Hence, the consensus determination requires multiple alignments in high-coverage regions.

3. *Incomplete coverage*. This occurs when an algorithm is not able to assemble a given set of fragments into a single *contig*, a sequence in which the overlap between adjacent fragments is greater than or equal to a predefined threshold (cutoff parameter).

4. *Repeated regions*. Repeats are sequences that appear two or more times in the target DNA. Repeated regions have caused problems in many genome-sequencing projects, and none of the current assembly programs can handle them perfectly.

5. *Chimeras and contamination*. Chimeras arise when two fragments that are not adjacent or overlapping on the target molecule join into one fragment. Contamination occurs due to incomplete purification of the fragment from the vector DNA.

After the order is determined, the progressive alignment algorithm is applied to combine all the pairwise alignments obtained in the overlap phase.

Consensus Phase Deriving the DNA sequence from the layout. The most common technique used in this phase is to apply the majority rule in building the consensus. To measure the quality of a consensus, we can look at the distribution of the coverage. *Coverage* at a base position is defined as the number of fragments at that position. It is a measure of the redundancy of the fragment data, and it denotes the number of fragments, on average, in which a given nucleotide in the target DNA is expected to appear. It is computed as the number of bases read from fragments over the length of the target DNA [37]:

$$\text{coverage} = \frac{\sum_{i=1}^{n} \text{length of the fragment } i}{\text{target sequence length}} \tag{16.1}$$

where n is the number of fragments. The higher the coverage, the fewer the gaps, and the better the result.

16.3.2 DNA Fragment Assembly Using Metaheuristics

Let us give some details about the most important issues of our implementation and how we have used metaheuristics to solve the DNA fragment assembly problem. First, we address common details such as the solution representation or the fitness function, and then we describe the specific features of each algorithm. The methods used are a genetic algorithm [38], a nontraditional GA called the

CHC method [39], scatter search [9], and simulated annealing [4] (for more detail about these algorithms, we refer the reader to the book by Glover and Kochenberger [3].

Common Issues

1. *Solution representation*. We use the permutation representation with integer number encoding. A permutation of integers represents a sequence of fragment numbers where successive fragments overlap. The solution in this representation requires a list of fragments assigned with a unique integer ID. For example, eight fragments will need eight identifiers: 0, 1, 2, 3, 4, 5, 6, 7. The permutation representation requires special operators to make sure that we always get legal (feasible) solutions. To maintain a legal solution, the two conditions that must be satisfied are: (1) All fragments must be presented in the ordering, and (2) no duplicate fragments are allowed in the ordering.

2. *Fitness function*. A fitness function is used to evaluate how good a particular solution is. In the DNA fragment assembly problem, the fitness function measures the multiple sequence alignment quality and finds the best scoring alignment. Our fitness function [18] sums the overlap score ($w(f, f1)$) for adjacent fragments ($f[i]$ and $f[i+1]$) in a given solution. When this fitness function is used, the objective is to maximize such a score. It means that the best individual will have the highest score, since the order proposed by that solution has strong overlap between adjacent fragments.

$$F1(l) = \sum_{i=0}^{n-2} w(f[i]f[i+1]) \tag{16.2}$$

3. *Program termination*. The program can be terminated in one of two ways. We can specify the maximum number of evaluations to stop the algorithm, or we can stop the algorithm when the solution is no longer improved.

GA Details

1. *Population size*. We use a fixed-size population of random solutions.

2. *Recombination operator*. Two or more parents are recombined to produce one or more offspring. The purpose of this operator is to allow partial solutions to evolve in different individuals and then combine them to produce a better solution. It is implemented by running through the population and for each individual, deciding whether it should be selected for crossover using a parameter called *crossover rate* (P_c). For our experimental runs, we use the order-based crossover (OX). This operator first copies the fragment ID between two random positions in Parent1 into the offspring's corresponding positions. We then copy the rest of the fragments from Parent2 into the offspring in the relative order presented in Parent2. If the fragment ID is already present in the offspring, we skip

that fragment. The method preserves the feasibility of every tentative solution in the population.

3. *Mutation operator*. This operator is used for the modification of single individuals. The reason we need a mutation operator is for the purpose of maintaining diversity in the population. Mutation is implemented by running through the entire population and for each individual, deciding whether or not to select it for mutation, based on a parameter called the *mutation rate* (P_m). For our experimental runs, we use the swap mutation operator and invert the segment mutation operator. The first operator randomly selects two positions from a permutation and then swaps the two fragment positions. The second one also selects two positions from a permutation and then inverts the order of the fragments in partial permutation defined by the two random positions (i.e., we swap two edges in the equivalent graph). Since these operators do not introduce any duplicate number in the permutation, the solution they produce is always feasible.

4. *Selection operator*. The purpose of the selection is to weed out bad solutions. It requires a population as a parameter, processes the population using the fitness function, and returns a new population. The level of the selection pressure is very important. If the pressure is too low, convergence becomes very slow. If the pressure is too high, convergence will be premature to a local optimum.

In this chapter we use the ranking selection mechanism [40], in which the GA first sorts individuals based on fitness and then selects the individuals with the best fitness score until the population size specified is reached. Preliminary results favored this method out of a set of other selection techniques analyzed. Note that the population size will grow whenever a new offspring is produced by crossover or mutation operators.

CHC Details

1. *Incest prevention*. The CHC method has a mechanism of *incest prevention* to avoid recombination of similar solutions. Typically, the Hamming distance is used as a measure of similarity, but this one is unsuitable for permutations. In the experiments we consider that the distance between two solutions is the total number of edges minus the number of common edges.

2. *Recombination*. The crossover that we use in our CHC creates a single child by preserving the edges that parents have in common and then randomly assigning the remaining edges to generate a legal permutation.

3. *Population restart*. Whenever the population converges, the population is partially randomized for a restart. The restart method uses as a template the best individual found so far, creating new individuals by repeatedly swapping edges until a specific fraction of edges differ from those of the template.

SS Details

1. *Initial population creation*. There exist several ways to get an initial population of good and disperse solutions. In our experiments, the solutions for the

population were generated randomly to achieve a certain level of diversity. Then we apply the *improvement method* to these solutions to get better solutions.

2. *Subsets generation and recombination operator*. It generates all two-element subsets and then it applies the recombination operator to them. For our experimental runs, we use the order-based crossover that was explained before.

3. *Improvement method*. We apply a hillclimber procedure to improve the solutions. The hillclimber is a variant of Lin and Kernighan's two-opt [41]. Two positions are selected randomly, and then the subpermutation is inverted by swapping the two edges. Whenever an improvement is found, the solution is updated; the hillclimber continues until it achieves a predetermined number of swap operations.

SA Details

1. *Cooling scheme*. The cooling schedule controls the values of the temperature parameter. It specifies the initial value and how the temperature is updated at each stage of the algorithm:

$$T_k = \alpha T_{k-1} \tag{16.3}$$

In this case we use a decreasing function (Equation 16.3) controlled by the α factor, where $\alpha \in (0, 1)$.

2. *Markov chain length*. The number of the iterations between two consecutive changes of the temperature is given by the parameter *Markov chain length*, whose name alludes to the fact that the sequence of solutions accepted is a Markov chain (a sequence of states in which each state depends only on the preceding one).

3. *Move operator*. This operator generates a new neighbor from the current solution. For our experimental runs, we use the edge swap operator. This operator randomly selects two positions from a permutation and inverts the order of the fragments between these two fragment positions.

16.3.3 Experimental Analysis

A target sequence with accession number BX842596 (GI 38524243) was used in this work. It was obtained from the NCBI website (http://www.ncbi.nlm.nih.gov). It is the sequence of a *Neurospora crassa* (common bread mold) BAC and is 77,292 base pairs long. To test and analyze the performance of our algorithm, we generated two problem instances with GenFrag [42]. GenFrag takes a known DNA sequence and uses it as a parent strand from which to generate fragments randomly according to the criteria (mean fragment length and coverage of parent sequence) supplied by the user. The first problem instance, 842596_4, contains 442 fragments with average fragment length 708 bps and coverage 4. The second problem instance, 842596_7, contains 773 fragments with average fragment length 703 bps and coverage 7. We evaluated the results in terms of the number of contigs assembled.

We use a GA, a CHC, an SS, and an SA to solve this problem. To allow a fair comparison among the results of these heuristics, we have configured them to perform a similar computational effort (the maximum number of evaluations for any algorithm is 512,000). Since the results of these algorithms vary depending on the different parameter settings, we performed a complete analysis previously to study how the parameters affect the performance of the algorithms. A summary of the conditions for our experimentation is given in Table 16.2. We have performed statistical analyses to ensure the significance of the results and to confirm that our conclusions are valid (all the results are different statistically).

Table 16.3 shows all the results and performance with all data instances and algorithms described in this work. The table shows the fitness of the best solution obtained (b), the average fitness found (f), average number of evaluations (e), and average time in seconds (t). We do not show the standard deviation because the fluctuations in the accuracy of different runs are rather small, showing that the algorithms are very robust (as proved by the ANOVA results). The best results are shown in boldface type.

TABLE 16.2 Parameters and Optimum Solutions of the Problem

Common parameters	
Independent runs	30
Cutoff	30
Max. evaluations	51,2000
Genetic algorithms	
Population size	1024
Crossover	OX (0.8)
Mutation	Edge swap (0.2)
Selection	Ranking
CHC	
Population size	50
Crossover	Specific (1.0)
Restart	Edge swap (30%)
Scatter search	
Initial population size	15
Reference set	8 (5 + 3)
Subset generation	All two-element subsets (28)
Crossover	OX (1.0)
Improvement	Edge swap (100 iterations)
Simulated annealing	
Move operator	Edge swap
Markov chain length	Total number evaluations/100
Cooling scheme	Proportional ($\alpha = 0.99$)

TABLE 16.3 Results for the Two Instances

	38524243_4				38524243_7			
Algorithm	*b*	*f*	*e*	*t*	*b*	*f*	*e*	*t*
Genetic	92,772	88,996	508,471	32.62	108,297	104,330	499,421	85.77
CHC	61,423	54,973	487,698	65.33	86,239	81,943	490,815	162.29
Scatter search	94,127	90,341	**51,142**	27.83	262,317	254,842	**52,916**	66.21
Simulated annealing	**225,744**	**223,994**	504,850	**7.92**	**416,838**	**411,818**	501,731	**12.52**

Let us discuss some of the results found in this table. First, for the two instances, it is clear that the SA outperforms the other algorithms from every point of view. In fact, SA obtains better fitness values than do the previous best known solutions [20]. Also, its execution time is the lowest. The reason is that the SA operates on a single solution, while the remaining methods are population based, and in addition, they execute time-consuming operators (especially the crossover operation).

The CHC is the worst algorithm in both execution time and solution quality. Its longer runtime is due to the additional computations needed to detect the converge of the population or to detect incest mating. CHC is not able to solve the DNA fragment assembly problem adequately and perhaps requires a local search (as proposed by Karpenko et al. [16]) to reduce the search space.

The SS obtains better solutions than the GA, and it also achieves these solutions in a shorter runtime. This means that a structured application of operator and explicit reference search points are a good idea for this problem.

The computational effort to solve this problem (considering only the number of evaluations) is similar for all the heuristics with the exception of the SS, because its most time-consuming operation is the improvement method that does not perform complete evaluations of the solutions. This result indicates that all the algorithms examine a similar number of points in the search space, and the difference in the solution quality is due to how they explore the search space. For this problem, trajectory-based methods such as simulated annealing are more effective than population-based methods. Thus, the resulting ranking of algorithms from best to worst is SA, SS, GA, and finally, CHC.

Table 16.4 shows the final number of contigs computed in every case, a contig being a sequence in which the overlap between adjacent fragments is greater than a threshold (cutoff parameter). Hence, the optimum solution has a single contig. This value is used as a high-level criterion to judge the entire quality of the results, since, as we said before, it is difficult to capture the dynamics of the problem in a mathematical function. These values are computed by applying a final step of refinement with a greedy heuristic popular in this application [19]. We have found that in some (extreme) cases it is possible that a solution with a better fitness than other solution generates a larger number of contigs (worse solution). This is the reason for still needing more research to get a more accurate mapping from fitness to contig number. The values of this table, however, confirm

TABLE 16.4 Final Best Contigs

Algorithm	38524243_4	38524243_7
GA	6	4
CHC	7	5
SS	6	4
SA	**4**	**2**

again that the SA method outperform the rest clearly, the CHC obtains the worst results, and the SS and GA obtain a similar number of contigs.

16.4 SHORTEST COMMON SUPERSEQUENCE PROBLEM

The shortest common supersequence problem (SCSP) is a classical problem from the realm of string analysis. Roughly speaking, the SCS problem amounts to finding a minimal-length sequence $S \in \Sigma^*$ of symbols from a certain alphabet Σ such that every sequence in a certain set $L \in 2^{\Sigma^*}$ can be generated from S by removing some symbols of the latter. The resulting combinatorial problem is enormously interesting from the point of view of bioinformatics [43] and bears a close relationship to the sequence alignment and microarray production, among other tasks.

Unfortunately, the SCS problem has been shown to be hard under various formulation and restrictions, resulting not just in a NP-hard problem, but also in a non-FPT problem [44], so practical resolution is probably unaffordable by conventional exact techniques. Therefore, several heuristics and metaheuristics have been defined to tackle it. Before detailing these, let us first describe the SCSP more formally.

16.4.1 Description of the Problem

We write $|s|$ for the length of sequence s ($|s_1 s_2 \cdots s_n| = n$) and $|\Sigma|$ for the cardinality of set Σ. Let ϵ be the empty sequence ($|\epsilon| = 0$). We use $s \rhd \alpha$ for the total number of occurrences of symbol α in sequence s:

$$s_1 s_2 \cdots s_n \rhd \alpha = \sum_{1 \le i \le n, s_i = \alpha} 1 \tag{16.4}$$

We write αs for the sequence obtained by appending the symbol α in front of sequence s. Deleting the symbol α from the front of sequence s is denoted $s|_\alpha$. We also use the | symbol to delete a symbol from the front of a set of strings: $\{s_1, \ldots, s_m\}|_\alpha = \{s_1|_\alpha, \ldots, s_m|_\alpha\}$.

Sequence s is said to be a supersequence of r (denoted as $s \succ r$) if $r = \epsilon$ or if $s|_{s_1}$ is a supersequence of $r|_{s_1}$. Plainly, $s \succ r$ implies that r can be embedded in s, meaning that all symbols in r are present in s in exactly the same order, although not necessarily consecutive. We can now state the SCSP as follows: An instance $I = (\Sigma, L)$ for the SCSP is given by a finite alphabet Σ and a set L of m sequences $\{s_1, \ldots, s_m\}$, $s_i \in \Sigma^*$. The problem consists of finding a sequence s of minimal length that is a supersequence of each sequence in L ($s \succ s_i, \forall s_i \in L$ and $|s|$ is minimal). A particularly interesting situation for bioinformatics arises when the sequences represent molecular data (i.e., sequences of nucleotides or amino acids).

16.4.2 Heuristics and Metaheuristics for the SCSP

One of the simplest and most effective algorithms for the SCSP is *majority merge* (MM). This is a greedy algorithm that constructs a supersequence incrementally by adding the symbol most frequently found at the front of the sequences in L and removing these symbols from the corresponding strings. Ties can be broken randomly, and the process is repeated until all sequences in L are empty. A drawback of this algorithm is its myopic functioning, which makes it incapable of grasping the global structure of strings in L. In particular, MM misses the fact that the strings can have different lengths [45]. This implies that symbols at the front of short strings will have more chances to be removed, since the algorithm still has to scan the longer strings. For this reason, it is less urgent to remove those symbols. In other words, it is better to concentrate first on shortening longer strings. This can be done by assigning a weight to each symbol, depending on the length of the string in whose front it is located. Branke et al. [45] propose using precisely this string length as a weight (i.e., the weight of a symbol is the sum of the length of all strings at whose front it appears). This modified heuristic is termed *weighted majority merge* (WMM), and its empirical evaluation indicates that it can outperform MM on some problem instances in which there is no structure or where the structure is deceptive [45,46].

The limitations of these simple heuristics have been dealt with via the use of more sophisticated techniques. One of the first metaheuristic approaches to the SCSP is due to Branke et al. [45]. They consider several evolutionary algorithm approaches, in particular a GA that uses WMM as a decoding mechanism. More precisely, the GA evolves weights (metaweights, actually) that are used to further refine the weights computed by the WMM heuristic. The latter algorithm is then used to compute tentative supersequences on the basis of these modified weights. This procedure is similar to the EA used by Cotta and Troya [47] for the multidimensional knapsack problem, in which a greedy heuristic was used to generate solutions, and weights were evolved in order to modify the value of objects. A related approach was presented by Michel and Middendorf [48] based on ACO. Pheromone values take the role of weights, and ants take probabilistic decisions based on these weights. An interesting feature of this ACO approach is the fact that pheromone update is not done on an individual basis (i.e., just on the symbols being picked each time), but globally (i.e., whenever a symbol is picked, the pheromone of previous symbols that allowed that symbol to be picked is also increased).

More recent approaches to the SCSP are based on EAs [46]. These EAs range from simple direct approaches based on penalty functions, to more complex approaches based on repairing mechanisms or indirect encodings:

1. *Simple EA*. Solutions are represented as a sequence of symbols corresponding to a tentative supersequence. Fitness is computed as the length of the supersequence plus a penalty term in case a solution does not contain a valid supersequence. This penalty term is precisely the length of the solution provided by the MM heuristic for the remaining sequences:

$$F(s, L) = \begin{cases} 0 & \text{if } \forall i : s_i = \epsilon \\ 1 + F(s', L|_\alpha) & \text{if } \exists i : s_i \neq \epsilon \text{ and } s = \alpha s' \\ |\text{MM}(L)| & \text{if } \exists i : s_i \neq \epsilon \text{ and } s = \epsilon \end{cases} \tag{16.5}$$

Standard recombination and mutation operators can be used in this EA.

2. *EA + repairing*. Based on the simple EA, a repair function ρ is used to complete the supersequence with the symbols indicated by the MM heuristic:

$$\rho\,(s, L) = \begin{cases} s & \text{if } \forall i : s_i = \epsilon \\ \rho(s', L) & \text{if } \exists i : s_i \neq \epsilon \text{ and } \neg\exists i : s_i = \alpha s_i' \text{ and } s = \alpha s' \\ \alpha\rho(s', L|_\alpha) & \text{if } \exists i : s_i = \alpha s_i' \text{ and } s = \alpha s' \\ \text{MM}(L) & \text{if } \exists i : s_i \neq \epsilon \text{ and } s = \epsilon \end{cases} \tag{16.6}$$

Notice that this function not only turns infeasible solutions into feasible ones, but also improves feasible solutions by removing unproductive steps.

3. *Indirect EAs*. In the line of the GRASP-like decoding mechanism defined by Cotta and Fernández [49], the EA evolves sequences of integers that denote the rank of the symbol that has to be picked at each step, on the basis of the ranking provided by a simple heuristic such as MM or WMM.

The experimental results [46] indicate that the EA with repairing is better than the other two approaches for small alphabet sizes (e.g., in the case of nucleotide sequences). For larger alphabets, the EA with indirect encoding is slightly better. Further improvements are obtained from the inclusion of a local search technique within the EA, thus resulting in a MA [15]. Local search is done by removing symbols from the supersequence and retaining the deletion if the resulting solution is valid. This local search procedure is costly, so it has to be used with caution. To be precise, it is shown that partial Lamarckism rates (around 1 or 5%) provide the best trade-off between computational cost and quality of results.

A completely different metaheuristic, beam search (BS), has also been proposed for the SCSP [50]. This technique approaches the incremental construction of k (a parameter) solutions via a pseudo-population-based greedy mechanism inspired in branch and bound and the breadth-first traversal of search trees. This is a fully deterministic approach (save for tie breaking) that is shown to provide moderately good solutions to the problem. Several improvements to this basic technique have been proposed. On the one hand, the hybridization of BS and MAs has also been attempted [50]. On the other hand, a probabilistic version of BS (PBS) has been described [51]. Both approaches have been shown to be superior to the basic technique and are very competitive in general. PBS is remarkable for its success in small sequences, but it starts to suffer scalability problems when the length of sequences increases.

16.4.3 Experimental Analysis

An experimental comparison has been made between the most competitive algorithms described before: namely, the MA-BS hybrid [50] and both the simple heuristics MM and WMM. The instances considered for the experimentation comprise both DNA sequences ($|\Sigma| = 4$) and protein sequences ($|\Sigma| = 20$). In the first case we have taken two DNA sequences of the SARS coronavirus from a genomic database (http://gel.ym.edu.tw/sars/genomes.html); these sequences are 158 and 1269 nucleotides long. As to the protein sequences, we have considered three of them, extracted from Swiss-Prot (http://www.expasy.org/sprot/):

- *Oxytocin*. Quite important in pregnant women, this protein causes contraction of the smooth muscle of the uterus and the mammary gland. The sequence is 125 amino acids long.
- *p53*. This protein is involved in the cell cycle and acts as tumor suppressor in many tumor types; the sequence is 393 amino acids long.
- *Estrogen*. Involved in the regulation of eukaryotic gene expression, this protein affects cellular proliferation and differentiation; the sequence is 595 amino acids long.

In all cases, problem instances are constructed by generating strings from the target sequence by removing symbols from the latter with probability p. In our experiments, problem instances comprise 10 strings, and $p = \{5\%, 10\%, 20\%\}$, thus representing increasingly difficult optimization problems.

All algorithms are run for 600 seconds on a Pentium-4 2.4-GHz 512-Mbyte computer. The particular parameters of the MA component are population size = 100, $p_X = 0.9$, $p_M = 1/L$, $p_{LS} = 0.01$, uniform crossover, tournament selection, and steady-state replacement. As to the BS component, it considers $k = 10{,}000$. The results (averaged for 20 runs) are shown in Figures 16.2 and 16.3.

As can be seen, the conventional greedy heuristics cannot compete with the hybrid approach. Notice that the results of the latter are very close to the putative optimal (the original sequence to be reconstructed). Actually, it can be seen that the hybrid MA-BS algorithm manages to find this original solution in all runs. As reported by Blum et al. [51], the PBS algorithm is also capable of performing satisfactorily for most of these problem instances but is affected by the increased dimensionality of the problem in the SARS 1269 instance, for high values of p. MA-BS is somewhat more robust in this sense.

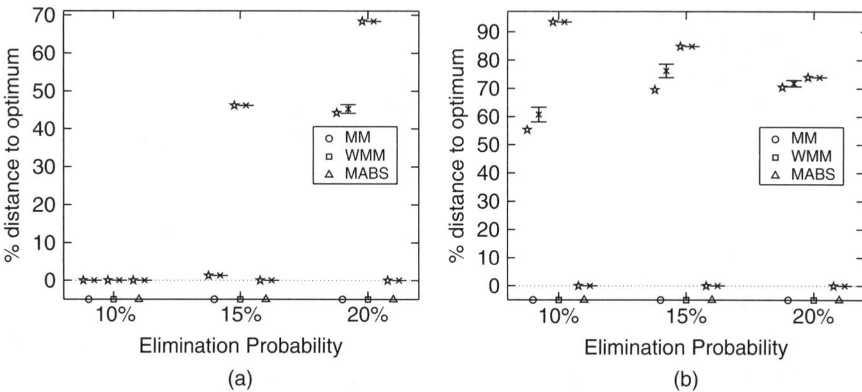

Figure 16.2 Results of MM, WMM, and MA-BS for the (a) SARS158 and (b) SARS1269 instances. Each group of bars indicates the results of the corresponding algorithm (MM, WMM, and MA-BS) for three different values of p (i.e., 5%, 10%, 20%). The star indicates the best solution found by the corresponding algorithm, the cross marks the mean, and the bars indicate the standard deviation.

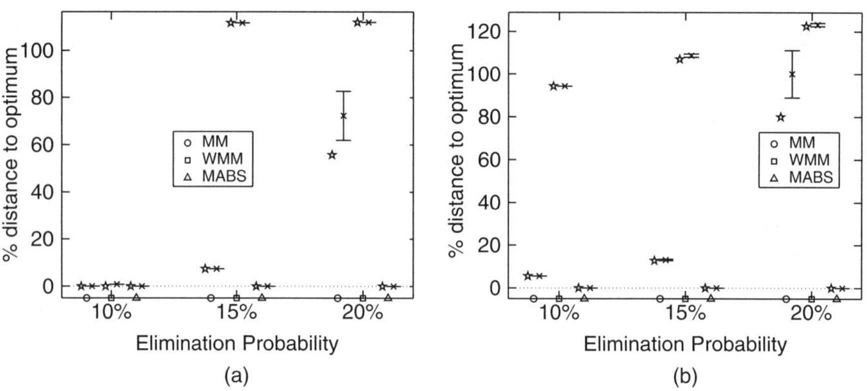

Figure 16.3 Results of MM, WMM, and MA-BS for the (a) P53 and (b) estrogen instances; all techniques manage to find the optimal solution in most runs for the oxytocin instance. The exceptions are MM and WMM for the 20% case.

16.5 CONCLUSIONS

The increase in biological data and the need to analyze and interpret them have opened several important research lines in computation science since the application of computational approaches allows us to facilitate an understanding of various biological process. In this chapter we have shown how these problems can be solved with metaheuristics. In particular, we have tackled two difficult problems: the DNA fragment assembly problem and the shortest common supersequence problem.

Both the DNA FAP and the SCSP are very complex problems in computational biology. We tackled the first problem with four sequential heuristics: three population based (GA, SS, and CHC), and one trajectory based (SA). We have observed that the latter outperforms the rest of the methods (which implies more research in trajectory-based methods in the future). It obtains much better solutions than the others do, and it runs faster. This local search algorithm also outperformed the best known fitness values in the literature. As to the second, we have shown that the best approach described in the literature is the hybrid of memetic algorithms and beam search defined by Gallardo et al. [50]. This hybrid method vastly outperforms conventional greedy heuristics on both DNA sequences and protein sequences.

Acknowledgments

The authors are partially supported by the Ministry of Science and Technology and FEDER under contract TIN2005-08818-C04-01 (the OPLINK project). The second and third authors are also supported under contract TIN-2007-67134.

REFERENCES

1. J. Cohen. Bioinformatics: an introduction to computer scientists. *ACM Computing Surveys*, 36:122–158, 2004.
2. C. Blum and A. Roli. Metaheuristics in combinatorial optimization: overview and conceptual comparison. *ACM Computing Surveys*, 35(3):268–308, 2003.
3. F. Glover and A. G. Kochenberger. Handbook of Metaheuristics. International Series in Operations Research and Management Science. Springer-Verlag, New York, Jan. 2003.
4. S. Kirkpatrick, C. D. Gelatt, and M. P. Vecchi. Optimization by simulated annealing. *Science*, 220(4598):671–680, 1983.
5. F. Glover. Tabu search: I. *ORSA, Journal of Computing*, 1:190–206, 1989.
6. N. Mladenovic and P. Hansen. Variable neighborhood search. *Computers and Operations Research*, 24:1097–1100, 1997.
7. T. Bäck, D. B. Fogel, and Z. Michalewicz, eds. *Handbook of Evolutionary Computation*. Oxford University Press, New York, 1997.
8. M. Dorigo and T. Stützle. *Ant Colony Optimization*. MIT Press, Cambridge, MA, 2004.
9. F. Glover, M. Laguna, and R. Martí. Fundamentals of scatter search and path relinking. *Control and Cybernetics*, 39(3):653–684, 2000.
10. P. Moscato and C. Cotta. A gentle introduction to memetic algorithms. In F. Glover and G. Kochenberger, eds., *Handbook of Metaheuristics*. Kluwer Academic, Norwell, MA, 2003, pp. 105–144.
11. P. Moscato, C. Cotta, and A. Mendes. Memetic algorithms. In G. C. Onwubolu and B. V. Babu, eds., *New Optimization Techniques in Engineering*. Springer-Verlag, New York, 2004, pp. 53–85.

12. P. Moscato and C. Cotta. Memetic algorithms. In T. González, ed., *Handbook of Approximation Algorithms and Metaheuristics*. Taylor & Francis, London, 2007.

13. C. Notredame and D. G. Higgins. SAGA: sequence alignment by genetic algorithm. *Nucleic Acids Research*, 24:1515–1524, 1996.

14. C. Zhang and A. K. Wong. A genetic algorithm for multiple molecular sequence alignment. *Bioinformatics*, 13(6):565–581, 1997.

15. C. Cotta. Memetic algorithms with partial Lamarckism for the shortest common supersequence problem. In J. Mira and J. R. Álvarez, eds., *Artificial Intelligence and Knowledge Engineering Applications: A Bioinspired Approach*, vol. 3562 of *Lecture Notes in Computer Science*. Springer-Verlag, New York, 2005, pp. 84–91.

16. O. Karpenko, J. Shi, and Y. Dai. Prediction of MHC class II binders using the ant colony search strategy. *Artificial Intelligence in Medicine*, 35(1–2):147–156, 2005.

17. T. K. Rasmussen and T. Krink. Improved hidden Markov model training for multiple sequence alignment by a particle swarm optimization–evolutionary algorithm hybrid. *Biosystems*, 72(1–2):5–17, 2003.

18. R. Parsons, S. Forrest, and C. Burks. Genetic algorithms, operators, and DNA fragment assembly. *Machine Learning*, 21:11–33, 1995.

19. L. Li and S. Khuri. A comparison of DNA fragment assembly algorithms. In *Proceedings of the International Conference on Mathematics and Engineering Techniques in Medicine and Biological Sciences*, 2004, pp. 329–335.

20. E. Alba, G. Luque, and S. Khuri. Assembling DNA fragments with parallel algorithms. In B. McKay, ed., *Proceedings of the 2005 Congress on Evolutionary Computation*, Edinburgh, UK, 2005, pp. 57–65.

21. P. Meksangsouy and N. Chaiyaratana. DNA fragment assembly using an ant colony system algorithm. In *Proceedings of the 2003 Congress on Evolutionary Computation*, vol. 3. IEEE Press, New York 2003, pp. 1756–1763.

22. A. Kel, A. Ptitsyn, V. Babenko, S. Meier-Ewert, and H. Lehrach. A genetic algorithm for designing gene family-specific oligonucleotide sets used for hybridization: the g protein-coupled receptor protein superfamily. *Bioinformatics*, 14:259–270, 1998.

23. V. G. Levitsky and A. V. Katokhin. Recognition of eukaryotic promoters using a genetic algorithm based on iterative discriminant analysis. *In Silico Biology*, 3(1):81–87, 2003.

24. K. Deb, S. Agarwal, A. Pratap, and T. Meyarivan. A fast elitist non-dominated sorting genetic algorithm for multi-objective optimization: NSGA-II. In *Proceedings of the Parallel Problem Solving from Nature VI*, 2000, pp. 849–858.

25. C. Cotta, A. Mendes, V. Garcia, P. Franca, and P. Moscato. Applying memetic algorithms to the analysis of microarray data. In G. Raidl et al., eds., *Applications of Evolutionary Computing*, vol. 2611 of *Lecture Notes in Computer Science*. Springer-Verlag, New York, 2003, pp. 22–32.

26. A. Mendes, C. Cotta, V. Garcia, P. França, and P. Moscato. Gene ordering in microarray data using parallel memetic algorithms. In T. Skie and C.-S. Yang, eds., *Proceedings of the 2005 International Conference on Parallel Processing Workshops*, Oslo, Norway. IEEE Press, New York, 2005, pp. 604–611.

27. X. Xiao, E. R. Dow, R. Eberhart, Z. B. Miled, and R. J. Oppelt. Gene clustering using self-organizing maps and particle swarm optimization. In *Proceedings of the 6th IEEE International Workshop on High Performance Computational Biology*, Nice, France, Apr. 2003.

28. C. Notredame, L. Holm, and D. G. Higgins. COFFEE: an objective function for multiple sequence alignments. *Bioinformatics*, 14(5):407–422, 1998.

29. G. Fogel and D. Corne. *Evolutionary Computation in Bioinformatics*. Morgan Kaufmann, San Francisco, CA, 2002.

30. C. Cotta. Protein structure prediction using evolutionary algorithms hybridized with backtracking. In J. Mira and J. R. Álvarez, eds., *Artificial Neural Nets Problem Solving Methods*, vol. 2687 of *Lecture Notes in Computer Science*. Springer-Verlag, New York, 2003, pp. 321–328.

31. J. J. Gray, S. Moughon, C. Wang, O. Schueler-Furman, B. Kuhlman, C. A. Rohl, and D. Baker. Protein–protein docking with simultaneous optimization of rigid-body displacement and side-chain conformations. *Journal of Molecular Biology*, 331(1):281–299, 2003.

32. R. Santana, P. Larrañaga, and J. A. Lozano. Protein folding in 2-dimensional lattices with estimation of distribution algorithms. In J. M Barreiro, F. Martín-Sánchez, V. Maojo, and F. Sanz, eds., *Proceedings of the 5th Biological and Medical Data Analysis International Symposium*, vol. 3337 of *Lecture Notes in Computer Science*. Springer-Verlag, New York, 2004, pp. 388–398.

33. P. O. Lewis. A genetic algorithm for maximum likelihood phylogeny inference using nucleotide sequence data. *Molecular Biology and Evolution*, 15:277–283, 1998.

34. C. Cotta and P. Moscato. Inferring phylogenetic trees using evolutionary algorithms. In J. J. Merelo et al., eds., *Parallel Problem Solving from Nature VII*, vol. 2439 of *Lecture Notes in Computer Science*. Springer-Verlag, New York, 2002, pp. 720–729.

35. C. Cotta. Scatter search with path relinking for phylogenetic inference. *European Journal of Operational Research*, 169(2):520–532, 2006.

36. J. E. Gallardo, C. Cotta, and A. J. Fernández. Reconstructing phylogenies with memetic algorithms and branch and bound. In S. Bandyopadhyay, U. Maulik, and J. T. L. Wang, eds., *Analysis of Biological Data: A Soft Computing Approach*. World Scientific, Hackensack, NJ, 2007, pp. 59–84.

37. J. Setubal and J. Meidanis. Fragment assembly of DNA. In *Introduction to Computational Molecular Biology*. University of Campinas, Brazil, 1997, pp. 105–139.

38. J. H. Holland. *Adaptation in Natural and Artificial Systems*. University of Michigan Press, Ann Arbor, MI, 1975.

39. L. J. Eshelman. The CHC adaptive search algorithm: how to have safe search when engaging in nontraditional genetic recombination. In *Foundations of Genetic Algorithms*, Vol. 1. Morgan Kaufmann, San Francisco, CA, 1991, pp. 265–283.

40. D. Whitely. The GENITOR algorithm and selection pressure: why rank-based allocation of reproductive trials is best. In J. D. Schaffer, ed., *Proceedings of the 3rd International Conference on Genetic Algorithms*. Morgan Kaufmann, San Francisco, CA, 1989, pp. 116–121.

41. S. Lin and B. W. Kernighan. An effective heuristic algorithm for TSP. *Operations Research*, 21:498–516, 1973.

42. M. L. Engle and C. Burks. Artificially generated data sets for testing DNA fragment assembly algorithms. *Genomics*, 16, 1993.

43. M. T. Hallet. An integrated complexity analysis of problems from computational biology. Ph.D. thesis, University of Victoria, 1996.

44. R. Downey and M. Fellows. *Parameterized Complexity*. Springer-Verlag, New York, 1998.

45. J. Branke, M. Middendorf, and F. Schneider. Improved heuristics and a genetic algorithm for finding short supersequences. *OR-Spektrum*, 20:39–45, 1998.

46. C. Cotta. A comparison of evolutionary approaches to the shortest common supersequence problem. In J. Cabestany, A. Prieto, and D. F. Sandoval, eds., *Computational Intelligence and Bioinspired Systems*, vol. 3512 of *Lecture Notes in Computer Science*. Springer-Verlag, New York, 2005, pp. 50–58.

47. C. Cotta and J. M. Troya. A hybrid genetic algorithm for the 0–1 multiple knapsack problem. In G. D. Smith, N. C. Steele, and R. F. Albrecht, eds., *Artificial Neural Nets and Genetic Algorithms*, Vol. 3. Springer-Verlag, New York, 1998, pp. 251–255.

48. R. Michel and M. Middendorf. An ant system for the shortest common supersequence problem. In D. Corne, M. Dorigo, and F. Glover, eds., *New Ideas in Optimization*. McGraw-Hill, London, 1999, pp. 51–61.

49. C. Cotta and A. Fernández. A hybrid GRASP–evolutionary algorithm approach to Golomb ruler search. In X. Yao et al., eds., *Parallel Problem Solving from Nature VIII*, vol. 3242 of *Lecture Notes in Computer Science*. Springer-Verlag, New York, 2004, pp. 481–490.

50. J. E. Gallardo, C. Cotta, and A. J. Fernández. Hybridization of memetic algorithms with branch-and-bound techniques. *IEEE Transactions on Systems, Man, and Cybernetics, Part B*, 37(1):77–83, 2007.

51. C. Blum, C. Cotta, A. J. Fernández, and J. E. Gallardo. A probabilistic beam search approach to the shortest common supersequence problem. In C. Cotta and J. van Hemert, eds., *Evolutionary Computation in Combinatorial Optimization*, vol. 4446 of *Lecture Notes in Computer Science*. Springer-Verlag, New York, 2007, pp. 36–47.

CHAPTER 17

Optimal Location of Antennas in Telecommunication Networks

G. MOLINA, F. CHICANO, and E. ALBA

Universidad de Málaga, Spain

17.1 INTRODUCTION

Mobile communications is a major area in the industry of the twenty-first century. As customers get used to having both mobility and connectivity, these types of services are required more and more. Mobile communications require the use of a mobile device by the end user, the presence of a network accessible by the mobile device from any place the user has to be, and a backbone network that manages the connections and communications between users. Also, ad hoc and sensor networks need to define a cluster responsible for communications to take place, in a dynamic and novel way of assigning data transfer functionalities to a given terminal. Recently, numerous companies have entered this area and began to compete to offer the best services at the lowest cost. Therefore, a great number of issues have arisen as problems to be solved in order to optimize the features of the service.

In this chapter we present one of the major issues found in this domain, the design of the access network, and propose a set of optimization techniques that can solve it efficiently and effectively. These techniques can be employed for static regular cellular phone networks and for dynamic networks with no infrastructure (such as ad hoc or sensor networks); hence, results could be of very high interest in academia and industry.

A generally accepted method of building an access network is to divide the terrain to be covered into small cells, each of which can be covered by a single transmitter conveniently located in a base station (BS); this solution is known as a *radio network*. The problem we solve is how to achieve maximum coverage of the terrain in order to obtain a valuable service for the customer (ideally, the coverage should be complete) by placing the lowest number of transmitters, so

Optimization Techniques for Solving Complex Problems, Edited by Enrique Alba, Christian Blum, Pedro Isasi, Coromoto León, and Juan Antonio Gómez

that the cost of the service remains competitive. This is equivalent to selecting the optimal positions for placing the transmitters, and this problem is known as the *radio network design* (RND) *problem*.

In this chapter we compare several algorithms on the same large set of instances to highlight their different advantages for the RND problem. We define three academic problem classes that differ in the type of transmitter modeled. For each of these classes we test our algorithms on several instances. This test is intended to give some insight into the performances of the algorithms and provide knowledge for their parameter tuning. In addition, we define a real-world-like instance based on the city of Malaga to test the algorithms under realistic constraints and complexity.

The chapter is organized as follows. A quick review of the state of the art in optimization for a radio network design is given in Section 17.2. In Section 17.3 a formal description of the RND problem is presented. In Section 17.4 we describe briefly the optimization techniques employed. In Sections 17.5 and 17.6 we solve the set of basic instances and the real-world instance, respectively. Finally, some conclusions are drawn in Section 17.7.

17.2 STATE OF THE ART

A large number of problems have had to be solved in order to bring telecommunication services to a wide public. Most technical and technological problems were faced during the early ages of this industry, and this has led to the current telecommunication means. The industry is currently reaching its maturity, and a new type of problem have to be solved: optimization problems. Now that technological issues have become secondary, the main goal for companies is to obtain the most efficient ways to administrate their resources so as to get the maximum profits. A large variety of optimization problems are being faced in the telecommunications field, and a wide variety of techniques have proven to be useful in those tasks. Recent work in this area has obtained good results in real-world scenarios, which encourages further research. We present in this section a review of the state of the art (to the authors' knowledge) in the field of RND; this review is summed up in Table 17.1.

Celli et al. [14] applied a genetic algorithm (GA) to optimize telecommunication networks, with the objective of building minimum cost networks that satisfy a set of requirements. With the set of network nodes and end users given as inputs, the algorithm must determine which nodes are multiplexers, which are exchangers, and what links are established between nodes. The results obtained by the GA outperformed those obtained by a standard heuristic algorithm. A well-known clustering problem, the location area management, was solved by Ali [4]. With the aim of clustering a cellular mobile radio network so as to obtain the minimum amount of roaming information (which depends on the number of cells in a frontier and the number of users traveling between different clusters at any time), an iterative algorithm was developed and applied. Meguerdichian

TABLE 17.1 Parallel Metaheuristics Applied to Network Design Problems

Author(s)	Year	Related Optimization Problem	Metaheuristic
Celli et al. [14]	1995	Network design	Genetic algorithm
Meunier et al. [22]	2000	Position and configuration of mobile base stations	Multiobjective evolutionary algorithm
Calégari et al. [12]	2001	Antenna placement–hitting set problem	Genetic algorithm
Watanabe et al. [24]	2001	Antenna arrangement problem	Multiobjective evolutionary algorithm
Alba et al. [3]	2002	Antenna placement	Sequential and parallel genetic algorithm
Ali [4]	2002	Location area management	Iterative algorithm
Kim and Jung [18]	2003	Configuration of base stations	Genetic algorithm
Amaldi et al. [5]	2003	3G network design	Greedy, taboo search
Maple et al. [20]	2004	3G network planning	Genetic algorithm
Cahon et al. [8]	2004	Position and configuration of mobile base stations	Multiobjective genetic algorithm
Créput et al. [15]	2005	Network mesh generation	Hybrid evolutionary strategy
Alba and Chicano [2]	2005	Antenna placement	Genetic algorithm

et al. [21] proposed some algorithms that determine the coverage of an ad hoc sensor network based on graph techniques. The methods proposed obtain an estimation of the best- and worst-case coverage of the network with optimal polynomial effort.

The RND problem has also received a great deal of attention. Watanabe et al. [24] worked out a parallel evolutionary multiobjective approach for deciding antenna placement and configuration in cellular networks. The authors presented two parallel models for multiobjective GAs applied to the problem: the master–slave with local cultivation genetic algorithm (MSLC) and the divided range multiobjective genetic algorithm (DRMOGA). The MSLC algorithm is based on the standard master–slave approach, but the evolutionary operators are carried out on the slaves using a two-individual population, and the evolution follows the minimal generation gap model. DRMOGA is a standard distributed island model that uses domain decomposition. The empirical analysis compares models proposed with both a multiobjective genetic algorithm (MOGA) [17] and a standard distributed GA. They show that MSLC gets the best results for Pareto front covering and nondominated individuals, while establishing that DRMOGA results are affected by the number of subpopulations: The number of nondominated individuals decreases when the number of subpopulations grows.

In the same line of work, Meunier et al. [22] presented a parallel implementation of a GA with a multilevel encoding deciding the activation of sites, the number and type of antennas, and the parameters of each BS. Two modified versions of the classical genetic operators, geographical crossover and multilevel mutation, are introduced. The fitness evaluation utilizes a ranking function, similar to Fonseca and Fleming's MOGA algorithm [17], and a sharing technique is employed to preserve diversity among solutions. In addition, a linear penalization model is used to handle the constraint considered (a minimal value for the covered area). A master–slave parallel implementation is presented for solving high-dimensional problems in reasonable times, with each slave processing a part of the geographical working area. The algorithm is evaluated with a large and realistic highway area generated by France Telecom. The authors analyze the convenience of using the sharing strategy proposed instead of concentrating on a small part of the Pareto front, showing that better Pareto front sampling is obtained in the first case.

In a later work, Cahon et al. [8] solve the same problem with a multiobjective GA. They use the three parallel/distributed GA models implemented in the ParadisEO (parallel and distributed evolving objects) framework: the *island (a)synchronous cooperative model*, the *parallel evaluation of the solution model*, and the *distributed evaluation of a single solution model* [9]. Working on a cluster of 40 Pentium III PCs, the Pareto fronts obtained for the test instances studied confirmed the robustness and efficiency of the island model for solving the problem. In addition, since the fitness evaluation process demands high computational effort, the problem is suitable for applying the parallel and distributed evaluation models. The computational efficiency analysis showed that the parallel evaluation model follows almost-linear speedup behavior. The distributed evaluation model scales superlinearly up to 10 processors and then follows a logarithmic decay.

Calégari et al. [10–13] developed a distributed GA to find the optimal placement of antennas. In ref. 12 the authors compare a greedy technique, a Darwinian algorithm, and a parallel GA (PGA). The PGA uses a bit string representation for codifying the entire set of possible antenna locations and a parametric fitness function evaluating the covered area as a function of a parameter that can be tuned in order to obtain acceptable service ratio values. Experiments were performed on two real-life cases: Vosges (rural scenario) and Geneva (urban scenario). On average, the PGA and the greedy technique show the same solution quality. But when an optimal solution is known, it can be found using PGA, whereas the greedy approach usually falls in local optima. Alba et al. [3] tackled the same problem with sequential and parallel GAs over an artificial instance. In a later work, Alba and Chicano [2] performed a deep study of the parallel approach evaluating the influence of the number of possible locations, the number of processors, and the migration rates. They found a sublinear speedup and concluded that the isolation of the subpopulations is beneficial for the search.

Maple et al. [20] used a PGA to solve a network planning problem, consisting of determining the optimum placement for base stations in third-generation mobile networks. They proposed a multiobjective approach, employing several objective functions for considering multiple network design factors. The model evaluates the network capacity (attempting to maximize the maximum number of users permitted in a cell), considers intracell and intercell interference, use known propagation models for coverage (attempting to maximize the covering radius of a cell), and incorporates the design cost calculation (attempting to minimize the base placement cost). Using a common strategy in telecommunication network operation, the authors employ a binary representation that selects a subset of sites for BSs from a finite set of possible locations. When a site is selected for use, the GA is also used to determine the antenna height and its transmission power. For dealing with the massive solution space and the complex fitness calculation process, they propose a parallel GA following the coarse-grain subpopulation model. The authors did not present numerical results for the optimization problem, stating that the research was "currently being undertaken" to implement the algorithm on a 30-node Beowulf cluster.

Créput et al. [15] proposed a parallel evolutionary strategy for dimensioning a cellular network to cover a city, addressing the problem of evaluating the optimal number and location of BSs needed for satisfying quality of service (QoS) and traffic requirements. They use a geometric approach for facing the adaptive meshing (AM) process, where a pattern of regular hexagonal cells transform themselves and adapt their shapes according to traffic density, geometrical constraints, and other parameters. For solving the problem with relatively low computational effort, the authors propose the hybrid islands evolutionary strategy (HIES), combining a hillclimbing local search procedure with a subpopulation distributed evolutionary mechanism. A high-level crossover-and-mutation schema and an elitist selection operator are used to avoid local minima reached in the local search.

In the particular approach presented in this chapter, each subpopulation or island is limited to one individual, so the HIES proposal is similar to memetic algorithms [23] incorporating a geographical isolation distribution for individuals, as in a cellular PGA. All three HIES operators (local search, crossover, and macromutation) are stochastic procedures, designed specifically for the problem to solve. The AM is an intrinsically multiobjective problem. However, Créput et al. used a linear aggregative fitness function considering the objective (minimization of the total number of BSs) and four constraints related to the resource distribution optimization, the regularity of cell geometry, the number of visible cells, and the elimination of overloaded cells. In the experimental evaluation, the authors considered four test instances, including a real-life scenario (city of Lyon, France) and three problems specifically built for representing typical application cases. Results show satisfactory meshing patterns, producing well-contoured meshes on a map while eliminating overloaded cells. The authors worked with several values for the population-size parameter ($5 < \text{population size} < 80$) in their experiments, but they did not distribute the algorithm on several machines or in a multiprocessor computer. They refer to their distributed algorithm as the parallel version. It has the ability of HIES to improve its performance as population size increases. The authors state that it is able to achieve highly adapted individuals using a moderate number of generations. Although the parallel version increases the population size, it allows us to obtain better results than do versions using lower population sizes, using a similar number of function evaluations. These results suggest that there is room to improve the HIES computational efficiency, executing on a multiple machine cluster, given that performing the simulations required from 5 to 20 hours of execution time for the test scenarios studied.

Kim and Jung [18] focused their work on the parameterization of the BSs in a given area. They proposed an iterative algorithm to solve the corresponding optimization problem. This algorithm partitions the problem geographically according to domain areas, then solves the resulting subproblems iteratively using a GA. A signal-to-noise criterion (determined by free-space propagation model) is employed to calculate the coverage regardless of signal interference, for the sake of simplicity. Their technique was tested against a global GA for the entire problem (without partition) and a random search technique, outperforming both.

The radio network design problem for UMTS was studied by Amaldi et al. [5]. The third generation for mobile telecommunications requires a different approach since its features allow for more flexibility in its use. The cell capacity is not limited a priori—resources are shared all over the network—and the main limitation is interference; therefore, a capacity study had to be made. Hata's propagation model was used to deal with real-world-like instances over a rectangular service area where a set S of candidate sites is defined, and another set TP of test points is randomly determined. Two types of power control mechanisms are considered [power-based and signal-to-interference-ratio (SIR)-based], and two types of optimization techniques are employed: greedy procedures (direct and reverse) and taboo search (TS). The experimental results showed that TS performs better than greedy procedures, although the differences were not big.

17.3 RADIO NETWORK DESIGN PROBLEM

The radio coverage problem amounts to covering an area with a set of transmitters. The part of an area that is covered by a transmitter is called a *cell*. A cell is usually disconnected. In the following we assume that the cells and the area considered are discretized; that is, they can be described as a finite collection of geographical locations (e.g., taken from a geo-referenced grid). The computation of cells may be based on sophisticated wave propagation models, on measurements, or on draft estimations. In any case, we assume that cells can be computed and returned by an ad hoc function.

Let us consider the set L of all potentially covered locations and the set M of all potential transmitter locations. Let G be the graph $(M \cup L, E)$, where E

is a set of edges such that each transmitter location is linked to the locations it covers, and let the vector $\vec{x}$ be a solution to the problem where $x_i \in \{0, 1\}$ and $i \in [1, |M|]$. The value x_i is 1 or 0, depending on whether or not a transmitter is being used in the corresponding site. As the geographical area needs to be discretized, the potentially covered locations are taken from a grid, as shown in Figure 17.1.

Throughout this chapter we consider different versions of the RND problem, which will differ in the types of antennas that might be placed in every location. There will be simple versions with antennas that will require no parameters to determine their coverage, and more complex versions in which antennas will require some parameters (i.e., direction) to determine the area covered by it.

Searching for the minimum subset of transmitters that covers a maximum surface of an area comes to searching for a subset $M' \subseteq M$ such that $|M'|$ is minimum and such that $|\text{Neighbors}(M', E)|$ is maximum, where

$$\text{Neighbors}(M', E) = \{u \in L \mid \exists v \in M', (u, v) \in E\} \tag{17.1}$$

$$M' = \{t \in M \mid x_t = 1\} \tag{17.2}$$

The problem we consider resembles the unicost set covering problem (USCP), which is known to be NP-hard. The radio coverage problem differs, however, from the USCP in that the goal is to select a subset of transmitters that ensures *good* coverage of the area and does not ensure *total* coverage. The difficulty of our problem arises from the fact that the goal is twofold, no part being secondary. If minimizing was the primary goal, the solution would be trivial: $M' = \emptyset$. If maximizing the number of covered locations was the primary goal, the problem would be the USCP. An objective function $f(\vec{x})$ to combine the two goals has

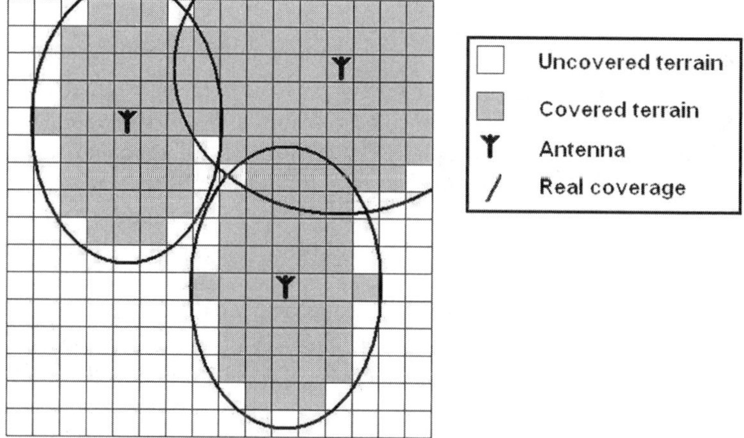

Figure 17.1 Three candidate transmitter locations and their associated covered cells on a grid.

been proposed [11]:

$$f(\vec{x}) = \frac{\text{Coverage}(\vec{x})^{\alpha}}{|M'(\vec{x})|} \tag{17.3}$$

where

$$\text{Coverage}(\vec{x}) = 100 \times \frac{|\text{Neighbors}(M', E)|}{|\text{Neighbors}(M, E)|} \tag{17.4}$$

The parameter $\alpha > 0$ can be tuned to favor the cover rate item with respect to the number of transmitters. If we set $\alpha = 1$, the algorithm will not distinguish between a solution with a single antenna producing coverage C and another with $N >> 1$ antennas producing coverage $N \times C$. This defeats the purpose of RND since the algorithm would not be searching for solutions that produce high coverage in an efficient way, but only for efficient solutions regardless of the coverage obtained. Therefore, we have to set $\alpha > 1$ in order to guide the search toward solutions with high cover rates. As Calégari et al. did [11], we use $\alpha = 2$.

17.4 OPTIMIZATION ALGORITHMS

In this section we describe briefly the three techniques used to solve RND problem instances: simulated annealing (SA), genetic algorithm (GA), and cross generational elitist selection, heterogeneous recombination, and cataclysmic mutation (CHC).

17.4.1 Simulated Annealing

Simulated annealing is a trajectory-based optimization technique [7] proposed by Kirkpatrick et al. in [19]. The pseudocode for this algorithm is shown in Algorithm 17.1. The algorithm works iteratively to keep a single tentative solution S_a at any time. In every iteration, a new solution S_n is generated from the preceding one, S_a, and either replaces it or not depending on an acceptance criterion. The acceptance criterion works as follows: Both the old (S_a) and the new (S_n) solutions have an associated quality value, determined by an objective function (also called a *fitness function*). If the new solution is better than the old one, it will replace it. If it is worse, it replaces it with probability P. This probability depends on the difference between their quality values and a control parameter T named *temperature*. This acceptance criterion provides a way of escaping from local optima. The mathematical expression for the probability P is

$$P = \frac{2}{1 + e^{\text{fitness}(S_a) - \text{fitness}(S_n)/T}} \tag{17.5}$$

Algorithm 17.1 Pseudocode of SA

```
t ← 0;
Initialize(T,Sa)
Evaluate(Sa)
while not EndCondition(t,Sa) do
  while not CoolingCondition(t) do
    Sn ← ChooseNeighbor(Sa)
    Evaluate(Sn)
    if Accept(Sa,Sn,T) then
       Sa ← Sn
    end if
    t ← t+1
  end while
  Cooldown(T)
end while
```

As iterations go on, the value of the temperature parameter is reduced following a cooling schedule, thus biasing SA toward accepting only better solutions. In this work we employ the geometric rule $T(n+1) = \alpha \cdot T(n)$, where $0 < \alpha < 1$, and the cooling is performed every k interations (k is the *Markov chain length*).

17.4.2 Genetic Algorithm

Genetic algorithms belong to the wide family of evolutionary algorithms [6]. They appear for the first time as a widely recognized optimization method as a result of the work of John Holland in the early 1970s, particularly his 1975 book. A standard GA is a population-based technique [7] that uses a selection operator to pick solutions from the population, a crossover and a mutation operator to produce new solutions from them, and a replacement operator to choose individuals for the next population. The pseudocode for this algorithm is shown in Algorithm 17.2.

Our implementation of the GA uses a ranking method for parent selection and elitist replacement for the next population, that is, the best individual of the current population is included in the next one. The crossover method is a two-point crossover and the mutation method selects at random some percentage of the positions and sets a new random value on them.

17.4.3 Cross-Generational Elitist Selection, Heterogeneous Recombination, and Cataclysmic Mutation

The last algorithm we propose for solving the RND problem is Eshelman's CHC [39], a kind of evolutionary algorithm (EA) [6]. Like other EAs, CHC works with a set of solutions (*population*) at any time. The pseudocode for this algorithm is shown in Algorithm 17.3. At each step a new set of solutions is produced by selecting pairs of solutions from the parent population P_a and recombining

Algorithm 17.2 Pseudocode of GA

```
t ← 0
Initialize(Pa)
Evaluate(Pa)
while not EndingCondition(t,Pa) do
  Parents ← SelectionParents(Pa)
  Offspring ← Crossover(Parents)
  Mutate(Offspring)
  Evaluate(Offspring)
  Pn ← Replacement(Offspring,Pa)
  t ← t+1
  Pa ← Pn
end while
```

Algorithm 17.3 Pseudocode of CHC

```
t ← 0
Initialize(Pa,convergence count)
Evaluate(Pa)
while not EndingCondition(t,Pa) do
  Parents ← SelectionParents(Pa)
  Offspring ← HUX(Parents)
  Evaluate(Offspring)
  Pa ← ElitistSelection(Offspring,Pa)
  if not Modified(Pa,Pn) then
     convergence count ← convergence count-1
     if convergence count = = 0 then
         Pa ← Restart(Pa)
         Initialize(convergence count)
     end if
  end if
  t ← t+1
  Pa ← Pn
end while
```

them. An incest prevention criterion prevents individuals that are too similar to each other to mate, and recombination is made using a special procedure known as HUX. This procedure copies first the parents into the offspring, then randomly exchanges half of the diverging information between the offspring. This method has been designed to preserve the maximum amount of diversity in the population since no new diversity is introduced during the iteration (because there is no mutation operator). The next population is formed by selecting the best individuals among the parents and the offspring (elitism).

In a normal execution population convergence is achieved, so the normal behavior of the algorithm should be to stall on it. A special mechanism is used to introduce new diversity when this happens: the *restart* mechanism. Upon restarting, all the solutions except the very best ones (or only the best) are modified significantly through a high-rate mutation.

17.5 BASIC PROBLEMS

As an initial approach to the RND problem, some academic problem instances are defined. These instances are used as benchmarks to compare the optimization algorithms. They are designed inspired by the problem model employed by Calégari et al. [11]. As stated previously, a discrete model is employed to represent the terrain area where the radio network has to be deployed. This model consists of a square grid with 287×287 points. This discretization is helpful when calculating the coverage obtained by a given radio network.

We assume that all the antennas forming a radio network are equal and offer the same coverage, which can be determined using the corresponding antenna model. We have defined three different antenna models, hence three different problems. The first has an associated coverage cell of square shape with the antenna located at the center; this shape is employed for comparison means with previous works [1]. We refer to this problem as RNDsqr. The second one produces a circular coverage cell with the antenna located at the centre; this is the simplest model for a realistic antenna coverage (the omnidirectional antenna). We refer to this problem as RNDcirc. The third model used is a directive antenna that produces a sectorial coverage cell; this is a model close to the antennas that are actually employed in cellular radio networks, such as the global system for mobile (GSM) network. We refer to this problem as RNDsect.

The first two models are designed to offer the same covered area (although with different shapes). The square-shaped cell of the first model covers a 41×41-point area in the terrain grid, whereas the circular cell of the second model is a 23-point radius discretized circle in the terrain grid. With this, the two antenna models cover 1681 and 1701 grid points, respectively (i.e., roughly 2% of the complete terrain area). The third model is defined after the second model; therefore, the radius is maintained, but the cell is then one-sixth of the circle. Figure 17.2 shows a graphical representation of the three models.

In RNDsect, to make up for the coverage decrease of a single antenna, we allow three antennas to be placed in any given site. Each antenna has an associated direction where its coverage cell will be; there are six available directions, equally spatiated 60°. Thus, the codification of a candidate solution does not only select the set of geographical sites but also has to specify the three directions for the directive antennas. In this way, any site selected produces half the coverage of a circular cell. We have considered two variants for this problem:

1. *Simple*. The area covered by the three transmitters from one location must form a solid half-circle.

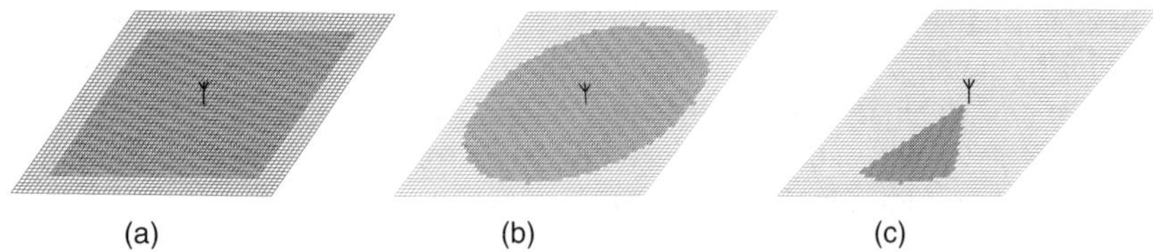

Figure 17.2 Antenna models used: (a) square cell; (b) circular cell; (c) sectorial cell.

2. *Complex*. The three transmitters from one location can point in any direction (out of six possible ones) as long as any two of them do not point in the same direction.

In RNDsqr, the optimal solution consists of 49 transmitters located in the area in order to form a regular 7×7 grid structure, achieving 100% terrain coverage. In RNDcirc the optimal solution is formed by 52 transmitters forming a regular hexagonal network, but only 90% coverage can be reached. In this case the efficiency of the solution is clearly lower because circular cells, unlike square cells, require that to obtain high coverages, some overlap exists. The graphical representations of these optimal solutions are shown in Figure 17.3. For the third problem the hexagonal network is maintained, since the sectorial cell has been defined after the circular cell.

In all the instances solved, the optimal solution is always offered as a possible solution. The list of available solution sites (M) contains the locations of all the sites belonging to the optimal solution, and a set of random locations with uniform distribution to complete the search space of the problem. The optimization algorithm has to select all the locations of the optimal solution and none of the others from M to solve the instance optimally. Since this is a first approach to the problem and the instances defined are not too complex, every experiment

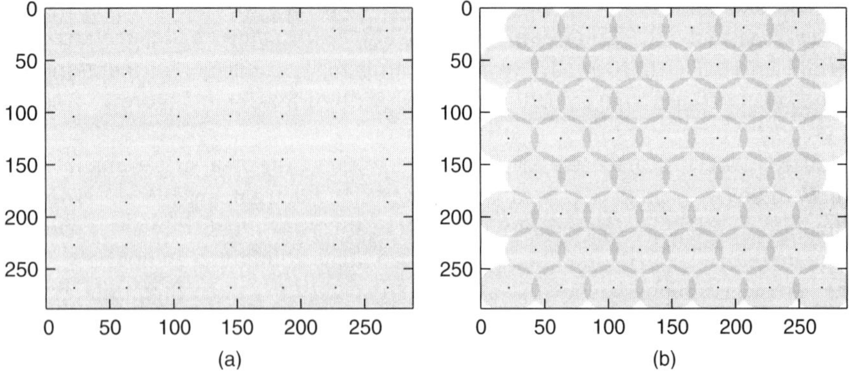

Figure 17.3 Optimal solution for the basic problem using (a) square- and (b) circular-cell antennas.

TABLE 17.2 Parameter Settings of the Optimization Algorithms

SA		GA		CHC	
Markov chain length	50	Population size	512	Population size	100
Cooling factor	0.999	Offspring	1	Crossover problem	0.8
Initial temperature	1.05	Crossover problem	0.8	Min. Hamming distance	25%
Mutation rate	1/instance size	Mutation rate	1/instance size	Restart mutation rate	35%

consists of 50 independent runs to ensure the statistical confidence of the results. Table 17.2 shows the parameter configurations used for the optimization algorithms. These configurations have been obtained using an empirical parameter tuning. In addition, a statistical analysis is performed for each experiment. The result is shown in column A as a + or a − sign; the first indicates that the analysis proved, with 95% confidence level, that the differences between the best algorithm found and the rest of the algorithms evaluated are statistically significant, the latter indicating that no significant differences were found. In each case the procedure for generating the statistical information presented in the tables is the following. First, a Kolmogorov–Smirnov test is performed to check whether or not the variables are normal. If they are, an ANOVA I test is performed; otherwise, we perform a Kruskal–Wallis test. After that, we do a multiple comparison test.

17.5.1 Results for RNDsqr

We have tested the three optimization algorithms on five instances with different sizes (complexity). The size of an instance is determined by the number of available location sites. In our experiments the smallest instance contains 149 available locations and the largest, 349. Independent empirical parameter tuning is performed for each algorithm and each instance. The experiments are run until the optimal solution is found [see Figure 17.3(a)]; that solution scores a fitness value of 204.082. The computational efforts measured for the algorithms are presented in Table 17.3 (in number of solution evaluations). The table also shows previous results from the literature [1].

All the optimization algorithms were able to find the optimum with almost complete effectivity (all the hit ratios were above 98%), but CHC clearly proved to be the most efficient algorithm to solve this RND problem. The statistical analysis guarantees the statistical significance of the results for all the instances. The number of evaluations required by CHC is 13,350 for the instance of size 149, which is only 15.4% of the number of evaluations required by SA, the second-best algorithm, and 9.4% of those performed by GA. Furthermore, CHC scales better that the rest of the algorithms since it requires 70,220 evaluations for the instance of size 349, only 8.7 and 1.9% of the evaluations performed by SA and GA, respectively.

TABLE 17.3 Computational Effort of the Algorithms for RNDsqr (Number of Evaluations)

Instance	SA	GA	CHC	dssGA8[1]	A
149	86,760	141,900	13,350	785,900	+
199	197,000	410,500	24,650	1,467,000	+
249	334,100	987,100	39,030	2,481,000	+
299	638,000	1,892,000	54,080	2,998,000	+
349	810,800	3,612,000	70,220	4,710,000	+

17.5.2 Result for RNDcirc

For circular cells (i.e., omnidirectional antennas) we have repeated the experiments described earlier to appreciate the influence that cell geometry may have on the algorithms performances. Figure 17.3 (b) shows the optimal solution for this problem, and Table 17.4 presents the computational efforts measured for this problem. In this problem the optimum scores a fitness value of 164.672. Since this problem has been defined by the authors, there are no existing results in the literature to compare to.

The second problem is tougher to solve for all the algorithms. To solve RNDcirc, CHC requires at least 30% more evaluations than to solve RNDsqr, while GA requires at least 45.6% more evaluations. SA is able to solve the smallest instance with slightly less effort (83,180 evaluations instead of 86,760) but needs at least 33% more evaluations for the remaining four instances. This problem is also less scalable than RNDsqr: In the worst case (the instance of size 349) CHC, SA, and GA require 238.5, 656.8, and 453.4% extra evaluations, respectively. The results also confirm that CHC is the best-suited algorithm for solving RND. When using omnidirectional antennas, CHC is consistently capable of finding the optimal solution with significantly less effort than for the other two algorithms. The number of evaluations performed by CHC ranges from 17,360 to 237,700 for instances of size 149 to 349. Compared to SA and GA for the five instances ordered by increasing complexity, CHC requires only from 20.9% down to 3.9% of the number of evaluations performed by SA and from 8.4% down to 1.2%

TABLE 17.4 Computational Effort of the Algorithms for RNDcirc (Number of Evaluations)

Instance	SA	GA	CHC	A
149	83,180	206,600	17,360	+
199	262,300	1,152,000	46,960	+
249	913,600	3,354,000	85,770	+
299	2,946,000	8,081,000	151,200	+
349	6,136,000	19,990,000	237,700	+

of the ones performed by GA, hence having the best scalability of the three techniques.

17.5.3 Result for RNDsect

The third problem is by far more complex than the previous two. Not only does the set of locations have to be chosen, but also the direction for the antennas. There are six possibilities in the simple variant of the problem, but in the complex one the number of possibilities rises to 20. Therefore, to keep the experiments within reasonable time lengths, we decided to add some restrictions.

First, only an instance containing 149 candidate sites will be solved. This is due to the fact that the search space grows exponentially by a factor of 7 for the simple variant, or 21 in the complex variant, versus a factor of only 2 in the two previous problems (they are binary problems). If we consider the number of feasible solutions existing in the solution space and divide it by the number of solutions producing the desired layout, we have:

- $7^{149} = 8.31 \times 10^{125}$ for the simple version of the problem.
- $21^{149} = 1.02 \times 10^{197}$ for the complex version of the problem.

The solution spaces for these two problems have the same dimension than those of binary problems with 418 and 654 location sites, respectively. Additionally, we restrict the study to CHC, as it has proved to be the most reliable algorithm for solving RND problems. The available location sites list is defined using the predefined candidate sites from the optimal solution of the previous instance (omnidirectional antennas), but a second list of sites with locations very close to these is added. This is done so that two geographically close sites may be selected, and by placing three directional antennas in each, the original circular cell can be obtained (as long as the antennas are given adequate directions).

The executions are now run until a predefined number of solution evaluations is met, and the fitness values obtained are averaged over the total number of independent executions. This differs from previous experiments, where the executions were run until the optimum was found. The value chosen for the number of evaluations is 1 million. There is a second variation regarding the fitness function: The number of transmitters used is replaced with the number of location sites selected (there are three transmitters per location). Although this does not affect the search behavior (it multiplies the fitness value by a constant factor of 3), it makes the results be more intuitive since the fitness values are closer to the previous values. In theory, sectorial cells have half the efficiency of circular cells (they cover exactly a half-circle), so an equivalent solution should produce half the fitness (requiring double the number of sites to obtain the same coverage). Given that the optimal fitness was 164.672 for omnidirectional antennas, we should expect an optimum solution for this problem to produce a fitness value of approximately 82.336. As in the previous case (RNDcirc), this problem has been defined by the authors.

The results in Table 17.5 show that CHC performs remarkably well for RNDsect despite the high complexity. The best solutions found (see Figure 17.4) are very close to the expected optimum, a hexagonal network. Some little errors can be noticed in the complex version solution, but the network follows mostly the hexagonal optimal pattern. The fitness values are better for the simple version, which was expected, but the values for the complex version are only 7% lower on average. From these results we can state that CHC is still well suited for this problem and can solve a simple instance with 1 million solution evaluations or less, but will probably require a little more effort to solve a complex instance. Furthermore, the fitness obtained for the simple version (84.884) is actually higher than the one expected (82.336). This happens because sectorial cells, unlike circular ones, allow us to cover the terrain frame efficiently. This can be noticed by looking at the best solutions for RNDcirc [Figure 17.3(b)] and RNDsect [Figure 17.4(a)]: The first solution has eight wide uncovered areas in the horizontal edges that cannot be covered efficiently using omnidirectional antennas, while the second solution has none.

TABLE 17.5 Results of the Study for CHC Using Directive Transmitters

Problem	Simple Version	Complex Version
Best fitness	85.328	80.693
Average fitness	84.884	78.787
Worst fitness	84.628	76.211

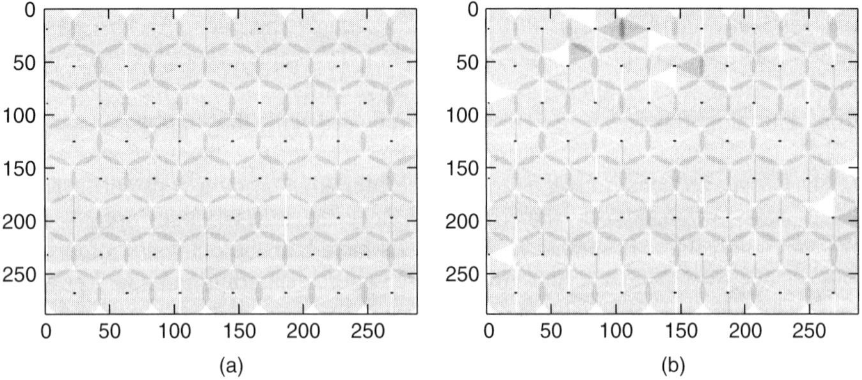

Figure 17.4 Best solutions found for the basic problem using sectorial cell antennas: (a) simple version; (b) complex version.

17.6 ADVANCED PROBLEM

Real-world radio networks are mostly deployed on urban scenarios to provide coverage for a set of services (GSM, UMTS, etc). An urban scenario has some characteristics that make it different from the basic cases studied in Section 17.5. In a city, the antennas may be located only at specific sites, such as rooftops or other high places; at the same time, there are restricted places such as hospitals or schools where antennas may not be placed. A typical urban scenario is non-homogeneous and will have regions with more buildings than others, and may also have rivers, parks, or some other building-free places.

We have developed a model for a real-world instance based on the city of Malaga (Spain) [see Figure 17.5(a)], where the available 1000 candidate location sites are distributed corresponding to the density of buildings that are suitable for installing antennas over the city area. A discrete 450×300 grid model is used to represent the terrain [Figure 17.5(b)]. Each point of the grid represents a surface area of 15×15 square meters. In this instance, the maximum coverage that can be attained—using all the transmitters—is 95.52%.

The antennas used in real radio networks are directive, because they offer very advantageous features in terms of interference immunity and hence allow a greater frequency reuse. A standard hexagonal cell receives its radio coverage from three directive antennas located at three of its vertices. Each of those antennas has a coverage angle of 120°, and the combined coverages of the three offer complete coverage of the cell [see Figure 17.6(a)]. Nevertheless, when a ground site is selected to place an antenna, actually three directive antennas are installed pointing in different directions—equally spatiated 120°—thus producing a full-coverage circle around the site [see Figure 17.6(b)]. Therefore, we have chosen for this problem a circular cell model for the antennas used, where the circle does not represent a logical cell in the hexagonal cell network, but the coverage produced by a selected ground site. For this problem the coverage cell will have a radius of 30 grid points, or 450 meters. This value is close to the real one in mobile telephony.

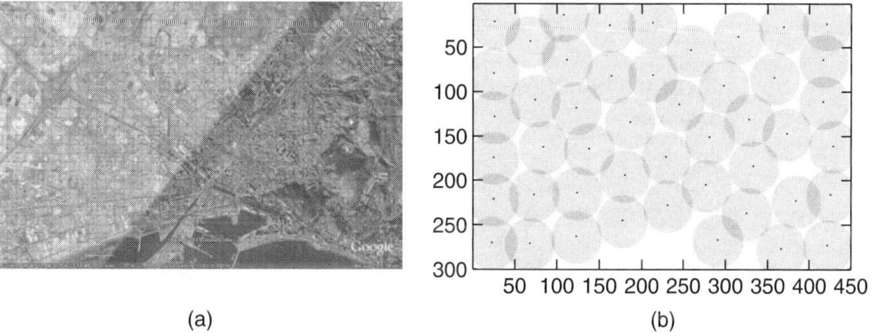

Figure 17.5 (a) Map of the city of Malaga; (b) best solution found with CHC.

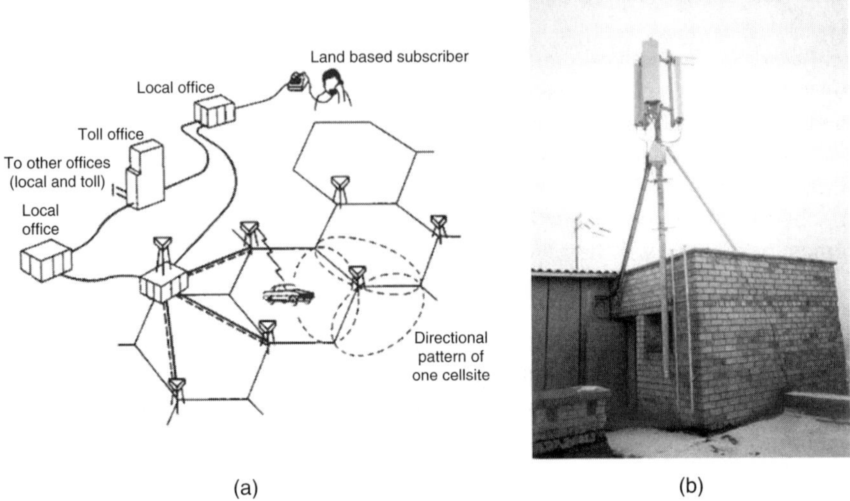

Figure 17.6 (a) GSM cellular network; (b) GSM base station transceiver on a building roof.

The CHC algorithm is used to solve this problem instance, and 30 independent executions are run. The optimal solution is not known beforehand; therefore, the stopping criterion is set as a number of solution evaluations. Since this problem is of higher complexity than the corresponding basic instance (RNDcirc), we set a higher number of evaluations than the one that was necessary to solve the former: 5 million evaluations. The best solution found so far is stored after every 25,000 evaluations to allow a trace of the evolution of the algorithm and to be able to determine how many evaluations may suffice if a certain fitness value is desired. The fitness values obtained at five selected checkpoints (25,000, 100,000, 500,000, 1 million, and 5 million evaluations) are shown in Table 17.6; for each checkpoint we show the average fitness and the highest and lowest fitnesses of the best solutions found among the 30 independent executions.

TABLE 17.6 Fitness Results for the Malaga Instance

Evals.	CHC		
($\times 10^3$)	Average	Best	Worst
25	139.444	146.740	130.793
100	153.157	156.120	148.051
500	160.408	162.945	157.453
1000	162.070	164.181	158.561
5000	163.278	164.703	160.262

The best solution found for this problem (see Figure 17.5) achieves 87.92% terrain coverage and uses 47 antennas. There are three main uncovered areas: one on the bottom of the terrain and the other two in the top and the right side. The first one corresponds to the sea, and the other two correspond to mountains. The coverage obtained can be augmented (it can rise as high as 95.52%), although probably at the cost of placing more antennas. As stated in Section 17.3, the fitness function (Equation 17.3) can be tuned to set the relative importance of the coverage and the number of antennas placed. If higher coverage solutions are desired, the parameter α can be set to a value higher than 2.

If we observe the evolution of the best solutions found with the number of evaluations, we notice a quick increase in the fitness in the first checkpoints, and a convergence of the fitness for the last checkpoints. Between the last two checkpoints (1 million and 5 million evaluations), the increase of 400% in computer effort results in a fitness value only 0.75% higher on average. Between the third and fourth checkpoints (500,000 and 1 million evaluations) there is a fitness increase of 1.04%, with an increase of only 100% of the computer effort. This suggests that 1 million solution evaluations performed produces a good balance between effort and solution quality, and more evaluations should be performed only when an especially high-quality solution is desired.

17.7 CONCLUSIONS

In this chapter we have presented the radio network design (RND) problem. This problem consists of selecting the set of geographical locations to place radio antennas (or transmitters) to provide a given region with radio coverage. The objectives of this design are to obtain the largest covered terrain while using the smallest number of antennas required. This problem can be found in cellular telephony (GSM, UMTS) and bears a high resemblance to problems found in new emerging technologies, such as wireless sensor networks.

We have defined a discrete terrain model and three different types of transmitter. A RND problem has been defined for each of the antenna models: RNDsqr for antennas with square coverage cells, RNDcirc for antennas with circular coverage cells, and RNDsect for antennas with sectorial-shaped coverage cells. Three optimization algorithms—simulated annealing (SA), genetic algorithm (GA) and cross-generational elitist selection, heterogeneous recombination, and cataclysmic mutation (CHC)—have been used to solve the problem. CHC has largely outperformed the other two techniques in the three basic scenarios for all the instance sizes.

A real-world instance has been defined to further test the performance of the CHC algorithm. Taking the city of Malaga as a reference, a set of 1000 candidate sites has been defined for placing a radio network of omnidirectional antennas. The CHC algorithm proves to be capable of solving this new instance. The coverage obtained in the best solutions found by the algorithm is lower than the maximum coverage that can be achieved; a tuning of the fitness function may be used to guide the searching process toward solutions with higher cover rates (and higher number of antennas). Currently, an intense research effort is being carried for the Malaga instance. Several groups of research from different universities in Spain are applying a plethora of optimization techniques to this problem to establish the state of the art for this problem instance. Future lines of interest may include the use of more realistic antenna coverage and terrain models, requiring the tuning of the BS's parameters, frequency assignment issues, and multiple-coverage degree requirements.

Acknowledgments

This chapter has been partially funded by the Spanish Ministry of Education and Science and by European FEDER under contract TIN2005-08818-C04-01 (the OPLINK project). Guillermo Molina is supported by grant AP2005-0914 from the Spanish government.

REFERENCES

1. E. Alba. Evolutionary algorithms for optimal placement of antennae in radio network design. In *Proceedings of the 18th International Parallel and Distributed Processing Symposium*, Apr. 26–30, 2004, p. 168.
2. E. Alba and F. Chicano. On the behavior of parallel genetic algorithms for optimal placement of antennae in telecommunications. *International Journal of Foundations of Computer Science*, 16(2):343–359, Apr. 2005.
3. E. Alba, C. Cotta, F. Chicano, and A. J. Nebro. Parallel evolutionary algorithms in telecommunications: two case studies. In *Proceedings of the Congreso Argentino de Ciencias de la Computación* (*CACIC'02*), Buenos Aires, Argentina, 2002.
4. S. Z. Ali. Design of location areas for cellular mobile radio networks. *IEEE 55th Vehicular Technology Conference*, 3:1106–1110, 2002.
5. E. Amaldi, A. Capone, and F. Malucelli. Planning UMTS base station location: optimization models with power control and algorithms. *IEEE Transactions on Wireless Communications*, 2(5):939–952, Sept. 2003.
6. T. Bäck. *Evolutionary Algorithms in Theory and Practice: Evolution Strategies, Evolutionary Programming, Genetic Algorithms*. Oxford University Press, New York, 1996.
7. C. Blum and A. Roli. Metaheuristics in combinatorial optimization: overview and conceptual comparison. *ACM Computing Surveys*, 35(3):268–308, 2003.
8. S. Cahon, N. Melab, and E.-G. Talbi. ParadisEO: a framework for the reusable design of parallel and distributed metaheuristics. *Journal of Heuristics*, 10(3):357–380, 2004.
9. S. Cahon, N. Melab, E.-G. Talbi, and M. Schoenauer. ParaDisEO-based design of parallel and distributed evolutionary algorithms. *Artificial Evolution*, 2003, pp. 216–228.
10. P. Calégari, F. Guidec, and P. Kuonen. A parallel genetic approach to transceiver placement optimisation. In *Proceedings of the SIPAR Workshop'96: Parallel and Distributed Systems*, Oct. 1996, pp. 21–24.

11. P. Calégari, F. Guidec, P. Kuonen, and D. Kobler. Parallel island-based genetic algorithm for radio network design. *Journal of Parallel and Distributed Computing*, 47(1):86–90, 1997.

12. P. Calégari, F. Guidec, P. Kuonen, and F. Nielsen. Combinatorial optimization algorithms for radio network planning. *Theoretical Computer Science*, 263(1–2):235–265, 2001.

13. P. Calégari, F. Guidec, P. Kuonen, and D. Wagner. Genetic approach to radio network optimization for mobile systems. In *Proceedings of the 47th Vehicular Technology Conference*, vol. 2, Phoenix, AZ. IEEE Computer Society Press, Los Alamitos, CA, May 1997, pp. 755–759.

14. G. Celli, E. Costamagna, and A. Fanni. Genetic algorithms for telecommunication network optimization. In *Proceedings of the IEEE International Conference on Systems, Man and Cybernetics: Intelligent Systems for the 21st Century*, vol. 2; Oct. 22–25, 1995, pp. 1227–1232.

15. J. Créput, A. Koukam, T. Lissajoux, and A. Caminada. Automatic mesh generation for mobile network dimensioning using evolutionary approach. *IEEE Transactions on Evolutionary Computation*, 9(1):18–30, 2005.

16. L. J. Eshelman. The CHC adaptive search algorithm: how to have safe search when engaging in nontraditional genetic recombination. In *Foundations of Genetic Algorithms*. Morgan Kaufmann, San Francisco, CA, 1991, pp. 265–283.

17. C. M. Fonseca and P. J. Fleming. Genetic algorithms for multiobjective optimization: formulation, discussion and generalization. In *Genetic Algorithms: Proceedings of the 5th International Conference*. Morgan Kaufmann, San Francisco, CA, 1993, pp. 416–423.

18. Y. S. Kim and H.-M. Jung. Efficient radio network optimization. In *Proceedings of the 57th IEEE Semiannual Vehicular Technology Conference*, vol. 3; Apr. 22–25, 2003, pp. 1546–1549.

19. S. Kirkpatrick, C. D. Gelatt, and M. P. Vecchi. Optimization by simulated annealing. *Science*, 4598(220):671–680, May 1983.

20. C. Maple, L. Guo, and J. Zhang. Parallel genetic algorithms for third generation mobile network planning. In *Proceedings of the International Conference on Parallel Computing in Electrical Engineering* (*PARELEC'04*), 2004, pp. 229–236.

21. S. Meguerdichian, F. Koushanfar, M. Potkonjak, and M. B. Srivastava. Coverage problems in wireless ad-hoc sensor networks. In *Proceedings of the 20th Annual Joint Conference of the IEEE Computer and Communications Societies* (*INFOCOM'01*), vol. 3; 2001, pp. 1380–1387.

22. H. Meunier, E.-G. Talbi, and P. Reininger. A multiobjective genetic algorithm for radio network optimization. In *Proceedings of the 2000 Congress on Evolutionary Computation*, La Jolla, CA, 2000, pp. 317–324.

23. P. Moscato. Memetic algorithms: a short introduction. In *New Ideas in Optimization*, McGraw-Hill, Maidenhead, UK, 1999, pp. 219–234.

24. S. Watanabe, T. Hiroyasu, and M. Mikiand. Parallel evolutionary multi-criterion optimization for mobile telecommunication networks optimization. In *Proceedings of the Evolutionary Methods for Design, Optimisation and Control with Applications to Industrial Problems Conference (EUROGEN'01)*, Athens, Greece, Sept. 19–21, 2001, pp. 167–172.

CHAPTER 18

Optimization of Image-Processing Algorithms Using FPGAs

M. A. VEGA
Universidad de Extremadura, Spain

A. GÓMEZ
Centro Extremeño de Tecnologias Avanzadas, Spain

J. A. GÓMEZ and J. M. SÁNCHEZ
Universidad de Extremadura, Spain

18.1 INTRODUCTION

At present, FPGAs (field-programmable gate arrays) are very popular devices in many different fields. In particular, FPGAs are a good alternative for many real applications in image processing. Several systems using programmable logic devices have been designed, showing the utility of these devices for artificial-vision applications [1]. Whereas other papers display the results of implementing image-processing techniques by means of "standard" FPGAs or reconfigurable computing systems [2]. In this chapter we present the implementation and optimization of 16 image-processing operations by means of reconfigurable hardware. Our FPGA implementations use several parallelism techniques, include other optimizations as we will see, and have high operation frequencies. Thanks to all this, they obtain very good results. In fact, our FPGA-based implementations are even more than 149 times faster than the corresponding software implementation.

The chapter is organized as follows. In Section 18.2 we give a brief background about the use of FPGAs for optimizing image-processing algorithms. In Section 18.3 we describe the main characteristics of our FPGA-based image-processing operations. In Section 18.4 we give more details about the types of algorithms we use in our hardware implementations. Then we present a

Optimization Techniques for Solving Complex Problems, Edited by Enrique Alba, Christian Blum, Pedro Isasi, Coromoto León, and Juan Antonio Gómez

number of comparisons between the software and FPGA-based implementations, and finally, conclusions are given in Section 18.6.

18.2 BACKGROUND

Vision has been the center of attention for researchers from the beginnings of computing since it is the most notorious perception mechanism for human beings. Research into the emulation of visual capability by means of computers has been developed to such an extent that, at present, vision plays a very important role in many application fields, from medical imaging to factory automation in robotics; and it is a key discipline in the general objective of allowing machines to "understand" the world.

Generally, a computer vision system extracts quantitative information from an image through the following steps: image acquisition, manipulation, understanding, and decision making (actuation), the last three which can be performed by a computer. On the other hand, the image manipulation and understanding steps are carried out by means of image-processing techniques. Any situation requiring enhancement, restoration, or analysis of a digital image is a good candidate for these techniques.

The main challenge is that computer vision systems are normally used in real-time applications. The use of a general-purpose computer allows a simple verification of an algorithm, but usually not real-time processing. Then very different technologies have been used to build computer vision applications, going from parallel architectures to specific-purpose processors, or even programmable logic devices.

Parallel computer vision architectures are based on the fact that computer vision implies the execution of data-intensive and computing-intensive applications that require the use of parallel mechanisms. Different computer vision systems have used parallel architectures with different interconnection structures: lineal array, mesh, pyramid, and so on. These structures show good performance at specific stages of a computer vision operation chain; but no structure by itself is in general suited for complete applications. Also, despite the large number of researchers and the advances in parallel architectures and programming languages over the last decades, parallel processing is still an immature science. Programming a parallel computer is still considered difficult, and the performance obtained by parallel computers is often way below their theoretical peak performance.

Parallel architectures for computer vision are used primarily in research environments or in prototyping applications; they are not cost-effective for many commercial applications. The VLSI technology explosion has allowed the emergence of a wide number of specific-purpose computer vision processors feasible for commercial applications. In some cases they are ASICs dedicated to specific applications, and in other cases they are planned as standard components of multiple-application systems. We can classify the specific-purpose processors used in computer vision into the following types: ASSPs (application-specific standard products), ASIPs (application-specific instruction set processors), DSPs (digital signal processors), and NSPs (native signal processors). The main problem with these devices, which are used at specific stages of a computer vision chain, is that they are not suitable for solving complete computer vision algorithms. Even more important, these devices lack the flexibility required for adapting to the requirements of any application.

The trend in computer vision hardware is to increase the system flexibility as much as possible. For instance, parallel computers have evolved into systems capable of reconfiguring their interconnections to adapt to the most efficient topology for each particular application; and some specific-purpose processors show a certain configurability degree to adapt their data paths to different word widths. The emergence of high-capacity programmable logic devices (mainly FPGAs) has made a reality of the reconfigurable computing paradigm, where hardware can be totally configured at the processor level to reach some specific, even changing, computation requirements. Reconfigurable computing has gathered strength over the past years, producing the appearance of custom computing machines (CCMs), a technology able to provide high computational performance in a large number of uses, including computer vision applications. A custom computing machine consists of a host processor, such as a microprocessor, connected to programmable hardware that implements the computationally complex part of a program. Thanks to reconfigurable computing, computer vision developers have the benefits of hardware execution speed (exploiting algorithms' inherent parallelism) and software flexibility, together with a better price/performance ratio than that of other technologies.

Image processing has always been a focus of interest for programmable logic developers. Several systems have been designed using programmable logic devices, showing the utility for computer vision applications, while other works display the results of implementing specific image-processing techniques by means of "standard" FPGAs or reconfigurable computing systems. However, reconfigurable computing is still a research area, and although many hardware developments exist in this discipline, so far no complete system has proven to be competitive enough to establish a significant commercial presence. For more detailed background on the use of FPGAs to optimize image-processing algorithms, readers are referred to an article by Vega-Rodríguez et al. [2].

18.3 MAIN FEATURES OF FPGA-BASED IMAGE PROCESSING

We have used the Celoxica RC1000 board (http://www.celoxica.com) with a Xilinx Virtex XCV2000E-8 FPGA (http://www.xilinx.com) to implement all the image-processing operations, designing the circuits by means of the Handel-C language and the Celoxica DK2 platform. Thanks to DK2, the EDIF files have been obtained for the various circuits. These EDIF files were used later in the

TABLE 18.1 Features for All the Image Processing Algorithms

Operation	Resource Use [Slices (%)]	Maximum Frequency (MHz)
H	18,986 (98)	43.737
CI	65 (1)	71.093
BCE	840 (4)	62.664
BS	861 (4)	68.143
HSS	5,827 (30)	51.800
MF	5,767 (30)	54.031
LPF	19,198 (99)	47.794
HPF	19,198 (99)	48.405
LF	19,198 (99)	43.446
VGF	19,198 (99)	39.107
HGF	19,198 (99)	45.867
DGF	19,198 (99)	39.331
BE	7,422 (38)	50.924
BD	7,422 (38)	50.924
GSE	19,198 (99)	42.192
GSD	19,198 (99)	44.088

Xilinx ISE 6.2i platform to generate the corresponding configuration files (.bit) for the FPGA.

Table 18.1 presents the main characteristics of the various hardware circuits implemented: histogram (H), complement image (CI), binary contrast enhancement or thresholding (BCE), brightness slicing (BS), histogram sliding and stretching or brightness/contrast adjustment (HSS), median filter (MF), low-pass filter (LPF), high-pass filter (HPF), Laplacian filter (LF), vertical gradient filter (VGF), horizontal gradient filter (HGF), diagonal gradient filter (DGF), binary erosion (BE), binary dilation (BD), gray-scale erosion (GSE), and gray-scale dilation (GSD). More details about these operations are given in a book by Baxes [3]. As we can observe, operations working pixel by pixel use fewer FPGA resources (except in the histogram operation). This is due to the low computation cost of this operation type, since these operations only have to visit all the image pixels (or simply the image color table) and update the pixel values according to the values provided by a LUT.

In addition to the number of slices occupied, Table 18.1 includes the percentage of resources used (knowing that the Virtex-E 2000 FPGA has 19,200 slices). In all the operations we have striven to use the maximum number of FPGA resources performing replications, since this results in a shorter execution time (parallelism use). As for the operation frequency, all the operations admit at least 39 MHz and surpass 70 MHz in some cases. This high frequency allows us to obtain better performance.

18.4 ADVANCED DETAILS

Hardware modules are "programmed" by creating Handel-C models which are designed, simulated, and refined using the DK tool, and finally, synthesized using the Xilinx ISE platform. In each case, every module is implemented using the architecture that is best suited for real-time operation (high-speed execution) and for the FPGA device we use. We have performed very diverse optimizations in each module: use of techniques of parallelism such as replication and pipelining, optimization of the multipliers by means of adder trees in the convolution modules, reduction of the sorting and selection network in the median filter and the gray-scale morphological operations, search for high regularity, reutilization of common resources, and so on. Due to the great number of image-processing operations that we have implemented, in the following subsections we explain the implementation only for the most relevant image-processing operations. You can obtain more details in an article by Vega-Rodríguez et al. [4].

18.4.1 Image Convolution

Convolution is a very important operation in the field of image processing, and it has been used in a great number of applications [5,6]. Among these applications is its use for artificial vision highlights. The main challenge is that artificial vision systems are normally used in real-time applications. For this reason, the implementation of convolution by means of reconfigurable hardware is an important issue [7,8]. In fact, some proposals even use techniques such as simulated annealing and genetic programming to find the best implementation for convolution [9]:

$$O(x, y) = \sum_{i=-k}^{k} \sum_{j=-k}^{k} I(x + y, y + j)w(i, j) \tag{18.1}$$

Convolution is an operation of linear spatial filtering. For this operation, to apply the mask on a given position means to multiply each weight of the mask for the pixel of the image on which it is, and to add the resulting products. This sum is used as the value of the pixel for the output image. Therefore, the convolution of an image I with a square mask w of width $n = 2k + 1$ will generate an image O that could be defined by means of the weighted average of Figure 18.1 and Equation (18.1) [3]. The choice of the mask weights is what determines the action of spatial filtering to carry out, which can be a low-pass, high-pass, gradient, Laplacian, and so on, filter. On each convolution circuit we have carried out several important optimizations. As an example to explain these optimizations, we study which would be the resulting circuit to implement a low-pass filter. A popular mask for this filter is shown in Equation (18.2) [3]. In this case, the corresponding circuit could be implemented as shown in Figure 18.2 [3]. In this figure we observe the necessary nine dividers and the summing network (with eight adders). To avoid floating-point operations, we have normalized the

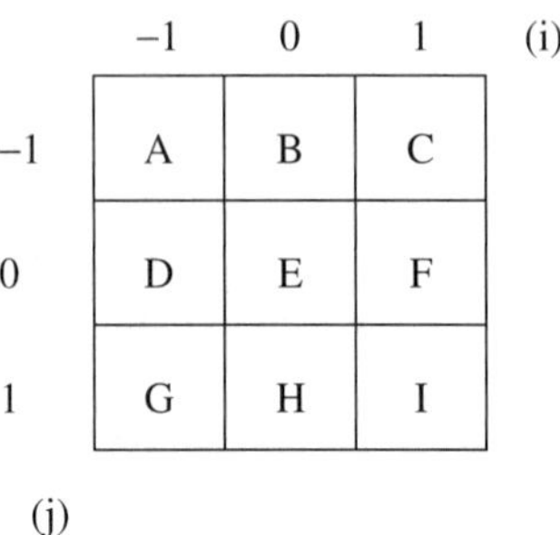

Figure 18.1 Example of a 3 × 3 convolution mask w.

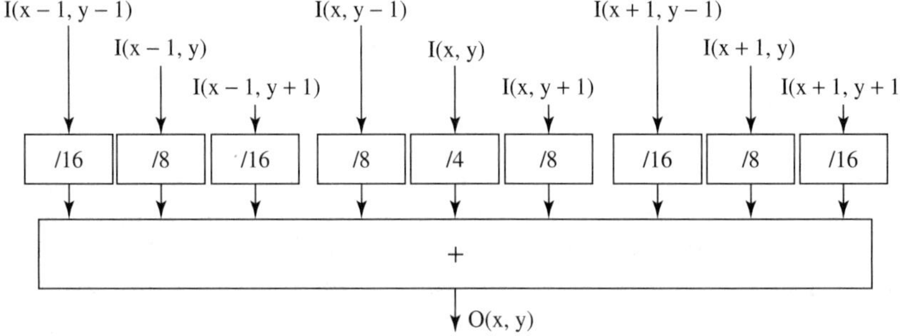

Figure 18.2 Possible implementation of a low-pass filter.

mask coefficients to integer values. In this way the mask in Equation (18.2) is equivalent to the one shown in Equation (18.3), which can be implemented by means of five multiplication operations and only one division.

$$W = \begin{pmatrix} \frac{1}{16} & \frac{1}{8} & \frac{1}{16} \\ \frac{1}{8} & \frac{1}{4} & \frac{1}{8} \\ \frac{1}{16} & \frac{1}{8} & \frac{1}{16} \end{pmatrix} \tag{18.2}$$

$$W = \begin{pmatrix} 1 & 2 & 1 \\ 2 & 4 & 2 \\ 1 & 2 & 1 \end{pmatrix} \times \frac{1}{16} \tag{18.3}$$

First, we apply an optimization to the divider-multipliers. The optimization of the divider-multipliers in Figure 18.2 is carried out in two aspects. On the one hand, the implementation of a division whose divider is 16 can be performed by shifts (in this case, four positions on the right) instead of using a divider circuit. On the other hand, optimization of the multipliers is based on the use of adder trees. As an example, if we need the multiplication $y = 51x$, we can decompose it in $y = 32x + 16x + 2x + x$. Products that are a power of 2 can be implemented by means of shifts, or even more, by means of the use of connections with zeros (direct connections to ground).

Figure 18.3 illustrates the resulting circuit after this optimization (the number of bits in each adder and connection is indicated). In this figure, each product has been carried out by means of direct connections to ground. The final circuit ">>" symbolizes the right shift of 4 bits for dividing by 16. However, in our final architecture, not only the divider-multipliers have been optimized, but also the data paths, with the purpose of using fewer bits in each operation and connection, and therefore to improve system performance and reduce the necessary resources, as well as obtaining a more regular architecture. Figure 18.4 shows the resulting circuit for a low-pass filter after optimizing the data paths in the circuit of Figure 18.3. In this figure we also show the number of bits in each adder and connection so that we can compare this circuit with the one in Figure 18.3. In conclusion, we can see an important reduction in the number of bits, besides a more regular architecture (better parallelism exploitation, more common resources, etc). Evidently, both optimizations (divider-multipliers and data paths) will be more important when the convolution to be implemented is more complex.

Note that the fact of multiplying the result of the central column by 2 instead of using the initial coefficients [2,4,2] allows us to implement the three columns using the same hardware scheme (coefficients [1,2,1]), therefore simplifying the circuit design. Furthermore, this reduces the number of multiplications, because we multiply only the column result, not its three coefficients. Finally, it is necessary to divide the result obtained by 16, using shifts. Even more notably, the shift

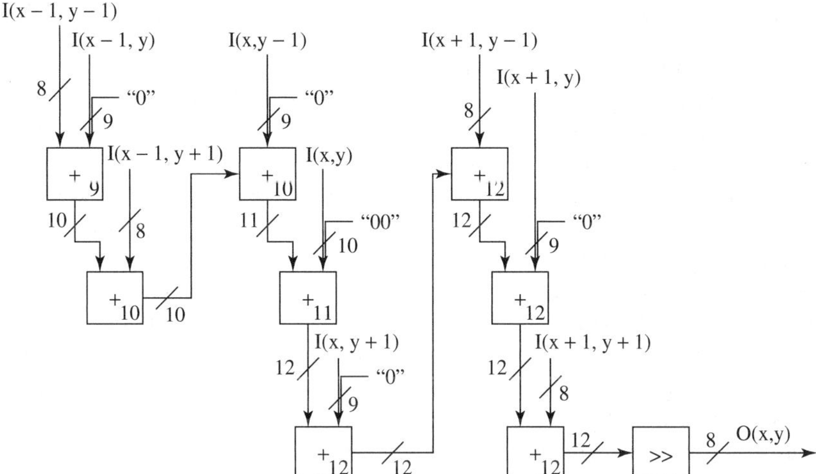

Figure 18.3 Low-pass filter after applying divider-multiplier optimization.

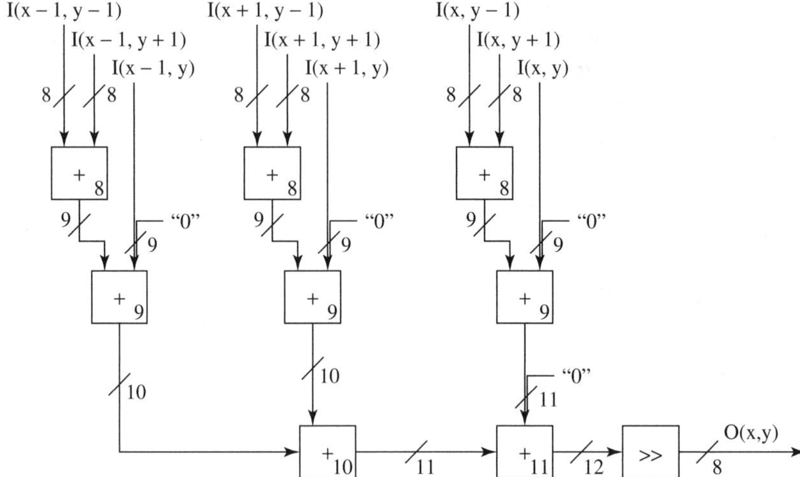

Figure 18.4 Implementation of a low-pass filter after optimization.

operation (the shift register) can be avoided by using appropriate connections. For example, in this case we have to discard the last 4 bits of the result.

Finally, using the statement *par* and function *vectors* of Celoxica Handel-C, we have replicated the resulting circuit (Figure 18.4 for a low-pass filter) in order to apply the corresponding convolution simultaneously on 148 pixel neighborhoods (the exact number of replications depends on the corresponding convolution, ranging from 138 to 148). In this way we take advantage of the inherent neighborhood parallelism, accelerating the operation 148 times.

18.4.2 Binary Morphology

Mathematical morphology [10] can be used to carry out very different tasks of image analysis, including elimination of noise, image enhancement, extraction of features, thinning/thickening, and shape recognition. In fact, it has been used with success in biomedical image analysis, robotics vision, industrial inspection, and other applications of artificial vision (several examples are explained in ref. 11). Morphology can be characterized as being computationally intensive, so it is difficult to implement it in real-time applications with general-purpose processors. This is a reason to study its FPGA implementation.

There are two forms of morphological processing: binary and gray scale [12]. For both forms, two basic operations exist: erosion and dilation. Erosion reduces the size of objects uniformly in relation to their background. Dilation, the dual operation, increases the size of objects uniformly. Morphological operators are implemented by means of nonlinear spatial filtering [3].

Binary morphological processing establishes the premises and methodology of image morphological processing. Binary morphological operators operate with binary images such as those created using thresholding. Therefore, we say that pixels possess logical values instead of brightness values. A black pixel is usually said to be in the 0-state and a white pixel is said to possess the 1-state. Here erosion and dilation combine the values of the pixels logically with a structuring element (or morphological mask). This structuring element is similar to the mask used in a convolution but has logical values instead of weighting values. Each logical value can have a 0- or 1-state. According to morphological mathematics, the binary erosion of an image I with regard to a structuring element w is expressed by the subtraction of Minkowski:

$$E = I \ominus w = \{x \in E / \exists a \in I, \exists b \in w, x = a - b\} \tag{18.4}$$

Similarly, the binary dilation of an image I with regard to a structuring element w is expressed as.

$$D = I \oplus w = \{x \in D / \exists a \in I, \exists b \in w, x = a + b\} \tag{18.5}$$

These mathematical expressions can be clarified by means of the use of morphological masks and the equation [3]

$$\begin{aligned}
O(x, y) &= 1 \text{ or } 0 \text{ (predefined) IF} \\
\{A = I(x-1, y-1) \text{ and } B &= I(x, y-1) \text{ and } C = I(x+1, y-1) \text{ and} \\
D = I(x-1, y) \text{ and } E &= I(x, y) \text{ and } F = I(x+1, y) \text{ and} \\
G = I(x-1, y+1) \text{ and } H &= I(x, y+1) \text{ and } I = I(x+1, y+1)\} \\
&\text{ELSE } O(x, y) = \text{opposed state}
\end{aligned} \tag{18.6}$$

Assuming 3×3 structuring elements (as in Figure 18.1), in a binary morphological operation each pixel in the image and its eight neighbors are compared logically to the values of the structuring element. When the nine input pixels are identical to their respective mask values, the resulting pixel value is set to a predefined logical value 1 for erosion and 0 for dilation. If one or more input pixels differ from their respective mask values, the resulting value is set to the opposite state. Equation (18.6) presents the general equation for erosion and dilation, assuming an input image I and an output image O.

In the erosion case, the structuring element only has elements with a 1-state. For the dilation, the mask only possesses elements with a 0-state. Erosion and dilation are considered primary morphological operations, being the base of many other morphological operations. For example, the opening operation is simply erosion followed by dilation:

$$O = I \circ w = (I \ominus w) \oplus w \tag{18.7}$$

Similarly, the closing operation (opposite of opening) is defined like dilation followed by erosion:

$$O = I \bullet w = (I \oplus w) \ominus w \qquad (18.8)$$

Due to the great similarity between both operations, binary erosion and dilation, they have been designed jointly in one hardware module. These operations have been implemented as shown in Figure 18.5, where $I(x+i, y+j)$ makes reference to each of the nine pixels covered by the 3×3 mask. As these operators are applied to binary images, each input pixel will be white (11111111 in binary) or black (00000000). To simplify, instead of using 8-bit pixels, in the implementation of these operators, 1-bit pixels are used. So in each input pixel, the most significant bit only is used. This bit will be 1 (white) or 0 (black). The use of only one bit for the pixels allows us to simplify the circuit, being able to use OR and AND gates.

Finally, the input *Operation*, also of 1 bit, indicates if an erosion (1) or a dilation (0) is being carried out. This input is used in a 2 : 1 multiplexer to obtain the correct output pixel $O(x, y)$. Notice that for saving resources the output of the multiplexer has been wired eight times. Thus, an 8-bit bus is obtained and the use of more complex logic to generate the values 0 and 255 is avoided. It is important to emphasize that this implementation (Figure 18.5) is better than the direct implementation of Equation 18.6 (with nine comparators of 8 bits).

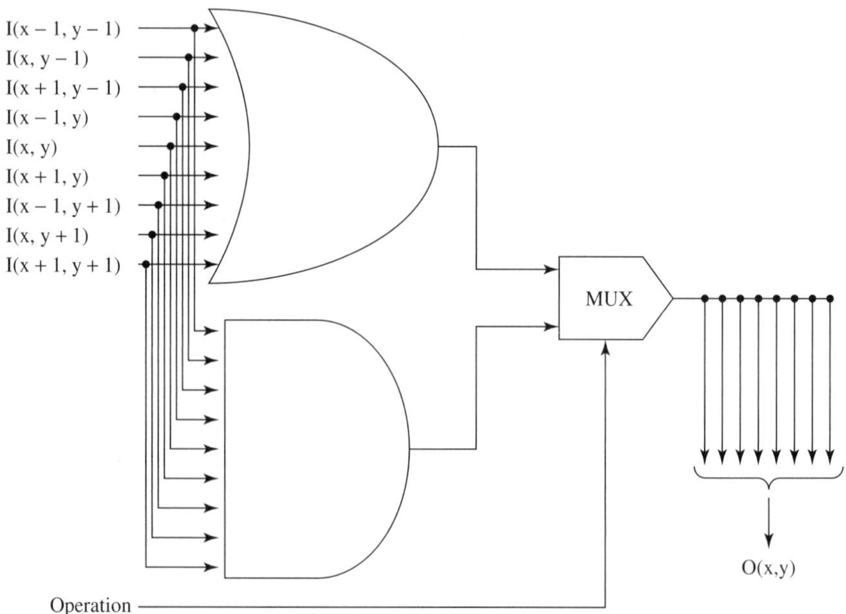

Figure 18.5 Implementation for binary erosion and dilation.

On the other hand, using the statement *par* and function *vectors* of Handel-C, we have replicated the circuit of Figure 18.5 in order to apply the corresponding operation simultaneously on 402 pixel neighborhoods, thereby accelerating the operation 402 times (use of inherent parallelism).

18.4.3 Gray-Scale Morphology

Gray-scale morphological operators are a continuation of binary techniques, allowing direct processing of gray-scale images. Mathematically, the erosion (E) and dilation (D) operations are

$$\begin{aligned} E &= (I \ominus w)(x, y) \\ &= \min\{I(x+y, y+j) - w(i, j)/(i, j) \in G, (x, y) + (i, j) \in F\} \qquad (18.9) \\ D &= (I \oplus w)(x, y) \\ &= \max\{I(x-y, y-j) + w(i, j)/(i, j) \in G, (x, y) - (i, j) \in F\} \qquad (18.10) \end{aligned}$$

respectively, where F and G are the index domains of I and w, respectively. $I(x, y)$ represents the value of the pixel at image I in the position (x, y). $w(i, j)$ indicates the value of the structuring element w in the position (i, j). These expressions can be clarified by means of the use of morphological masks and [3]

$$\begin{aligned} O(x, y) = \min/\max\{&A + I(x-1, y-1), B + I(x, y-1), \\ &C + I(x+1, y-1), D + I(x-1, y), E + I(x, y), \\ &F + I(x+1, y), G + I(x-1, y+1), H + I(x, y+1), \\ &I + I(x+1, y+1)\} \qquad (18.11) \end{aligned}$$

The central mask value and its eight neighbors are added to the corresponding input pixels. For the erosion case, the structuring element is formed by values that range from −255 to 0, while for the dilation the values range from 0 to 255. The output value is determined as the minimum/maximum (minimum for the erosion and maximum for the dilation) value of these nine sums.

As in binary morphology, the gray-scale erosion and dilation is the base of many other morphological operations: for example, opening and closing, which are defined like their binary equivalents. Gray-scale erosion and dilation have been designed by means of two different hardware modules. Although two hardware modules have been developed, these possess a similar scheme of operation. We explain this scheme now, and later we detail the differences between the two operations. Both erosion and dilation follow the general scheme shown in Figure 18.6, where $I(x+i, y+j)$ makes reference to each of the nine pixels (of 8 bits) covered by the 3×3 mask. The structuring element has values between 0 and 255 $(A, B, \ldots, I)$ to subtract (erosion) or to add (dilation) with the associate pixel.

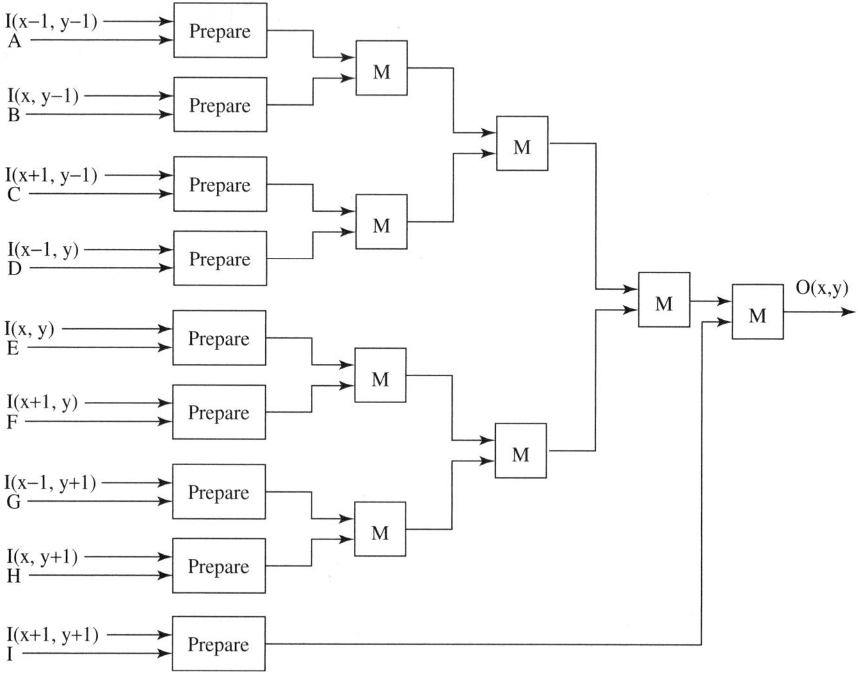

Figure 18.6 Implementation of gray-scale dilation and erosion.

In the case of erosion, Figure 18.7 illustrates the internal scheme for each of the circuits *Prepare* and M. For the circuit *Prepare*, the corresponding mask element Wn is subtracted from the pixel $I(x+i, y+j)$ using an 8-bit unsigned binary subtracter. Then the multiplexer allows us to control if the resulting value is negative, replacing it by 0. The outputs of the nine circuits *Prepare* are used inside a partial-sorting network that allows us to select the output that possesses the minimum value. For that, the sorting and selection network consists of eight nodes M.

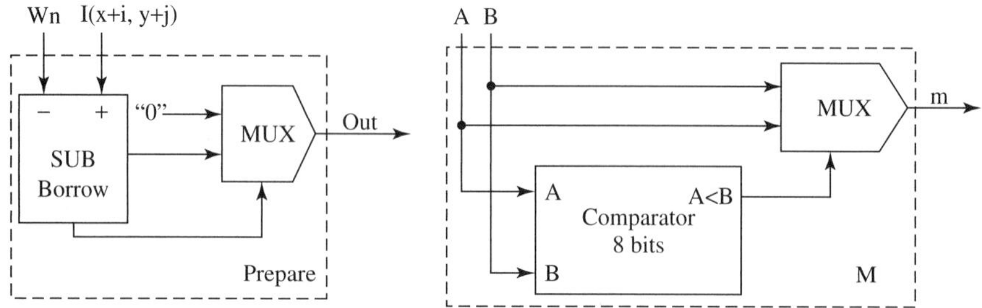

Figure 18.7 Circuits *Prepare* and *M* for the gray-scale erosion operation.

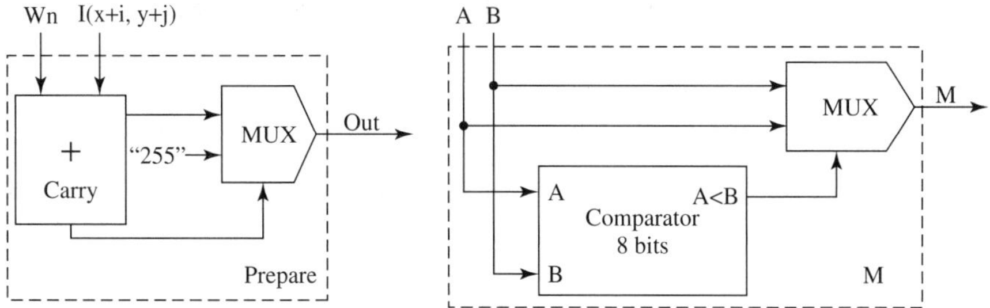

Figure 18.8 Circuits *Prepare* and *M* for the gray-scale dilation operation.

In the case of dilation, Figure 18.8 presents the internal scheme for the circuits *Prepare* and M. The corresponding mask element Wn is added to the pixel $I(x+i, y+j)$ using an 8-bit unsigned adder, and the multiplexer allows us to control if the resulting value is higher than 255. Afterward, the outputs of the nine circuits *Prepare* are used inside a partial sorting network that makes it possible to select the output that has the maximum value.

Finally, it is important to indicate that for these operations (gray-scale erosion and dilation) we have been able to replicate the corresponding circuit 148 times, therefore accelerating the operation 148 times (application of the operation simultaneously on 148 pixel neighborhoods).

18.5 EXPERIMENTAL ANALYSIS: SOFTWARE VERSUS FPGA

Table 18.2 presents several comparisons between the software and hardware (FPGA) versions of the different image-processing operations. The software versions are executed on a 1.7-GHz Pentium-4 with 768M bytes of RAM, and the hardware versions, on a Virtex-E 2000 FPGA. The software versions are written in C++ and compiled with the appropriate optimizations. All the results refer to images with 256 gray levels and they are expressed in seconds. These times also include the communication time, and they are the average response times after 10 executions of each experiment. In order not to influence the results, all the measurements in Table 18.2 have been made with the same workload in the server.

As we can observe in Table 18.2, FPGA implementation is always better than software implementation. Furthermore, the larger the image, the higher the benefits obtained by hardware (FPGA) execution, since a larger image implies greater use of the hardware computation capability, that is, greater exploitation of the inherent parallelism in image-processing operations.

The hardware implementation of all the pixel-by-pixel operations (except the histogram) always have the same execution time, because they take advantage of the existence of the image color table (all the 8-bit images have this color table),

TABLE 18.2 Comparisons Between the Software (Pentium) and Hardware (FPGA) Versions

	Images 600 × 600			Images 1024 × 768			Images 2032 × 1524		
Op.	SW	HW	Speedup	SW	HW	Speedup	SW	HW	Speedup
H	2.515	0.071	35.423	5.875	0.156	37.660	27.953	0.203	137.700
CI	0.016	0.001	16	0.026	0.001	26	0.078	0.001	78
BCE	0.015	0.001	15	0.028	0.001	28	0.078	0.001	78
BS	0.015	0.001	15	0.031	0.001	31	0.078	0.001	78
HSS	0.015	0.001	15	0.031	0.001	31	0.078	0.001	78
MF	3.464	0.711	4.872	7.593	1.754	4.329	34.328	5.536	6.201
LPF	2.531	0.064	39.547	5.461	0.148	36.899	26.195	0.461	56.822
HPF	2.468	0.063	39.175	5.437	0.139	39.115	26.344	0.458	57.468
LF	2.281	0.050	45.520	4.913	0.098	50.133	25.659	0.438	58.583
VGF	2.265	0.053	42.736	4.921	0.109	45.147	25.262	0.420	60.184
HGF	2.281	0.053	43.078	4.903	0.110	44.573	25.203	0.417	60.475
DGF	2.234	0.051	43.804	4.906	0.109	45.009	25.187	0.419	60.153
BE	2.468	0.020	123.400	5.431	0.041	132.463	26.188	0.175	149.219
BD	2.484	0.020	124.200	5.382	0.040	134.550	26.346	0.189	139.399
GSE	2.421	0.048	50.438	5.377	0.108	49.787	26.398	0.361	73.068
GSD	2.453	0.049	50.061	5.403	0.106	50.972	26.336	0.355	74.238

performing their operation on the color table, and their performance is therefore independent of image size.

On the one hand, we can notice clearly that the result obtained by median filter is not as good as for other operations. On the other hand, binary erosion and dilation demonstrate great speedup (more than 149 times faster).

18.6 CONCLUSIONS

Image processing has always been a focus of interest for programmable logic developers. However, the use of FPGAs in image processing is still a research area, and although many hardware developments exist, so far no complete system has been proven competitive enough to establish a significant commercial presence.

In this chapter we have described the FPGA-based implementation and optimization of many different image-processing algorithms. In fact, 16 hardware implementations have been studied, all clearly improving the performance obtained by software (some operations are more than 139 times faster). These results are possible because we have applied many different optimizations to our FPGA-based implementations. In particular, our implementations use several parallelism techniques, include other operation-specific optimizations, and have high operational frequencies. Thanks to this, very good execution times are obtained.

Acknowledgments

This work has been supported in part by the Spanish government under grant TIC2002-04498-C05-01 (the TRACER project) and grant TIN2005-08818-C04-03 (the OPLINK project).

REFERENCES

1. M. A. Vega-Rodríguez, J. M. Sánchez-Pérez, and J. A. Gómez-Pulido. Guest editors' introduction—special issue on FPGAs: applications and designs. *Microprocessors and Microsystems*, 28(5–6):193–196, 2004.
2. M. A. Vega-Rodríguez, J. M. Sánchez-Pérez, and J. A. Gómez-Pulido. Recent advances in computer vision and image processing using reconfigurable hardware. *Microprocessors and Microsystems*, 29(8–9):359–362, 2005.
3. G. Baxes. *Digital Image Processing: Principles and Applications*. Wiley, New York, 1994.
4. M. A. Vega-Rodríguez, A. Gómez-Iglesias, J. A. Gómez-Pulido, and J. M. Sanchez-Perez. Reconfigurable computing system for image processing via the Internet. *Microprocessors and Microsystems*, 31(8):498–515, 2007.
5. R. Haralick and L. Shapiro. *Computer and Robot Vision*. Addison-Wesley, Reading, MA, 1992.
6. E. Dougherty and P. Laplante. *Introduction to Real-Time Imaging*. SPIE Press, Bellingham, WA, 1995.
7. E. Jamro and K. Wiatr. FPGA implementation of addition as a part of the convolution. In *Proceedings of the EuroMicro Symposium on Digital System Design*, Warsaw, Poland. IEEE Computer Society Press, Los Alamitos, CA, 2001, pp. 458–465.
8. S. Perri, M. Lanuzza, P. Corsonello, and G. Cocorullo. SIMD 2-D convolver for fast FPGA-based image and video processors. In *Proceedings of the International Conference on Military and Aerospace Programmable Logic Devices*, Washington, DC, 2003.
9. E. Jamro and K. Wiatr. Genetic programming in FPGA implementation of addition as a part of the convolution. In *Proceedings of the EuroMicro Symposium on Digital System Design*, Warsaw, Poland. IEEE Computer Society Press, Los Alamitos, CA, 2001, pp. 466–473.
10. R. Haralick, S. Sternberg, and X. Zhuang. Image analysis using mathematical morphology. *IEEE Transactions on Pattern Analysis and Machine Intelligence*, 9(4):532–550, 1987.
11. H. Joo and R. Haralick. Understanding the Application of Mathematical Morphology to Machine Vision. In *Proceedings of the IEEE International Symposium on Circuits and Systems*, Portland, OR, vol. 2, 1989, pp. 977–982.
12. C. Giardina and E. Dougherty. *Morphological Methods in Image and Signal Processing*. Prentice Hall, Upper Saddle River, NJ, 1998.

CHAPTER 19

Application of Cellular Automata Algorithms to the Parallel Simulation of Laser Dynamics

J. L. GUISADO and F. JIMÉNEZ-MORALES
Universidad de Sevilla, Spain

J. M. GUERRA
Universidad Complutense de Madrid, Spain

F. FERNÁNDEZ
Universidad de Extremadura, Spain

19.1 INTRODUCTION

In this chapter we review the use of a biologically inspired heuristic technique—cellular automata (CA)—as a problem solver for one of the most paradigmatic complex systems: the laser. CAs are a class of mathematical system that can be used to model spatiotemporal phenomena, characterized by the discreteness of all of its variables: space, time, and normally state variables. An important property of CAs is their intrinsic parallel character. Therefore, they are specially well suited to be implemented very efficiently on parallel computers.

In this work we also exploit this property to carry out a parallel implementation of the CA model developed for laser dynamics. In addition, we study the performance and scalability of this parallel implementation and conclude that it is very satisfactory. In particular, we have described a CA-based algorithm which is an alternative to model the dynamics of lasers, normally modeled using differential equations. This approach can be very useful for modeling lasers in situations in which the differential equations are difficult to integrate, or even difficult to apply: lasers ruled by stiff differential equations, devices with complex boundary conditions, very small devices, and so on.

Optimization Techniques for Solving Complex Problems, Edited by Enrique Alba, Christian Blum, Pedro Isasi, Coromoto León, and Juan Antonio Gómez

In Section 19.2 we introduce some background information about the relevant subjects that are treated. Then in Section 19.3 we present the problem to be solved by the proposed algorithm: laser dynamics. We describe the proposed algorithm in detail in Section 19.4. In Section 19.5 we review some of the laser properties that are reproduced successfully by the CA model. Next, we describe a parallel implementation of the CA model and analyze its performance and scalability when executed on a small computer cluster. Finally, some conclusions and prospects for future work are noted in Section 19.7.

19.2 BACKGROUND

The concept of stimulated emision was raised by Albert Einstein in his 1917 paper "Zur Quantenmechanik der Strahlung" [1]. Use of this idea has permitted us to amplify radiation by propagation across a medium in which the population of an upper energy state is larger than the population of a low-energy state (population inversion). The first experimental device working as a microwave amplifier was reported in 1955 [2]. The extension of this concept to the optical domain was achieved by Maiman in an inverted ruby rod optically pumped by a flashlamp in 1960 [3].

Maiman called this phenomenon the *laser* (light amplification by stimulated emission of radiation) effect. A typical laser device needs a population inversion mechanism to enhance the upper-state population to be larger than that remaining in a lower-energy state. This mechanism is usually known as the pumping system. To enhance the effective amplification, the inverted medium is usually placed inside a Fabry–Perot resonator that provides feedback, making the amplified light bounce between the mirrors. In this form the laser behaves as a regenerative light oscillator, and transient, periodic, or chaotic oscillatory processes arise in it. The amplification gain is spread homogeneously or inhomogeneously over an interval of frequencies, frequently covering several resonator mode frequencies. In some cases the light in many of the covered resonator modes becomes amplified, frequently coupled in phase, causing a large variety of complex dynamic phenomena to take place. In the case of large-aperture lasers, a large number of transverse mode geometries may be developed by the system, and the dynamic behavior has a spatiotemporal character. In many cases the spatiotemporal dynamics of these lasers has been observed to be complex [4], but in homogeneously broadened gain, lasers may be fairly well reproduced by simple theoretical models [4]. In these models a single frequency field in resonant interaction with the populations of a couple of states in the medium is assumed. In this simplified model, field and matter obey the Maxwell–Bloch equations [5]. These equations couple the electromagnetic field with matter polarization and population inversion. The dynamics is influenced by the relaxation constants of field, atomic polarization, and population inversion, the lowest of the tree damping constants becoming dominant. Frequently, the polarization damping constant is the largest and the polarization becomes a slave of field and population inversion. By eliminating the polarization from the equations and under an assumption of homogeneous field spatial distribution, a couple of rate equations may describe some simple laser dynamics.

The Maxwell–Bloch equations are $2 + 1$ partial differential equations with damping constants sometimes several orders of magnitude different. Thus, integration of this system is sometimes a very rigid problem. For this reason, an alternative approach to modelization would be welcome. Cellular automata models may play an important role in this context, taking into account its intrinsic parallelization in computing. To test the feasibility of this approach, we have started from simulation of the simplest rate equations.

In this work we first describe a CA algorithm to model a simple laser device, and after reviewing some of the laser dynamics phenomena that are reproduced by the CA model, we develop a parallel implementation of the model to be executed on parallel computers. The very nature of cellular automata is strongly connected with parallel systems: The reason is the way a CA works, by simultaneous updating of a cell's states along the entire automaton. No sequential process should therefore be employed for that task, but instead, a parallel computation of new states for all the cells. Nevertheless, despite the intrinsic parallel nature of the model, researchers have usually applied sequential processes that simulate the parallel processes. The reason is the sequential nature of the computers generally used for simulations. Although this is an easy approach to employing CA systems, they would be of no practical use for solving real-world problems because of the computational load for large sizes of two- or three-dimensional CA.

The possibilities for running CA in parallel are two: The first consists of using available parallel computers to develop parallel CA; the second requires the design of specialized hardware aimed at CA execution. Although some attempts have been described for this second alternative, such as the CAM (cellular automata machine) computer [6], we focus here on available parallel, cluster, or grid deployment of CA. In fact, general-purpose parallel computers are well suited for scalable CA models, from the point of view of speedup, programmability, and portability. Two types of architectures are of interest: both single-instruction and multiple-data (SIMD) and also multiple-instruction, multiple-data (MIMD). For the implementation of CA on these computers, two principal approaches are available: using general-purpose parallel programming languages such as HPF, HPC++, or Linda, or employing a standard high-level sequential language combined with specific libraries allowing parallel applications to run, such as MPI (message-passing interface), PVM (parallel virtual machine), or OpenMP (open multiprocessing).

During the last decade, some attempts to introduce parallelism within CA have been described. Most of them were not intended to implement directly the inherently parallel CA internal working rules, which can easily be simulated in sequential fashion, but to improve speedup of the entire process by using many processors. The first attempts to parallelize CA were carried out by Resnick with the StarLogo system [7] and by Cannataro et al. [8], although many approaches and results were described later using parallel CAs, such as CAMEL [9], Nemo

[10], PECANS [11], DEVS [12], and P-CAM [13]. The topic has been reviewed by Talia [14].

19.3 LASER DYNAMICS PROBLEM

We start by modeling the simplest laser dynamics phenomena, which can be described in the most simplified but still realistic way by two coupled nonlinear rate equations [15]:

$$\frac{dn(t)}{dt} = KN(t)\,n(t) - \frac{n(t)}{\tau_c} \tag{19.1}$$

$$\frac{dN(t)}{dt} = R - \frac{N(t)}{\tau_a} - KN(t)\,n(t) \tag{19.2}$$

The first gives the variation in the number of laser photons $n(t)$ with time, proportional to the laser beam intensity. The term $+KN(t)n(t)$ represents an increase in the number of photons by stimulated emission (K is the coupling constant between the radiation and the population inversion). The term $-n(t)/\tau_c$ accounts for the decaying (or absorption) process exhibited by laser photons inside a laser cavity with a characteristic decay time τ_c. The second equation represents the temporal variation of the population inversion $N(t)$. The term $+R(t)$ represents the pumping of electrons with a pumping rate R to the upper laser level. The term $-N(t)/\tau_a$ introduces the decaying of electrons from the upper laser level to lower levels with a characteristic decay time τ_a. The product term $-KN(t)n(t)$ reflects decreasing population inversion by stimulated emission. The presence of the product term $KN(t)n(t)$ in each equation gives them a nonlinear nature. For small-amplitude fluctuations, its solutions can show relaxation oscillations in their evolution toward a steady state. For strong oscillations the two variables $n(t)$ and $N(t)$ are changing in a rapid and typically nonlinear way, and there does not seem to be a simple analytical solution [15,16].

A simplified but sufficiently realistic description of the laser system represented by these equations is the four-level laser system depicted in Figure 19.1, where we have represented some of the basic physical processes that play a role. Electrons are excited from the ground level up to level E_3 by some external pumping process. Population inversion is produced between levels E_1 and E_2. The reason is that the lifetimes of energy levels E_3 and E_1 are negligible compared to the lifetime of level E_2. As a result, electrons in levels E_3 and E_1 decay very fast, but level E_2 is metastable. Stimulated emission occurs when an electron in level E_2 decays down to level E_1, stimulated by the presence of a stimulator photon with energy $E = E_2 - E_1$. Two processes not represented in Figure19.1 are also very important: absorption of electrons in level E_2 (which decay to lower levels due to different processes not related to stimulated emission) and absorption of laser photons, a fraction of which disappear because they leave the laser cavity through the semireflecting mirror or are absorbed by the material.

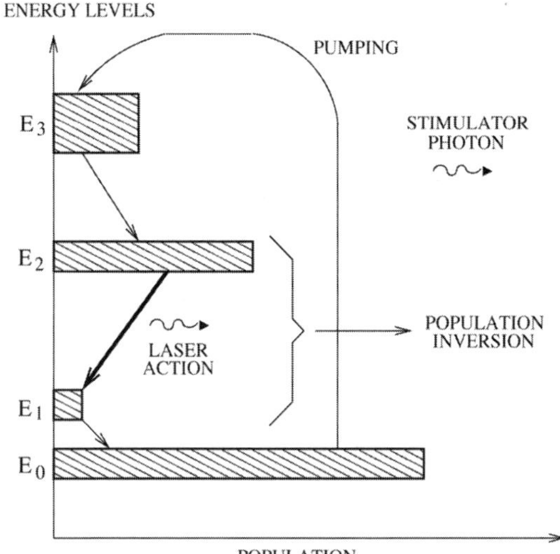

Figure 19.1 Some of the basic physical processes in a four-level laser system, a simplified but still realistic description of many real laser systems.

19.4 ALGORITHMIC PROPOSAL

In this section we describe the algorithm that we have used to simulate laser dynamics, proposed originally by some of the present authors [16]. The algorithm is based on a two-dimensional, partially probabilistic multivariable CA that simulates a transverse section of the active medium in a laser system. The defining characteristics of the CA are described below.

19.4.1 Cellular Space

The cellular space is a two-dimensional square lattice that contains $N_c = L \times L$ cells. Periodic boundary conditions are used.

19.4.2 States of the Cells

Two variables are associated with each cell of the CA: $a_{ij}(t)$ and $c_{ij}(t)$. The first, $a_{ij}(t)$, represents the state of the electron in cell $\{ij\}$ (row i and column j) at time t: When $a_{ij}(t) = 0$, the electron is in the ground state, and when $a_{ij}(t) = 1$, the electron is in the upper laser state. Also, $c_{ij}(t) \in \{0, 1, 2, \ldots, M\}$ represents the number of laser photons in cell $\{ij\}$ at time t. This number is bounded by an upper value M which must be chosen large enough to avoid the saturation of the system. The state variables represent "bunches" of real photons and electrons. Its values are connected to the number of photons and electrons in the real system by a normalization constant.

NW	N	NE
W	C	E
SW	S	SE

Figure 19.2 Moore neighborhood.

19.4.3 Neighborhood

Each cell interacts locally with a number of surrounding cells belonging to a *Moore neighborhood*: the cell itself (C), its four nearest neighbors located at the north (N), south (S), east (E), and west (W) positions, and the four next-nearest neighbors, located at the northeast (NE), southeast (SE), southwest (SW), and northwest (NW) positions, as shown in Figure 19.2.

19.4.4 Transition Rules

The evolution of the system is computed using the transition rules, which specify the state of each cell at time step $t + 1$, depending on its state and the state of the cells included in its neighborhood at time step t. Rules represent the physical processes working at the microscopic level in the laser system. The application of the transition rules is the main operation of a CA algorithm. In our case the overall structure of the CA laser model algorithm is shown in Algorithm 19.1, which corresponds to the main program. After initializing the system, the transition rules

Algorithm 19.1 Pseudocode Diagram for the CA Laser Model

```
Initialize system
Input data
for time step = 1 to maximum time step do
  for each cell in the array do
     Apply stimulated emission rule (Algorithm 19.2)
     Apply photon decay, electron decay, and
         pumping rules (Algorithm 19.3)
     Apply noise photons creation rule (Algorithm 19.4)
  end for
  Calculate populations after this time step
  Optional additional calculations on intermediate results
end for
Final calculations
Output results
```

are applied to each CA cell inside a time loop. Subroutines used for computation of the transition rules are shown in Algorithms 19.2 to 19.5.

Our CA model uses five transition rules:

1. *Stimulated emission rule* (Algorithms 19.2 and 19.3). If the electronic state of a cell has a value of $a_{ij}(t) = 1$ at time t and the sum of the values of the laser photons states in the nine neighboring cells is larger than a certain threshold θ (which in our simulations has been taken to be 1), then at time $t + 1$ a new photon will be emitted in that cell: $c_{ij}(t + 1) = c_{ij}(t) + 1$, and the electron will decay to the ground level: $a_{ij}(t + 1) = 0$. All the cells of the CA must be updated in parallel. To this end, changes from this rule are computed using a temporal matrix c'_{ij}. After the rule has been applied to all the cells of the CA, the values of c_{ij} are updated with the contents of c'_{ij}.

2. *Photon decay rule* (Algorithm 19.4). Each photon is destroyed τ_c time steps after being created. In particular, (tlc_{ijk}) represents the number of time steps that will have to elapse until a particular photon located in cell $\{ij\}$ (at row i and column j) is destroyed, where k distinguishes between the different photons that can occupy the same cell. When a photon is created, $tlc_{ijk} = \tau_c$. After that, 1 is substracted from tlc_{ijk} at each time step and the photon will be destroyed when $tlc_{ijk} = 0$.

3. *Electron decay rule* (Algorithm 19.4; this algorithm computes three rules: photon decay, electron decay, and pumping). After an electron is excited from the ground level to the upper laser level, it will decay to the ground level again after τ_a time steps if it has not yet decayed by stimulated emission. In particular, (tla_{ij}) represents the number of time steps that will have to elapse until a particular electron located in cell $\{ij\}$ decays to the ground level. When the electron is excited initially, $tla_{ij} = \tau_a$. After that, 1 is substracted from tla_{ij} at each time step and the electron will decay to the ground level again when $tla_{ij} = 0$.

4. *Pumping rule* (Algorithm 19.4). If the electronic state of a cell $\{ij\}$ has a value of $a_{ij}(t) = 0$ at time t, then at time $t + 1$ that state will have a value of $a_{ij}(t + 1) = 1$ with a pumping probability λ.

5. *Noise photons creation* (Algorithm 19.5). A small number of laser photons in randomly chosen positions is introduced at each time step to reproduce spontaneous emission and thermal contributions, responsible for the initial laser startup. To this end, for a small number of randomly chosen cells $\{ij\}$ ($< 0.01\%$ of total), $c_{ij}(t + 1) = c_{ij}(t) + 1$ is applied.

19.5 EXPERIMENTAL ANALYSIS

In this section we present a review of some of the experimental results found in the simulations carried out using this model. As shown originally [16–19], the CA model of laser dynamics can reproduce different aspects of the phenomenology of laser systems. The behavior of the system depends on three parameters: the

Algorithm 19.2 Stimulated Emission Rule

```
{List of variables:}
{Lx: length of the CA lattice in the x direction }
{Ly: length of the CA lattice in the y direction }
{k: index to distinguish different photons in the same cell}
{aij: state of the electron at cell (i, j)}
{cij: number of laser photons at cell (i, j)}
{c'ij: auxiliary variable to calculate new values of cij}
{τc: (maximum) lifetime of laser photons}
{tlaij: current lifetime of excited electron at cell (i, j)}
{tlcijk: current lifetime of photon number k at cell (i, j)}
{θ: threshold for the number of photons in
    neighborhood that can produce a stimulated emission}
{M: maximum number of laser photons in a cell}
for j = 0 to Ly − 1 do
   for i = 0 to Lx − 1 do {CA lattice loop}
     if aij = 1 then
        if neighbors (*cij, Lx, Ly, i, j) > θ then
           {neighbors function: Alg. 19.5}
           {Look for first value of k for which tlcijk = 0}
           k ← 1
           while tlcijk ≠ 0 and k [[leq]] M do
             k ← k + 1
           end while
           if k < = M then
              aij ← 0
              tlaij ← 0
              c'ij ← c'ij + 1
              tlcijk ← τc + 1
              {τc + 1 is assigned because 1 is subtracted in
                   the decay loop}
           end if
        end if
     end if
   end for
end for
{Update value of c matrix with contents of c' matrix}
for j = 0 to Ly − 1 do
   for i = 0 to Lx − 1 do {CA lattice loop}
      cij ← c'ij
   end for
end for
```

Algorithm 19.3 Function to Calculate the Sum of Photons in State 1 in the Moore Neighborhood, Implementing Periodic Boundary Conditions

```
{function: neighbors}
{input parameters: *c_ij, L_x, L_y, i, j}
{output: sum}
a_1 ← i − 1
if a_1 < 0 then
    a_1 ← L_x + a_1
end if
a_2 ← i + 1
if a_2 ≥ L_x then
    a_2 ← a_2 − L_x
end if
    a_3 ← j − 1
if a_3 < 0 then
    a_3 ← L_y + a_3
end if
a_4 ← j + 1
if a_4 ≥ L_y then
    a_4 ← a_4 − L_y
end if
Σ ← c_{a1 j} + c_{a2 j} + c_{i a3} + c_{i a4} + c_{a1 a3} + c_{a1 a4} + c_{a2 a3} + c_{a2 a4} + c_ij
{returns sum}
```

pumping probability (λ), the lifetime of photons (τ_c), and the lifetime of excited electrons (τ_a). In a simulation, an initial state is provided [$a_{ij}(0) = 0,\ c_{ij}(0) = 0,\ \forall ij$, except for a small fraction, 0.01%, of noise photons present] and then the system is allowed to evolve for a number of time steps. In each step we measure two macroscopic magnitudes: the total number of laser photons, $n(t)$, and the total number of electrons in the upper laser state (population inversion), $N(t)$:

$$n(t) = \sum_{i=1}^{L_x} \sum_{j=1}^{L_y} c_{\text{ij}}(t) \qquad N(t) = \sum_{i=1}^{L_x} \sum_{j=1}^{L_y} a_{\text{ij}}(t) \tag{19.3}$$

A characteristic feature of laser systems is that laser action happens only when the pumping probability is over a threshold value. This property is reproduced correctly by the CA model [16], and the dependence of this threshold value on the other two system parameters (lifetimes τ_a and τ_c) is found to be in good agreement with laser behavior, as shown in Figure 19.3. Depending on the values of their three characteristic parameters, lasers exhibit two main distinctive behaviors in their time evolution: a constant or an oscillatory behavior [16,18]. As shown in Figure 19.4, the model reproduces these two types of behavior: The time evolution obtained from the simulations is similar to that exhibited by laser

Algorithm 19.4 Photon Decay, Electron Decay, and Pumping Rules

```
{List of variables:}
{L_x: length of the CA lattice in the x direction}
{L_y: length of the CA lattice in the y direction}
{k: index to distinguish different photons in the same cell}
{a_ij: state of the electron at cell (i, j)}
{c_ij: number of laser photons at cell (i, j)}
{c'_ij: auxiliary variable to calculate new values of c_ij}
{τ_a: (maximum) lifetime of excited electrons}
{τ_c: (maximum) lifetime of laser photons}
{tla_ij: current time of life of excited electron at cell
    (i, j)}
{tlc_ijk: current time of life of photon number k at cell
    (i, j)}
{M: maximum number of laser photons in a cell}
{λ: pumping probability}
{ξ: auxiliary variable}
for j = 0 to L_y − 1 do
  for i = 0 to L_x − 1 do {CA lattice loop}
    if c_ij > 0 then {Apply photon decay rule}
      for k = 1 to M do
        {Subtract 1 to every photon's lifetime}
        if tlc_ijk > 0 then
          tlc_ijk ← tlc_ijk − 1
          if tlc_ijk = 0 then {One photon decays}
            c_ij ← c_ij − 1
            c'_ij = c_ij
          end if
        end if
      end for
    end if
    if a_ij = 1 then {Apply electron decay rule}
      {Subtract 1 to time of life of every excited
          electron}
      tla_ij ← tla_ij − 1
      if tla_ij = 0 then
        {One electron decays}
        a_ij ← 0
      end if
    else if a_ij = 0 then {Apply pumping rule}
      {Generate random number in (0, 1) interval}
      ξ ← random_number(0, 1)
      if ξ < λ then {λ: pumping probability}
        {One electron is pumped}
```

Algorithm 19.4 (*Continued*)

```
            a_ij ← 1
            tla_ij ← τ_a
         end if
      end if
   end for
end for
```

Algorithm 19.5 Noise Photons Creation Rule

```
{Introduce n_n number of photons in random positions}
for n = 0 to n_n − 1 do
   {Generate two random integers in (0, size − 1) interval}
   i ← random_number(0, L_x − 1)
   j ← random_number(0, L_y − 1)
   {Look for first value of k for which tlc_ijk = 0}
   k ← 1
   while tlc_ijk ≠ 0 and k ≤ M do
     k ← k + 1
   end while
   if k ≤ M then
      {Create new photon}
      c'_ij ← c'_ij + 1
      tlc_ijk ← τ_c
   end if
end for
```

systems, described, for example, by Siegman [15]. A lattice size of 400 $\times$ 400 cells was used for this figure.

In addition, the CA model exhibits another type of complex behavior in which irregular oscillations with fluctuations on a wide range of time scales appear (see ref. 18), as shown in Figure 19.5. This regime could correspond to a chaotic state, as found in the dynamics of many lasers, but this is still under investigation. Also, the dependence on system parameters of the type of behavior exhibited in the time evolution of the system is in good qualitative agreement with the laser behavior [16], as shown in Figure 19.6. In this figure we show a contour plot of a magnitude called *Shannon's entropy* of the distribution of the number of laser photons, for a fixed value of $\tau_c = 10$ time steps and obtained using simulations with a 200 $\times$ 200 lattice.

This magnitude is a good indicator of the presence of oscillations in the time evolution of the number of laser photons (for a precise definition and discussion, see, e.g., ref. 17). In this plot, R is the laser pumping rate and R_t is the threshold laser pumping rate, which are linearly related to the pumping probability λ and

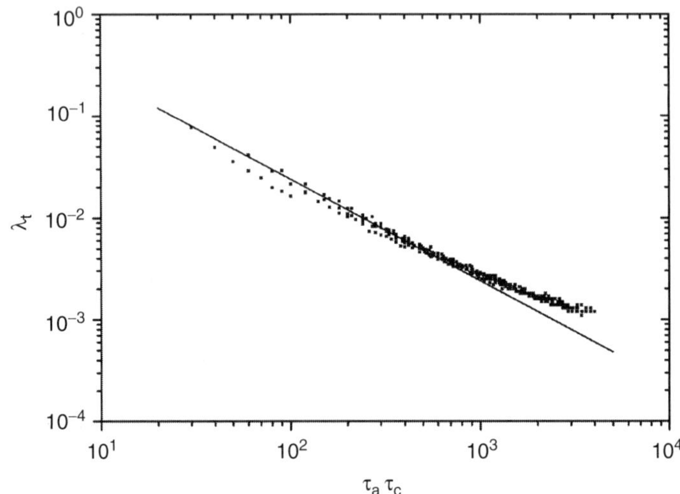

Figure 19.3 Dependence of the threshold pumping probability λ_t from the CA laser model on the product of the characteristic lifetimes τ_a and τ_c (measured in time steps), plotted on a logarithmic scale. The solid line is the laser behavior predicted by the standard laser rate equations, and the dots are the results of the simulations.

the threshold pumping probability λ_t that appear in the CA model, so that $R/R_t = \lambda/\lambda_t$. Points a, b, and c show the values of the parameters that correspond to Figures 19.4 and 19.5: a corresponds to constant behavior [Figure 19.4(a)], b to oscillatory behavior [Figure 19.4(b)], and c to a regime with irregular oscillations (Figure 19.5). High values of Shannon's entropy (dark zones) correspond to oscillatory behavior and low values (bright zones) to nonoscillatory response. The predictions of the standard laser rate equations are indicated by the black line: Areas of oscillatory behavior should appear above and to the right of this curve, and constant behavior should appear in the remaining areas. There is good qualitative agreement between the predictions and the results of the simulations indicated by Shannon's entropy, as the high values of this magnitude appear above and to the right of the black line, and their contour resembles the shape of this line.

19.6 PARALLEL IMPLEMENTATION OF THE ALGORITHM

Earlier we described a CA algorithm employed to simulate laser dynamics and presented some of its experimental results. It was shown that a very simple coarse-grained CA model can reproduce the laser behavior in a qualitative way. But a finer-grained CA model is needed to simulate a specific laser device quantitatively. In particular, it could reproduce more details of that specific device (such as complicated boundary conditions) and have a granularity closer to that of the

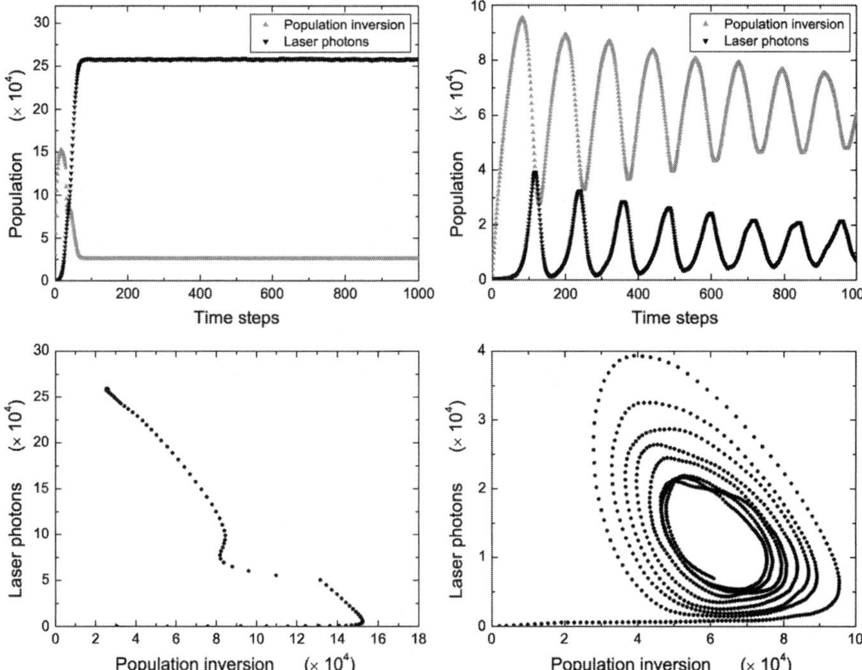

Figure 19.4 Simulation results for two different sets of values of the system parameters depicting the time evolution of the two macroscopic magnitudes, number of laser photons $n(t)$ and population inversion $N(t)$. They show how the model reproduces the two main characteristic behaviors exhibited by lasers. Upper: $n(t)$ and $N(t)$ versus time. Lower: evolution in a phase space with $n(t)$ versus $N(t)$. Left: Constant behavior; parameters: $\{\lambda = 0.192,\ \tau_c = 10,\ \tau_a = 30\}$. After an initial transient, the system goes to a fixed point. Right: Oscillatory behavior; parameters: $\{\lambda = 0.0125,\ \tau_c = 10,\ \tau_a = 180\}$. The system follows a spiral toward a steady-state limit point.

real macroscopic system. Moreover, to reproduce the shape of the laser device, a three-dimensional version of the CA model is needed. For these purposes a very large lattice size is needed, and the resulting algorithm needs a prohibitively large runtime for a sequential computer. Therefore, a parallel implementation is needed. In the present section we describe a parallel implementation of the previous CA model and study its performance and scalability running on a small computer cluster. These results were introduced in refs. 19 and 20.

The CA model has been parallelized for running on parallel computers with distributed memory using the message-passing paradigm. The *parallel virtual machine* (PVM) implementation of this paradigm has been used because we were interested in a later study of the model using dynamic load-balancing mechanisms developed for it specifically. Parallelization has been done using the master–slave programming model, as shown in Figure 19.7. Workload has been allocated with

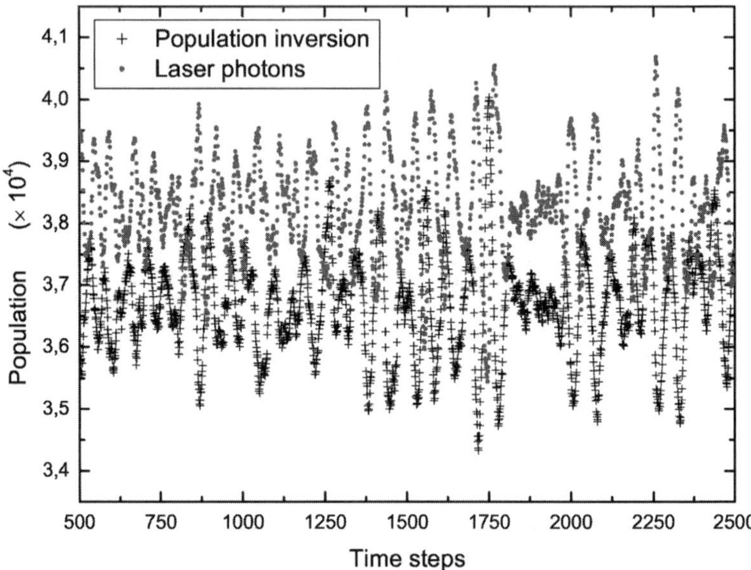

Figure 19.5 Regime with irregular oscillations for $\lambda = 0.031$, $\tau_c = 10$, $\tau_a = 180$. The number of laser photons and population inversion are plotted versus time after a transient of 500 time steps. Lattice size: 400 $\times$ 400 cells.

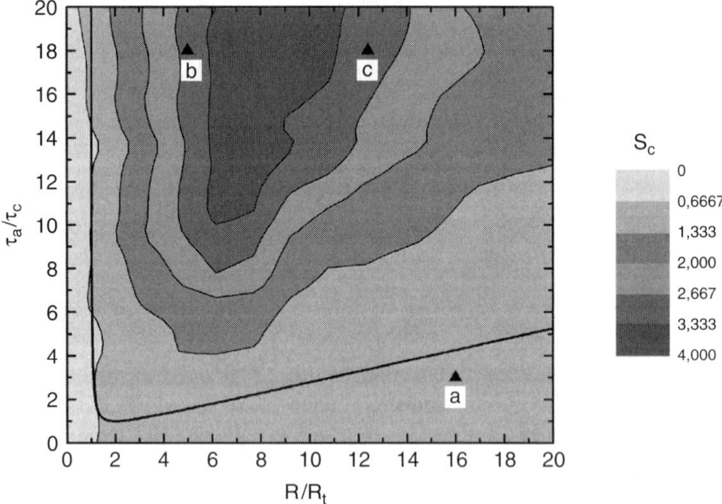

Figure 19.6 Contour plot of Shannon's entropy of the distribution of the number of laser photons obtained from the simulations with a fixed value of $\tau_c = 10$ time steps. This plot shows that there is good qualitative agreement between dependence on system parameters of the type of behavior exhibited by the system, as obtained from the simulations, and the laser behavior, delimited by the black line.

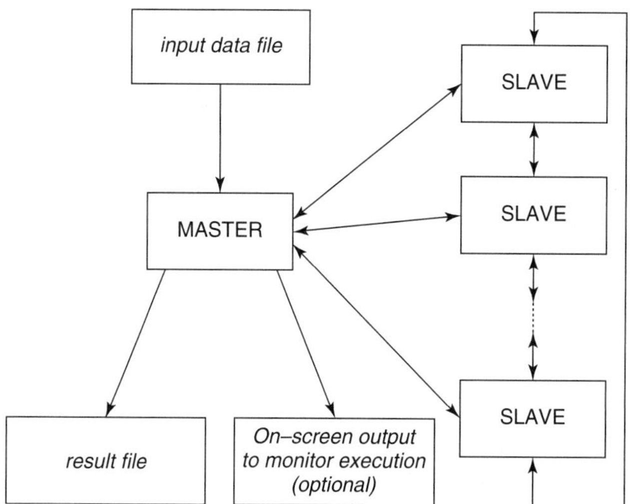

Figure 19.7 Block diagram of the parallel implementation of the CA model of laser dynamics, showing processes running on different processors (boxes in bold type represent different processors), communications between them (bold lines), and data flows.

data decomposition methodology: Identical tasks are computed with different portions of the data. A *master program* performs the initialization of the system, divides the CA lattice in p partitions of equal size, and sends each to a *slave program* that runs on a different processor. The particular tasks carried out by the master and slave programs are:

- Master program
  1. Reading input data (system size, number of partitions, parameter values, number of time steps) and initialization
  2. Spawning slave programs
  3. Partitioning the initial data of the automaton
  4. Sending common information and initial data to each slave
  5. Collection of results from slaves at each time step
  6. Termination of slave programs
  7. Calculations performed using collected data
  8. Outputting final data to external files
  9. Timing functions to measure performance
- Slave program
  1. Reception of common information and initial data from master
  2. Time evolution computation for the assigned partition: application of CA evolution rules

3. Exchange of state of the boundary cells with slave programs computing the neighboring partitions
4. Computation of intermediate results and their communication to the master program

A one-dimensional domain decomposition has been used: The CA is divided vertically into parallel stripes (subdomains), and each is assigned to a different processor. Two additional columns of ghost cells have been included at the left and right sides of each subdomain. They store the state of neighboring cells, which belong to a different subdomain that will be needed to apply the transition rules to the cells of the original subdomain. For the transition rules of our CA model, the only state information needed from neighboring cells is the photon state $c_{ij}(t)$, so this is the only information that must be communicated from the neighboring subdomains. Each slave program is responsible for computing the time evolution on its assigned partition.

We have measured the performance and scalability of the parallel implementation by running simulations on the cluster Abacus from the University of Extremadura, a Beowulf-type cluster composed of 10 nodes with an Intel Pentium-4 processor, six with a clock frequency of 2.7 GHz and four with 1.8 GHz, communicated by a fast Ethernet switch with 100 Mbps of bandwith. To avoid indeterminism in the results due to the heterogeneity of the cluster, for simulations with one to six nodes, slave programs have always been run on the "fast" (2.7-GHz) machines, and for simulations with seven to 10 nodes, additional "slow" (1.8-GHz) machines have been used to complete the required number of nodes. The master program has always been run on the master node of the cluster (1.8 GHz).

To measure the performance of the parallel implementation, we have run the same experiment for three different system sizes using a different number of partitions (each assigned to a slave program running on a different processor). The resulting runtime measures are plotted in Figure 19.8, showing a significant decrease in the number of processors. The only exception is the change from six to seven processors, where an increase is registered due to the assigning strategy that has been used: Only fast nodes are assigned to jobs with six or fewer processors, and for jobs with more than six processors, some slow processors have to be used.

The performance of the parallel application can be evaluated using *speedup* (S_p) [21], which indicates how much faster a parallel algorithm is than a corresponding sequential algorithm. It can be defined as the ratio of the runtime of the sequential version of the program running on one processor of the parallel computer (T_1) to the runtime of the parallel version running on m processors of the same computer (T_m):

$$S_p(m) = \frac{T_1}{T_m} \tag{19.4}$$

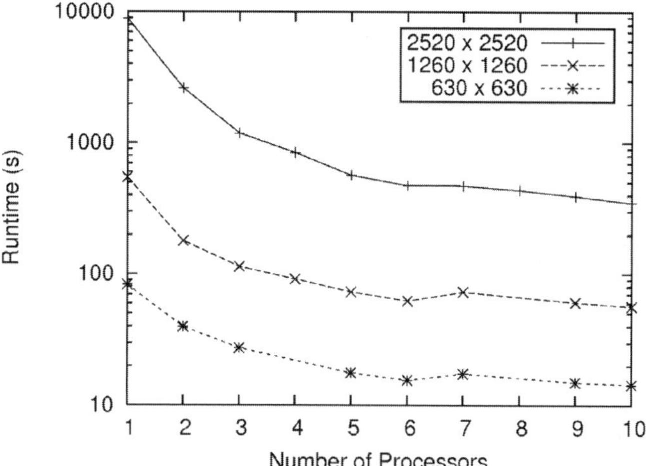

Figure 19.8 Runtime of the experiments, using a logarithmic scale, for different numbers of partitions of the whole CA, each running on a different processor. Measurements for three different system sizes are shown.

In Figure 19.9 we have shown the speedup obtained for parallel implementation of our CA model [19] for three different system sizes, compared to *linear speedup*, represented by the line $y = x$, which could be defined as the ideally optimal speedup. For the smallest system size, very good performance has been obtained. For the other two system sizes, still better performance figures are obtained: in fact, *superlinear speedup* (speedup higher than linear). The reasons are finite memory effects on the memory hierarchy: For very large system sizes, the physical memory of one processor is not enough and swap memory must be used, thus considerably increasing the runtime for the sequential version of the program and obtaining a speedup value higher than lineal. Because of this circumstance the calculation for very large system sizes (e.g., for a detailed three-dimensional simulation) may not be affordable on a single PC (for the prohibitively large runtime needed due to the use of swap memory) but be feasible on a cluster, in which the system is partitioned so that each node needs less memory and does not have to use swap memory.

Figure 19.10 is a Gantt chart representing the various types of tasks executed for each node and the messages transferred between different nodes versus time. The activity of the master node, which executes only the master program, is represented above and the activity from the six slave nodes executing the slave program is represented below it. Two different periods can be recognized: computation periods, represented by dark gray horizontal rectangles, in which each slave node calculates the CA state for the next time step on its subdomain, and the communication periods, represented by white horizontal rectangles, in which the photon state values of the cells located in the borders of each subdomain are

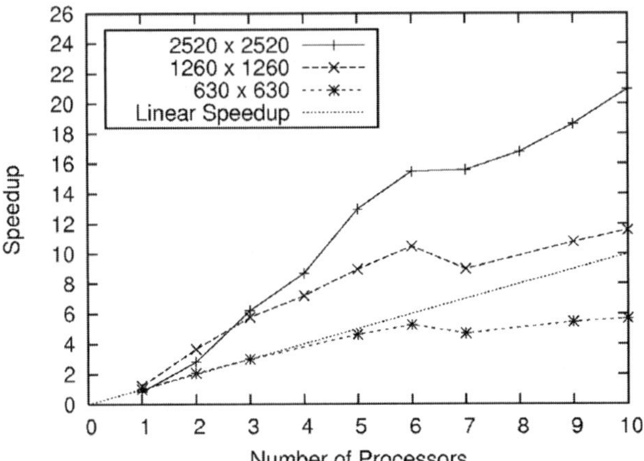

Figure 19.9 Speedup obtained for the parallel implementation with respect to the sequential program for different numbers of processors and for three different system sizes. For comparison, the ideally optimal linear speedup has been shown. Very good performance is obtained for a moderate system size (630 × 630 cells) and a superlinear speedup for larger system sizes.

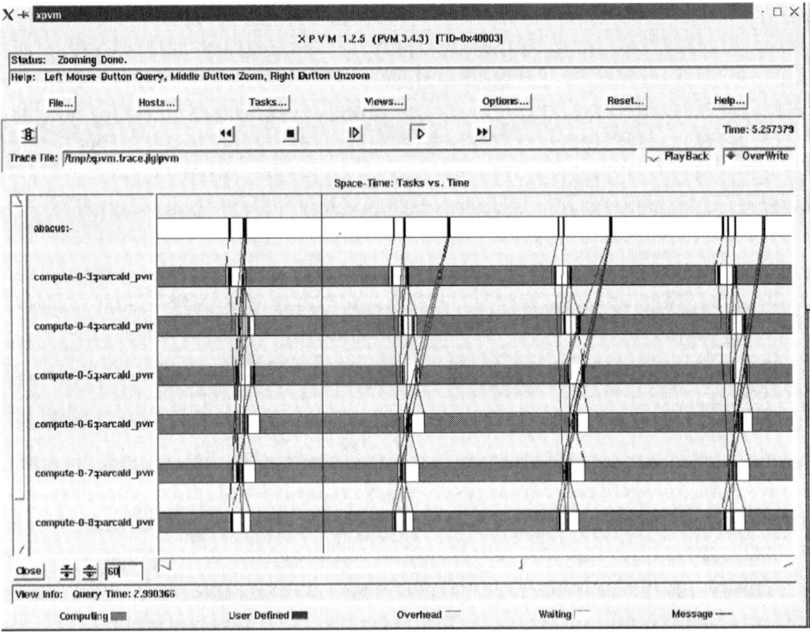

Figure 19.10 Gantt chart with the tasks executed by each cluster node and the messages passed between different nodes versus time once the calculation phase has started. It shows that the application is running with a high computation-to-communication ratio, on the order of 10, and therefore it is exploiting the parallel computational power of the machine.

communicated to the slave node responsible for the neighboring subdomain, and the total number of electrons and photons in each subdomain are sent to the master node. These communications are carried out by exchanging messages, represented by the thin lines joining execution points from different nodes. Computation periods are much longer than communication periods, as can be deduced by the length of the horizontal rectangles. The average computation-to-communication ratio for the slave nodes, obtained by calculating the ratio between the average horizontal length of their dark gray and white rectangles, is on the order of 10. This indicates that the application is taking good advantage of the parallelization on the computer cluster [19].

Finally, it is interesting to ask if the parallel implementation of the model is scalable for clusters of this order of magnitude. Following Dongarra et al. [22], an application is said to be scalable if when the number of processors and the problem size are increased by a factor of x, the running time remains the same. To analyze this question, we have run the same experiment, increasing the system size and the number of processors by the same factor. The results are shown in Figure 19.11. In an ideal case, the running time should be the same for all cases. In our case, only a small excess (from 2 to 5%) of runtime compared to the optimal value was obtained, showing that the parallel implementation scales well on a small computer cluster [19].

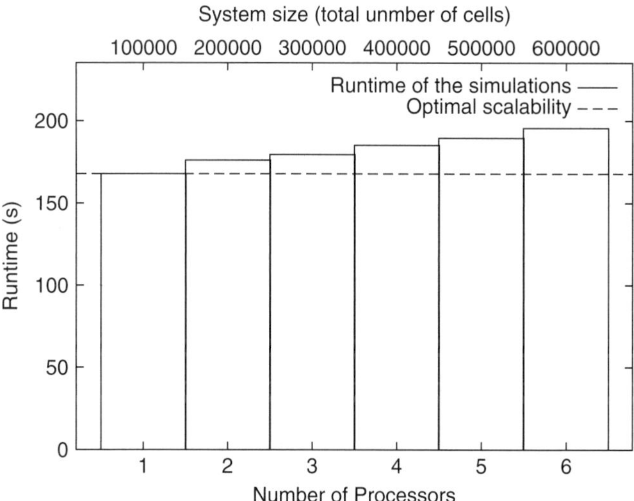

Figure 19.11 Scalability of the combination parallel application–parallel computer can be analized by comparing the runtimes obtained for the same experiment when increasing the system size and the number of processors by the same factor. For an optimal ideal scalability, the same runtime (horizontal straight line) would be obtained. Here, a small excess in runtime from the optimal value (from 2 to 5%) is obtained. It shows that the parallel implementation scales well at this level of parallelization.

19.7 CONCLUSIONS

In this chapter we have reviewed the use of a cellular automata-based algorithm to simulate the time evolution of a complex system: laser dynamics. In addition, we have described how to implement this algorithm in parallel for computer cluster environments, and we have studied the performance and scalability of such an implementation on a small cluster. As a result, we can conclude that the CA algorithm described is a useful technique as an alternative to the standard description of laser dynamics for certain situations. Furthermore, it is feasible to run large fine-grained or three-dimensional simulations of specific real laser systems with the CA model using the parallel implementation on computer clusters, simulations that are not possible on a single-processor sequential computer for their large runtime and memory requirements.

Once the feasibility of running large simulations of the CA model is proved, some of the steps that can be taken to continue this research line are: developing a three-dimensional model with more realistic boundary conditions, verifying that it reproduces the behavior of specific real laser devices, and studying the feasibility of extending the present parallel model to parallel computing environments beyond cluster computing, such as grid computing.

Acknowledgments

This work was partially supported by the Spanish MEC and FEDER under contract TIN2005-08818-C04-03 (the OPLINK project).

REFERENCES

1. A. Einstein. Zur Quantenmechanik der Strahlung. *Physikalische Zeitschrift*, 18:121–128, 1917.
2. J. P. Gordon, H. J. Zeiger, and C. H. Townes. The maser, new type of microwave amplifier, frequency standard, and spectrometer. *Physical Review*, 99:1264–1274, 1955.
3. T. H. Maiman. Optical and microwave-optical experiments in ruby. *Physical Review Letters*, 4:564–566, 1960.
4. E. Cabrera, S. Melle, O. Calderón, and J. M. Guerra. Evolution of the correlation between orthogonal polarization patterns in broad area lasers. *Physical Review Letters*, 97:233902, 2006.
5. L. A. Lugiato, G. L. Oppo, J. R. Tredicce, L. M. Narducci, and M. A. Pernigo. Instabilities and spatial complexity in a laser. *Journal of the Optical Society of America*, 7:1019, 1990.
6. T. Toffoli and N. Margolus. *Cellular Automata Machines: A New Environment for Modelling*. MIT Press, Cambridge, MA, 1987.
7. M. Resnick. *Turtles, Termites, and Traffic Jams*. MIT Press, Cambridge, MA, 1994.

8. M. Cannataro, S. Di Gregorio, R. Rongo, W. Spataro, G. Spezzano, and D. Talia. A parallel cellular automata environment on multicomputers for computational science. *Parallel Computing*, 21(5):803–823, 1995.
9. G. Spezzano, D. Talia, S. Di Gregorio, R. Rongo, and W. Spataro. A parallel cellular tool for interactive modeling and simulation. *IEEE Computational Science and Engineering*, 3(3):33–43, 1996.
10. D. Hutchinson, L. Kattner, M. Lanthier, A. Maheshwari, D. Nussbaum, D. Roytenberg, and J. R. Sack. Parallel neighbourhood modeling: research summary. In *Proceedings of the 8th Annual ACM Symposium on Parallel Algorithms and Architectures*, 1996, pp. 204–207.
11. L. Carotenuto, F. Mele, M. Furnari, and R. Napolitano. PECANS: a parallel environment for cellular automata modeling. *Complex Systems*, 10(1):23–42, 1996.
12. B. Zeigler, Y. Moon, D. Kim, and G. Ball. The DEVS environment for high-performance modeling and simulation. *IEEE Computational Science and Engineering*, 4(3):61–71, 1997.
13. A. Schoneveld and J. F. de-Ronde. P-CAM: a framework for parallel complex systems simulations. *Future Generation Computer Systems*, 16(2):217–234, 1999.
14. D. Talia. Cellular processing tools for high-performance simulation. *IEEE Computer*, 33(9):44–52, 2000.
15. A. E. Siegman. *Lasers*. University Science Books, Mill Valley, CA, 1986.
16. J. L. Guisado, F. Jiménez-Morales, and J. M. Guerra. Cellular automaton model for the simulation of laser dynamics. *Physical Review E*, 67(6):066708, 2003.
17. J. L. Guisado, F. Jiménez-Morales, and J. M. Guerra. Application of Shannon's entropy to classify emergent behaviors in a simulation of laser dynamics. *Mathematical and Computer Modelling*, 42:847–854, 2005.
18. J. L. Guisado, F. Jiménez-Morales, and J. M. Guerra. Computational simulation of laser dynamics as a cooperative phenomenon. *Physica Scripta*, T118:148–152, 2005.
19. J. L. Guisado, F. Jiménez-Morales, and F. Fernández de Vega. Cellular automata and cluster computing: an application to the simulation of laser dynamics. *Advances in Complex Systems*, 10(1):167–190, 2007.
20. J. L. Guisado, F. Fernández de Vega, and K. Iskra. Performance analysis of a parallel discrete model for the simulation of laser dynamics. In *Proceedings of the International Conference on Parallel Processing Workshops*. IEEE Computer Society Press, Los Alamitos, CA, 2006, pp. 93–99.
21. I. Foster. *Designing and Building Parallel Programs*. Addison-Wesley, Reading, MA, 1995.
22. J. Dongarra, I. Foster, G. Fox, W. Gropp, K. Kennedy, L. Torczon, and A. White. *Sourcebook of Parallel Computing*. Morgan Kaufmann, San Francisco, CA, 2003.

CHAPTER 20

Dense Stereo Disparity from an Artificial Life Standpoint

G. OLAGUE
Centro de Investigación Cientifica y de Educación Superior de Ensenada, México

F. FERNÁNDEZ
Universidad de Extremadura, Spain

C. B. PÉREZ
Centro de Investigación Cientifica y de Educación Superior de Ensenada, México

E. LUTTON
Institut National de Recherche en Informatique et en Automatique, France

20.1 INTRODUCTION

Artificial life is devoted to the endeavor of understanding the general principles that govern the living state. The field of artificial life [1] covers a wide range of disciplines, such as biology, physics, robotics, and computer science. One of the first difficulties of studying artificial life is that there is no generally accepted definition of life. Moreover, an analytical approach to the study of life seems to be impossible. If we attempt to study a living system through gradual decomposition of its parts, we observe that the initial property of interest is no longer present. Thus, there appears to be no elementary living thing: Life seems to be a property of a collection of components but not a property of the components themselves. On the other hand, classical scientific fields have proved that the traditional approach of gradual decomposition of the whole is useful in modeling the physical reality. The aim of this chapter is to show that the problem of matching the contents of a stereoscopic system could be approached from an artificial life standpoint. Stereo matching is one of the most active research areas in computer vision. It consists of determining which pair of pixels, projected on at least two images, belong to the same physical three-dimensional point.

Optimization Techniques for Solving Complex Problems, Edited by Enrique Alba, Christian Blum, Pedro Isasi, Coromoto León, and Juan Antonio Gómez

The correspondence problem has been approached using sparse, quasidense, and dense stereo matching algorithms. Sparse matching has normally been based on sparse points of interest to achieve a three-dimensional reconstruction. Unfortunately, most modeling and visualization applications need dense reconstruction rather than sparse point clouds. To improve the quality of image reconstruction, researchers have turned to the dense surface reconstruction approach.

The correspondence problem has been one of the primary subjects in computer vision, and it is clear that these matching tasks have to be solved by computer algorithms [3]. Currently, there is no general solution to the problem, and it is also clear that successful matching by computer can have a large impact on computer vision [4,5]. The matching problem has been considered the most difficult and most significant problem in computational stereo. The difficulty is related to the inherent ambiguities being produced during the image acquisition concerning the stereo pair: for example, geometry, noise, lack of texture, occlusions, and saturation. *Geometry* concerns the shapes and spatial relationships between images and the scene. *Noise* refers to the inevitable variations of luminosity, which produces errors in the image formation due to a number of sources, such as quantization, dark current noise, or electrical processing. *Texture* refers to the properties that represent the surface or structure of an object. Thus, *lack of texture* refers to the problem of unambiguously describing those surfaces or objects with similar intensity or gray values. *Occlusions* can be understood as those areas that appear in only one of the two images due to the camera movement. Occlusion is the cause of complicated problems in stereo matching, especially when there are narrow objects with large disparity and optical illusion in the scene. *Saturation* refers to the problem of quantization beyond the dynamic range in which the image sensor normally works. Dense stereo matching is considered an ill-posed problem. Traditional dense stereo methods are limited to specific precalibrated camera geometries and closely spaced viewpoints. Dense stereo disparity is a simplification of the problem in which the pair is considered to be rectified and the images are taken on a linear path with the optical axis perpendicular to the camera displacement. In this way the problem of matching two images is evaluated indirectly through a univalued function in disparity space that best describes the shape of the surfaces in the scene. There is one more problem with the approach to studying dense stereo matching using the specific case of dense stereo disparity. In general, the quality of the solution is related to the contents of the image pair, so the quality of the algorithms depends on the test images. In our previous work we presented a novel matching algorithm based on concepts from artificial life and epidemics that we call the *infection algorithm*.

The goal of this work is to show that the quality of the algorithm is comparable to the state of the art published in computer vision literature. We decided to test our algorithm with the test images provided at the Middlebury stereo vision web page [11]. However, the problem is very difficult to solve and the comparison is image dependent. Moreover, the natural vision system is an example of a system in which the visual experience is a product of a collection of components but not a property of the components. The infection algorithm presented by

Olague et al. [8] uses an epidemic automaton that propagates the pixel matches as an infection over the entire image with the purpose of matching the contents of two images. It looks for the correspondences between real stereo images following a susceptible–exposed–infected–recovered (SEIR) model that leads to fast labeling. SEIR epidemics refer to diseases with incubation periods and latent infection. The purpose of the algorithm is to show that a set of local rules working over a spatial lattice could achieve the correspondence of two images using a guessing process. The algorithm provides the rendering of three-dimensional information permitting visualization of the same scene from novel viewpoints. Those new viewpoints are obviously different from the initial photographs. In our past work, we had four different epidemic automata in order to observe and analyze the behavior of the matching process. The best results that we have obtained were those related to cases of 47 and 99%. The first case represents geometrically a good image with a moderate percentage of computational effort saving. The second case represents a high percentage of automatically allocated pixels, producing an excellent percentage of computational effort saving, with acceptable image quality.

Our current work aims to improve the results based on a new algorithm that uses concepts from evolution, such as inheritance and mutation. We want to combine the best of both epidemic automata to obtain high computational effort saving with excellent image quality. Thus, we are proposing to use knowledge based on geometry and texture to decide during the algorithm which epidemic automaton is based more firmly on neighborhood information. The benefit of the new algorithm is shown in Section 20.3 through a comparison with previous results. Both the new and the preceding algorithms use local information such as the zero-mean normalized cross-correlation, geometric constraints (i.e., epipolar geometry, orientation), and a set of rules applied within the neighborhood. Our algorithm manages global information, which is encapsulated through the epidemic cellular automaton and the information about texture and edges in order to decide which automata is more appropriate to apply.

The chapter is organized as follows. The following section covers the nature of the correspondence problem. In Section 20.2 we introduce the new algorithm, explaining how the evolution was applied to decide between two epidemic automata. Finally, Section 20.3 shows the results of the algorithm, illustrating the behavior, performance, and quality of the evolutionary infection algorithm. In the final section we state our conclusions.

20.1.1 Problem Statement

Computational stereo studies how to recover the three-dimensional characteristics of a scene from multiple images taken from different viewpoints. A major problem in computational stereo is how to find the corresponding points between a pair of images, which is known as the *correspondence problem* or *stereo matching*. The images are taken by a moving camera in which a unique three-dimensional physical point is projected into a unique pair of image points. A pair of points should correspond to each other in both images. A correlation measure can be used as a similarity criterion between image windows of fixed size. The input is a stereo pair of images, I_l (left) and I_r (right). The correlation metric is used by an algorithm that performs a search process in which the correlation gives the measure used to identify the corresponding pixels on both images. In this work the infection algorithm attempts to maximize the similarity criterion within a search region. Let p_l, with image coordinates (x, y), and p_r, with image coordinates (x', y'), be pixels in the left and right image, $2W + 1$ the width (in pixels) of the correlation window, $\overline{(I_l(x, y))}$ and $\overline{(I_r(x, y))}$ the mean values of the images in the windows centered on p_l and p_r, $R(p_l)$ the search region in the right image associated with p_l, and $\phi(I_l, I_r)$ a function of both image windows. The ϕ function is defined as the zero-mean normalized cross-correlation in order to match the contents of both images:

$$\phi(I_l, I_r) = \frac{\sum_{i,j\in[-W,W]}[AB]}{\sqrt{\sum_{i,j\in[-W,W]} A^2 \sum_{i,j\in[-W,W]} B^2]}} \tag{20.1}$$

where

$$A = (I_l(x+i, y+j) - \overline{I_l(x, y)})$$
$$B = (I_r(x'+i, y'+j) - \overline{I_r(x', y')})$$

However, stereo matching has many complex aspects that turn the problem intractable. In order to solve the problem, a number of constraints and assumptions are exploited which take into account occlusions, lack of texture, saturation, or field of view. Figure 20.1 shows two images taken at the EvoVisión laboratory that we used in the experiments. The movement between the images is a translation with a small rotation along the x, y, and z axes, respectively: $T_x =$ 4.91mm, $T_y = 114.17$mm, $T_z = 69.95$mm, $R_x = 0.84^\circ$, $R_y = 0.16^\circ$, $R_z = 0.55^\circ$. Figure 20.1 also shows five lattices that we have used in the evolutionary infection algorithm. The first two lattices correspond to the images acquired by the stereo rig. The third lattice is used by the epidemic cellular automaton in order to process the information that is being computed. The fourth lattice corresponds to the reprojected image, while the fifth lattice (Canny image) is used as a database in which we save information related to contours and texture. In this work we are interested in providing a quantitative result to measure the benefit of using the infection algorithm. We decided to apply our method to the problem of dense two-frame stereo matching. For a comprehensive discussion on the problem, we refer the reader to the survey by Scharstein and Szeliski [11]. We perform our experiments on the benchmark Middlebury database. This database includes four stereo pairs: *Tsukuba*, *Sawtooth*, *Venus*, and *Map*. It is important to mention that the Middlebury test is limited to specific camera geometries and closely spaced viewpoints. Dense stereo disparity is a simplification of the problem in which the pair is considered to be rectified, and the images are taken on a linear

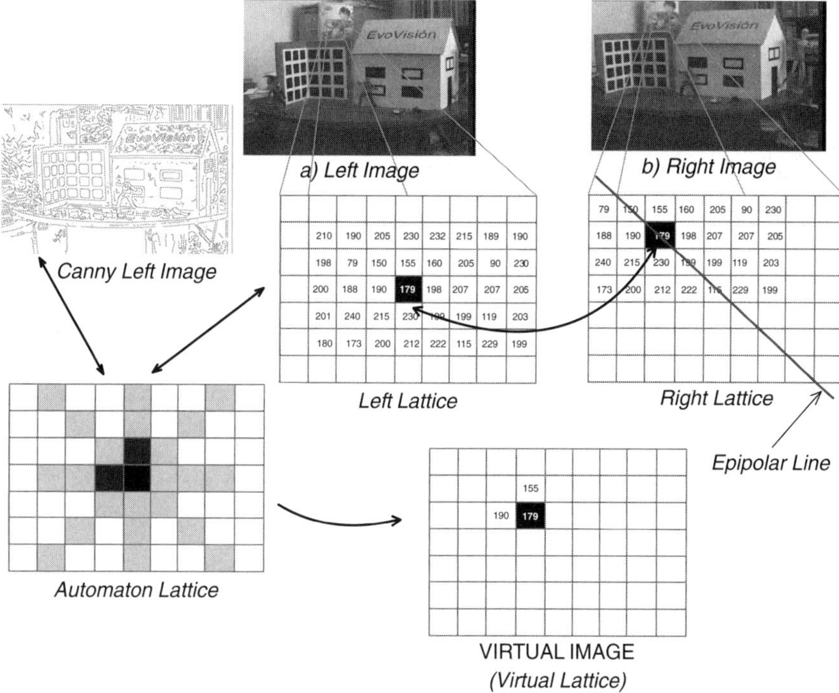

Figure 20.1 Relationships between each lattice used by the infection algorithm. A pixel in the left image is related to the right image using a cellular automaton, canny image, and virtual image during the correspondence process. Gray represents the sick (explored) state, and black represents the healthy (not-explored) state.

path with the optical axis perpendicular to the camera displacement. In this way, the problem of matching is evaluated indirectly through a univalued function in disparity space that best describes the shape of the surfaces. The performance of our algorithm has been evaluated with the methodology proposed in [11]: The error is the percentage of pixels far from the true disparity by more than one pixel.

20.2 INFECTION ALGORITHM WITH AN EVOLUTIONARY APPROACH

The infection algorithm is based on the concept of natural viruses used to search for correspondences between real stereo images. The purpose is to find all existing corresponding points in stereo images while saving on the number of calculations and maintaining the quality of the data reconstructed. The motivation to use what we called the infection algorithm is based on the following: When we observe a scene, we do not observe everything in front of us. Instead, we focus our attention on some regions that retain our interest in the scene. As a result, it is not necessary to analyze each part of the scene in detail. Thus, we pretend to "guess" some parts of the scene through a process of propagation based on artificial epidemics.

The search process of the infection algorithm is based on a set of transition rules that are coded as an epidemic cellular automaton. These rules allow the development of global behaviors. A mathematical description of the infection algorithm has been provided by Olague et al.[9]. In this chapter we have introduced the idea of evolution within the infection algorithm using the concepts of inheritance and mutation in order to achieve a balance between exploration and exploitation. As we can see, the idea of evolution is rather different from that in traditional genetic algorithms. Concepts such as an evolving population are not considered in the evolutionary infection algorithm. Instead, we incorporate aspects such as inheritance and mutation to develop a dynamic matching process. To introduce the new algorithm, let us define some notation:

- *Cellular automata*. A cellular automaton is a continuous map $G\colon S^L \rightarrow S^L$ which commutes with

$$\sigma_i \qquad 1 \leq i \leq d \tag{20.2}$$

  This definition is not, however, useful for computations. Therefore, we consider an alternative characterization. Given a finite set S and d-dimensional shift space S^L, consider a finite set of transformations, $N \subseteq L$. Given a function $f\colon S^N \rightarrow S$, called a *local rule*, the global cellular automaton map is

$$G_f(c)_v = f(c_v + N) \tag{20.3}$$

  where $v \in L$, $c \in S^Z$, and $v + N$ are the translates of v by elements in N.
- *Epidemic cellular automata*. Our epidemic cellular automaton can be introduced formally as a quadruple $E = (S, d, N, f)$, where $S = 5$ is a finite set composed of four states and the wild card $(*)$, $d = 2$ a positive integer, $N \subset Z^d$ a finite set, and $f_i : S^N \rightarrow S$ an arbitrary set of (local) functions, where $i = \{1, \ldots, 14\}$. The global function $G_f\colon S^L \rightarrow S^L$ is defined by $G_f(c)_v = f(c_v + N)$.
  It is also useful to note that S is defined by the following sets:
  - $S = \{\alpha_1, \varphi_2, \beta_3, \varepsilon_0, *\}$ is a finite alphabet.
  - $S_f = \{\alpha_1, \beta_3\}$ is the set of final output states.
  - $S_0 = \{\varepsilon_0\}$ is called the initial input state.

Our epidemic cellular automaton has four states, defined as follows. Let α_1 be the explored (sick) state, which represents cells that have been infected by the virus (it refers to pixels that have been computed in order to find their matches); ε_0 be the not-explored (healthy) state, which represents cells that have not been infected by the virus (it refers to pixels that remain in the initial state); β_3 be the automatically allocated (immune) state, which represents cells that cannot be infected by the virus [this state represents cells that are immune to the disease (it refers to pixels that have been confirmed by the algorithm in order to allocate a pixel match automatically)]; and φ_2 be the proposed (infected) state, which represents cells that have acquired the virus with a probability of recovering from the disease (it refers to pixels that have been "guessed" by the algorithm in order to decide later the best match based on local information).

In previous work we defined four different epidemic cellular automata, from which we detect two epidemic graphs that provide singular results in our experiments (see Figure 20.2). One epidemic cellular automaton produces 47% effort saving; the other, 99%. These automata use a set of transformations expressed by a set of rules grouped within a single graph. Each automaton transforms a pattern of discrete values over a spatial lattice. Entirely different behavior is achieved by changing the relationships between the four states using the same set of rules. Each rule represents a relationship that produces a transition based on local information. These rules are used by the epidemic graph to control the global behavior of the algorithm. In fact, the evolution of cellular automata is governed typically not by a function expressed in closed form, but by a "rule table" consisting of a list of the discrete states that occur in an automaton together with the values to which these states are to be mapped in one iteration of the algorithm.

The goal of the search process is to achieve a good balance between two rather different epidemic cellular automata in order to combine the benefits of each automaton. Our algorithm not only finds a match within the stereo pair, but provides an efficient and general process using geometric and texture information. It is efficient because the final image combines the best of each partial image within the same amount of time, and it is general because the algorithm could be used with any pair of images with little effort at adaptation.

Our algorithm attempts to provide a remarkable balance between the exploration and exploitation of the matching process. Two cellular automata were selected because each provides a particular characteristic from the exploration

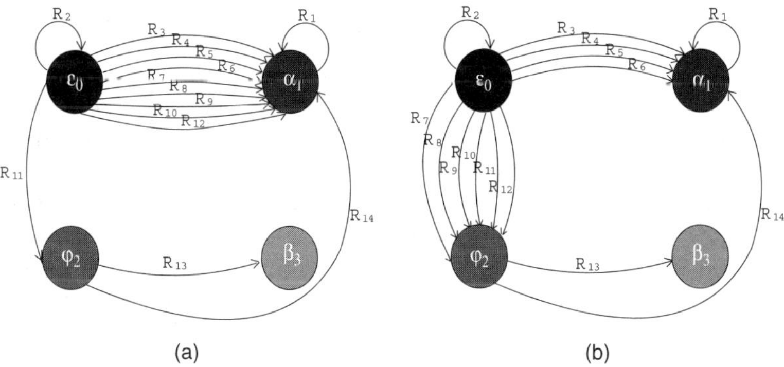

Figure 20.2 Evolutionary epidemic graphs used in the infection algorithm to obtain (a) 47% and (b) 99% of computational savings.

and exploitation standpoint. The 47% epidemic cellular automaton, called A, provides a strategy that exploits the best solution. Here, *best solution* refers to areas where matching is easier to find. The 99% epidemic cellular automaton, called B, provides a strategy that explores the space when matching is difficult to achieve.

The pseudocode for the evolutionary infection algorithm is depicted in Algorithm 20.1. The first step consists of calibrating both cameras. Knowing the calibration for each camera, it is possible to compute the spatial relationship between the cameras. Then two sets of rules, which correspond to 47 and 99%, are coded to decide which set of rules will be used during execution of the algorithm. The sets of rules contain information about the configuration of the pixels in the neighborhood. Next, we built a lattice with the contour and texture information, which we called a canny left image. Thus, we iterate the algorithm as long as the number of pixels with immune and sick states is different between times t and $t + 1$. Each pixel is evaluated according to a decision that is made based on three criteria:

1. The decision to use A or B is weighted considering the current pixels evaluated in the neighborhood, so inheritance is incorporated within the algorithm.
2. The decision is also made based on the current local information (texture and contour). Thus, environmental adaptation is contemplated as a driving force in the dynamic matching process.
3. A probability of mutation that could change the decision as to using A or B is computed. This provides the system with the capability of stochastically adapting the dynamic search process.

Thus, while the number of immune and sick cells does not change between times t and $t + 1$, the algorithm searches for the set of rules that better match the constraints. An action is then activated which produces a path and sequence around the initial cells. When the algorithm needs to execute a rule to evaluate a pixel, it calculates the corresponding epipolar line using the fundamental matrix information. The correlation window is defined and centered with respect to the epipolar line when the search process is begun. This search process provides a nice balance between exploration and exploitation. The exploration process occurs when the epidemic cellular automaton analyzes the neighborhood around the current cell in order to decide where there is a good match. Once we find a good match, a process of exploitation occurs to guess as many point matches as possible. Our algorithm not only executes the matching process, but also takes advantage of the geometric and texture information to achieve a balance between the results of the 47 and 99% epidemic cellular automata. This allows a better result in texture quality, as well as geometrical shape, also saving computational effort. Figure 20.2 shows the two epidemics graphs, 47% and 99%, in which we can appreciate that the differences between the graphs are made by changing the relationships among the four states. Each relationship is represented as a transition based on a local rule, which as a set is able to control the global

Algorithm 20.1 Pseudocode for the Evolutionary Infection Algorithm

```
BEGIN
Calculate projection matrices MM1 and MM2 of the left and
    right images.
Calculate fundamental matrix (F) left to right.
Coding of rule matrices 47 and 99%.
Process left image with the canny edge detector.
Initiate epidemic automata to healthy (not-explored) state.
Corner detection of left image with canny edge detector.
Obtain the left image with canny edge detector.
WHILE the number of immune and sick states are different at t
    and t+1.
DO
 FOR col = 6 to N
 FOR row = 6 to M
 neighbors = countKindNeighbor(row,col)
 pos = statesPosition(row,col)
 stateCell = automata[row][col].cell
 contEdge = findEdgeProbability(row,col)
 IF (contEdge > threshold)
    action = findRule(stateCell, neighbors, pos, Mrules47)
 ELSE
  action = findRule(stateCell, neighbors, pos, Mrules99)
 ENDIF
 SWITCH(action)
  1:EXPLORED
  2:PROPOSED
  3:Analyze the inclination of the epipolar lines.
  IF inclination is at least equal to the inclination
      of three neighbors
     AUTOMATICALLY ALLOCATED
  ELSE
   EXPLORED
  ENDIF
 ENDSWITCH
 ENDFOR
ENDDO
Virtual image is obtained.
END
```

behavior of the algorithm. Next, we explain how each rule works according to the foregoing classification.

20.2.1 Transitions of the Epidemic Automata

Each epidemic graph has 14 transition rules that we divide in three classes: basic rules, initial structure rules, and complex structure rules. Each rule could be represented as a predicate that encapsulates an action allowing a change of state on the current cell based on their neighborhood information. The basic rules relate the obvious information between the initial and explored states. The initial structure rules consider only the spatial set of relationships within the close neighborhood. The complex structure rules consider not only the spatial set of relationships within the close neighborhood, but also those within the external neighborhood. The transitions of our epidemic automata are based on a set of rules (Table 20.1) coded according to a neighborhood that is shown in Figure 20.3. The basic rules correspond to rules 1 and 2. The initial structure is formed by the rules 3, 4, 5, and 6. Finally, the complex structure rules correspond to the rest of the rules. Up to this moment, the basic and initial structure rules have not been changed. The rest of the rules are modified to produce different behaviors and a certain percentage of computational effort saving. The 14 epidemic rules related to the case of 47% are explained as follows.

- *Rules 1 and 2*. The epidemic transitions of these rules represent two obvious actions (Figures 20.4 and 20.5). First, if no information exists in the close neighborhood, no change is made in the current cell. Second, if the central cell was already sick (explored), no change is produced in the current cell.
- *Rules 3 to 6*. The infection algorithm begins the process with the nucleus of infection around the entire image (Figure 20.6). The purpose of creating

TABLE 20.1 Summary of the 14 Rules Used in the Infection Algorithm

$R_1 : CC(\alpha_1) \rightarrow ACCION(\alpha_1)$

$R_2 : CC(\epsilon_0) \bigwedge LU_{\epsilon_0} \bigwedge CU_{\epsilon_0} \bigwedge RU_{\epsilon_0} \bigwedge LR_{\epsilon_0} \bigwedge RR_{\epsilon_0} \bigwedge LD_{\epsilon_0} \bigwedge CD_{\epsilon_0} \bigwedge RD_{\epsilon_0} \rightarrow ACCION(\epsilon_0)$

$R_3 : CC(\epsilon_0) \bigwedge LR(\epsilon_0) \bigwedge RR(\epsilon_0) \bigwedge LD(\epsilon_0) \bigwedge CD(\epsilon_0) \bigwedge RD(\alpha_1) \rightarrow ACCION(\alpha_1)$

$R_4 : CC(\epsilon_0) \bigwedge CU(\epsilon_0) \bigwedge RU(\alpha_1) \bigwedge RR(\alpha_1) \bigwedge CD(\epsilon_0) \rightarrow ACCION(\alpha_1)$

$R_5 : CC(\epsilon_0) \bigwedge LU(\alpha_1) \bigwedge CU(\epsilon_0) \bigwedge LR(\alpha_1) \bigwedge CD(\epsilon_0) \rightarrow ACCION(\alpha_1)$

$R_6 : CC(\epsilon_0) \bigwedge LU(\alpha_1) \bigwedge CU(\alpha_1) \bigwedge RU(\alpha_1) \bigwedge LR(\epsilon_0) \bigwedge RR(\epsilon_0) \rightarrow ACCION(\alpha_1)$

$R_7 : CC(\epsilon_0) \bigwedge EXP(3) \bigwedge EXT(3) \rightarrow ACCION(\alpha_1)$

$R_8 : CC(\epsilon_0) \bigwedge EXP(3) \bigwedge AUT(3) \rightarrow ACCION(\alpha_1)$

$R_9 : CC(\epsilon_0) \bigwedge CU(\epsilon_0) \bigwedge LR(\alpha_1) \bigwedge RR(\epsilon_0) \bigwedge LD(\alpha_1) \bigwedge CD(\varphi_2) \rightarrow ACCION(\alpha_1)$

$R_{10} : CC(\epsilon_0)CU(\epsilon_0) \bigwedge LR(\epsilon_0) \bigwedge RR(\alpha_1) \bigwedge CD(\varphi_2) \bigwedge RD(\alpha_1) \rightarrow ACCION(\alpha_1)$

$R_{11} : CC(\epsilon_0)EXP(3) \rightarrow ACCION(\varphi_2)$

$R_{12} : CC(\epsilon_0)PROP(3) \rightarrow ACCION(\alpha_1)$

$R_{13} : CC(\varphi_2)EXP(3) \rightarrow ACCION(\beta_3)$

$R_{14} : CC(\varphi_2)PROP(3) \rightarrow ACCION(\alpha_1)$

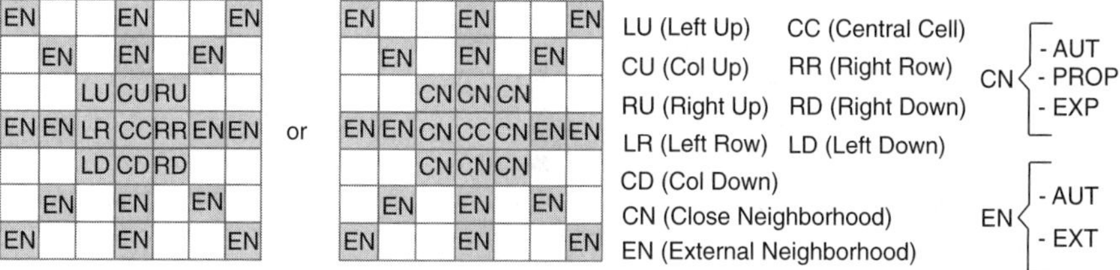

Figure 20.3 Layout of the neighborhood used by the cellular automata.

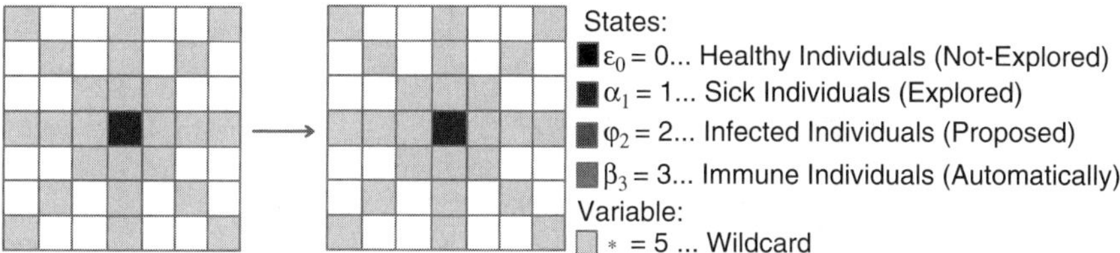

Figure 20.4 Rule 1: Central cell state does not change if the pixel is already evaluated.

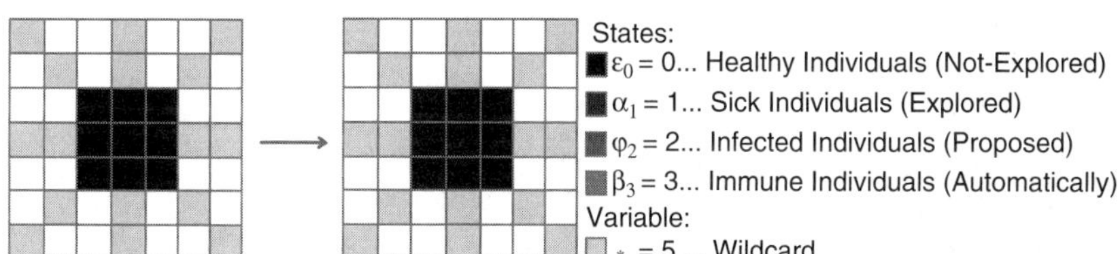

Figure 20.5 Rule 2: Central cell state does not change if there is a lack of information in the neighborhood.

an initial structure in the matching process is to explore the search space in such a way that the information is distributed in several processes. Thus, propagation of the matching is realized in a higher number of directions from the central cell. We use these rules only during the beginning of the process.

- *Rules 7 and 8*. These rules assure the evaluation of the pixels in a region where immune (automatically allocated) individuals exist (Figures 20.7 and 20.8). The figure of rule 8 is similar to rule 7. The main purpose is to control the quantity of immune individuals within a set of regions.
- *Rules 9 and 10*. These transition rules avoid the linear propagation of infected (proposed) individuals (Figure 20.9). Rules 9 and 10 take into account the information of the close neighborhood and one cell of the external neighborhood.

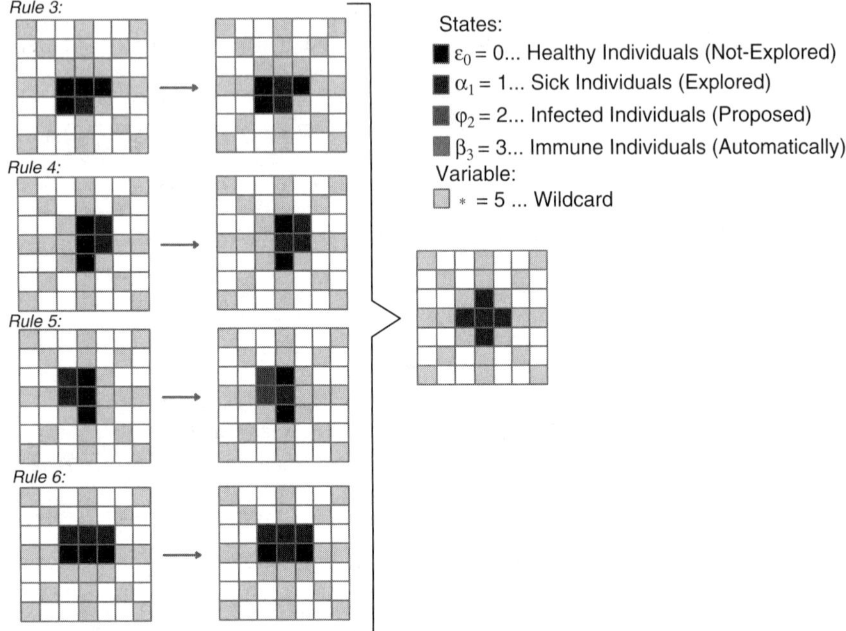

Figure 20.6 Transition of rules 3, 4, 5, and 6, creating the initial structure for the propagation.

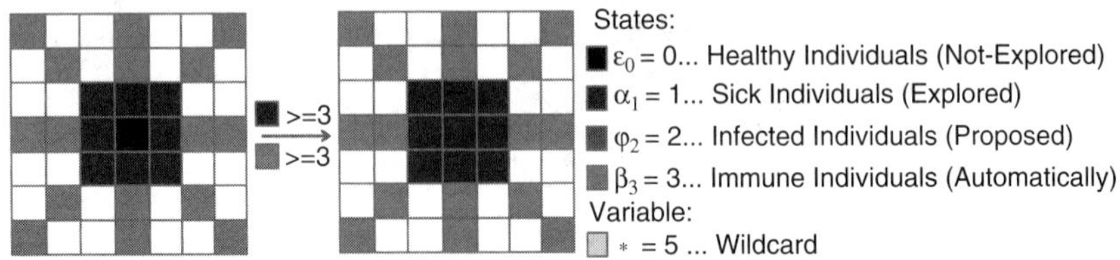

Figure 20.7 Rule 7: This transition indicates the necessity to have at least three sick (explored) individuals in the close neighborhood and three immune (automatically allocated) individuals on the external neighborhood to change the central cell.

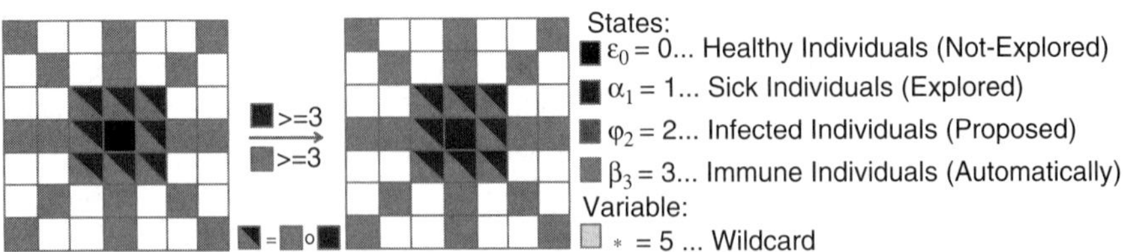

Figure 20.8 Rule 8: This epidemic transition indicates that it is necessary to have at least three sick (explored) pixels in the close neighborhood and at least three immune pixels in the entire neighborhood to change the central cell.

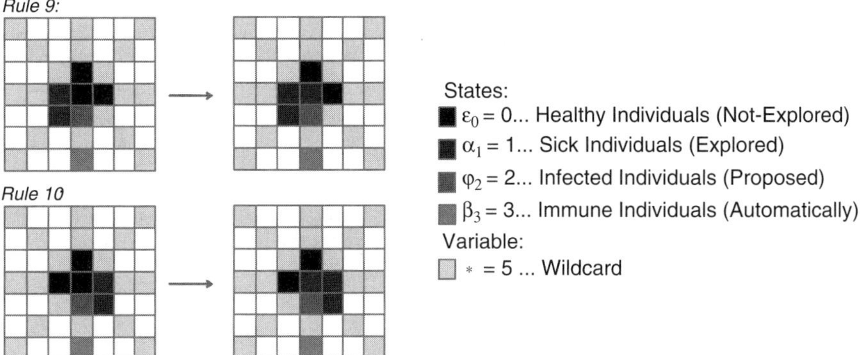

Figure 20.9 Rules 9 and 10 avoid the linear propagation of the infected pixels.

- *Rule 11*. This rule generates the infected (proposed) individuals to obtain later a higher number of the immune (automatically allocated) individuals (Figure 20.10). If the central cell is on the healthy state (not-explored) and there are at least three sick individuals (explored) in the close neighborhood, then the central cell is infected (proposed).
- *Rules 12 and 14*. The reason for these transitions is to control the infected (proposed) individuals (Figures 20.11 and 20.12). If we have at least three infected individuals in the close neighborhood, the central cell is evaluated.
- *Rule 13*. This rule is one of the most important epidemic transition rules because it indicates the computational effort saving of individual pixels during the matching process (Figure 20.13). If the central cell is infected (proposed) and there are at least three sick (explored) individuals in the close neighborhood, then we guess automatically the corresponding pixel in the right image without computation. The number of sick (explored) individuals can be changed according to the desired percentage of computational savings.

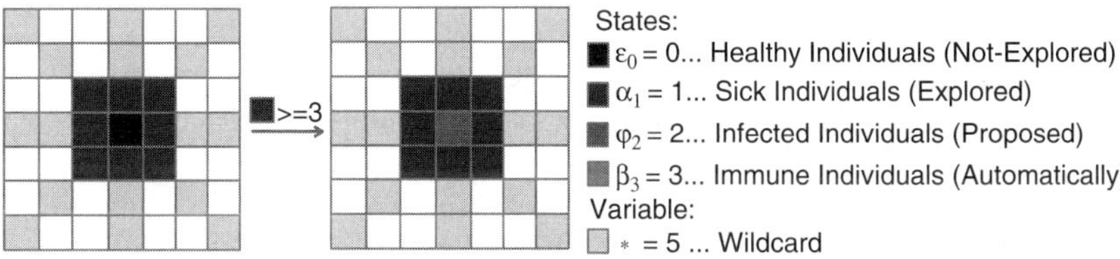

Figure 20.10 Rule 11: This epidemic transition rule represents the quantity of infected (proposed) individuals necessary to obtain the immune (automatically allocated) individuals. In this case, if there are three sick (explored) individuals in the close neighborhood, the central cell changes to an Infected (proposed) state.

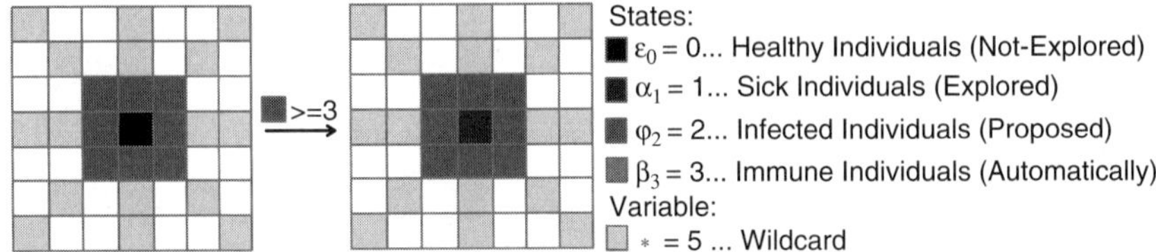

Figure 20.11 Rule 12: This transition controls the infected (proposed) individuals within a region. It requires at least three infected pixels around the central cell.

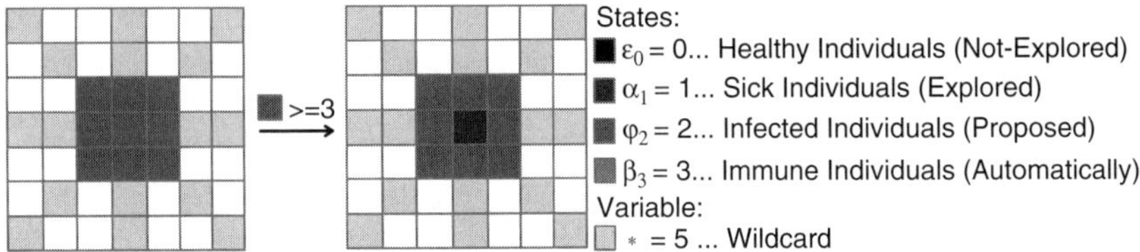

Figure 20.12 Rule 14: This epidemic transition controls the infected (proposed) individuals in different small regions of the image. If the central cell is in an infected (proposed) state, the central cell is evaluated.

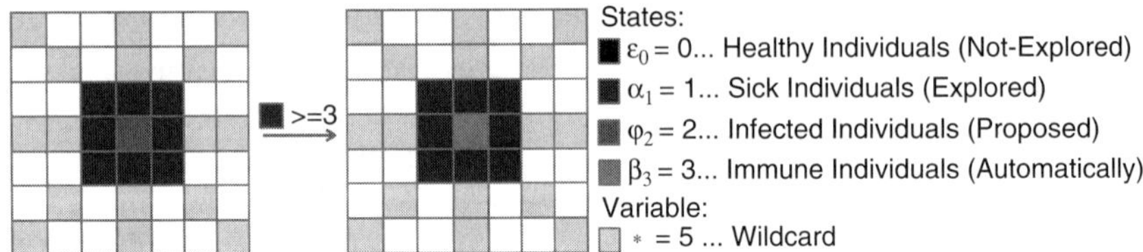

Figure 20.13 Rule 13: This transition indicates when the infected (proposed) individuals will change to immune (automatically allocated) individuals.

20.3 EXPERIMENTAL ANALYSIS

We have tested the infection algorithm with an evolutionary approach on a real stereo pair of images. The infection algorithm was implemented under the Linux operating system on an Intel Pentium-4 at 2.0 GHz with 256 Mbytes of RAM. We have used libraries programmed in C++, designed especially for computer vision, called VxL (Vision x Libraries). We have proposed to improve the results obtained by the infection algorithm through the implementation of an evolutionary approach using inheritance and mutation operations. The idea was to combine the best of both epidemic automata, 47 and 99%, to obtain high computational effort saving together with an excellent image quality. We used knowledge based

Figure 20.14 Results of different experiments in which the rules were changed to contrast the epidemic cellular automata: (a) final view with 47% savings; (b) final view with 70% savings; (c) final view with 99% savings; (d) final view with evolution of 47% and 99% epidemic automata.

on geometry and texture to decide during the correspondence process which epidemic automaton should be applied during evolution of the algorithm.

Figure 20.14 shows a set of experiments where the epidemic cellular automaton was changed to modify the behavior of the algorithm and to obtain a better virtual image. Figure 20.14(a) is the result of obtaining 47% of computational effort savings, Figure 20.14(b) is the result of obtaining 70% of computational effort savings, and Figure 20.14(c) shows the result to obtain 99% of computational effort savings. Figure 20.14(d) presents the results that we obtain with the new algorithm. Clearly, the final image shows how the algorithm combines both epidemic cellular automata. We observe that the geometry is preserved with a nice texture reconstruction. We also observe that the new algorithm spends about the same time employed by the 70% epidemic cellular automaton, with a slightly better texture result. Figure 20.15(a) shows the behavior of the evolutionary infection algorithm that corresponds to the final result of Figure 20.14(d). Figure 20.15(b) describes the behavior of the two epidemic cellular automata during execution of the correspondence process.

We decided to test the infection algorithm with the standard test used in the computer vision community [11]. Scharstein and Szeliski have set up test data to use it as a test bed for quantitative evaluation and comparison of different stereo algorithms. In general, the results of the algorithms that are available for comparison use subpixel resolution and a global approach to minimize the disparity.

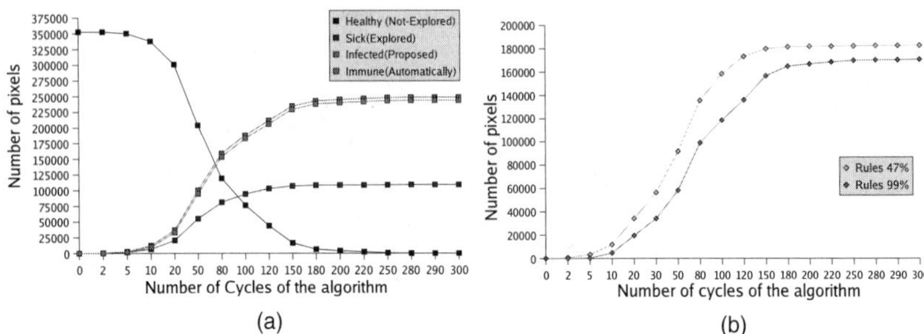

Figure 20.15 Evolution of (a) the states and (b) the epidemic cellular automata to solve the dense correspondence matching.

The original images can be obtained in gray-scale and color versions. We use the gray-scale images even if this represent a drawback with respect to the final result. Because we try to compute the best possible disparity map, we apply a 0% saving. The results are comparable to other algorithms that make similar assumptions: gray-scale image, window-based approach, and pixel resolution [2,12,15]. In fact, the infection algorithm is officially in the Middlebury database.

To improve the test results, we decided to enhance the quality of the input image with an interpolation approach [6]. According to Table 20.2 (note that the *untex.* column of the *Map* image is missing because this image does not have untextured regions), these statistics are collected for all unoccluded image pixels (shown in column *all*), for all unoccluded pixels in the untextured regions (shown in column *untex.*), and for all unoccluded image pixels close to a disparity discontinuity (shown in column *disc.*). In this way we obtain results showing that the same algorithm could be ameliorated if the resolution of the original images is improved. However, the infection algorithm was realized to explore the field

TABLE 20.2 Results on the Middlebury Database

Algorithm	*all*	*untex.*	*disc.*
Tsukuba			
Original	8.90(37)	7.64(37)	42.48(40)
First interp.	7.95(36)	8.55(37)	30.24(38)
Sawtooth			
Original	5.79(39)	4.87(39)	34.20(40)
First interp.	3.59(34)	1.31(30)	24.24(37)
Venus			
Original	5.33(34)	6.60(32)	41.30(39)
First interp.	4.41(34)	5.48(32)	32.94(38)
Map			
Original	3.33(37)	—	32.43(39)
First interp.	1.42(25)	—	17.57(36)

of artificial life using the correspondence problem. Therefore, the final judgment should also be made from the standpoint of the artificial life community. In the future we expect to use the evolutionary infection algorithm in the search for novel vantage viewpoints.

20.4 CONCLUSIONS

In this chapter we have shown that the problems of dense stereo matching and dense stereo disparity could be approached from an artificial life standpoint. We believe that the complexity of the problem reported in this research and its solution should be considered as a rich source of ideas in the artificial life community. A comparison with a standard test bed provides enough confidence that this type of approach can be considered as part of the state of the art. The best algorithms use knowledge currently not used in our implementation. This provides a clue for future research in which some hybrid approaches could be proposed by other researchers in the artificial life community.

Acknowledgments

This research was funded by CONACyT and INRIA under the LAFMI project 634-212. The second author is supported by scholarship 0416442 from CONACyT.

REFERENCES

1. C. Adami. *Introduction to Artificial Life*. Springer-Verlag, New York, 1998.
2. S. Birchfield and C. Tomasi. Depth discontinuities by pixel-to-pixel stereo. Presented at the International Conference on Computer Vision, 1998.
3. M. Z. Brown, D. Burschka, and G. D. Hager. Advances in computational stereo. *IEEE Transactions on Pattern Analysis and Machine Intelligence*, 25(8):993–1008, 2003.
4. J. F. Canny. A computational approach to edge detection. *IEEE Transactions on Pattern Analysis and Machine Intelligence*, 8(6):679–698, 1986.
5. G. Fielding and M. Kam. Weighted matchings for dense stereo correspondence. *Pattern Recognition*, 33(9):1511–1524, 2000.
6. P. Legrand and J. Levy-Vehel. Local regularity-based image denoising. In *Proceedings of the IEEE International Conference on Image Processing*, 2003, pp. 377–380.
7. Q. Luo, J. Zhou, S. Yu, and D. Xiao. Stereo matching and occlusion detection with integrity and illusion sensitivity. *Pattern Recognition Letters*, 24(9–10):1143–1149, 2003.
8. G. Olague, F. Fernández, C. B. Pérez, and E. Lutton. The infection algorithm: an artificial epidemic approach to dense stereo matching. In X. Yao et al., eds., *Parallel Problem Solving from Nature VIII*, vol. 3242 of *Lecture Notes in Computer Science*. Springer-Verlag, New York, 2004, pp. 622–632.

9. G. Olague, F. Fernández, C. B. Pérez, and E. Lutton. The infection algorithm: an artificial epidemic approach for dense stereo matching. *Artificial Life*, 12(4):593–615, 2006.
10. C. B. Pérez, G. Olague, F. Fernández, and E. Lutton. An evolutionary infection algorithm for dense stereo correspondence. In *Proceedings of the 7th European Workshop on Evolutionary Computation in Image Analysis and Signal Processing*, vol. 3449 of Lecture Notes in Computer Science. Springer-Verlag, New York, 2005, pp. 294–303.
11. D. Scharstein and R. Szeliski. A taxonomy and evaluation of dense two-frame stereo correspondence algorithms. *International Journal of Computer Vision*, 47(1):7–42, 2004.
12. J. Shao. Combination of stereo, motion and rendering for 3D footage display. Presented at the IEEE Workshop on Stereo and Multi-Baseline Vision, Kauai, Hawaii, 2001.
13. M. Sipper. *Evolution of Parallel Cellular Machines*. Springer-Verlag, New York, 1997.
14. J. Sun, N. N. Zheng, and H. Y. Shum. Stereo matching using belief propagation. *IEEE Transactions on Pattern Analysis and Machine Intelligence*, 25(7):787–800, 2003.
15. C. Sun. Fast stereo matching using rectangular subregioning and 3D maximum-surface techniques. Presented at the IEEE Computer Vision and Pattern Recognition 2001 Stereo Workshop. *International Journal of Computer Vision*, 47(2):99–117, 2002.
16. C. L. Zitnick and T. Kanade. A cooperative algorithm for stereo matching and occlusion detection. *IEEE Transactions on Pattern Analysis and Machine Intelligence*, 22(7):675–684, 2000.

CHAPTER 21

Exact, Metaheuristic, and Hybrid Approaches to Multidimensional Knapsack Problems

J. E. GALLARDO, C. COTTA, and A. J. FERNÁNDEZ

Universidad de Málaga, Spain

21.1 INTRODUCTION

In this chapter we review our recent work on applying hybrid collaborative techniques that integrate branch and bound (B&B) and memetic algorithms (MAs) in order to design effective heuristics for the multidimensional knapsack problem (MKP). To this end, let us recall that branch and bound (B&B) [102] is an exact algorithm for finding optimal solutions to combinatorial problems that works basically by producing convergent lower and upper bounds for the optimal solution using an implicit enumeration scheme. A different approach to optimization is provided by evolutionary algorithms (EAs) [2–4]. These are powerful heuristics for optimization problems based on principles of natural evolution: namely, adaptation and survival of the fittest. Starting from a *population* of randomly generated *individuals* (representing solutions), a process consisting of *selection* (promising solutions are chosen from the population), *reproduction* (new solutions are created by combining selected ones), and *replacement* (some solutions are replaced by new ones) is repeated. A *fitness function* measuring the quality of the solution is used to guide the process.

A key aspect of EAs is robustness, meaning that they can be deployed on a wide range of problems. However, it has been shown that some type of domain knowledge has to be incorporated into EAs for them to be competitive with other domain-specific optimization techniques [5–7]. A very successful approach to achieving this knowledge augmentation is to use memetic algorithms (MAs) [8–11], which integrate domain-specific heuristics into the EA. In this chapter,

Optimization Techniques for Solving Complex Problems, Edited by Enrique Alba, Christian Blum, Pedro Isasi, Coromoto León, and Juan Antonio Gómez

hybridizations of MAs with B&B are presented. The hybridizations are aimed at combining the search capabilities of both algorithms in a synergistic way.

The chapter is organized as follows. In the remainder of this section, general descriptions of branch and bound and memetic algorithms are given. In Section 21.2 the *multidimensional knapsack problem* is described along with branch and bound and a memetic algorithm to tackle it. Then in Section 21.3 we present our hybrid proposals integrating both approaches. Subsequently, in Section 21.4 we show and analyze the empirical results obtained by the application of each of the approaches described on different instances of the benchmark. Finally, in Section 21.5 we provide conclusions and outline ideas for future work.

21.1.1 Branch-and-Bound Algorithm

The branch-and-bound algorithm (B&B) is a very general exact technique, proposed by Land and Doig [12], that can be used to solve combinatorial optimization problems. B&B is basically an enumeration approach that prunes the nonpromising regions of the search space [102]. For this purpose, the algorithm uses two key ingredients: a *branching rule*, which provides a way of dividing a region of the search space into smaller regions (ideally, partitions of the original one), and a way to calculate an *upper bound* (assuming a maximization problem) on the best solution that can be attained in a region of the search space.

A very general pseudocode for this algorithm is shown in Algorithm 21.1. The algorithm maintains the best solution found so far (*sol* in the pseudocode), usually termed the *incumbent solution*. B&B starts by considering the entire search space of the problem (S). This is split into several subspaces (denoted by set C in the

Algorithm 21.1 Branch-and-Bound Algorithm

```
open := S
sol := null {whose value is assumed −∞}
while open ≠ ∅ do
 select s from open
 open := open \ {s}
 C := branch on s
 for c ∈ C do
   if is Solved(c) then
      if value(c) > value(sol) then
         sol := c
      end if
   else if Upper Bound(c) > value(sol) then
     open := open ∪ {c}
   end if
 end for
end while
return sol
```

algorithm) using the branching rule. If a subspace is solved trivially and its value improves on the incumbent solution, the latter is updated. Otherwise, a bound for each subspace is calculated. If the bound of one subspace is worse than the incumbent solution, this part of the search space can safely be eliminated (this is called *pruning*). Otherwise, the process goes on until all subspaces are either solved or pruned, and the final value of the incumbent solution is returned as an optimal solution to the problem.

The efficiency of the algorithm relies on the effectiveness of the branching rule and the accuracy of the bounds. If these ingredients of the algorithm are not well defined, the technique degenerates into an exhaustive inefficient search. On one side, an effective branching rule will avoid revisiting the same regions of the search space. One the other side, tight bounds will allow more pruning. Usually, it is difficult to find tight bounds that are not too expensive computationally, and a balance has to be found between the amount of pruning achieved and the computational complexity of the bounds. In any case, these techniques cannot practically be used on large instances for many combinatorial optimization problems (COPs).

Note that the search performed by the algorithm can be represented as a tree traversed in a certain way (which depends on the order in which subspaces are introduced and extracted in set *open*). If a depth-first strategy is used, the memory required grows linearly with the depth of the tree; hence, large problems can be considered. However, the time consumption can be excessive. On the other hand, a best-first strategy minimizes the number of nodes explored, but the size of the search tree (i.e., the number of nodes kept for latter expansion) will grow exponentially in general. A third option is to use a breadth-first traversal (i.e., every node in a level is explored before moving on to the next). In principle, this option would have the drawbacks of the preceding two strategies unless a heuristic choice is made: to keep at each level only the best nodes (according to some *quality* measure). This implies sacrificing exactness but provides a very effective heuristic search approach. The term *beam search* (BS) has been coined to denote this strategy [13,14]. As such, BS algorithms are incomplete derivatives of B&B algorithms and are thus heuristic methods. A generic description of BS is depicted in Algorithm 21.2. Essentially, BS works by extending every partial solution from a set $\mathcal{B}$ (called the *beam*) in all possible ways. Each new partial solution so generated is stored in a set $\mathcal{B}'$. When all solutions in $\mathcal{B}$ have been processed, the algorithm constructs a new beam by selecting the best up to k_{bw} (called the *beam width*) solutions from $\mathcal{B}'$. Clearly, a way of estimating the quality of partial solutions, such as an upper bound or an heuristic, is needed for this.

One context in which B&B is frequently used is integer programming (IP) [15]. IP is a generalization of linear programming (LP) [16], so let us first describe the latter problem. In LP there are *decision variables*, *constraints*, and an *objective function*. Variables take real values, and both the objective function and the constraints must be expressed as a series of linear expressions in the decision variables. The goal is to find an assignment for the variables maximizing (or minimizing) the objective function. Linear programs can be used to model many

Algorithm 21.2 Beam Search Algorithm

```
sol := null {whose value is assumed −∞}
B := { () }
while B ≠ ∅ do
  B' := ∅
  for s ∈ B do
    for c ∈ Children of(s) do
      if is Completed(c) then
         if value(c) > value(sol) then
            sol := c
         end if
      else if Upper Bound(c) > sol then
       B' := B' ∪ { c}
      endif
    end for
  end for
  B := select best k_bw nodes from B'
end while
return sol
```

practical problems, and nowadays, solvers can find optimal solutions to problems with hundred of thousands of constraints and variables in reasonable time. The first algorithm proposed to solve LPs was the *simplex method* devised by Dantzig in 1947 [17], which performs very well in practice, although its performance is exponential in the worst case. Afterward, other methods that run in polynomial time have been proposed, such as *interior points* methods [18]. In practice, the simplex algorithm works very well in most cases; other approaches are only useful for very large instances.

An integer program is a linear program in which some or all variables are restricted to take integer values. This *small* change in the formulation increases considerably the number of problems that can be modeled, but also increases dramatically the difficulty of solving problems. One approach to solving IPs is to use a B&B algorithm that uses solutions to *linear relaxations* as bounds. These LP relaxations are obtained by removing the integral requirements on decision variables and can be solved efficiently using any LP method. Assuming a maximization problem, it follows that a solution to its LP relaxation is an upper bound for the solution of the original problem. In order to branch, the search space is usually partitioned into two parts by choosing an integer variable x_i that has fractional value v_i^f in the LP-relaxed solution, and introducing two branches with the additional constraints $x_i \leq \lfloor v_i^f \rfloor$ and $x_i \geq \lceil v_i^f \rceil$. One common heuristic is to choose as a branching variable the one with fractional part closest to 0.5, although commercial IP solvers use more sophisticated techniques.

21.1.2 Memetic Algorithms

The need to exploit problem knowledge in heuristics has been shown repeatedly in theory and in practice [5–7, 19]. Different attempts have been made to answer this need; *memetic algorithms* [8–11] (MAs) have probably been among the most successful to date [20].

The adjective *memetic* comes from the term *meme*, coined by Dawkins [21] to denote an analogy to the *gene* in the context of cultural evolution. As evolutionary algorithms, MAs are also population-based metaheuristics. The main difference is that the components of the population (called *agents* in the MAs terminology) are not passive entities. The agents are active entities that cooperate and compete in order to find improved solutions.

There are many possible ways to implement MAs. The most common implementation consists of combining an EA with a procedure to perform a local search that is usually done after evaluation, although it must be noted that the MA paradigm does not simply reduce itself to this particular scheme. According to Eiben and Smith [4], Figure 21.1 shows places were problem-specific knowledge can be incorporated. Some of the possibilities are:

- During the initialization of the population, some heuristic method may be used to generate high-quality initial solutions. One example of this type of technique is elaborated in this chapter: use of a variant of a B&B algorithm to initialize the population of a MA periodically, with the aim of improving its performance.
- Recombination or mutation operators can be designed intelligently so that specific problem knowledge is used to improve offspring.

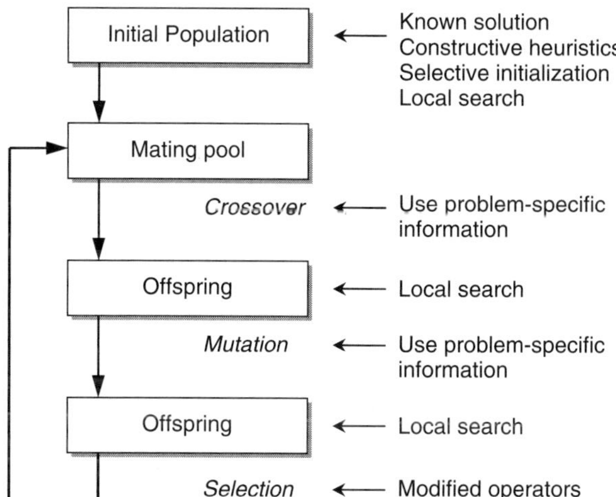

Figure 21.1 Places to incorporate problem knowledge within an evolutionary algorithm, according to Eiben and Smith [4].

- Problem knowledge can be incorporated in the genotype-to-phenotype mapping present in many EAs, such as when repairing an nonfeasible solution. This technique is used in the MA designed by Chu and Beasley [22] for the MKP described in Section 21.2.3.

In addition to other domains, MAs have proven very successful across a wide range of combinatorial optimization problems, where they are state-of-the-art approaches for many problems. For a comprehensive bibliography, the reader is referred to [23, 24].

21.2 MULTIDIMENSIONAL KNAPSACK PROBLEM

In this section we review B&B algorithms and a memetic algorithm that have been proposed to tackle the MKP. Let us first introduce the problem.

21.2.1 Description of the Problem

The *multidimensional knapsack problem* (MKP) is a generalization of the classical *knapsack problem* (KP). In the KP, there is a knapsack with an upper weight limit b and a collection of n items with different values p_j and weights r_j. The problem is to choose the collection of items that gives the highest total value without exceeding the weight limit of the knapsack.

In the MKP, m knapsacks with different weight limits b_i must be filled with the same items. Furthermore, these items have a different weight r_{ij} for each knapsack i. The objective is to find a set of objects with maximal profit such that the capacity of the knapsacks is not exceeded. Formally, the problem can be formulated as:

$$\text{maximize:} \quad \sum_{j=1}^{n} p_j x_j \tag{21.1}$$

$$\text{subject to:} \quad \sum_{j=1}^{n} r_{ij} x_j \leq b_i \qquad i = 1, \ldots, m \tag{21.2}$$

$$x_j \in \{0, 1\} \qquad j = 1, \ldots, n \tag{21.3}$$

where x is an $n$ary binary vector such that $x_j = 1$ if the ith object is included in the knapsacks, and 0 otherwise.

Each of the m constraints in Equation 21.2 is called a knapsack constraint. The problem can be seen as a general statement of any 0–1 integer programming problem with nonnegative coefficients. Many practical problems can be formulated as an instance of the MKP: for example, the capital budgeting problem, project selection and capital investment, budget control, and numerous loading problems (see, e.g., ref. 25).

Although the classical knapsack problem is weakly NP-hard and admits a fully polynomial time approximation scheme (FPTAS) [26], this is not the case for the MKP, which is much more difficult, even for $m = 2$, and does not admit a FPTAS unless P = NP [27]. On the other hand, there exists a polynomial time approximation scheme (PTAS) with a running time of $O(n^{\lceil m/\varepsilon \rceil - m})$ [28,29], which implies a nonpolynomial increase in running time with respect to the accuracy sought.

21.2.2 Branch-and-Bound Algorithms for the MKP

As stated earlier, a complete solution for an instance of the MKP with n objects will be represented by an $n$ary binary vector v such that $v_i = 1$ if the ith object is included in the solution, or 0 otherwise. The natural way of devising a B&B algorithm for the MKP is to start with a totally unspecified solution (a solution for which the inclusion or exclusion of any object is unknown), and to progressively fix the state of each object. This will lead to a branching rule that generates two new nodes from each partial node in the B&B tree by including or excluding a new object in the corresponding solution.

A way of estimating an upper bound for partial solutions is needed so that nonpromising partial solutions can be eliminated from the B&B queue. For this purpose we consider two possibilities. In the first, when the jth object is included in the solution, the lower bound for this partial solution is increased with the corresponding profit p_j (and the remaining available space is decreased by r_{ij} in each knapsack i), whereas the upper bound is decreased by p_j when the item is excluded. The second possibility is to carry out a standard LP exploration of the search tree for this type of problem, as described in Section 21.1.1 (see also ref. 30). More precisely, the linear relaxation of each node is solved (i.e., the problem corresponding to unfixed variables is solved, assuming that variables can take fractional values in the interval [0,1]) using linear programming (LP) techniques such as the simplex method [17]. If all variables take integral values, the subproblem is solved. This is not generally the case, though, and some variables are noninteger in the LP-relaxed solution; in the latter situation, the variable whose value is closest to 0.5 is selected, and two subproblems are generated, fixing this variable at 0 or 1, respectively. The LP-relaxed value of the node is used as its upper bound, so that nodes whose value is below the best-known solution can be pruned from the search tree.

21.2.3 Chu and Beasley's MA for the MKP

In many studies (e.g., refs. 22 and 31–35) the MKP has been tackled using EAs. Among these, the EA developed by Chu and Beasley [22] remains a cutting-edge approach to solving the MKP. This EA uses the natural codification of solutions: namely, binary n-dimensional strings $\vec{x}$ representing the incidence vector of a subset S of objects on the universal set O [i.e., $(x_j = 1) \Leftrightarrow o_j \in S$].

Of course, infeasible solutions might be encoded in this way, and this has to be considered in the EA. Typically, these situations can be solved in three ways:

(1) allowing the generation of infeasible solutions and penalizing accordingly, (2) using a repair mechanism for mapping infeasible solutions to feasible solutions, and (3) defining appropriate operators and/or problem representation to avoid the generation of infeasible solutions. Chu and Beasley's approach is based on the second solution, that is, a Lamarckian repair mechanism is used (see ref. 36 for a comparison of Lamarckian and Baldwinian repair mechanisms for the MKP). To do so, an initial preprocessing of the problem instance is performed off-line. The goal is to obtain a heuristic precedence order among variables: They are ordered by decreasing *pseudoutility* values: $u_j = p_j / \sum_{i=1}^{m} a_i r_{ij}$, where we set the surrogate multipliers a_i to the dual variable values of the solution of the LP relaxation of the problem (see ref. 22 for details). Variables near the front of this ordered list are more likely to be included in feasible solutions (and analogously, variables near the end of the list are more likely to be excluded from feasible solutions). More precisely, whenever a nonfeasible solution is obtained, variables are set to zero in increasing order of pseudoutility until feasibility is restored. After this, feasible solutions are improved by setting variables to 1 in decreasing order of pseudoutility (as long as no constraint is violated). In this way, the repairing algorithm can actually be regarded as a (deterministic) local improvement procedure, and hence this EA certainly qualifies as a MA. Since this MA just explores the feasible portion of the search space, the fitness function can readily be defined as $f(\vec{x}) = \sum_{j=1}^{n} p_j x_j$.

21.3 HYBRID MODELS

In this section we present hybrid models that integrate an MA with the B&B method. Our aim is to combine the advantages of the two approaches while avoiding (or at least minimizing) their drawbacks when working alone. First, in the following subsection, we discuss briefly some related literature regarding the hybridization of exact techniques and metaheuristics.

21.3.1 Hybridizing Exact and Metaheuristic Techniques

Although exact and heuristics methods are two very different ways of tacking COPs, they can be combined in hybrid approaches with the aim of combining their characteristics synergistically. This has been recognized by many researchers, and as a result, many proposals have emerged in recent years. Some authors have reviewed and classified the literature on this topic.

Dumitrescu and Stutzle's [37] classification of algorithms combines local search and exact methods. They focus on hybrid algorithms in which exact methods are used to strengthen local search.

Fernandez and Lourenço's [38] extensive compendium of exact and metaheuristics hybrid approaches for different COPs includes mixed-integer programming, graph coloring, frequency assignment, partitioning, maximum independent sets, maximum clique, traveling salesman, vehicle routing, packing, and job shop scheduling problems.

Puchinger and Raidl's [39] classification of exact and metaheuristics hybrid techniques is possibly the most general. There are two main categories in their classification: *collaborative* and *integrative combinations*. In the following we describe each one in detail.

Collaborative Combinations This class includes hybrid algorithms in which the exact and metaheuristic methods exchange information but neither is a subordinate of the other. As both algorithms have to be executed, two cases can be considered:

1. *Sequential execution*, in which one of the algorithms is completely executed before the other. For example, one technique can act as a kind of preprocessing for the other, or the result of one algorithm can be used to initialize the other.
2. *Parallel or intertwined execution*, where the two techniques are executed simultaneously, either in parallel (i.e., running at the same time on different processors) or in an intertwined way by alternating between the two algorithms.

One example in the first group is the hybrid algorithm proposed by Vasquez and Hao [40] to tackle the MKP, which combines linear programming and tabu search. Their central hypothesis is that the neighborhood of a solution to a relaxed version of a problem contains high-quality solutions to the original problem. They use a two-phase algorithm that first solves exactly a relaxation of the problem, and afterward explores its neighborhood carefully and efficiently. For the first phase, they use the simplex method; for the second they use a tabu search procedure. This hybrid algorithm produces excellent results for large and very large instances of a problem.

An example of parallel execution is presented by Denzinger and Offermann [41]. It is based on a multiagent model for obtaining cooperation between search systems with different search paradigms. The basic schema consisted of teams with a number of agents subjected to the same search paradigm that exchanged a certain class of information. To demonstrate the feasibility of the approach, a GA- and B&B-based system for a job shop scheduling problem was described. Here, the GA and B&B agents only exchanged information in the form of solutions, whereas the B&B agents could also exchange information in the form of closed subtrees. As a result of the cooperation, better solutions were found given a timeout.

In this chapter we present proposals in the second group: namely, hybridizations of a B&B variant (e.g., using *depth-first* strategy or a beam search algorithm) and a memetic algorithm that are executed in an intertwined way.

Integrative Combinations Here, one technique uses the other for some purpose; the first acts as a master while the second behaves as a subordinate of the first. Again, two cases may be considered. The first consists of incorporating an exact algorithm into a metaheuristic. One example is the MA for the MKP by Chu and Beasley [22] (described in detail in Section 21.2.3).

Cotta and Troya [42] presented a framework for hybridization along the lines initially sketched by Cotta et al. [43] (i.e., based on using the B&B algorithm as a recombination operator embedded in the EA). This hybrid operator is used for recombination: It intelligently explores the possible children of solutions being recombined, providing the best possible outcome. The resulting hybrid algorithm provided better results than those of pure EAs in several problems where a full B&B exploration was not practical on its own.

Puchinger et al. [44] presented another attempt to incorporate exact methods in metaheuristics. This work considered different heuristics algorithms for a real-world glass-cutting problem, and a combined GA and B&B approach was proposed. The GA used an order-based representation that was decoded with a greedy heuristic. Incorporating B&B in the decoding for occasional (with a certain probability) locally optimizing subpatterns turned out to increase the solution quality in a few cases.

The other possibility for integrative combinations is to incorporate a metaheuristics into an exact algorithm. One example is a hybrid algorithm that combines genetic algorithms and integer programming B&B approaches to solve the MAX-SAT problems described by French et al. [45]. This hybrid algorithm gathered information during the run of a linear programming–based B&B algorithm and used it to build the population of an EA population. The EA was eventually activated, and the best solution found was used to inject new nodes in the B B search tree. The hybrid algorithm was run until the search tree was exhausted; hence, it is an exact approach. However, in some cases it expands more nodes than does the B B algorithm alone. A different approach has been put forth recently [46] in which a metaheuristic is used for strategic guidance of an exact B&B method. The idea was to use a genetic programming (GP) model to obtain improved node selection strategies within B&B for solving mixed-integer problems. The information collected from the B&B after operating for a certain amount of time is used as a training set for GP, which is run to find a node selection strategy more adequate for the specific problem. Then a new application of the B&B used this improved strategy. The idea was very interesting, although it is not mature and requires further research to obtain more general conclusions.

Cotta et al. [43] used a problem-specific B&B approach to the traveling salesman problem based on 1-trees and Lagrangean relaxation [47], and made use of an EA to provide bounds to guide the B&B search. More specifically, two different approaches to integration were analyzed. In the first model, the genetic algorithm played the role of master and the B&B was incorporated as a slave. The primary idea was to build a hybrid recombination operator based on the B&B philosophy. More precisely, B&B was used to build the best possible tour within the (Hamiltonian) subgraph defined by the union of edges in the parents. This recombination procedure was costly but provided better results than did a blind edge recombination. The second model proposed consisted of executing the B&B algorithm in parallel with a certain number of EAs, which generated a number of different high-quality solutions. The diversity provided by the independent EAs contributed to making edges suitable to be part of the optimal solution that were probably included in some individuals, and nonsuitable edges were unlikely to be taken into account. Despite the fact that these approaches showed promise, the work described by Cotta et al. [43] showed only preliminary results.

21.3.2 Hybrid Proposals

One way to integrate evolutionary techniques and B&B models is through *direct collaboration*, which consists of letting both techniques work alone in parallel (i.e., letting both processes perform independently), that is, at the same level. The two processes will share the solution. There are two ways of obtaining the benefit of this parallel execution:

1. The B&B can use the lower bound provided by the EA to purge the problem queue, deleting those problems whose upper bound is smaller than the one obtained by the EA.
2. The B&B can inject information about more promising regions of the search space into the EA population to guide the EA search.

In our hybrid approach, a single solution is shared among the EA and B&B algorithms that are executed in an interleaved way. Whenever one of the algorithms finds a better approximation, it updates the solution and yields control to the other algorithm.

Two implementation of this scheme are considered. In the first [48], the hybrid algorithm starts by running the EA to obtain a first approximation to the solution. In this initial phase, the population is initialized randomly and the EA executed until the solution is not improved for a certain number of iterations. This approximation can later be used by the B&B algorithm to purge the problem queue. No information from the B&B algorithm is incorporated in this initial phase of the EA, to avoid the injection of high-valued building blocks that could affect diversity, polarizing further evolution.

Afterward, the B&B algorithm is executed. Whenever a new solution is found, it is incorporated into the EA population (replacing the worst individual), the B&B phase is paused, and the EA is run to stabilization. Periodically, pending nodes in the B&B queue are incorporated into the EA population. Since these are partial solutions and the EA population consists of full solutions, they are completed and corrected using the repair operator. The intention of this transfer is to direct the EA to these regions of the search space. Recall that the nodes in the queue represent the subset of the search space still unexplored. Hence, the EA is used to find probably good solutions in this region. Upon finding an improved lower bound (or upon stabilization of the EA, in case no improvement is found), control is returned to the B&B, hopefully with an improved lower bound. This process is repeated until the search tree is exhausted or a time limit is reached. The hybrid is then an anytime algorithm that provides both a quasioptimal solution and an indication of the maximum distance to the optimum. One interesting peculiarity of BS is that it works by extending in parallel a set of different partial solutions in several possible ways. For this reason, BS is a particularly suitable tree search method to be used in a hybrid collaborative framework, as it can be used to provide periodically diverse promising partial solutions to a population-based search method such as an MA. We have used exactly this approach in our second implementation [49,50], and a general description of the resulting algorithm is given in Algorithm 21.3.

The algorithm begins by executing BS for l_0 levels of the search tree. Afterward, the MA and BS are interleaved until a termination condition is reached. Every time the MA is run, its population is initialized using the nodes in the BS queue. Usually, the size of the BS queue will be larger than the MA population size, so some criteria must be used to select a subset from the queue. For instance, these criteria may be selecting the best nodes according to some measure of quality or selecting a subset from the BS queue that provides high diversity. Note that nodes in the BS queue represent schemata (i.e., they are partial solutions in which some genes are fixed but others are indeterminate), so they must first be converted to full solutions in a problem-dependent way. This is another aspect that must be considered when instantiating a general schema for various combinatorial problems.

Upon stabilization of the MA, control is returned to the B&B algorithm. The lower bound for the optimal solution obtained by the MA is then compared to the current incumbent in the B&B, updating the latter if necessary. This may lead to new pruned branches in the BS tree. Subsequently, BS is executed for descending l levels of the search tree. This process is repeated until the search tree is exhausted or a time limit is reached.

Algorithm 21.3 General Description of the Hybrid Algorithm

```
for l0 levels do
  run BS
end for
repeat
  select popsize nodes from problem queue
  initialize MA population with selected nodes
  run MA
  if MA solution better than BS solution then
     let BS solution := MA solution
  end if
  for l levels do
     run BS
  end for
until timeout or tree-exhausted
return BS solution
```

21.4 EXPERIMENTAL ANALYSIS

We tested our algorithms with problems available at the OR library [51] maintained by Beasley. We took two instances per problem set. Each problem set is characterized by a number m of constraints (or knapsacks), a number n of items, and a *tightness ratio*, $0 \leq \alpha \leq 1$. The closer to 0 the tightness ratio is, the more constrained the instance. All algorithms were coded in C, and all tests were carried out on a Pentium-4 PC (1700 MHz and 256 Mbytes of main memory).

21.4.1 Results for the First Model

In this section we describe the results obtained by the first hybrid model described in Section 21.3.2. The B&B algorithm explores the search tree in a depth-first way and uses the first simple upper bound described in Section 21.2.2. A single execution for each instance was performed for the B&B methods, whereas 10 runs were carried out for the EA and hybrid algorithms. The algorithms were run for 600 seconds in all cases. For the EA and the hybrid algorithm, population size was fixed at 100 individuals, initialized with random feasible solutions. Mutation probability was set at 2 bits per string, recombination probability at 0.9, a binary tournament selection method was used, and a standard uniform crossover operator was chosen.

The results are shown in Figure 21.2, where the relative distances to the best solution found by any of the algorithms are shown. As can be seen, the hybrid algorithm outperforms the original algorithms in most cases. Figure 21.3 shows the online evolution of the lower bound for the three algorithms in two instances.

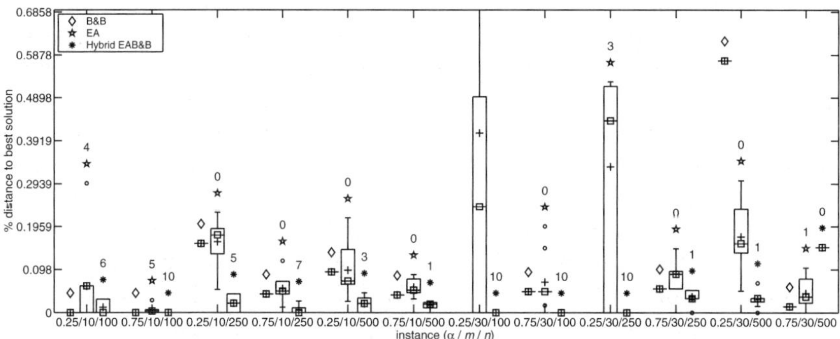

Figure 21.2 Results of the B&B algorithm, the EA, and the first hybrid model for problem instances of different number of items (n), knapsacks (m), and the tightness ratio (α). Box plots show the relative distance to the best solution found by the three algorithms. In all box plots, the figure above indicates the number of runs (out of 10) leading to the best solution. A + sign indicates the mean of the distribution, whereas a □ marks its median. Boxes comprise the second and third quartiles of the distribution. *Outliers* are indicated by small circles in the plot.

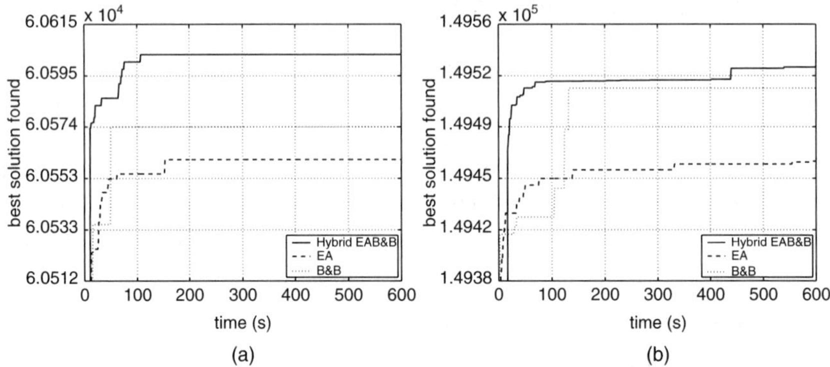

Figure 21.3 Temporal evolution of the lower bound in the three algorithms for a problem instance with (a) $\alpha = 0.75$, $m = 30$, $n = 100$, and (b) $\alpha = 0.75$, $m = 30$, $n = 250$. Curves are averaged for the 10 runs in the case of the EA and the hybrid algorithm.

Notice how the hybrid algorithm yields consistently better results all over the run. This confirms the goodness of the hybrid model as an anytime algorithm.

21.4.2 Results for the Second Model

We solved the same problems using the EA, a beam search (BS) algorithm ($k_{\text{bw}} = 100$), and the second hybrid algorithm. The upper bound was obtained solving the LP relaxation of the problem in all cases. A single execution for each instance was performed for the BS method, whereas 10 independent runs per instance were carried out for the EA and hybrid algorithms. The algorithms were run for 600 seconds in all cases. For the EA and hybrid algorithm, the size of the population was fixed at 100 individuals, initialized with random feasible solutions. With the aim of maintaining some diversity in the population, duplicated individuals were not allowed. The crossover probability was set at 0.9, binary tournament selection was used, and a standard uniform crossover operator was chosen. Execution results for the BS algorithm, the EA, and the hybrid model are shown in Figure 21.4. As can be seen, the hybrid algorithm provides better results for the largest problem instances regardless of the tightness ratio. For the smallest problem instances, the EA performs better. This may be due to the lower difficulty of the latter instances; the search overhead of switching from the EA to the B&B may not be worth while in this case. The hybrid algorithm only begins to be advantageous in larger instances, where the EA faces a more difficult optimization scenario. Notice also that the hybrid algorithm is always able to provide a solution better than or equal to the one provided by BS.

Figure 21.5 shows the evolution of the best value found by the various algorithms for two specific problem instances. Note that the hybrid algorithm always provides better results here than the original ones, especially in the case of the more constrained instance ($\alpha = 0.25$).

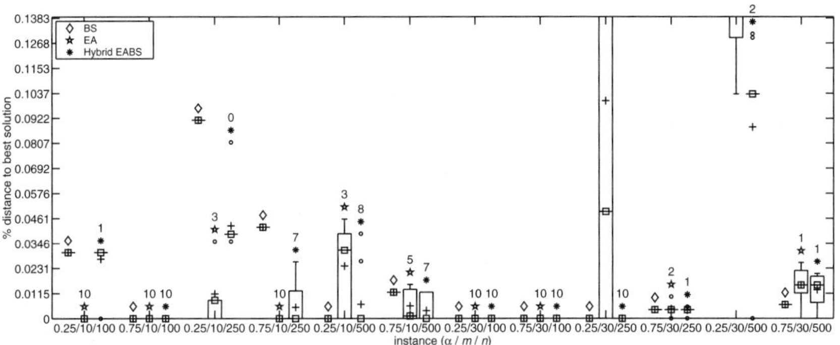

Figure 21.4 Results of the BS algorithm, the EA, and the second hybrid model for problem instances of different number of items (n), knapsacks (m), and tightness ratio (α).

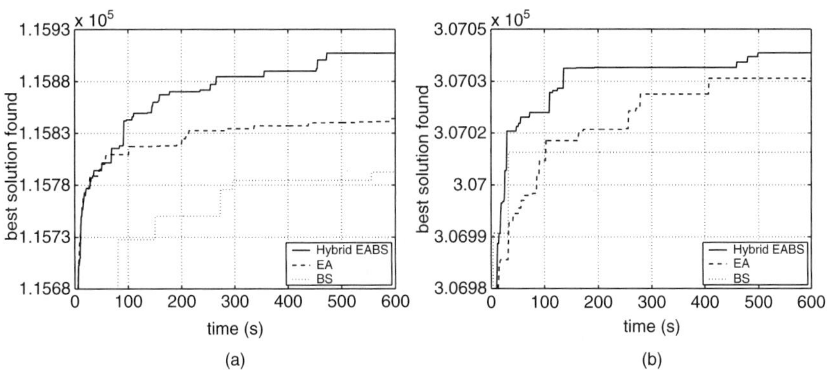

Figure 21.5 Evolution of the best solution in the evolutionary algorithm (EA), beam search, and the hybrid algorithm during 600 seconds of execution: (a) $\alpha = 0.25$, $m = 30$, $n = 500$; (b) $\alpha = 0.75$, $m = 10$, $n = 500$. In both cases, curves are averaged for 10 runs for the EA and the hybrid algorithm.

21.5 CONCLUSIONS

We have presented hybridizations of an EA with a B&B algorithm. The EA provides lower bounds that the B&B can use to purge the problem queue, whereas the B&B guides the EA to look into promising regions of the search space. The resulting hybrid algorithm has been tested on large instances of the MKP problem with encouraging results: The hybrid EA produces better results than the constituent algorithms at the same computational cost. This indicates the synergy of this combination, thus supporting the idea that this is a profitable approach to tackling difficult combinatorial problems.

Acknowledgments

The authors are partially supported by the Ministry of Science and Technology and FEDER under contract TIN2005-08818-C04-01 (the OPLINK project). The first and third authors are also supported under contract TIN2007-67134.

REFERENCES

1. E. L. Lawler and D. E. Wood. Branch and bounds methods: a survey. *Operations Research*, 4(4):669–719, 1966.
2. T. Bäck. *Evolutionary Algorithms in Theory and Practice*. Oxford University Press, New York, 1996.
3. T. Bäck, D. B. Fogel, and Z. Michalewicz. *Handbook of Evolutionary Computation*. Oxford University Press, New York, 1997.
4. A. E. Eiben and J. E. Smith. *Introduction to Evolutionary Computation*. Springer-Verlag, New York, 2003.
5. L. Davis. *Handbook of Genetic Algorithms*. Van Nostrand Reinhold, New York, 1991.
6. D. H. Wolpert and W. G. Macready. No free lunch theorems for optimization. *IEEE Transactions on Evolutionary Computation*, 1(1):67–82, 1997.
7. J. Culberson. On the futility of blind search: an algorithmic view of "no free lunch." *Evolutionary Computation*, 6(2):109–128, 1998.
8. P. Moscato. Memetic algorithms: a short introduction. In D. Corne, M. Dorigo, and F. Glover, eds., *New Ideas in Optimization*, McGraw-Hill, Maidenhead, UK, 1999, pp. 219–234.
9. P. Moscato and C. Cotta. A gentle introduction to memetic algorithms. In F. Glover and G. Kochenberger, eds., *Handbook of Metaheuristics*. Kluwer Academic, Norwell, MA, 2003, pp. 105–144.
10. P. Moscato, A. Mendes, and C. Cotta. Memetic algorithms. In G. C. Onwubolu and B. V. Babu, eds., *New Optimization Techniques in Engineering*. Springer-Verlag, New York, 2004, pp. 53–85.
11. N. Krasnogor and J. Smith. A tutorial for competent memetic algorithms: model, taxonomy, and design issues. *IEEE Transactions on Evolutionary Computation*, 9(5):474–488, 2005.
12. A. H. Land and A. G. Doig. An automatic method for solving discrete programming problems. *Econometrica*, 28:497–520, 1960.
13. A. Barr and E. A. Feigenbaum. *Handbook of Artificial Intelligence*. Morgan Kaufmann, San Francisco, CA, 1981.
14. P. Wiston. *Artificial Intelligence*. Addison-Wesley, Reading MA, 1984.
15. G. L. Nemhauser and L. A. Wolsey. *Integer and Combinatorial Optimization*. Wiley, New York, 1988.
16. D. R. Anderson, D. J. Sweeney, and T. A. Williams. *Introduction to Management Science: Quantitative Approaches to Decision Making*. West Publishing, St. Paul, MN, 1997.
17. G. B. Dantzig. Maximization of a linear function of variables subject to linear inequalities. In T. C. Koopmans, ed., *Activity Analysis of Production and Allocation*. Wiley, New York, 1951, pp. 339–347.

18. N. Karmakar. A new polynomial-time algorithm for linear programming. *Combinatorica*, 4:373–395, 1984.
19. W. E. Hart and R. K. Belew. Optimizing an arbitrary function is hard for the genetic algorithm. In R. K. Belew and L. B. Booker, eds., *Proceedings of the 4th International Conference on Genetic Algorithms*, San Mateo CA. Morgan Kaufmann, San Francisco, CA, 1991, pp. 190–195.
20. W. E. Hart, N. Krasnogor, and J. E. Smith. *Recent Advances in Memetic Algorithms*, vol. 166 of Studies in Fuzziness and Soft Computing. Springer-Verlag, New York, 2005.
21. R. Dawkins. *The Selfish Gene*. Clarendon Press, Oxford, UK, 1976.
22. P. C. Chu and J. E. Beasley. A genetic algorithm for the multidimensional knapsack problem. *Journal of Heuristics*, 4:63–86, 1998.
23. P. Moscato and C. Cotta. Memetic algorithms. In T. F. Gonzalez, ed., *Handbook of Approximation Algorithms and Metaheuristics*. Chapman & Hall/CRC Press, Boca Raton, FL, 2007, Chap. 27.
24. Memetic Algorithms' home page. http://www.densis.fee.unicamp.br/~moscato/memetic_home.html, 2002.
25. H. Salkin and K. Mathur. *Foundations of Integer Programming*. North-Holland, Amsterdam, 1989.
26. O. H. Ibarra and C. E. Kim. Fast approximation for the knapsack and sum of subset problems. *Journal of the ACM*, 22(4):463–468, 1975.
27. B. Korte and R. Schrader. On the existence of fast approximation schemes. In O. L. Mangasarian, R. R. Meyer, and S. Robinson, eds., *Nonlinear Programming 4*. Academic Press, New York, 1981, pp. 415–437.
28. A. Caprara, H. Kellerer, U. Pferschy, and D. Pisinger. Approximation algorithms for knapsack problems with cardinality constraints. *European Journal of Operational Research*, 123:333–345, 2000.
29. D. Lichtenberger. An extended local branching framework and its application to the multidimensional knapsack problem. Diploma thesis, Institut für Computergrafik und Algorithmen, Technischen Universität Wien, 2005.
30. E. Balas and H. Martin. Pivot and complement: a heuristic for 0–1 programming. *Management Science*, 26(1):86–96, 1980.
31. C. Cotta and J. M. Troya. A hybrid genetic algorithm for the 0–1 multiple knapsack problem. In G. D. Smith, N. C. Steele, and R. F. Albrecht, eds., *Artificial Neural Nets and Genetic Algorithms 3*. Springer-Verlag, New York, 1998, pp. 251–255.
32. J. Gottlieb. Permutation-based evolutionary algorithms for multidimensional knapsack problems. In J. Carroll, E. Damiani, H. Haddad, and D. Oppenheim, eds., *Proceedings of the ACM Symposium on Applied Computing*, Villa Olmo, Como, Italy 2000. ACM Press, New York, 2000, pp. 408–414.
33. S. Khuri, T. Bäck, and J. Heitkötter. The zero/one multiple knapsack problem and genetic algorithms. In E. Deaton, D. Oppenheim, J. Urban, and H. Berghel, eds., *Proceedings of the 1994 ACM Symposium on Applied Computation*, Phoenix, AZ,ACM Press, New York, 1994, pp. 188–193.
34. G. R. Raidl. An improved genetic algorithm for the multiconstraint knapsack problem. In *Proceedings of the 5th IEEE International Conference on Evolutionary Computation*, Anchorage, AK, 1998, pp. 207–211.

35. G. R. Raidl and J. Gottlieb. Empirical analysis of locality, heritability and heuristic bias in evolutionary algorithms: a case study for the multidimensional knapsack problem. *Technical Report TR 186–1–04–05*. Institute of Computer Graphics and Algorithms, Vienna University of Technology, 2004.

36. H. Ishibuchi, S. Kaige, and K. Narukawa. Comparison between Lamarckian and Baldwinian repair on multiobjective 0/1 knapsack problems. In C. Coello Coello, A. Hernández Aguirre, and E. Zitzler, eds., *Evolutionary Multi-criterion Optimization Proceedings of the 3rd International Conference on (EMO'05)*, Guanajuato, Mexico, vol. 3410 of *Lecture Notes in Computer Science*. Springer-Verlag, New York, 2005, pp. 370–385.

37. I. Dumitrescu and T. Stutzle. Combinations of local search and exact algorithms. In G. L. Raidl et al., eds., *Applications of Evolutionary Computation*, vol. 2611 of *Lecture Notes in Computer Science*. Springer-Verlag, New York, 2003, pp. 211–223.

38. S. Fernandes and H. Lourenço. Hybrid combining local search heuristics with exact algorithms. In F. Almeida et al., eds., *Procedimiento V Congreso Español sobre Metaheurísticas, Algoritmos Evolutivos y Bioinspirados*, Puerto de la Cruz, Tenerife, Spain, 2007, pp. 269–274.

39. J. Puchinger and G. R. Raidl. Combining metaheuristics and exact algorithms in combinatorial optimization: a survey and classification. In J. Mira and J. R. Álvarez, eds., *Artificial Intelligence and Knowledge Engineering Applications: A Bioinspired Approach*, Las Palmas, Canary Islands, Spain, vol. 3562 of *Lecture Notes in Computer Science*. Springer-Verlag, New York, 2005, pp. 41–53.

40. M. Vasquez and J. K. Hao. A hybrid approach for the 0–1 multidimensional knapsack problem. In *Proceedings of the International Joint Conference on Artificial Intelligence*, Seattle, WA. Morgan Kaufmann, San Francisco, CA, 2001, pp. 328–333.

41. J. Denzinger and T. Offermann. On cooperation between evolutionary algorithms and other search paradigms. In *Proceedings of the 6th IEEE International Conference on Evolutionary Computation*, Washington, DC. IEEE Press, Piscataway, NJ, 1999, pp. 2317–2324.

42. C. Cotta and J. M. Troya. Embedding branch and bound within evolutionary algorithms. *Applied Intelligence*, 18(2):137–153, 2003.

43. C. Cotta, J. F. Aldana, A. J. Nebro, and J. M. Troya. Hybridizing genetic algorithms with branch and bound techniques for the resolution of the TSP. In D. W. Pearson, N. C. Steele, and R. F. Albrecht, eds., *Artificial Neural Nets and Genetic Algorithms 2*, Springer-Verlag, New York, 1995, pp. 277–280.

44. J. Puchinger, G. R. Raidl, and G. Koller. Solving a real-world glass cutting problem. In J. Gottlieb and G. R. Raidl, eds., *Proceedings of the 4th European Conference on Evolutionary Computation in Combinatorial Optimization*, Coimbra, Portugal, vol. 3004 of *Lecture Notes in Computer Science*. Springer-Verlag, New York, 2004, pp. 165–176.

45. A. P. French, A. C. Robinson, and J. M. Wilson. Using a hybrid genetic-algorithm/branch and bound approach to solve feasibility and optimization integer programming problems. *Journal of Heuristics*, 7(6):551–564, 2001.

46. K. Kostikas and C. Fragakis. Genetic programming applied to mixed integer programming. In M. Keijzer, U. O'Reilly, S. M. Lucas, E. Costa, and T. Soule, eds., *Proceedings of the 7th European Conference on Genetic Programming*, Coimbra, Por-

tugal, vol. 3003 of *Lecture Notes in Computer Science*. Springer-Verlag, New York, 2004, pp. 113–124.

47. A. Volgenant and R. Jonker. A branch and bound algorithm for the symmetric traveling salesman problem based on the 1-tree relaxation. *European Journal of Operational Research*, 9:83–88, 1982.
48. J. E. Gallardo, C. Cotta, and A. J. Fernández. Solving the multidimensional knapsack problem using an evolutionary algorithm hybridized with branch and bound. In J. Mira and J. R. Álvarez, eds., *Proceedings of the 1st International Work-Conference on the Interplay Between Natural and Artificial Computation*, Las Palmas, Canary Islands, Spain, vol. 3562 of *Lecture Notes in Computer Science*, Springer-Verlag, New York, 2005.
49. J. E. Gallardo, C. Cotta, and A. J. Fernández. A hybrid model of evolutionary algorithms and branch-and-bound for combinatorial optimization problems. In *Proceedings of the 2005 Congress on Evolutionary Computation*, Edinburgh, UK. IEEE Press, Piscataway, NJ, 2005, pp. 2248–2254.
50. J. E. Gallardo, C. Cotta, and A. J. Fernández. On the hybridization of memetic algorithms with branch-and-bound techniques. *IEEE Transactions on Systems, Man and Cybernetics, Part B*, 37(1):77–83, 2007.
51. J. E. Beasley. OR-library: distributing test problems by electronic mail. *Journal of the Operational Research Society*, 41(11):1069–1072, 1990.

CHAPTER 22

Greedy Seeding and Problem-Specific Operators for GAs Solution of Strip Packing Problems

C. SALTO
Universidad Nacional de La Pampa, Argentina

J. M. MOLINA and E. ALBA
Universidad de Málaga, Spain

22.1 INTRODUCTION

The two-dimensional strip packing problem (2SPP) is present in many real-world applications in the glass, paper, textile, and other industries, with some variations on its basic formulation. Typically, the 2SPP consists of a set of M rectangular pieces, each defined by a width w_i and a height h_i ($i = 1, \ldots, M$), which have to be packed in a larger rectangle, the *strip*, with a fixed width W and unlimited length. The objective is to find a layout of all the pieces in the strip that minimizes the required strip length, taking into account that the pieces have to be packed with their sides parallel to the sides of the strip, without overlapping. This problem is NP-hard [11].

In the present study some additional constrains are imposed: pieces must not be rotated and they have to be packed into three-stage level packing patterns. In these patterns, pieces are packed by horizontal levels (parallel to the bottom of the strip). Inside each level, pieces are packed bottom-left-justified, and when there is enough room in the level, pieces with the same width are stacked one above the other. Three-stage level patterns are used in many real applications in the glass, wood, and metal industries, which is the reason for incorporating this restriction in the problem formulation. We have used evolutionary algorithms (EAs) [4,15], in particular genetic algorithms (GAs), to find a pattern of minimum

Optimization Techniques for Solving Complex Problems, Edited by Enrique Alba, Christian Blum, Pedro Isasi, Coromoto León, and Juan Antonio Gómez

length. GAs deal with a population of tentative solutions, and each one encodes a problem solution on which genetic operators are applied in an iterative manner to compute new higher-quality solutions progressively.

In this chapter a hybrid approach is used to solve the 2SPP: A GA is used to determine the order in which the pieces are to be packed, and a placement routine determines the layout of the pieces, satisfying the three-stage level packing constraint. Taking this GA as our basic algorithm, we investigate the advantages of using genetic operators incorporating problem-specific knowledge such as information about the layout of the pieces against other classical genetic operators. To reduce the trim loss in each level, an additional final operation, called an adjustment operator, is always applied to each offspring generated. Here we also investigate the advantages of seeding the initial population with a set of greedy rules, including information related to the problem (e.g., piece width, piece area), resulting in a more specialized initial population. The main goal of this chapter is to find an improved GA to solve larger problems and to quantify the effects of including problem-specific operators and seeding into algorithms, looking for the best trade-off between exploration and exploitation in the search process.

The chapter is organized as follows. In Section 22.2 we review approaches developed to solve two-dimensional packing problems. The components of the GA are described in Section 22.3. In Section 22.4 we present a detailed description of the new operators derived. Section 22.5 covers the greedy generation of the initial population. In Section 22.6 we explain the parameter settings of the algorithms used in the experimentation. In Section 22.7 we report on algorithm performance, and in Section 22.8 we give some conclusions and analyze future lines of research.

22.2 BACKGROUND

Heuristic methods are commonly applied to solve the 2SPP. In the case of level packing, three strategies [12] have been derived from popular algorithms for the one-dimensional case. In each case the pieces are initially sorted by decreasing height and packed by levels following the sequence obtained. Let i denote the current piece and l the last level created:

1. *Next-fit decreasing height* (NFDH) *strategy*. Piece i is packed left-justified on level l, if it fits. Otherwise, a new level ($l = l + 1$) is created and i is placed left-justified into it.
2. *First-fit decreasing height* (FFDH) *strategy*. Piece i is packed left-justified on the first level where it fits, if any. If no level can accommodate i, a new level is initialized as in NFDH.
3. *Best-fit decreasing height* (BFDH) *strategy*. Piece i is packed left-justified on that level, among those where it fits, for which the unused horizontal space is minimal. If no level can accommodate i, a new level is initialized.

The main difference between these heuristics lies in the way that a level to place the piece in is selected. NFDH uses only the current level, and the levels are considered in a greedy way. Meanwhile, either FFDH or BFDH analyzes all levels built to accommodate the new piece. Modified versions of FFDH and BFDH are incorporated into the mechanism of a special genetic operator in order to improve the piece distribution, and a modified version of NFDH is used to build the packing pattern corresponding to a possible solution to the three-stage 2SPP, like the one used in refs. 18 and 21.

Exact approaches are also used [7,14]. Regarding the existing surveys of metaheuristics in the literature, Hopper and Turton [10,11] review the approaches developed to solve two-dimensional packing problems using GAs, simulated annealing, tabu search, and artificial neural networks. They conclude that EAs are the most widely investigated metaheuristics in the area of cutting and packing. In their survey, Lodi et al. [13] consider several methods for the 2SPP and discuss mathematical models. Specifically, the case where items have to be packed into rows forming levels is discussed in detail.

Few authors restrict themselves to problems involving guillotine packing [5,16] and n-stage level packing [18,19,21,23], but Puchinger and Raidl [18,19] deal in particular with the bin packing problem, where the objective is to minimize the number of bins needed (raw material with fixed dimensions) to pack all the pieces, an objective that is different from, but related to, the one considered in the present work.

22.3 HYBRID GA FOR THE 2SPP

In this section we present a steady-state GA for solving the 2SPP. This algorithm creates an initial population of μ solutions in a random (uniform) way and then evaluates these solutions. The evaluation uses a layout algorithm to arrange the pieces in the strip to construct a feasible packing pattern. After that, the population goes into a cycle in which it undertakes evolution. This cycle involves the selection of two parents by binary tournament and the application of some genetic operators to create a new solution. The newly generated individual replaces the worst one in the population only if it is fitter. The stopping criterion for the cycle is to reach a maximum number of evaluations. The best solution is identified as the best individual ever found that minimizes the strip length required.

A chromosome is a permutation $\pi = (\pi_1, \pi_2, \ldots, \pi_M)$ of M natural numbers (piece identifiers) that defines the input for the layout algorithm, which is a modified version of the NFDH strategy. This heuristic—in the following referred to as *modified next-fit* (MNF)—gets a sequence of pieces as its input and constructs the packing pattern by placing pieces into stacks (neighbor pieces with the same width are stacked one above the other) and then stacking into levels (parallel to the bottom of the strip) bottom-left-justified in a greedy way (i.e., once a new stack or a new level is begun, previous stacks or levels are never reconsidered). See Figure 22.1 for an illustrative example. A more in-depth explanation of the MNF procedure has been given by Salto et al. [23].

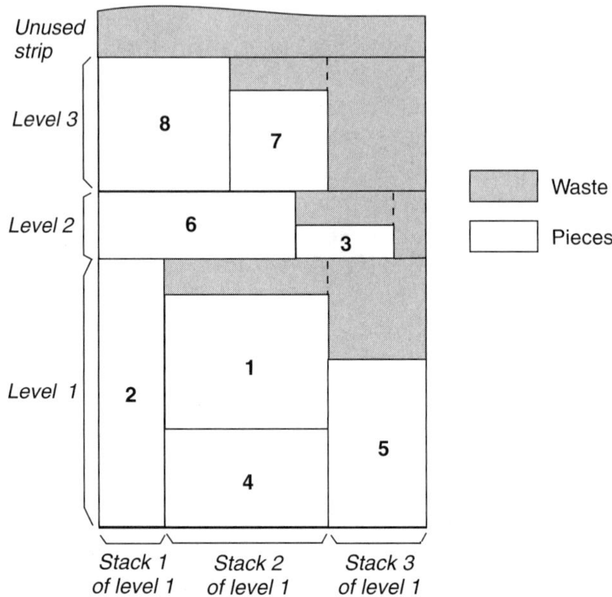

Figure 22.1 Packing pattern for the permutation (2 4 1 5 6 3 8 7).

The fitness of a solution π is defined as the strip length needed to build the corresponding layout, but an important consideration is that two packing patterns could have the same length—so their fitness will be equal—although from the point of view of reusing the trim loss, one of them can actually be better because the trim loss in the last level (which still connects with the remainder of the strip) is greater than the one present in the last level in the other layout. Therefore, we have used the following more accurate fitness function:

$$F(\pi) = strip.length - \frac{l.waste}{strip.length * W} \tag{22.1}$$

where $strip.length$ is the length of the packing pattern corresponding to the permutation π, and $l.waste$ is the area of reusable trim loss in the last level l.

22.4 GENETIC OPERATORS FOR SOLVING THE 2SPP

In this section we describe the recombination and mutation operators used in our algorithms. Moreover, a new operator, adjustment, is introduced.

22.4.1 Recombination Operators

We have studied five recombination operators. The first four operators have been proposed in the past for permutation representations. Three operators focus on combining the order or adjacency information from the two parents, taking into account the position and order of the pieces: partial-mapped crossover (PMX) [8], order crossover (OX) [6], and cycle crossover (CX) [17]. The fourth, edge recombination (EX) [25], focuses on the links between pieces (edges), preserving the linkage of a piece with other pieces.

The fifth operator, called *best inherited level recombination* (BILX), is a new one (introduced in ref. 23 as BIL) tailored for this problem. This operator transmits the best levels (groups of pieces that define a level in the layout) of one parent to the offspring. In this way, the inherited levels could be kept or could even capture some piece from their neighboring levels, depending on how compact the levels are. The BILX operator works as follows. Let nl be the number of levels in one parent, parent_1. In a first step the waste values of all nl levels of parent_1 are calculated. After that, a probability of selection, inversely proportional to its waste value, is assigned to each level and a number $(nl/2)$ of levels are selected from parent_1. The pieces π_i belonging to the levels selected are placed in the first positions of the offspring, and then the remaining positions are filled with the pieces that do not belong to those levels, in the order in which they appear in the other parent, parent_2.

Actually, the BILX operator differs from the operator proposed by Puchinger and Raidl [18] in that our operator transmits the best levels of one parent and the remaining pieces are taken from the other parent, whereas Puchinger and Raidl recombination-pack all the levels from the two parents, sorted according to decreasing values, and repair is usually necessary to guarantee feasibility (i.e., to eliminate piece duplicates and to consider branching constraints), thus becoming a more complex and expensive mechanism.

22.4.2 Mutation Operators

We have tested four mutation operators, described below. These operators were introduced in a preliminary work [23].

The *piece exchange* (PE) operator randomly selects two pieces in a chromosome and exchanges their positions. *Stripe exchange* (SE) selects two levels of the packing pattern at random and exchanges (swaps) these levels in the chromosome. In the *best and worst stripe exchange* (BW_SE) mutation, the best level (the one with the lowest trim loss) is relocated at the beginning of the chromosome, while the worst level is moved to the end. These movements can help the best level to capture pieces from the following level and the worst level to give pieces to the previous level, improving the fitness of the pattern. Finally, the *last level rearrange* (LLR) mutation takes the first piece π_i of the last level and analyzes all levels in the pattern—following the MNF heuristic—trying to find a place for that piece. If this search is not successful, piece π_i is not moved at all. This process is repeated for all the pieces belonging to the last level. During this process, the piece positions inside the chromosome are rearranged consistently.

22.4.3 Adjustment Operators

Finally, we proceed to describe a new operator, which was introduced in a preliminary work [23] as an adjustment operator, the function of which is to improve the solution obtained after recombination and mutation. There are two versions of this operator: one (*MFF_Adj*) consists of the application of a modified version of the FFDH heuristic, and the other (*MBF_Adj*) consists of the application of a modified version of the BFDH heuristic.

MFF_Adj works as follows. It considers the pieces in the order given by the permutation π. The piece π_i is packed into the first level it fits, in an existing stack or in a new one. If no spaces were found, a new stack containing π_i is created and packed into a new level in the remaining length of the strip. The process is repeated until no pieces remain in π.

MBF_Adj proceeds in a similar way, but with the difference that a piece π_i is packed into the level (among those that can accommodate it) with less trim loss, on an existing stack or on a new one. If no place is found to accommodate the piece, a new stack is created containing π_i and packed into a new level. These steps are repeated until no piece remains in the permutation π, and finally, the resulting layout yields a suitable reorganization of the chromosome.

22.5 INITIAL SEEDING

The performance of a GA is often related to the quality of its initial population. This quality consists of two important measures: the average fitness of the individuals and the diversity in the population. By having an initial population with better fitness values, better final individuals can be found faster [1,2,20]. Besides, high diversity in the population inhibits early convergence to a locally optimal solution. There are many ways to arrange this initial diversity. The idea in this work is to start with a seeded population created by following some building rules, hopefully allowing us to reach good solutions in the early stages of the search. These rules include some knowledge of the problem (e.g., piece size) and incorporate ideas from the BFDH and FFDH heuristics [12]. Individuals are generated in two steps. In the first step, chromosomes are sampled randomly from the search space with a uniform distribution. After that, each of them is modified by one rule, randomly selected, with the aim of improving the piece location inside the corresponding packing pattern.

The rules for the initial seeding are listed in Table 22.1. These rules are proposed with the aim of producing individuals with improved fitness values and for introducing diversity into the initial population. Hence, sorting the pieces by their width will hopefully increase the probability of stacking pieces and produce denser levels. On the other hand, sorting by height will generate levels with less wasted space above the pieces, especially when the heights of the pieces are very similar. Rules 11 and 12 relocate the pieces with the goal of reducing the trim loss inside a level. Finally, rules 7 to 10 have been introduced to increase the initial diversity. As we will see, these rules are useful not

TABLE 22.1 Rules to Generate the Initial Population

Rule	Description
1	Sorts pieces by decreasing width.
2	Sorts pieces by increasing width.
3	Sorts pieces by decreasing height.
4	Sorts pieces by increasing height.
5	Sorts pieces by decreasing area.
6	Sorts pieces by increasing area.
7	Sorts pieces by alternating between decreasing width and height.
8	Sorts pieces by alternating between decreasing width and increasing height.
9	Sorts pieces by alternating between increasing width and height.
10	Sorts pieces by alternating between increasing width and decreasing height.
11	The pieces are reorganized following a modified BFDH heuristic.
12	The pieces are reorganized following a modified FFDH heuristic.

only for initial seeding; several can be used as a simple greedy algorithm for a local search to help during the optimization process. The MBF_Adj operator is an example based on rule 11, and MFF_Adj operator is an example based on rule 12.

22.6 IMPLEMENTATION OF THE ALGORITHMS

Next we comment on the actual implementation of the algorithms to ensure that this work is replicable in the future. This work is a continuation of previous work [22,23], extended here by including all possible combinations of recombination (PMX, OX, CX, EX, and BILX), mutation (PE, SE, BW_SE, and LLR), and adjustment operators (MFF_Adj and MBF_Adj). Also, these GAs have been studied using different methods of seeding the initial population. All these algorithms have been compared in terms of the quality of their results. In addition, analysis of the algorithm performance was addressed in order to learn the relationship between fitness values and diversity.

The population size was set at 512 individuals. By default, the initial population is generated randomly. The maximum number of iterations was fixed at 2^{16}. The recombination operators were applied with a probability of 0.8, and the mutation probability was set at 0.1. The adjustment operator was applied to all newly generated solutions. These parameters (e.g., population size, stop criterium, probabilities) were chosen after an examination of some values used previously with success (see ref. 21).

The algorithms were implemented inside MALLBA [3], a C++ software library fostering rapid prototyping of hybrid and parallel algorithms. The platform was an Intel Pentium-4 at 2.4 GHz and 1 Gbyte RAM, linked by fast Ethernet, under SuSE Linux with -GHz, 2.4, 4-Gbyte kernel version.

We have considered five randomly generated problem instances with M equal to 100, 150, 200, 250, and 300 pieces and a known global optimum equal to 200 (the minimum length of the strip). These instances belong to the subtype of level packing patterns, but the optimum value does not correspond to a three-stage guillotine pattern. They were obtained by an implementation of a data set generator, following the ideas proposed by Wang and Valenznela [24]. The length-to-width ratio of all M rectangles is set in the range $1/3 \leq l/w \leq 3$. These instances are available to the public at http://mdk.ing.unlpam.edu.ar/~lisi/2spp.htm.

22.7 EXPERIMENTAL ANALYSIS

In this section we analyze the results obtained by the different variants of the proposed GA acting on the problem instances selected.

For each algorithm variant we have performed 30 independent runs per instance using the parameter values described in Section 22.6. To obtain meaningful conclusions, we have performed an analysis of variance of the results. When the results followed a normal distribution, we used a t-test for the two-group case and an ANOVA test to compare differences among three or more groups (multiple comparison test). We have considered a level of significance of $\alpha = 0.05$ to indicate a 95% confidence level in the results. When the results did not follow a normal distribution, we used the nonparametric Kruskal–Wallis test (multiple comparison test) to distinguish meaningful differences among the means of the results for each algorithm.

Also, the evaluation considers two important measures for any search process: the capacity for generating new promising solutions and the pace rate of the progress in the surroundings of the best-found solution (fine tuning or intensification). For a meaningful analysis, we have considered the average fitness values and the entropy measure (as proposed in ref. 9), which is computed as follows:

$$\text{entropy} = \frac{\sum_{i=1}^{M} \sum_{j=1}^{M} (n_{ij}/\mu) \ln(n_{ij/\mu})}{M \ln M} \tag{22.2}$$

where n_{ij} represents the number of times that a piece i is set into a position j in a population of size μ. This function takes values in $[0 \cdots 1]$, and a value of 0 indicates that all the individuals in the population are identical.

The figures in Tables 22.2 to 22.6 below stand for the best fitness value obtained (column *best*) and the average objective values of the best feasible solutions found, along with their standard deviations (column $avg{\pm}\sigma$). The minimum *best* values are printed in bold. Moreover, Tables 22.4 and 22.6 include the average number of evaluations needed to reach the best value (column $eval_b$), which represents the numerical effort. Table 22.3 also shows the mean times (in seconds) spent in the search for the best solution (column T_b) and in the full search (column T_t).

TABLE 22.2 Experimental Results for the GA with All Recombination and Mutation Operators

		PMX		OX		BILX		EX		CX	
Inst.	Mutation	*best*	$avg \pm \sigma$	*best*	$avg \pm \sigma$	*best*	$avg \pm \sigma$	*best*	$avg \pm \sigma$	*best*	$avg \pm \sigma$
100	PE	255.44	270.48 ± 6.44	244.76	261.67 ± 8.77	**232.73**	243.63 ± 5.85	247.74	273.01 ± 9.73	263.54	279.16 ± 7.69
	SE	254.64	263.86 ± 5.90	241.70	260.00 ± 8.98	**234.66**	240.21 ± 4.81	266.50	283.94 ± 7.87	271.49	291.68 ± 10.81
	LLR	261.71	281.14 ± 8.16	248.68	265.45 ± 7.69	**237.71**	249.56 ± 5.51	288.57	304.68 ± 9.59	315.54	332.78 ± 8.63
	BW_SE	255.65	272.38 ± 9.52	245.59	260.28 ± 7.53	**234.71**	247.44 ± 6.02	271.03	294.02 ± 12.07	307.55	321.51 ± 8.07
150	PE	286.52	301.46 ± 7.31	280.51	292.82 ± 8.62	**241.80**	251.90 ± 5.51	284.56	312.80 ± 12.00	298.04	320.38 ± 11.58
	SE	271.73	291.00 ± 10.88	279.53	294.99 ± 8.47	**238.43**	247.94 ± 5.30	300.47	331.23 ± 17.08	322.38	340.61 ± 10.25
	LLR	288.54	317.63 ± 13.31	274.55	289.99 ± 7.18	**241.82**	253.96 ± 5.40	336.59	366.07 ± 16.30	393.37	411.69 ± 10.38
	BW_SE	291.54	306.74 ± 11.50	270.70	290.45 ± 8.50	**243.73**	253.45 ± 5.82	319.64	361.27 ± 17.13	371.67	394.68 ± 11.15
200	PE	283.79	300.10 ± 10.60	276.44	291.38 ± 7.12	**245.79**	254.00 ± 4.64	289.34	310.98 ± 12.56	290.68	317.73 ± 11.05
	SE	266.52	290.28 ± 10.48	274.27	287.63 ± 7.38	**243.52**	250.94 ± 4.82	296.49	319.52 ± 11.65	303.53	327.98 ± 12.81
	LLR	277.79	310.87 ± 12.72	280.49	291.65 ± 7.64	**246.67**	257.10 ± 4.87	322.03	358.33 ± 14.50	374.46	398.99 ± 7.74
	BW_SE	282.65	302.62 ± 11.10	262.17	283.16 ± 8.46	**248.77**	259.62 ± 4.88	293.56	340.11 ± 17.87	341.51	380.50 ± 14.55
250	PE	292.71	316.25 ± 10.06	292.44	310.11 ± 8.28	**243.86**	253.95 ± 5.91	318.62	332.68 ± 10.58	319.63	334.42 ± 8.34
	SE	288.32	304.95 ± 10.47	289.40	302.68 ± 5.94	**241.68**	249.32 ± 4.19	312.29	332.68 ± 10.83	319.29	336.40 ± 9.71
	LLR	294.27	324.28 ± 11.90	288.53	308.31 ± 10.67	**241.81**	257.72 ± 7.06	348.53	378.55 ± 13.81	377.60	411.80 ± 9.58
	BW_SE	303.81	324.53 ± 10.56	279.51	300.69 ± 10.17	**242.87**	257.22 ± 6.96	320.61	366.73 ± 18.65	346.57	386.20 ± 13.56
300	PE	325.12	339.06 ± 8.95	297.51	324.80 ± 11.10	**253.74**	263.27 ± 6.46	325.71	349.15 ± 11.88	339.41	355.93 ± 10.69
	SE	294.16	322.09 ± 11.81	294.45	322.50 ± 12.13	**241.04**	258.10 ± 6.10	330.23	353.56 ± 11.72	337.41	364.88 ± 12.81
	LLR	317.71	347.10 ± 13.29	301.72	323.64 ± 10.43	**251.78**	264.34 ± 7.46	379.75	411.89 ± 14.52	411.54	445.72 ± 9.70
	BW_SE	309.55	335.86 ± 12.28	294.50	313.86 ± 9.88	**257.37**	263.51 ± 4.91	344.25	385.13 ± 19.70	360.64	423.50 ± 16.34

TABLE 22.3 Experimental Results for the GA with All Recombination and Mutation Operators (*cont.*)

		PMX			OX			BILX			EX			CX		
Inst.	Mutation	$eval_b$	T_b	T_t	$eval_b$	T_b	T_t	$eval_b$	T_b	T_t	$eval_b$	T_b	T_t	$eval_b$	T_b	T_t
100	PE	61,141	39	42	61,825	40	42	31,731	24	50	63,263	41	43	62,852	37	38
	SE	41,232	28	45	53,755	36	44	13,604	11	51	26,132	18	45	10,256	6	40
	LLR	15,496	11	45	55,181	39	47	6,804	5	53	7,367	5	47	1,077	0	43
	BW_SE	23,270	17	47	49,034	35	47	7,217	6	56	10,270	8	48	1,875	1	44
150	PE	62,792	67	70	61658	67	71	33,775	42	82	63,602	68	70	64,162	63	64
	SE	54,339	63	76	57396	66	75	19,784	26	85	21,307	24	74	15,657	16	68
	LLR	22,463	27	80	58532	72	81	8,028	10	87	9,869	12	79	1,076	1	77
	BW_SE	29,006	37	84	55586	71	83	7,162	11	105	9,839	12	81	1,791	2	77
200	PE	60,996	96	103	62,825	100	104	42,648	81	124	63,409	98	101	62,294	88	92
	SE	50,312	86	111	58,343	100	112	27,593	55	128	20,901	35	108	12,942	19	99
	LLR	25,198	45	115	58,190	105	119	10,609	21	131	9,417	17	116	932	1	113
	BW_SE	28,761	54	124	59,000	109	122	9,101	21	153	11,679	20	120	2,084	3	113
250	PE	62,056	136	144	61,874	139	147	44,282	119	176	63,122	135	140	63,326	125	129
	SE	50,732	123	158	59,214	143	158	29,226	84	185	25,063	58	151	17,537	37	139
	LLR	28,083	72	167	55,743	144	170	11,150	32	187	8,203	20	162	804	1	161
	BW_SE	26,316	70	174	57,419	152	174	9,368	31	220	11,123	27	166	2,462	5	160
300	PE	63,365	185	192	61276	183	196	31,716	115	236	63,512	178	184	63,694	166	170
	SE	53,980	174	211	58,020	186	211	31,656	122	248	26,184	80	198	16,933	47	184
	LLR	25,826	88	221	57,446	196	225	8,447	33	252	7,132	22	214	851	2	209
	BW_SE	30,331	108	236	56,889	202	233	9,582	42	286	14,390	46	218	2,418	6	212

TABLE 22.4 Experimental Results for the GA with and Without the Adjustment Operators

	GA			GA + MFF_Adj			GA + MBF_Adj		
Inst.	*best*	$avg \pm \sigma$	$eval_b$	*best*	$avg \pm \sigma$	$eval_b$	*best*	$avg \pm \sigma$	$eval_b$
100	234.66	240.21 ± 4.81	13,603.70	**212.78**	216.10 ± 1.45	21,949.80	213.70	216.88 ± 2.15	21,601.87
150	238.43	247.94 ± 5.30	19,784.47	**210.89**	216.02 ± 1.77	38,386.70	212.64	216.11 ± 1.64	33,481.23
200	243.52	250.94 ± 4.82	27,593.10	**207.69**	211.00 ± 1.81	42,446.00	207.75	211.35 ± 1.69	39,837.30
250	241.68	249.32 ± 4.19	29,226.27	211.48	213.19 ± 1.16	37,566.00	**210.83**	213.54 ± 1.37	40,941.67
300	241.04	258.10 ± 6.10	31,656.37	**207.24**	211.25 ± 1.84	38,092.50	208.65	211.60 ± 1.44	41,586.00
Mean	239.86	249.30	24,372.78	**210.02**	213.51	35,688.20	210.71	213.90	35,489.61

TABLE 22.5 Experimental Results for the GA Using Different Seeding Methods

	$M = 100$			$M = 150$			$M = 200$			$M = 250$			$M = 300$		
Alg.	avg_i	$best$	$avg \pm \sigma$	avg_i	$best$	$avg \pm \sigma$	avg_i	$best$	$avg \pm \sigma$	avg_i	$best$	$avg \pm \sigma$	avg_i	$best$	$avg \pm \sigma$
GA	417.3	234.7	240.2 ±4.8	504.8	238.4	247.9 ±5.3	485.4	243.5	250.9 ±4.8	488.5	241.7	249.3 ±4.2	531.4	241.0	258.1 ±6.1
GA_1	353.3	229.7	241.8 ±9.4	408.0	231.7	243.7 ±6.0	381.7	223.7	240.1 ±6.1	354.6	230.8	237.5 ±3.5	409.6	234.7	248.0 ±5.8
GA_2	298.2	229.7	235.6 ±5.5	368.7	236.7	245.9 ±4.1	352.2	229.6	240.3 ±4.6	331.2	227.7	237.9 ±4.5	345.3	241.2	246.0 ±3.3
GA_3	325.9	229.6	233.8 ±5.0	274.1	226.4	229.2 ±1.9	294.7	222.5	228.0 ±2.5	284.2	222.4	228.1 ±2.4	273.3	227.8	236.9 ±3.4
GA_4	282.9	226.7	228.8 ±0.8	240.4	226.2	227.8 ±0.9	244.5	221.4	225.3 ±2.0	239.7	221.6	224.8 ±1.5	239.1	224.8	230.6 ±2.7
GA_5	381.3	240.6	251.9 ±5.4	427.8	257.6	271.7 ±7.0	415.8	249.6	261.5 ±5.3	377.3	242.6	255.2 ±5.4	401.1	255.6	268.2 ±5.9
GA_6	371.9	249.5	260.0 ±5.4	377.8	257.6	267.9 ±4.1	359.1	241.4	253.2 ±5.9	357.1	248.6	258.9 ±4.6	375.1	272.6	286.6 ±5.3
GA_7	396.7	236.7	244.6 ±6.5	486.7	239.6	258.4 ±7.8	452.3	236.6	249.6 ±6.6	412.2	243.3	251.6 ±4.7	454.4	260.3	273.6 ±5.6
GA_8	299.2	229.7	238.1 ±3.6	274.7	241.5	246.0 ±1.7	275.2	233.7	238.3 ±1.6	262.8	231.2	234.6 ±1.6	259.6	241.6	244.7 ±1.5
GA_9	384.5	233.7	246.3 ±6.2	499.8	245.6	258.0 ±6.1	460.1	237.3	250.4 ±5.1	450.0	246.6	255.8 ±5.9	486.9	259.6	272.2 ±5.0
GA_{10}	354.0	237.3	246.6 ±6.7	353.5	255.7	272.6 ±7.1	369.0	241.3	251.3 ±5.5	321.2	240.7	253.4 ±5.9	336.1	269.7	275.6 ±3.7
GA_{11}	277.9	229.3	235.2 ±3.0	290.2	232.8	241.6 ±4.0	278.5	227.7	233.7 ±3.0	274.0	226.7	232.6 ±2.8	283.4	229.4	234.4 ±3.1
GA_{12}	281.6	226.3	238.1 ±4.3	294.1	237.7	246.2 ±4.3	282.7	229.7	236.4 ±3.6	277.0	229.6	235.2 ±3.0	288.5	232.6	241.0 ±3.8
GA_{Rseed}	340.3	222.8	230.2 ±2.3	368.8	226.1	228.6 ±1.7	358.6	220.7	224.8 ±2.7	339.5	218.7	224.2 ±1.8	360.1	222.6	230.7 ±2.9

TABLE 22.6 Experimental Results for the GA with Seeding and Adjustment

	GA_{Rseed}			$GA_{\text{Rseed}}FF$			$GA_{\text{Rseed}}BF$		
Inst.	*best*	$avg \pm \sigma$	$eval_b$	*best*	$avg \pm \sigma$	$eval_b$	*best*	$avg \pm \sigma$	$eval_b$
100	222.75	230.18 ± 2.28	8,787.53	**214.48**	217.09 ± 1.11	11,426.30	214.79	220.00 ± 1.76	13,803.21
150	226.10	228.56 ± 1.66	19,125.93	**213.54**	214.77 ± 0.69	8,821.35	213.82	216.11 ± 0.76	22,327.71
200	220.74	224.85 ± 2.67	19,018.50	**207.79**	211.11 ± 1.25	15,696.52	207.91	210.96 ± 1.52	18,951.38
250	218.68	224.23 ± 1.81	18,811.87	211.34	212.51 ± 0.65	15,266.35	**210.80**	212.92 ± 1.13	17,441.08
300	222.63	230.71 ± 2.94	24,025.80	209.93	211.91 ± 0.90	17,975.77	**206.41**	212.60 ± 2.21	35,747.48
Mean	222.18	227.70	17,954	211.42	213.48	13,837	**210.75**	214.52	21,654

22.7.1 Experimental Results with Recombination and Mutation Operators

Let us begin with the results of the computational experiments listed in Table 22.2. These results show clearly that the GA using the BILX operator outperforms GAs using any other recombination, in terms of solution quality, for all instances. In all runs, the best values reached by applying the BILX operator have a low cost and are more robust than the best values obtained with the rest of the recombination operators (see the $avg{\pm}\sigma$ column), and the differences are more evident as the instance dimension increases. Using the test of multiple comparison, we have verified that the differences among the results are statistically significant.

The BILX's outperformance is due to the fact that in some way, BILX exploits the idea behind building blocks, in this case defined as a level (i.e., a group of pieces) and tends to conserve good levels in the offspring produced during recombination. As a result, BILX can discover and favor compact versions of useful building blocks.

Regarding the average number of evaluations to reach the best value (see the $eval_b$ column in Table 22.3), BILX presents significant differences in means when compared with OX, the nearest successor regarding best values, but it does not present significant differences in means compared with EX and CX. To confirm these results, we used the test of multiple comparisons. The resulting ranking was as follows: BILX, CX, EX, PMX, and OX, from more efficient to less efficient operators.

On the other hand, the fastest approach in the search is that applying CX (see Table 22.3), due to its simplistic mechanism to generate offspring. The slowest is the GA using BILX, due to the search the best levels to transmit and to the search for pieces that do not belong to the levels transmitted. Using the test of multiple comparisons, we verified that PMX and OX have very similar average execution times, whereas the differences among the other recombination operators are significant. The ranking was CX, OX/PMX, EX, and BILX, from faster to slower operators. Although BILX exhibits the highest runtime, it is important to note that BILX is one of the fastest algorithms to find the best solution. In addition, it finds the best packing patterns; therefore, BILX has a good trade-off between time and quality of results. In conclusion, BILX is the most suitable recombination of those studied.

We turn next to analyzing the results obtained for the various mutation operators with the GA using BILX (see the BILX columns of Tables 22.2 and 22.3). On average, SE has some advantages regarding the best values. Here again we can infer that the use of levels as building blocks improves the results. The test of multiple comparisons of means shows that these differences are significant for all instances. Both BW_SE and LLR need a similar number of evaluations to reach their best values, but those values are half of the number of evaluations that SE needs and nearly a third of the number that PE needs. Despite this fast convergence, BW_SE and LLR performed poorly in solution qualities. The reason could be that those mutations make no alterations to the layout once a "good" solution is reached, whereas SE and PE always modify the layout, due to their random choices. Using the test of multiple comparisons of means, we verified that BW_SE and LLR use a similar average number of evaluations to reach the best value, while the differences between the other two mutation operators are significant. The ranking is BW_SE/LLR, SE, and PE.

With respect to the time spent in the search, we can observe that the SE and BW_SE operators do not have significantly different means, owing to the fact that their procedures differ basically in the level selection step. On the other hand, LLR is the slowest for all instances because of the search on all levels, trying to find a place for each piece belonging to the last level. As expected, PE is the fastest operator, due to its simplistic procedure. All these observations were corroborated statistically.

To analyze the trade-off between exploration and exploitation, the behavior of the various mutation operators is illustrated (for $M = 200$, but similar results are obtained with the rest of the instances) in Figures 22.2 and 22.3, which present the evolution of the population entropy and the average population fitness, respectively (y-axis) with respect to the number of evaluations (x-axis). From

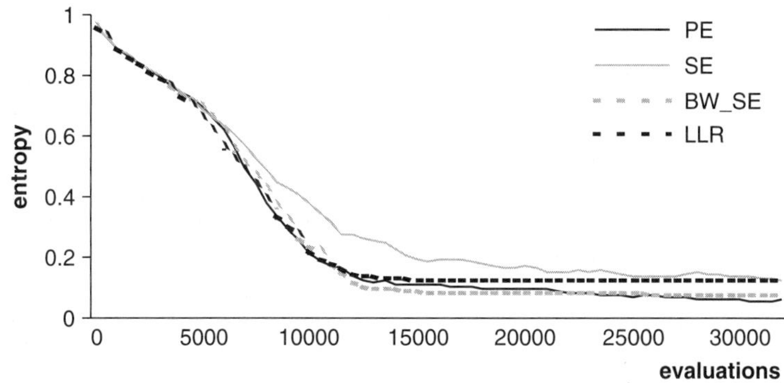

Figure 22.2 Population entropy for BILX and each mutation operator ($M = 200$).

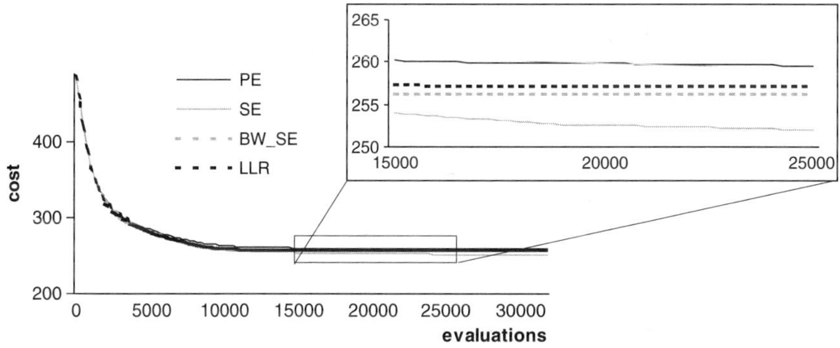

Figure 22.3 Average fitness for BILX and each mutation operator ($M = 200$).

these figures we can see that genotypic entropy is higher for the SE operator with respect to the others, while average population fitness is always lower for the SE. Therefore, the best trade-off between exploration and exploitation is obtained by SE.

To summarize this subsection, we can conclude that the combination of BILX and SE is the best one regarding good-quality results together with a low number of evaluations to reach their best values and also with a reasonable execution time.

22.7.2 Experimental Results with Adjustment Operators

To justify the use of adjustment operators, we present in this section the results obtained by three GAs using the BILX and SE operators (see Table 22.4): one without adjustment, and the others adding an adjustment operator (MFF_Adj or MBF_Adj).

We notice that the GA variants that use some of the adjustment operators significantly outperformed the plain GA in solution quality (corroborated statistically). The reason for this is the improvement obtained in the pieces layout after the application of these operators (which rearrange the pieces to reduce the trim loss inside each level). Nevertheless, it can be observed that the average best solution quality, in general, does not differ very much between the algorithms applying an adjustment (the average best values in the last row are quite similar), whereas the differences are significant in a t-test (p-values are lower than 0.05).

Figure 22.4 shows that the two algorithms using adjustments have a similar entropy decrease until evaluation 17,000; then GA+MFF_Adj maintains higher entropy values than those of GA+MBF_Adj. An interesting point is how the incorporation of adjustment operators allows us to maintain the genotypic diversity at quite high levels throughout the search process. From Figure 22.5 we

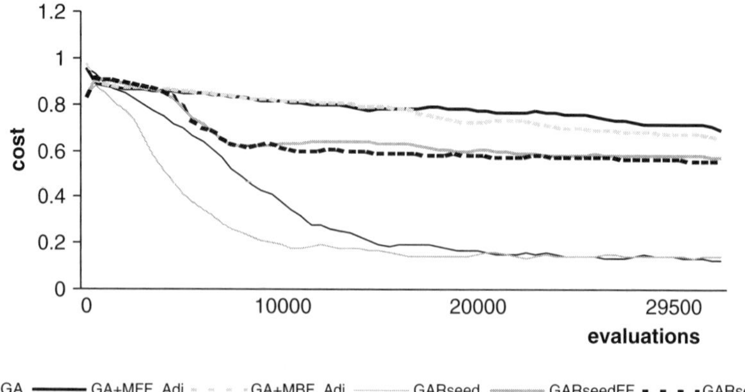

Figure 22.4 Population entropy for all algorithms ($M = 200$).

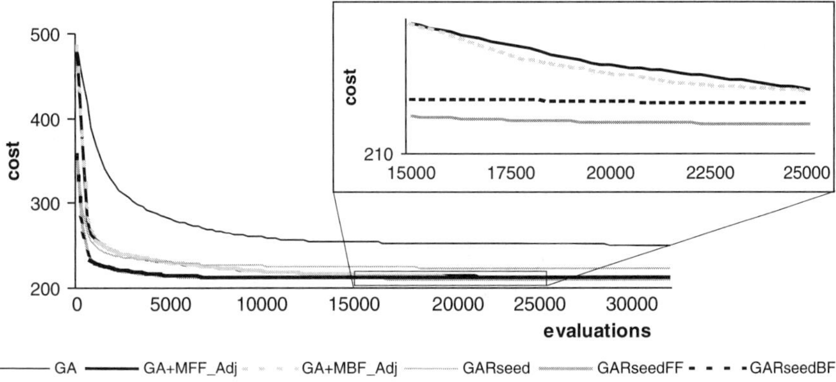

Figure 22.5 Average population fitness for all algorithms ($M = 200$).

observe that the curves of GAs using adjustment operators are close; the plain GA shows the worst performance. In conclusion, GA+MFF_Adj seems to provide the best trade-off between exploration and exploitation.

22.7.3 Studying the Effects of Seeding the Population

Up to this point we have studied GAs that start with a population of randomly sampled packing patterns. Next we analyze the results obtained when the population is initialized using problem-aware rules (see Section 22.5). We have studied GAs using the BILX and SE operators (because of the good results shown in the previous analysis) combined with three different methods of seeding the initial population: (1) by means of a random generation (*GA*), (2) by applying one rule from the Table 22.1 (GA_i where i stands for a rule number), and (3) by applying a rule from Table 22.1 but a randomly selected one for each individual in the population (GA_{Rseed}). The random generation is another rule applicable in this case.

Table 22.5 shows results obtained for the various methods of generating the initial population for all instances. In this table we include information about the average objective value of the initial population (column avg_i). Results indicate that any seeded GA starts the search process from better fitness values than the GA with randomly generated initial populations. Best initial populations, on average, are obtained using GA_4, but have poor genetic diversity (see the mean entropy values for the initial population in Figure 22.6). Rule 4 arranges the pieces by their height, so the pieces in each level have similar heights and the free space inside a level tends to be small. On the other hand, poor performance is obtained with both GA_7 and GA_9, which have the worst initial population means and also poor genetic diversity. Most of the seeded GAs present poor initial genetic diversity (below 0.5), except GA_{12}, GA_{11}, and GA_{Rseed}, with an initial diversity (near 1) close to that of one of the nonseeded GAs. GA_{12} and GA_{11} apply

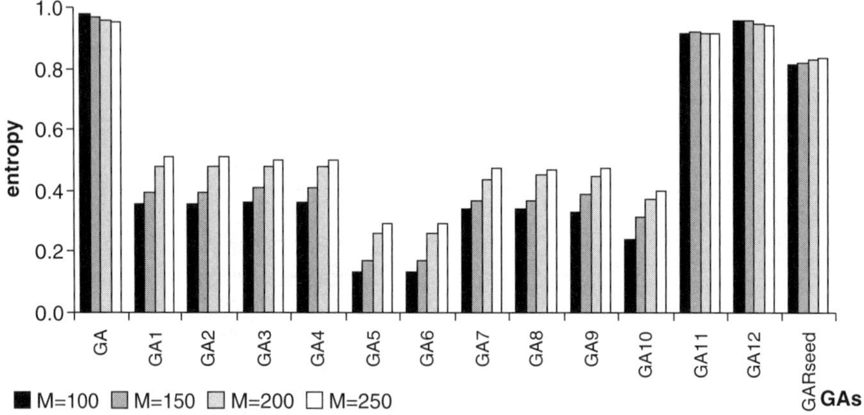

Figure 22.6 Average entropy of the initial population.

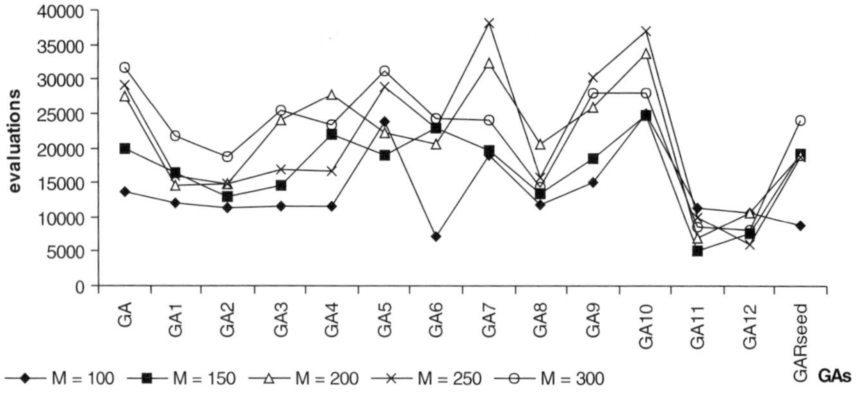

Figure 22.7 Mean number of evaluations to reach the best value for each instance.

modified FFDH and BFDH heuristics (respectively) to a randomly generated solution, and these heuristics do not take piece dimensions into account; hence, the original piece positions inside the chromosome suffer few modifications. On the other hand, GA_{Rseed} combines all the previous considerations with random generation of the piece positions, so high genetic diversity was expected (entropy value near 0.8), and the average objective value of its initial population is in the middle of the rule ranking.

As supposed, the GA_{Rseed} reduces the number of evaluations required to find good solutions (see Figure 22.7). To confirm these observations, we used the t-test, which indicates that the difference between GA_{Rseed} and GA is significant (p-values near 0). A neat conclusion of this study is that GA_{Rseed} significantly outperforms GA with a traditional initialization in all the metrics (the p-value is close to 0). This suggests that the efficiency of a GA could be improved simply by increasing the quality of the initial population.

Due to the good results obtained by the GA with the application of adjustment operators, we now compare the performance of GA_{Rseed} with that of two other algorithms: $GA_{\text{Rseed}}FF$ and $GA_{\text{Rseed}}BF$, which include in their search mechanism the MFF_Adj and MBF_Adj operators, respectively.

Table 22.6 shows the results for the three approaches. For every instance, the algorithm that finds the fittest individuals is always an algorithm using an adjustment operator. Here again, there are statistical differences among the results of each group (GA_{Rseed} vs. GA_{Rseed} applying some adjustment operator), but there are no differences inside each group. Looking at the average objective values of the best solutions found, $GA_{\text{Rseed}}FF$ produces best values and also small deviations. In terms of the average number of evaluations, Table 22.6 (column $eval_b$) shows that in general, the $GA_{\text{Rseed}}FF$ performs faster than the other algorithms.

Finally, comparing the best values reached by the algorithms using adjustment operators with those without those operators, we observe that the results of $GA_{\text{Rseed}}FF$ are of poorer quality than those obtained by a nonseeded GA applying an MFF_Adj operator (the differences are significant in a t-test, with p-values very close to 0), except for instances with $M = 200$ and $M = 300$. This suggests that the operators are effective in the search process when they are applied to populations with high genotypic diversity, as shown by the plain GA (i.e., the operator has more chances to rearrange pieces in the layout). On the other hand, the means of $GA_{\text{Rseed}}FF$ and $GA + MFF\_Adj$ do not present significant differences in a t-test.

Both GA_{Rseed} using adjustment operators have a similar decrease in the average population fitness and maintain a higher entropy than does GA_{Rseed} (see Figures 22.4 and 22.5), but the first algorithms were able to find the best final solutions in a resulting faster convergence. In fact, we can see that the seeded initial populations present a high entropy value (near 0.85; i.e., high diversity), but GA_{Rseed} using adjustment operators produces fitter individuals very quickly (close to evaluation, 2000).

22.8 CONCLUSIONS

In this chapter we have analyzed the behavior of improved GAs for solving a constrained 2SPP. We have compared some new problem-specific operators to traditional ones and have analyzed various methods of generating the initial population. The study, validated from a statistical point of view, analyzes the capacity of the new operators and the seeding in order to generate new potentially promising individuals and the ability to maintain a diversified population.

Our results show that the use of operators incorporating specific knowledge from the problem works accurately, and in particular, the combination of BILX recombination and SE mutation obtains the best results. These operators are based on the concept of building blocks, but here a building block is a group of pieces that defines a level in the phenotype. This marks a difference from some of the traditional operators, who randomly select the set of pieces to be interchanged. For the 2SPP, the absolute positions of pieces within the chromosome do not have relative importance.

Also, the incorporation of a new adjustment operator (in two versions) in the evolutionary process makes a significant improvement in the 2SPP algorithms results, providing faster sampling of the search space in all instances studied and exhibits a satisfactory trade-off between exploration and exploitation. This operator maintains the population diversity longer than in the case of plain GAs.

In addition, we observe an improvement in the GA performance by using problem-aware seeding, regarding both efficiency (effort) and quality of the solutions found. Moreover, the random selection of greedy rules to build the initial population works properly, providing good genetic diversity of initial solutions and a faster convergence without sticking in local optima, showing that the performance of a GA is sensitive to the quality of its initial population.

Acknowledgments

This work has been partially funded by the Spanish Ministry of Science and Technology and the European FEDER under contract TIN2005-08818-C04-01 (the OPLINK project). We acknowledge the Universidad Nacional de La Pampa and the ANPCYT in Argentina, from which we received continuous support.

REFERENCES

1. R. Ahuja and J. Orlin. Developing fitter genetic algorithms. *INFORMS Journal on Computing*, 9(3):251–253, 1997.
2. R. Ahuja, J. Orlin, and A. Tiwari. A greedy genetic algorithm for the quadratic assignment problem. *Computers and Operations Research*, 27(3):917–934, 2000.
3. E. Alba, J. Luna, L. M. Moreno, C. Pablos, J. Petit, A. Rojas, F. Xhafa, F. Almeida, M. J. Blesa, J. Cabeza, C. Cotta, M. Díaz, I. Dorta, J. Gabarró, and C. León. *MALLBA: A Library of Skeletons for Combinatorial Optimisation*, vol. 2400 of Lecture Notes in Computer Science, Springer-Verlag, New York, 2002, pp. 927–932.
4. T. Bäck, D. Fogel, and Z. Michalewicz. *Handbook of Evolutionary Computation*. Oxford University Press, New York, 1997.
5. A. Bortfeldt. A genetic algorithm for the two-dimensional strip packing problem with rectangular pieces. *European Journal of Operational Research*, 172(3):814–837, 2006.
6. L. Davis. *Handbook of Genetic Algorithms*. Van Nostrand Reinhold, New York, 1991.
7. S. P. Fekete and J. Schepers. On more-dimensional packing: III. Exact algorithm. *Technical Report ZPR97-290*. Mathematisches Institut, Universität zu Köln, Germany, 1997.

8. D. Goldberg and R. Lingle. Alleles, loci, and the TSP. In *Proceedings of the 1st International Conference on Genetic Algorithms*, 1985, pp. 154–159.
9. J. J. Grefenstette. Incorporating problem specific knowledge into genetic algorithms. In L. Davis, ed., *Genetic Algorithms and Simulated Annealing*. Morgan Kaufmann, San Francisco, CA, 1987, pp. 42–60.
10. E. Hopper. Two-dimensional packing utilising evolutionary algorithms and other meta-heuristic methods. Ph.D. thesis, University of Wales, Cardiff, UK, 2000.
11. E. Hopper and B. Turton. A review of the application of meta-heuristic algorithms to 2D strip packing problems. *Artificial Intelligence Review*, 16:257–300, 2001.
12. A. Lodi, S. Martello, and M. Monaci. Recent advances on two-dimensional bin packing problems. *Discrete Applied Mathematics*, 123:379–396, 2002.
13. A. Lodi, S. Martello, and M. Monaci. Two-dimensional packing problems: a survey. *European Journal of Operational Research*, 141:241–252, 2002.
14. S. Martello, S. Monaci, and D. Vigo. An exact approach to the strip-packing problem. *INFORMS Journal on Computing*, 15:310–319, 2003.
15. M. Michalewicz. *Genetic Algorithms + Data Structures = Evolution Programs*, 3rd rev. ed. Springer-Verlag, New York, 1996.
16. C. L. Mumford-Valenzuela, J. Vick, and P. Y. Wang. Heuristics for large strip packing problems with guillotine patterns: an empirical study. In *Metaheuristics: Computer Decision-Making*. Kluwer Academic, Norwell, MA, 2003, pp. 501–522.
17. I. Oliver, D. Smith, and J. Holland. A study of permutation crossover operators on the traveling salesman problem. In *Proceedings of the 2nd International Conference on Genetic Algorithms*, 1987, pp. 224–230.
18. J. Puchinger and G. Raidl. An evolutionary algorithm for column generation in integer programming: an effective approach for two-dimensional bin packing. In X. Yao et al, ed., *PPSN*, vol. 3242, Springer-Verlag, New York, 2004, pp. 642–651.
19. J. Puchinger and G. R. Raidl. Models and algorithms for three-stage two-dimensional bin packing. *European Journal of Operational Research*, 183(3):1304–1327, 2007.
20. C. R. Reeves. A genetic algorithm for flowshop sequencing. *Computers and Operations Research*, 22(1):5–13, 1995.
21. C. Salto, J. M. Molina, and E. Alba. Analysis of distributed genetic algorithms for solving cutting problems. *International Transactions in Operational Research*, 13(5):403–423, 2006.
22. C. Salto, J. M. Molina, and E. Alba. A comparison of different recombination operators for the 2-dimensional strip packing problem. In *Procedimiento Congreso Argentino de Ciencias de la Computación (CACIC'06)*, 2006, pp. 1126–1138.
23. C. Salto, J. M. Molina, and E. Alba. Evolutionary algorithms for the level strip packing problem. In *Proceedings of the International Workshop on Nature Inspired Cooperative Strategics for Optimization*, 2006, pp. 137–148.
24. P. Y. Wang and C. L. Valenzuela. Data set generation for rectangular placement problems. *European Journal of Operational Research*, 134:378–391, 2001.
25. D. Whitley, T. Starkweather, and D. Fuquay. Scheduling problems and traveling salesmen: the genetic edge recombination operator. In *Proceedings of the 3rd International Conference on Genetic Algorithms*, 1989, pp. 133–140.

CHAPTER 23

Solving the KCT Problem: Large-Scale Neighborhood Search and Solution Merging

C. BLUM and M. J. BLESA

Universitat Politècnica de Catalunya, Spain

23.1 INTRODUCTION

In recent years, the development of hybrid metaheuristics for optimization has become very popular. A hybrid metaheuristic is obtained by combining a metaheuristic intelligently with other techniques for optimization: for example, with complete techniques such as branch and bound or dynamic programming. In general, hybrid metaheuristics can be classified as either collaborative combinations or integrative combinations. *Collaborative combinations* are based on the exchange of information between a metaheuristic and another optimization technique running sequentially (or in parallel). *Integrative combinations* utilize other optimization techniques as subordinate parts of a metaheuristic. In this chapter we present two integrative combinations for tackling the k-cardinality tree (KCT) problem. The hybridization techniques that we use are known as *large-scale neighborhood search* and *solution merging*. The two concepts are related. In large-scale neighborhood search, a complete technique is used to search the large neighborhood of a solution to find the best neighbor. Solution merging is known from the field of evolutionary algorithms, where the union of two (or more) solutions is explored by a complete algorithm to find the best possible offspring solution. We refer the interested reader to the book by Blum et al. [5] for a comprehensive introduction to hybrid metaheuristics.

23.1.1 Background

The k-cardinality tree (KCT) problem was defined by Hamacher et al. [23]. Subsequently [17,26], the NP-hardness of the problem was shown. The problem has several applications in practice (see Table 23.1).

Optimization Techniques for Solving Complex Problems, Edited by Enrique Alba, Christian Blum, Pedro Isasi, Coromoto León, and Juan Antonio Gómez

TABLE 23.1 Applications of the KCT Problem

Application	References
Oil-field leasing	[22]
Facility layout	[18,19]
Open-pit mining	[27]
Matrix decomposition	[8,9]
Quorum-cast routing	[12]
Telecommunications	[21]

Technically, the KCT problem can be described as follows. Let $G(V, E)$ be an undirected graph in which each edge $e \in E$ has a weight $w_e \geq 0$ and each node $v \in V$ has a weight $w_v \geq 0$. Furthermore, we denote by $\mathcal{T}_k$ the set of all k-cardinality trees in G, that is, the set of all trees in G with exactly k edges. The problem consists of finding a k-cardinality tree $T_k \in \mathcal{T}_k$ that minimizes

$$f(T_k) = \left(\sum_{e \in E_{T_k}} w_e \right) + \left(\sum_{v \in V_{T_k}} w_v \right) \tag{23.1}$$

Henceforth, when given a tree T, E_T denotes the set of edges of T and V_T the set of nodes of T.

23.1.2 Literature Review

The literature treats mainly two special cases of the general KCT problem as defined above: (1) the edge-weighted KCT (eKCT) problem, where all node weights are equal to zero, and (2) the node-weighted KCT (nKCT) problem, where all edge weights are equal to zero. The eKCT problem was first tackled by complete techniques [12,20,28] and heuristics [12,15,16]. The best working heuristics build (in some way) a spanning tree of the given graph and subsequently apply a polynomial time dynamic programming algorithm [25] that finds the best k-cardinality tree in the given spanning tree. Later, research focused on the development of metaheuristics [6,11,29].

Much less research effort was directed at the nKCT problem. Simple greedy- as well as dual greedy-based heuristics were proposed in [16]. The first metaheuristic approaches appeared in [7,10].

23.1.3 Organization

The organization of this chapter is as follows. In Section 23.2 we outline the main ideas of two hybrid metaheuristics for the general KCT problem. First, we deal with an ant colony optimization (ACO) algorithm that uses large-scale neighborhood search as a hybridization concept. Subsequently, we present the main ideas of an evolutionary algorithm (EA) that uses a recombination operator based on solution merging. Finally, we present a computational comparison of the two approaches with state-of-the-art algorithms from the literature. The chapter finishes with conclusions.

23.2 HYBRID ALGORITHMS FOR THE KCT PROBLEM

The hybrid algorithms presented in this work have some algorithmic parts in common. In the following we first outline these common concepts, then specify the metaheuristics.

23.2.1 Common Algorithmic Components

First, both algorithms that we present are based strongly on tree construction, which is well defined by the definition of the following four components:

1. The graph $G' = (V', E')$ in which the tree should be constructed (here G' is a subgraph of a given graph G)
2. The required size $l \leq (V' - 1)$ of the tree to be constructed, that is, the number of required edges the tree should contain
3. The way in which to begin tree construction (e.g., by determining a node or an edge from which to start the construction process)
4. The way in which to perform each construction step

The first three aspects are operator dependent. For example, our ACO algorithm will use tree constructions that start from a randomly chosen edge, whereas the crossover operator of the EA starts a tree construction from a partial tree. However, the way in which to perform a construction step is the same in all tree constructions used in this chapter: Given a graph $G' = (V', E')$, the desired size l of the final tree, and the current tree T whose size is smaller than l, a construction step consists of adding exactly one node and one edge to T such that the result is again a tree. At an arbitrary construction step, let $\mathcal{N}$ (where $\mathcal{N} \cap V_T = \emptyset$) be the set of nodes of G' that can be added to T via at least one edge. (Remember that V_T denotes the node set of T.) For each $v \in \mathcal{N}$, let E_v be the set of edges that have v as an endpoint and that have their other endpoint, denoted by $v_{e,o}$, in T. To perform a construction step, a node $v \in \mathcal{N}$ must be chosen. This can be done in different ways. In addition to the chosen node v, the edge that minimizes

$$e_{\min} \leftarrow \operatorname{argmin}\{w_e + w_{v_{e,o}} \mid e \in E_v\} \tag{23.2}$$

is added to the tree. For an example, see Figure 23.1.

Second, both algorithms make use of a dynamic programming algorithm for finding the best k-cardinality tree in a graph that is itself a tree. This algorithm, which is based on the ideas of Maffioli [25], was presented by Blum [3] for the general KCT problem.

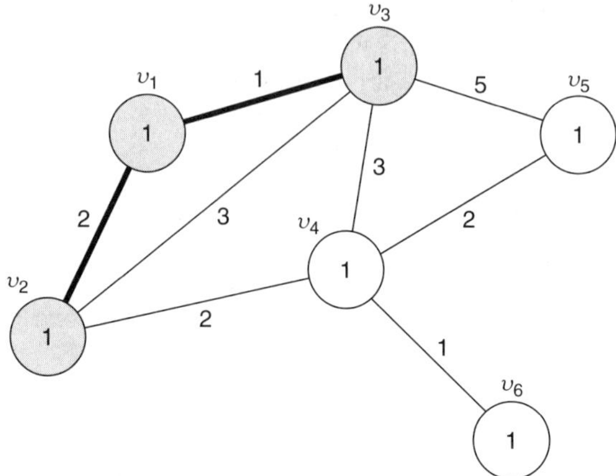

Figure 23.1 Graph with six nodes and eight edges. For simplicity, the node weights are all set to 1. Nodes and edges are labeled with their weights. Furthermore, we have given a tree T of size 2, denoted by shaded nodes and bold edges: $V_T = \{v_1, v_2, v_3\}$, and $E_T = \{e_{1,2}, e_{1,3}\}$. The set of nodes that can be added to T is therefore given as $\mathcal{N} = \{v_4, v_5\}$. The set of edges that join v_4 with T is $E_{v_4} = \{e_{2,4}, e_{3,4}\}$, and the set of edges that join v_5 with T is $E_{v_5} = \{e_{3,5}\}$. Due to the edge weights, $e_{\min}$ in the case of v_4 is determined as $e_{2,4}$, and in the case of v_5, as $e_{3,5}$.

23.2.2 ACO Combined with Large-Scale Neighborhood Search

ACO [13,14] emerged in the early 1990s as a nature-inspired metaheuristic for the solution of combinatorial optimization problems. The inspiring source of ACO is the foraging behavior of real ants. When searching for food, ants initially explore the area surrounding their nest in a random manner. As soon as an ant finds a food source, it carries to the nest some of the food found. During the return trip, the ant deposits a chemical pheromone trail on the ground. The quantity of pheromone deposited, which may depend on the quantity and quality of the food, will guide other ants to the food source. The indirect communication established via the pheromone trails allows the ants to find the shortest paths between their nest and food sources. This behavior of real ants in nature is exploited in ACO to solve discrete optimization problems using artificial ant colonies.

The hybridization idea of our implementation for the KCT problem is as follows. At each iteration, n_a artificial ants each construct a tree. However, instead of stopping the tree construction when cardinality k is reached, the ants continue constructing until a tree of size $l > k$ is achieved. To each of these l-cardinality trees is then applied Blum's dynamic programming algorithm [3] to find the best k-cardinality tree in this l-cardinality tree. This can be seen as a form of large-scale neighborhood search. The pheromone model $\mathcal{T}$ used by our ACO algorithm contains a pheromone value τ_e for each edge $e \in E$. After initialization of the variables T_k^{bs} (i.e., the best-so-far solution), T_k^{rb} (i.e., the restart-best

Algorithm 23.1 Hybrid ACO for the KCT Problem (HyACO)

```
INPUT a node and/or edge-weighted graph G, a cardinality
k < |V| - 1, and a tree size l with k ≤ l ≤ |V| - 1
T_k^bs ← NULL, T_k^rb ← NULL
cf ← 0, bs_update ← FALSE
forall e ∈ E do τ_e ← 0.5 end forall
while termination conditions not satisfied do
  for j = 1 to n_a do
    T_l^j ← ConstructTree(T,l)
    if (l > k) then T_k^j ← LargeScaleNeighborhoodSearch(T_l^j)
        end if
  end for
  T_k^ib ← argmin {f(T_k^1),...,f(T_k^na)}
  Update(T_k^ib,T_k^rb,T_k^bs)
  ApplyPheromoneValueUpdate(cf,bs_update,T,T_k^ib,T_k^rb,T_k^bs)
  cf ← ComputeConvergenceFactor(T,T_k^rb)
  if cf ≥ 0.99 then
    if bs_update = TRUE then
      forall e ∈ E do τ_e ← 0.5 end forall
      T_k^rb ← NULL
      bs_update ← FALSE
    else
      bs_update ← TRUE
    end if
  end if
end while
OUTPUT: T_k^bs
```

solution), and *cf* (i.e., the convergence factor), all the pheromone values are set to 0.5. At each iteration, after generating the corresponding trees, some of these solutions are used to update the pheromone values. The details of the algorithmic framework shown in Algorithm 23.1 are explained in the following.

`ConstructTree(T,l)` The construction of a tree works as explained in Section 23.2.1. It only remains to specify (1) how to start a tree construction, and (2) how to use the pheromone model for choosing among the nodes (and edges) that can be added to the current tree at each construction step. Each tree construction starts from an edge $e = (v_i, v_j)$ that is chosen probabilistically in proportion to the values $\tau_e/(w_e + w_{v_i} + w_{v_j})$. With probability $\mathbf{p}_{\text{det}}^{\text{ACO}}$, an ant chooses the node $v \in \mathcal{N}$ that minimizes $\tau_e \cdot (w_{e_{\min}} + w_v)$. Otherwise, v is chosen probabilistically in proportion to $\tau_e \cdot (w_{e_{\min}} + w_v)$. Two parameters have to be fixed: the probability $\mathbf{p}_{\text{det}}^{\text{ACO}}$ that determines the percentage of deterministic steps during the tree construction, and l, the size of the trees constructed by the ants. After tuning, we chose the setting ($\mathbf{p}_{\text{det}}^{\text{ACO}} = 0.85, l = k + 2s$) for the application to node-weighted grid graphs, and the setting ($\mathbf{p}_{\text{det}}^{\text{ACO}} = 0.95, l = k + 2s$) for the application to edge-weighted graphs, where $s = (|V| - 1 - k)/4$. The tuning process has been outlined in detail by Blum and Blesa [4].

`LargeScaleNeighborhoodSearch(`T_l^j`)` This procedure applies the dynamic programming algorithm [3] to an l-cardinality tree T_l^j (where j denotes the tree constructed by the jth ant). The algorithm returns the best k-cardinality tree $T_k{}^j$ embedded in T_l^j.

`Update(`T_k^{ib}`,`T_k^{rb}`,`T_k^{bs}`)` In this procedure, T_k^{rb} and T_k^{bs} are set to T_k^{ib} (i.e., the iteration-best solution) if $f(T_k^{ib}) < f(T_k^{rb})$ and $f(T_k^{ib}) < f(T_k^{bs})$, respectively.

`ApplyPheromoneUpdate(cf,`bs_update`,`$\mathcal{T}$`,`T_k^{ib}`,`T_k^{rb}`,`T_k^{bs}`)` As described by Blum and Blesa [6] for a standard ACO algorithm, the HyACO algorithm may use three different solutions for updating the pheromone values: (1) the restart-best solution T_k^{ib}, (2) the restart-best solution T_k^{rb}, and (3) the best-so-far solution T_k^{bs}. Their influence depends on the convergence factor cf, which provides an estimate about the state of convergence of the system. To perform the update, first an update value ξ_e for every pheromone trail parameter $\mathcal{T}_e \in \mathcal{T}$ is computed:

$$\xi_e \leftarrow \kappa_{ib} \cdot \delta(T_k^{ib}, e) + \kappa_{rb} \cdot \delta(T_k^{rb}, e) + \kappa_{bs} \cdot \delta(T_k^{bs}, e) \tag{23.3}$$

where κ_{ib} is the weight of T_k^{ib}, κ_{rb} the weight of T_k^{rb}, and κ_{bs} the weight of T_k^{bs} such that $\kappa_{ib} + \kappa_{rb} + \kappa_{bs} = 1.0$. The δ-function is the characteristic function of the set of edges in the tree; that is, for each k-cardinality tree T_k,

$$\delta(T_k, e) = \begin{cases} 1 & e \in E(T_k) \\ 0 & \text{otherwise} \end{cases} \tag{23.4}$$

Then the following update rule is applied to all pheromone values τ_e:

$$\tau_e \leftarrow \min\{\max\{\tau_{\min}, \tau_e + \rho \cdot (\xi_e - \tau_e)\}, \tau_{\max}\} \tag{23.5}$$

where $\rho \in (0, 1]$ is the evaporation (or learning) rate. The upper and lower bounds $\tau_{\max} = 0.99$ and $\tau_{\min} = 0.01$ keep the pheromone values in the range $(\tau_{\min}, \tau_{\max})$, thus preventing the algorithm from converging to a solution. After tuning, the values for ρ, κ_{ib}, κ_{rb}, and κ_{bs} are chosen as shown in Table 23.2.

TABLE 23.2 Schedule Used for Values ρ, κ_{ib}, κ_{rb}, and κ_{bs} Depending on cf (the Convergence Factor) and the Boolean Control Variable *bs_update*

	bs_update = FALSE			*bs_update* = TRUE
	$cf < 0.7$	$cf \in [0.7, 0.9)$	$cf \geq 0.9$	
ρ	0.05	0.1	0.15	0.15
κ_{ib}	2/3	1/3	0	0
κ_{rb}	1/3	2/3	1	0
κ_{bs}	0	0	0	1

`ComputeConvergenceFactor`($\mathcal{T}$, T_k^{rb}) This function computes, at each iteration, the convergence factor as

$$cf \leftarrow \frac{\sum_{e \in E(T_k^{rb})} \tau_e}{k \cdot \tau_{\max}} \tag{23.6}$$

where $\tau_{\max}$ is again the upper limit for the pheromone values. The convergence factor cf can therefore only assume values between 0 and 1. The closer cf is to 1, the higher is the probability to produce the solution T_k^{rb}.

23.2.3 Evolutionary Algorithm Based on Solution Merging

Evolutionary algorithms (EAs) [1,24] are used to tackle hard optimization problems. They are inspired by nature's capability to evolve living beings that are well adapted to their environment. EAs can be characterized briefly as computational models of evolutionary processes working on populations of individuals. Individuals are in most cases solutions to the tackled problem. EAs apply genetic operators such as *recombination* and/or *mutation* operators in order to generate new solutions at each iteration. The driving force in EAs is the *selection* of individuals based on their *fitness*. Individuals with higher fitness have a higher probability to be chosen as members of the next iteration's population (or as parents for producing new individuals). This principle is called *survival of the fittest* in natural evolution. It is the capability of nature to adapt itself to a changing environment that gave the inspiration for EAs.

The algorithmic framework of our hybrid EA approach to tackle the KCT problem is shown in Algorithm 23.2. In this algorithm, henceforth denoted by HyEA, T_k^{best} denotes the best solution (i.e., the best k-cardinality tree) found since the start of the algorithm, and T_k^{iter} denotes the best solution in the current population P. The algorithm begins by generating the initial population in the function `GenerateInitialPopulation`(*pop_size*). Then at each iteration the algorithm produces a new population by first applying a crossover operator in the function `ApplyCrossover(`P`)`, and then by subsequent replacement of the worst solutions with newly generated trees in the function `IntroduceNew-Material(`$\hat{P}$`)`. The components of this algorithm are outlined in more detail below.

Algorithm 23.2 Hybrid EA for the KCT Problem (HyEA)

```
INPUT: a node and/or edge-weighted graph G, and a
          cardinality k < |V| − 1
P ← GenerateInitialPopulation(pop_size)
T_k^best ← argmin f(T_k) | T_k ∈ P
while termination conditions are not met do
   P̂ ← ApplyCrossover(P)
   P ← IntroduceNewMaterial(P̂)
   T_k^iter ← argmin f(T_k) | T_k ∈ P
   if f(T_k^iter) < f(T_k^best) then
      T_k^best ← T_k^iter
   end if
end while
OUTPUT: T_k^best
```

`GenerateInitialPopulation` *(pop-size)* The initial population is generated in this method. It takes as input the size *pop_size* of the population, which was set depending on the graph size and the desired cardinality:

$$pop\_size \leftarrow \min\left\{\max\left\{10, \left\lfloor 5 \cdot \frac{|V|}{k+1} \right\rfloor\right\}, 100\right\} \tag{23.7}$$

The construction of each of the initial k-cardinality trees in graph G begins with a node that is chosen uniformly at random from V. At each further construction step is chosen, with probability $\mathbf{p}_{\text{det}}^{\text{EA}}$, the node v that minimizes $w_{e_{\min}} + w_v$ (see also Section 23.2.1). Otherwise (i.e., with probability $1 - \mathbf{p}_{\text{det}}^{\text{EA}}$), v is chosen probabilistically in proportion to $w_{e_{\min}} + w_v$. Note that when $\mathbf{p}_{\text{det}}^{\text{EA}}$ is close to 1, the tree construction is almost deterministic, and the other way around. After tuning by hand (see ref. 2 for more details) we chose for each tree construction a uniform value for $\mathbf{p}_{\text{det}}^{\text{EA}}$ from $[0.85, 0.99]$.

`ApplyCrossover(`P`)` At each algorithm iteration an offspring population $\hat{P}$ is generated from the current population P. For each k-cardinality tree $T \in P$, the following is done. First, tournament selection (with tournament size 3) is used to choose a crossover partner $T^c \neq T$ for T from P. In the following we say that two trees in the same graph are overlapping if and only if they have at least one node in common. In case T and T^c are overlapping, graph G^c is defined as the union of T and T^c; that is, $V_{G^c} = V_T \cup V_{T^c}$ and $E_{G^c} = E_T \cup E_{T^c}$. Then a spanning tree T^{sp} of G^c is constructed as follows. The first node is chosen uniformly at random. Each further construction step is performed as described above (generation of the initial population). Then the dynamic programming algorithm [3] is applied to T^{sp} for finding the best k-cardinality tree T^{child} that is contained in T^{sp}. Otherwise, that is, in case the crossover partners T and T^c are not overlapping, T is used as the basis for constructing a tree in G that contains both, T and T^c. This is done by extending T (with construction steps as outlined in Section 23.2.1) until the current tree under construction can be connected with T^c by at least one edge. In the case of several connecting edges, the edge $e = (v_i, v_j)$ that minimizes $w_e + w_{v_i} + w_{v_j}$ is chosen. Finally, we apply the dynamic programming algorithm [3] for finding the best k-cardinality tree T^{child} in the tree constructed. The better tree among T^{child} and T is added to the offspring population $\hat{P}$. Finally, note that both crossover cases can be seen as implementations of the solution merging concept.

`IntroduceNewMaterial(`$\hat{P}$`)` To avoid premature convergence of the algorithm, at each iteration this function introduces new material (in the form of newly constructed k-cardinality trees) into the population. The input of this function is the offspring population $\hat{P}$ generated by crossover. First, the function selects $X = \lfloor 100 - newmat \rfloor\%$ of the best solutions in $\hat{P}$ for the new population P. Then the remaining $100 - X\%$ of P are generated as follows: Starting from a node of G that is chosen uniformly at random, an l_{new}-cardinality tree $T_{l_{\text{new}}}$ (where $l_{\text{new}} \geq k$) is constructed by applying construction steps as outlined above. Here we chose

$$l_{\text{new}} \leftarrow k + \left\lfloor \frac{|V| - 1 - k}{3} \right\rfloor \tag{23.8}$$

and $newmat = 20$ (see ref. 2 for details). Then the dynamic programming algorithm [3] is used to find the best k-cardinality tree in $T_{l_{\text{new}}}$. This tree is then added to P.

23.3 EXPERIMENTAL ANALYSIS

We implemented both HyACO and HyEA in ANSI C++ using GCC 3.2.2 to compile the software. Our experimental results were obtained on a PC with Intel Pentium-4 processor (3.06 GHz) and 1 Gb of memory.

23.3.1 Application to Node-Weighted Grid Graphs

First, we applied HyACO and HyEA to the benchmark set of node-weighted grid graphs that was proposed in [10]. This set is composed of 30 grid graphs, that is, 10 grid graphs of 900 vertices (i.e., 30 times 30 vertices), 10 grid graphs of 1600 vertices, and 10 grid graphs of 2500 vertices. We compared our results to the results of the variable neighborhood decomposition search (denoted by VNDS) presented by Brimberg et al. [10]. This comparison is shown in Table 23.3. Instead of applying an algorithm several times to the same graph and cardinality, it is general practice for this benchmark set to apply the algorithm exactly once to each graph and cardinality, and then to average the results over graphs of the same type. The structure of Table 23.3 is as follows. The first table column indicates the graph type (e.g., grid graphs of size 30×30), the second column contains the cardinality, and the third table column provides the best known results (abbreviated as bkr). A best-known result is marked with a left–right arrow in case it was produced by one of our algorithms. Then for each of the three algorithms we provide the result (as described above) together with the average computation time that was spent to compute this result. Note that the computation time limit for HyACO and HyEA was approximately the same as the one that was used for VNDS [10]. The computer used to run the experiments [10] is about seven times slower than the machine we used. Therefore, we divided the computation time limits of VNDS by 7 to obtain our computation time limits.

Concerning the results, we note that in 9 out of 15 cases both HyACO and HyEA are better than VNDS. It is interesting to note that this concerns especially small to medium-sized cardinalities. For larger cardinalities, VNDS beats both HyACO and HyEA. Furthermore, HyEA is in 11 out of 15 cases better than HyACO. This indicates that even though HyACO and HyEA behave similarly to VNDS, HyEA seems to make better use of the dynamic programming algorithm than does HyACO. The computation times of the algorithms are of the same order of magnitude.

23.3.2 Application to Edge-Weighted Random Graphs

We also applied HyACO and HyEA to some of the edge-weighted random graph instances from ref. 29 (i.e., 10 instances with 1000, 2000, and 3000 nodes, respectively). The results are shown in Table 23.4 compared to the VNDS algorithm presented by Vrošević et al. [29]. Note that this VNDS is the current state-of-the-art algorithm for these instances. (The VNDS algorithms for node- and edge-weighted graphs are not the same. They are specialized for the corresponding special cases of the KCT problem.) The machine that was used to run VNDS is about 10 times slower than our machine. Therefore, we used as time limits for HyACO and HyEA one-tenth of the time limits of VNDS.

In general, the results show that the performance of all three algorithms is very similar. However, in 6 out of 15 cases our algorithms (together) improve over the performance of VNDS. Again, the advantage of our algorithms is for small and medium-sized cardinalities. This advantage drops when the size of the graphs increases. Comparing HyACO with HyEA, we note an advantage for HyEA, especially when the graphs are quite small. However, application to the largest graphs shows that HyACO seems to have advantages over HyEA in these cases, especially for rather small cardinalities.

23.4 CONCLUSIONS

In this chapter we have proposed two hybrid approaches to tackling the general k-cardinality tree problem. The first approach concerns an ant colony optimization algorithm that uses large-scale neighborhood search as a hybridization concept.

TABLE 23.3 Results for Node-Weighted Graphs [10]

Instance			VNDS		HyACO		HyEA	
Type	k	bkr	Result	Avg. Time	Result	Avg. Time	Result	Avg. Time
30 × 30	100	→ 8,203.50	8,571.90	24.00	**8,203.50**	65.99	8,206.50	63.43
	200	→ 17,766.60	17,994.40	88.00	17,850.10	89.29	**17,766.60**	97.99
	300	28,770.90	**28,770.90**	126.00	28,883.90	108.42	28,845.40	104.14
	400	42,114.00	**42,114.00**	80.00	42,331.90	133.78	42,282.70	117.80
	500	59,266.40	**59,266.40**	213.00	59,541.70	132.52	59,551.60	110.31
40 × 40	150	→ 17,461.70	18,029.90	112.00	17,527.10	211.78	**17,461.70**	229.39
	300	→ 37,518.50	38,965.90	114.00	37,623.80	277.70	**37,518.50**	297.82
	450	→ 60,305.60	61,290.10	261.00	60,417.00	270.27	**60,305.60**	269.80
	600	86,422 30	**86,422.30**	261.00	86,594.70	187.45	86,571.10	295.54
	750	117,654.00	**117,654.00**	303.00	118,570.00	217.23	118,603.50	260.62
50 × 50	250	→ 35,677.20	37,004.00	228.00	35,995.20	171.64	**35,677.20**	259.78
	500	→ 76,963.20	81,065.80	322.00	77,309.90	286.66	**76,963.20**	267.76
	750	→ 125,009.00	128,200.00	482.00	125,415.00	310.13	**125,009.00**	284.98
	1000	→ 181,983.00	182,220.00	575.00	**181,983.00**	316.63	182,101.50	313.55
	1250	250,962.00	**250,962.00**	681.00	253,059.00	335.35	252,683.10	303.51

TABLE 23.4 Results for Edge-Weighted Random Graphs [29]

Instance			VNDS		HyACO		HyEA	
Type	k	bkr	Result	Avg. Time	Result	Avg. Time	Result	Avg. Time
1000	100	→ 5,811.70	5,828.00	52.87	5,827.20	27.20	**5,811.70**	37.13
	200	→ 11,891.60	11,893.70	95.86	11,910.10	57.13	**11,891.60**	49.04
	300	→ 18,196.10	18,196.60	157.28	18,217.40	50.36	**18,196.10**	54.32
	400	24,734.00	**24,734.00**	174.35	24,757.80	76.68	24,741.80	74.73
	500	31,561.80	**31,561.80**	34.64	31,613.80	85.30	31,571.70	93.32
2000	200	→ 23,415.60	23,538.30	152.43	23,479.00	208.23	**23,415.60**	206.55
	400	→ 47,936.70	48,027.30	210.77	4,8030.4	247.95	**47,936.70**	286.83
	600	73,277.80	**73,277.80**	241.85	7,3392.9	301.19	73,344.00	365.52
	800	99,491.20	**99,491.20**	322.85	9,9801.4	301.74	99,595.50	384.56
	1000	126,485.00	**126,485.00**	352.35	127,325	300.67	126,920.20	457.36
3000	300	→ 35,160.30	35,186.10	225.32	**35,160.3**	103.93	35,354.10	84.19
	600	71,634.70	**71,634.70**	628.85	71,862.8	123.83	72,064.20	118.43
	900	109,463.00	**109,463.00**	1,134.97	110,094	108.29	110,036.80	138.33
	1200	148,826.00	**148,826.00**	915.30	149,839	120.39	149,453.50	160.94
	1500	189,943.00	**189,943.00**	785.00	191,627	180.49	190,526.10	184.71

The second approach is a hybrid evolutionary technique that makes use of solution merging within the recombination operator. Our algorihms were able to improve on many of the best known results from the literature. In general, the evolutionary techniques seem to have advantages for the ant colony optimization algorithm. In the future we will study other ways of using dynamic programming to obtain hybrid metaheuristics for a problem.

Acknowledgments

This work was supported by the Spanish MEC projects OPLINK (TIN2005-08818) and FORMALISM (TIN2007-66523), and by the *Ramón y Cajal* program of the Spanish Ministry of Science and Technology, of which Christian Blum is a research fellow.

REFERENCES

1. T. Bäck. *Evolutionary Algorithms in Theory and Practice*. Oxford University Press, New York, 1996.
2. C. Blum. A new hybrid evolutionary algorithm for the k-cardinality tree problem. In M. Keijzer et al., eds., *Proceedings of the Genetic and Evolutionary Computation Conference 2006 (GECCO'06)*. ACM Press, New York, 2006, pp. 515–522.
3. C. Blum. Revisiting dynamic programming for finding optimal subtrees in trees. *European Journal of Operational Research*, 177(1):102–115, 2007.
4. C. Blum and M. Blesa. Combining ant colony optimization with dynamic programming for solving the k-cardinality tree problem. In *Proceedings of the 8th International Work-Conference on Artificial Neural Networks, Computational Intelligence and Bioinspired Systems (IWANN'05)*, vol. 3512 of *Lecture Notes in Computer Science*. Springer-Verlag, New York, 2005, pp. 25–33.
5. C. Blum, M. Blesa, A. Roli, and M. Sampels, eds. *Hybrid Metaheuristics: An Emergent Approach to Optimization*, vol. 114 in *Studies in Computational Intelligence*. Springer-Verlag, New York, 2008.
6. C. Blum and M. J. Blesa. New metaheuristic approaches for the edge-weighted k-cardinality tree problem. *Computers and Operations Research*, 32(6):1355–1377, 2005.
7. C. Blum and M. Ehrgott. Local search algorithms for the k-cardinality tree problem. *Discrete Applied Mathematics*, 128:511–540, 2003.
8. R. Borndörfer, C. Ferreira, and A. Martin. Matrix decomposition by branch-and-cut. *Technical report*. Konrad-Zuse-Zentrum für Informationstechnik, Berlin, 1997.
9. R. Borndörfer, C. Ferreira, and A. Martin. Decomposing matrices into blocks. *SIAM Journal on Optimization*, 9(1):236–269, 1998.
10. J. Brimberg, D. Urošević, and N. Mladenović. Variable neighborhood search for the vertex weighted k-cardinality tree problem. *European Journal of Operational Research*, 171(1):74–84, 2006.

11. T. N. Bui and G. Sundarraj. Ant system for the k-cardinality tree problem. In K. Deb et al., eds., *Proceedings of the Genetic and Evolutionary Computation Conference (GECCO'2004)*, vol. 3102 of Lecture Notes in Computer Science. Springer-Verlag, New York, 2004, pp. 36–47.

12. S. Y. Cheung and A. Kumar. Efficient quorumcast routing algorithms. In *Proceedings of INFOCOM'94*. IEEE Press, Piscataway, NJ, 1994, pp. 840–847.

13. M. Dorigo. Optimization, learning and natural algorithms (in Italian). Ph.D. thesis, Dipartimento di Elettronica, Politecnico di Milano, Italy, 1992.

14. M. Dorigo, V. Maniezzo, and A. Colorni. Ant system: optimization by a colony of cooperating agents. *IEEE Transactions on Systems, Man, and Cybernetics, Part B*, 26(1):29–41, 1996.

15. M. Ehrgott and J. Freitag. K_TREE/K_SUBGRAPH: a program package for minimal weighted k-cardinality-trees and -subgraphs. *European Journal of Operational Research*, 1(93):214–225, 1996.

16. M. Ehrgott, J. Freitag, H. W. Hamacher, and F. Maffioli. Heuristics for the k-cardinality tree and subgraph problem. *Asia-Pacific Journal of Operational Research*, 14(1):87–114, 1997.

17. M. Fischetti, H. W. Hamacher, K. Jørnsten, and F. Maffioli. Weighted k-cardinality trees: complexity and polyhedral structure. *Networks*, 24:11–21, 1994.

18. L. R. Foulds and H. W. Hamacher. A new integer programming approach to (restricted) facilities layout problems allowing flexible facility shapes. *Technical Report 1992-3*. Department of Management Science, University of Waikato, New Zealand, 1992.

19. L. R. Foulds, H. W. Hamacher, and J. Wilson. Integer programming approaches to facilities layout models with forbidden areas. *Annals of Operations Research*, 81:405–417, 1998.

20. J. Freitag. Minimal k-cardinality trees (in German). Master's thesis, Department of Mathematics, University of Kaiserslautern, Germany, 1993.

21. N. Garg and D. Hochbaum. An $O(\log k)$ approximation algorithm for the k minimum spanning tree problem in the plane. *Algorithmica*, 18:111–121, 1997.

22. H. W. Hamacher and K. Joernsten. Optimal relinquishment according to the Norwegian petrol law: a combinatorial optimization approach. *Technical Report 7/93*. Norwegian School of Economics and Business Administration, Bergen, Norway, 1993.

23. H. W. Hamacher, K. Jörnsten, and F. Maffioli. Weighted k-cardinality trees. *Technical Report 91.023*. Dipartimento di Elettronica, Politecnico di Milano, Italy, 1991.

24. A. Hertz and D. Kobler. A framework for the description of evolutionary algorithms. *European Journal of Operational Research*, 126:1–12, 2000.

25. F. Maffioli. Finding a best subtree of a tree. *Technical Report 91.041*. Dipartimento di Elettronica, Politecnico di Milano, Italy, 1991.

26. M. V. Marathe, R. Ravi, S. S. Ravi, D. J. Rosenkrantz, and R. Sundaram. Spanning trees short or small. *SIAM Journal on Discrete Mathematics*, 9(2):178–200, 1996.

27. H. W. Philpott and N. Wormald. On the optimal extraction of ore from an open-cast mine. *Technical Report*. University of Auckland, New Zeland, 1997.

28. R. Uehara. The number of connected components in graphs and its applications. *IEICE Technical Report COMP99-10*. Natural Science Faculty, Komazawa University, Japan, 1999.
29. D. Urošević, J. Brimberg, and N. Mladenović. Variable neighborhood decomposition search for the edge weighted k-cardinality tree problem. *Computers and Operations Research*, 31:1205–1213, 2004.

CHAPTER 24

Experimental Study of GA-Based Schedulers in Dynamic Distributed Computing Environments

F. XHAFA and J. CARRETERO

Universitat Politècnica de Catalunya, Spain

24.1 INTRODUCTION

In this chapter we address the issue of experimental evaluation of metaheuristics for combinatorial optimization problems arising in dynamic environments. More precisely, we propose an approach to experimental evaluation of genetic algorithm–based schedulers for job scheduling in computational grids using a grid simulator. The experimental evaluation of metaheuristics is a very complex and time-consuming process. Moreover, ensuring significant statistical results requires, on the one hand, testing on a large set of instances to capture the most representative set of instances, and on the other, finding appropriate values of the search parameters of the metaheuristic that would be expected to work well in any instance of the problem.

Taking into account the characteristics of the problem domain under resolution is among the most important factors in experimental studies of the metaheuristics. One such characteristics is the static versus dynamic setting. In the static setting, the experimental evaluation and fine tuning of parameters is done through benchmarks of (static) instances. An important objective in this case is to run the metaheuristic a sufficient number of times on the same instance and using a fixed setting of parameters to compare the results with the state of the art. To avoid a biased result, parameters are fine-tuned using instances other than those used for reporting computational results. This is actually the most common scenario in evaluating metaheuristics for combinatorial optimization problems. It should be noted that among many issues, the generation of benchmarks of instances is very

Optimization Techniques for Solving Complex Problems, Edited by Enrique Alba, Christian Blum, Pedro Isasi, Coromoto León, and Juan Antonio Gómez

important here since we would like a benchmark to be as representative as possible of all problem instances. Examples of such benchmarks of static instances are those for the traveling salesman problem [28], quadratic assignment problems [6], min k-tree [18], and the 0–1 multiknapsack problem [12].

Simple random generation of instances, for example, could fail to capture the specifics of the problem. The interest is to obtain a *rich* benchmark of instances so that the performance and robustness of metaheuristics could be evaluated. One such example is the benchmark of static instances for independent job scheduling in distributed heterogeneous computing environments [5,20]. In this problem a set of independent jobs having their workload are to be assigned to a set of resources having their computing capacities. The authors of the benchmark used the expected time to compute (ETC) matrix model and generated 12 groups of 100 instances each, where each group tries to capture different characteristics of the problem. Essentially, different scenarios arise in distributed heterogeneous computing environments, due to the consistency of the computing capacity, the heterogeneity of resources, and the heterogeneity of the workload of jobs. For example, we could have scenarios in which resources have a high degree of heterogeneity and jobs have a low degree of heterogeneity. The aim of the benchmark is to represent real distributed computing environments as realistically as possible. This benchmark has been used for the evaluation of metaheuristic implementations in a static setting [5,23,24,25] and metaheuristic-based batch schedulers [29,30].

The state of the art in experimental evaluation of parameters for metaheuristics approaches in a dynamic setting is somewhat limited compared to studies carried out in the static setting. Dynamic optimization problems are becoming increasingly important, especially those arising from computational systems such as the problem of job scheduling. Such problems require the development of effective systems to experimentally study, evaluate, and compare various metaheuristics resolution techniques.

In this chapter we present the use of a grid simulator [31] to study genetic algorithms (GAs)– based schedulers. The main contribution of this work is an approach to *plug in* GA-based implementations for independent job scheduling in grid systems with the grid simulator. Unlike other approaches in the literature, in which the scheduling algorithms are part of the simulation tool, in our approach the simulator and the scheduler are decoupled. Therefore, although we use a GA-based scheduler, in fact, any other heuristics scheduler can be plugged into the simulator. Moreover, in this approach we are able to observe the performance of GA-based schedulers in a dynamic setting where resources and jobs vary over time. Also, using this approach we could tune the simulator to capture different real grid systems with the aim of evaluating performance behavior, in terms of both optimization criteria and the robustness of the GA-based scheduler.

The remainder of the chapter is organized as follows. In Section 24.2 we give a short overview of some related work on experimenting and on fine tuning of metaheuristics in static and dynamic setting. Independent job scheduling is defined in Section 24.3, and the GA-based scheduler for the problem is presented in Section 24.4. In Sections 24.5 and 24.6 we present the grid simulator and the interface for using the GA-based scheduler with the grid simulator, respectively. Some computational results are given in Section 24.7. We end the chapter with some conclusions and indications for future work.

24.2 RELATED WORK

Experimental analysis and fine tuning of metaheuristics have attracted the attention of many researchers, and a large body of literature exists for the static setting. In the general context of the experimental analysis of algorithms, Johnson [14] presented several fundamental issues. Although the approach is applicable to metaheuristics, due to its particularities, such as the resolution of large-scale problems, hybridization, convergence, use of adaptive techniques, and so on, the metaheuristics community has undertaken research for the experimental analysis and fine tuning of metaheuristics on its own. It should be noted, however, that most of the research work in the literature is concerned with experimenting on and fine tuning of metaheuristics in a static setting.

On the one hand, research efforts have been oriented toward the development of a methodology and rigorous basis for experimental evaluation of heuristics (e.g., Rardin and Uzsoy [22], Barr et al. [4]). On the other hand, concrete approaches to dealing in practice with the fine tuning of parameters such as the development of specific software (e.g., the CALIBRA [2]) or the use of statistics and gradient descent methods [9] and the use of self-adaptive procedures [17] have been proposed. These approaches are generic in purpose and could also be used in experimenting on and fine tuning of metaheuristics for scheduling in a static setting.

Due to increasing interest in dynamic optimization problems, recently many researchers have focused on experimental analysis in dynamic setting [15]. There is little research work on studying schedulers experimentally in dynamic environments. YarKhan and Dongarra [32] presented an experimental study of scheduling using simulated annealing in a grid environment. Based on experimental analysis, the authors report that a simulated annealing scheduler performs better, in terms of estimated execution time, than an ad hoc greedy scheduler. Zomaya and Teh [33] analyzed genetic algorithms for dynamic load balancing. The authors use a "window-sized" mechanism for varying the number of processors and jobs in a system. Madureira et al. [19] presented a GA approach for dynamic job shop scheduling problems that arise in manufacturing applications. In their approach, the objective was to adapt the planning of jobs due to random events occurring in the system rather than computing an optimum solution, and thus the approach deals with the design of a GA scheduler to react to random events occurring in the system. Klusácek et al. [16] presented an evaluation of local search heuristics for scheduling using a simulation environment.

Other approaches in the literature (Bricks [27], MicroGrid [26], ChicSim [21], SimGrid [8], GridSim [7], and GangSim [11]) use simulation as a primary objective and embed concrete scheduling policies rather than optimization approaches for scheduling in the simulator.

24.3 INDEPENDENT JOB SCHEDULING PROBLEM

Computational grids (CGs) are very suitable for solving large-scale optimization problems. Indeed, CGs join together a large computing capacity distributed in the nodes of the grid, and CGs are parallel in nature. Yet to benefit from these characteristics, efficient planning of jobs (hereafter, job and task are used indistinctly) in relation to resources is necessary. The scheduling process is carried out in a dynamic form; that is, it is carried out while jobs enter the system, and the resources may vary their availability. Similarly, it is carried out in running time to take advantage of the properties and dynamics of the system that are not known beforehand. The dynamic schedulers are therefore more useful than static schedulers for real distributed systems. This type of scheduling, however, imposes critical restrictions on temporary efficiency and therefore on its performance.

The simplest version of job scheduling in CGs is that of independent jobs, in which jobs submitted to the grid are independent and are not preemptive. This type of scheduling is quite common in real-life grid applications where independent users submit their jobs, in complete applications to the grid system, or in applications that can be split into independent jobs, such as in parameter sweep applications, software and/or data-intensive computing applications and data-intensive computing, and data mining and massive processing of data.

Problem Description Under the ETC Model Several approaches exist in the literature to model the job scheduling problem. Among them is the proposed expected time to compute (ETC) model, which is based on estimations or predictions of the computational load of each job, the computing capacity of each resource, and an estimation of the prior load of the resources.

This approach is common in grid simulation, where jobs and their workload, resources, and computing capacities are generating through probability distributions. Assuming that we dispose of the computing capacity of resources, the prediction of the computational needs (workload) of each job, and the prior load of each resource, we can then calculate the ETC matrix where each position $ETC[t][m]$ indicates the expected time to compute job t in resource m. The entries $ETC[t][m]$ could be computed by dividing the workload of job t by the computing capacity of resource m, or in more complex ways by including the associated migration cost of job t to resource (hereafter, resource and machine are used indistinctly) m, and so on.

The problem is thus defined formally as (a) a set of independent *jobs* that must be scheduled (any job has to be processed entirely in a unique resource); (b) a set of heterogeneous *machine* candidates to participate in the planning; (c) the *workload* (in millions of instructions) of each job; and (d) the *computing capacity* of each machine (in MIPS).

The number of jobs and resources may vary over time. Notice that in the ETC matrix model, information on MIPS and workload of jobs is not included. As stated above, this matrix can be computed from the information on machines and jobs. Actually, ETC values are useful to formulate the optimization criteria, or fitness of a schedule. Also, based on expected time to compute values, the ready times of machines (i.e., times when machines will have finished previously assigned jobs) can be defined and computed. Ready times of machines measure the previous workload of machines in the grid system.

Optimization Criteria Among the possible optimization criteria for the problem, the minimization of *makespan* (the time when the latest job finishes) and *flowtime* (the sum of finalization times of all the jobs) and the maximization of (average) utilization of grid resources can be defined. Next, we give their formal definition (see Equations 24.2 to 24.5); to this end we define first the completion time of machines, $completion[m]$, which indicates the time in which machine m will finalize the processing of the jobs assigned previously, as well as of those already planned for the machine; F_j denotes the finishing time of job j, $ready[m]$ is the ready time of machine m, $Jobs$ is the set of jobs, $Machines$ is the set of machines, and schedule is a schedule from the set S of all possible schedules:

$$\text{completion}[m] = \text{ready}[m] + \sum_{j \in \text{Jobs} \;—\; \text{schedule}(j)=m} \text{ETC}[j][m] \tag{24.1}$$

$$\text{makespan} = \max\{\text{completion}[i] | i \in \text{Machines}\} \tag{24.2}$$

$$\text{flowtime}[m] = \sum_{j \in \text{schedule}^{-1}(m)} F_j \tag{24.3}$$

$$\text{flowtime} = \sum_{m \in \text{Machines}} \text{flowtime}[m] \tag{24.4}$$

$$\text{avg\_utilization} = \frac{\sum_{\{i \in \text{Machines}\}} \text{completion}[i]}{\text{makespan} \cdot \text{nb\_machines}} \tag{24.5}$$

In the equations above, $Machines$ indicate the set of machines available. These optimization criteria can be combined in either a simultaneous or hierarchic mode. In the former, we try to optimize the criteria simultaneously, while in the later, the optimization is done according to a previously fixed hierarchy of the criteria according to the desired priority (see Section 24.7).

24.4 GENETIC ALGORITHMS FOR SCHEDULING IN GRID SYSTEMS

Genetic algorithms (GAs) [5] have proved to be a good alternative for solving a wide variety of hard combinatorial optimization problems. GAs are a population-based approach where individuals (*chromosomes*) represent possible solutions, which are successively evaluated, selected, crossed, mutated, and replaced by simulating the Darwinian evolution found in nature.

One important characteristic of GAs is the tendency of the population to converge to a fixed point where all individuals share almost the same genetic characteristics. If this convergence is accelerated by means of the selection and replacement strategy, good solutions will be obtained faster. This characteristic is interesting for the job scheduling problem in grid systems given that we might be interested to obtain a fast reduction in the makespan value of schedule.

GAs are high-level algorithms that integrate other methods and genetic operators. Therefore, to implement it for the scheduling problem, we used a standard template and designed the encodings, inner methods, operators, and appropriate data structures.

In this section we present briefly the GA-based scheduler [10,30], which has been extended with new versions for the purposes of this work. Two encodings and several genetic operators are presented; in the experimental study (see Section 24.7), those that showed the best performance are identified.

Schedule Encodings Two types of encoding of individuals, the direct representation and the permutation-based representation, were implemented. In the former, a schedule is encoded in a vector of size nb_jobs, where $schedule[i]$ indicates the machine where job i is assigned by the schedule. In the latter, each element in the encoding must be present only once, and it is a sequence $< S_1, S_2, \ldots, S_m >$, where S_i is the sequence of jobs assigned to machine m_i.

GA's Methods and Operators Next we describe briefly the methods and operators of the GA algorithm for the scheduling problem based on the encodings above.

1. *Generating the initial population*. In GAs, the initial population is usually generated randomly. Besides the random method, we have used two ad hoc heuristics: the longest job to fastest resource–shortest job to fastest resource (LJFR-SJFR) heuristic [1] and minimum completion time (MCT) heuristics [20]. These two methods are aimed at introducing more diversity to the initial population.

2. *Crossover operators*. We considered several crossover operators for the direct representation and other crossover operators for the permutation-based representation. In the first group, one-point crossover, two-point crossover, uniform crossover, and fitness-based crossover were implemented; in the second group we considered the partially matched crossover (PMX), cycle crossover (CX), and order crossover (OX).

3. *Mutation operators*. The mutation operators are oriented toward achieving load balancing of resources through movements and swaps of jobs in different resources. Thus, four mutation operators were defined: Move, Swap, Move&Swap, and Rebalancing mutation.

4. *Selection operators*. The selection of individuals to which the crossover operators will be applied is done using the following selection operators: select random, select best, linear ranking selection, binary tournament selection, and N-tournament selection.

5. *Replacement operators*. Several replacement strategies known in the literature were implemented: generational replacement, elitist generational replacement, replace only if better, and steady-state strategy.

The final setting of GA operators and parameters is done after a tuning process (see Section 24.7).

24.5 GRID SIMULATOR

In this section we describe the main aspects of HyperSim-G, a discrete event-based simulator for grid systems (the reader is referred to Xhafa et al. [31] for details). The simulator has been conceived, designed, and implemented as a tool for experimentally studying algorithms and applications in grid systems. There are many conditions that make simulation an indispensable means for experimental evaluation. Among them we could distinguish the complexity of the experimentation and the need to repeat and control it. Also, resources of the grid systems could have different owners who establish their local policies on use of resources, which could prevent free configuration of grid nodes.

24.5.1 Main Characteristics of the Grid Simulator

HyperSim-G simulator is based on a discrete event model; the behavior of the grid is simulated by discrete events occurring in the system. The sequence of events and the changes in the state of the system capture the dynamic behavior of the system.

The simulator offers many functionalities that allow us to simulate resources, jobs, communication with schedulers, recording data traces, and the definition of grid scenarios, among others. Thus, a complete specification of resource characteristics (computing capacity, consistency of computing, hardware/software specifics, and other parameters) can be done in the simulator. Regarding jobs submitted to the grid, a complete specification of job characteristics is possible. In addition to standard characteristics such as release time, jobs can also have their own requirements on resources that can solve them.

Parameters of the Simulator The simulator is highly parameterized, making it possible to simulate realistically grid systems. The parameters used in the simulator follow.

```
Usage: sim [options] Options:
  -n, --ttasks       <integer>    Total number of tasks (jobs)
  -b, --itasks       <integer>    Initial number of tasks
  -i, --iatime       <distrib>    Task mean interarrival time
  -w, --workload     <distrib>    Task mean workload (MI)
  -o, --ihosts       <integer>    Initial number of hosts
  -m, --mips         <distrib>    Host mean (MIPS)
  -a, --addhost      <distrib>    Add-host event distribution
  -d, --delhost      <distrib>    Del-host event distribution
  -f, --minhosts     <integer>    Minimum number of hosts
  -g, --maxhosts     <integer>    Maximum number of hosts
  -r, --reschedule                Reschedule uncompleted tasks
  -s, --strategy     <meta_p>     Scheduler meta policy
  -x, --hostselect   <host_p>     Host selection policy
  -y, --taskselect   <task_p>     Task selection policy
  -z, --activate     <wake_p>     Scheduler activation policy
  -l, --allocpolicy  <local_p>    Host local scheduling policy
  -1, --nruns        <integer>    Number of runs
  -2, --seed         <integer>    Random seed
  -3, --thrmach      <double>     Threshold of machine type
  -4, --thrtask      <double>     Threshold of task type
  -5, --inmbatch     <double>     Threshold for immediate
                                  and batch
  -t, --trace        <filename>   Enable trace on output file
  -h, --help                      Shows this help
```

Output of the Simulator For a setting of input parameters, the simulator outputs the following (statistical) values:

- `Makespan`: the finishing time of the latest job
- `Flowtime`: sum of finishing times of jobs
- `Total Potential Time`: sum of available times of machines
- `Total Idle Time`: sum of idle times of machines
- `Total Busy Time`: sum of busy time of machines
- `Total Lost Time`: sum of execution times lost due to resource dropping from the system
- `Host Utilization`: the quotient of host utilization
- `Number Hosts`: number of hosts (idle or not) available in the system
- `Number Free Hosts`: number of idle hosts available in the system
- `Global Queue Length`: number of jobs pending to schedule
- `Waiting Time`: time period between the time when a job enters the system and starts execution in a grid resource
- `Schedules Per Task`: number of times a job has been scheduled

- `Number Scheduler Calls`: number of times the scheduler has been invoked
- `Number Activations`: number of times scheduling took place (the scheduler is activated and job planning took place)
- `Scheduler Activation Interval`: time interval between two successive activations of the scheduler

24.5.2 Coupling of Schedulers with the Simulator

Unlike other simulation packages in which the scheduling policies are *embedded* in the simulator, in HyperSim-G the scheduling algorithms are completely separated from the simulator, which need not to know the implementation of the specific scheduling methods. This requirement regarding scheduling is achieved through a "refactoring" design and using new classes and methods, as shown in Figure 24.1.

HyperSim-G includes several classes for scheduling. We distinguish here the `Scheduler` class, which, in turn, uses four other classes: (a) `TaskSelectionPolicy`: this class implements job selection policies; for instance, `SelectAllTasks`; (b) `HostSelectionPolicy`: this class implements resource selection policies; for instance, `SelectAllHosts`, `SelectIdleHosts`; (c) `SchedulerActivacion`: this class implements different policies for launching the scheduler (e.g., `TimeIntervalActivacion`); and (d) the class `SchedulingPolicy`, which

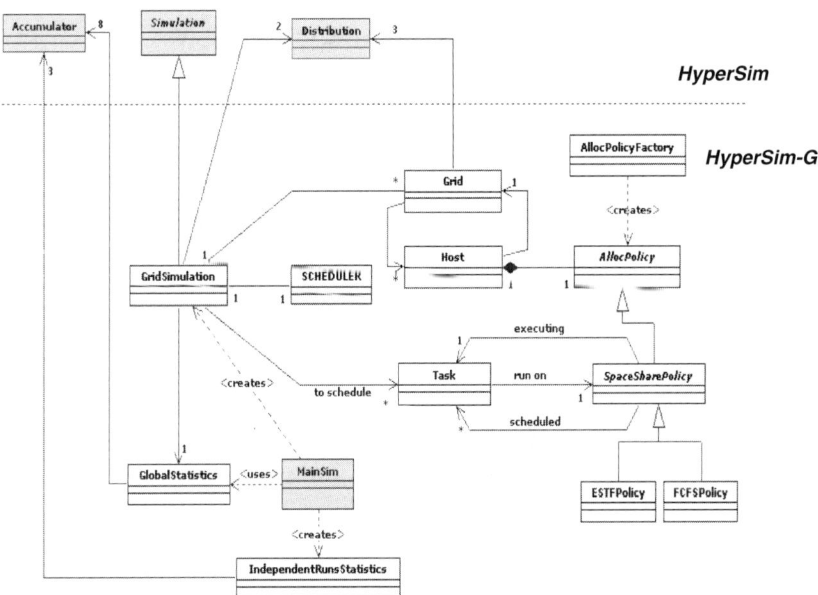

Figure 24.1 Scheduling entities in HyperSim-G.

serves as a basis for implementing different scheduling methods. It is through this class that any scheduling method could be *plugged in* the scheduler; the simulator need not know about the implementation of specific scheduling methods, such as GA-based schedulers.

24.6 INTERFACE FOR USING A GA-BASED SCHEDULER WITH THE GRID SIMULATOR

The GA-based scheduler is implemented using the skeleton design of Alba et al. [3]. However, it is not possible to plug in the scheduler to the simulator in a straightforward way since the skeleton interface and the SchedulingPolicy interface in the simulator cannot "recognize" each other. This is, in fact, the most common situation since the scheduler needs not to know about the simulator, and vice versa, the simulator needs not to know the implementation of the scheduler. Therefore, the simulator, as a stand-alone application, has been extended using the the adapter design pattern to support the use of external schedulers without changing the original definition of the simulator class.

GAScheduler Class This class inherits and implements the SchedulingPolicy interface. This class is able to "connect" the behavior of the GA-based scheduler with the simulator. In Figure 24.2 we show the class diagram using the adapter pattern.

Essentially, it is through the constructor of the `GAScheduler` and the virtual method `schedule` that the parameters and the problem instance are passed to any of the subclasses implementing the `SchedulingPolicy` interface. It should be noticed that the simulator will now act as a grid resource broker, keeps information on the system's evolution, and is in charge of planning the jobs dynamically as they arrive at the system.

Taking into account the dynamic nature of the grid is very important here. A new activation policy of the scheduler (`ResourceAndTimeActivation`) has been defined, which generates an event schedule: (a) each time a resource joins the grid or a resource is dropped from the grid; or (b) activate the schedule at each time interval to plan the jobs entered in the system during that time interval. Moreover, since we also aim to optimize the flowtime, a new local scheduling policy, called `SPTFPolicy` (shortest-processing-time-first policy), has been defined.

Use of the Simulator with the GA-Based Scheduler Currently, the simulator with the GA-based scheduler can be used from a command line by using the option `-s` (scheduling strategy) as follows:

```
-s GA_scheduler(t,s)
```

which instructs to the simulator to run the GA-based scheduler for t seconds in simultaneous optimization mode.

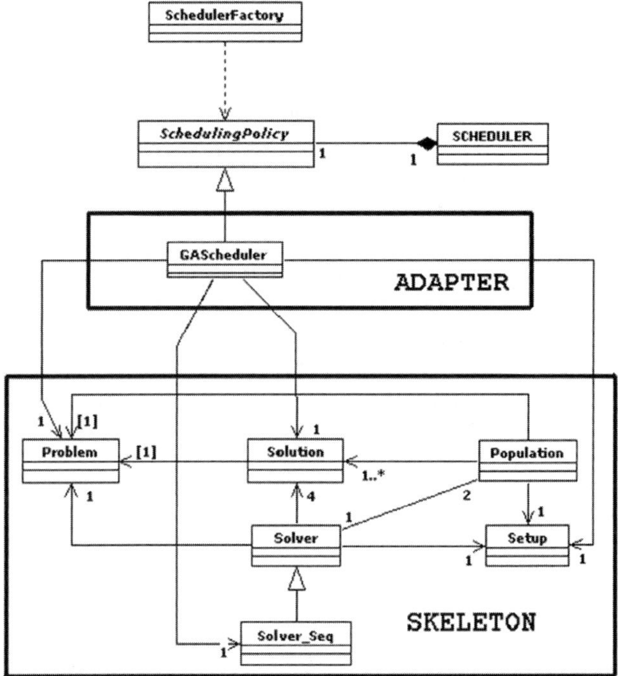

Figure 24.2 Adapter `GAScheduler` class.

24.7 EXPERIMENTAL ANALYSIS

In this section we present some computational results for evaluation of the performance of the GA-based schedulers used in conjunction with the grid simulator. Notice that the experimental study is conducted separately for each of the four GA-based schedulers: elitist generational GA with hierarchic optimization, elitist generational GA with simultaneous optimization, steady-state GA with hierarchic optimization, and steady-state GA with simultaneous optimization. The basic GA configuration shown in Table 24.1 is used for all of them.

24.7.1 Parameter Setting

Two cases have been considered for measuring the performance of the GA-based scheduler: the static and dynamic. In the first one, the planning of jobs is done statically; that is, no new jobs are generated, and the available resources are known beforehand. Essentially, this is using the simulator as a static benchmark of instances. The static case was helpful to fine-tune the parameters. In the dynamic case, which is the most relevant, both jobs and resources may vary over time.

For both static and dynamic cases, the performance of the GA-based scheduler is studied for four grid sizes: small, medium, large, and very large. The number of initial number of hosts and jobs in the system for each case is specified below for the static and dynamic cases, respectively.

GA Configuration We report here computational results for makespan and flowtime values obtained with two versions of GAs—the elitist generational GA and steady-state GA—and for each of them, both simultaneous and hierarchic optimization modes were considered.

The GA configuration used is shown in Table 24.1.

Notice that both GA-based schedulers are run for the same time, which in the static case is about 40 seconds and in the dynamic case is about 25 seconds. A total of 30 independent runs were performed; the results reported are averaged and the standard deviation is shown. The simulations were done in the same machine (Pentium-4 HT, 3.5 GHz, 1-Gbyte RAM, Windows XP) without other programs running in the background.

24.7.2 Parameters Used in a Static Setting

In the static setting the simulator is set using the parameter values [`sptf` stands for shortest-processing-time-first policy and `n(.,.)` denotes the normal distribution] in Table 24.2. Observe that in this case, the number of hosts and jobs is as follows: small (32 hosts, 512 jobs), medium (64 hosts, 1024 jobs), large (128 hosts, 2048 jobs), and very large (256 hosts, 4096 jobs).

In Tables 24.3, 24.4, and 24.5 we show the results for makespan, flowtime, and resource utilization for both versions of GA-based scheduler, and for each of them for both optimization modes. The confidence interval is also shown. A 95% CI means that we can be 95% sure that the range of makespan (flowtime,

TABLE 24.1 GA Configuration

	Elitist Generational	Steady State
nb_evolution_steps	5 * nbr_tasks	20 * nbr_tasks
population_size	$\lceil(\log_2(\text{nbr\_tasks}))^2 - \log_2(\text{nbr\_tasks})\rceil$	$4(\lceil\log_2(\text{nbr\_tasks})\rceil - 1)$
intermediate_population_size	population_size −2	population_size / 3
select_choice	SelectLinearRanking	
cross_choice	CrossCX	
cross_probability	0.80	1.00
mutate_choice	MutateRebalancing ($p_{m'} = 0.65$)	
mutate_probability	0.40	
mutate_extra_parameter	0.60	
replace_only_if_better	false	
replace_generational	false	
start_choice	MCT + LJFR_SJFR	
max_time_to_spend	40secs (static)/25secs (dynamic)	

TABLE 24.2 Setup of the Simulator in the Static Case

	Small	Medium	Large	Very Large
Init./total hosts	32	64	128	256
MIPS	n(1000, 175)			
Init./total tasks	512	1024	2048	4096
Workload	n(250000000, 43750000)			
Host selection	all			
Task selection	all			
Local policy	sptf			
Number of runs	30			

TABLE 24.3 Makespan Values (Arbitrary Time Units) for Two GA-Based Schedulers in a Static Setting

Makespan ± % C.I (0.95)	Small	Medium	Large	Very Large
Elitist generational (hierarchic)	3,975,630.270 ± 0.5714 %	3,986,741.953 ± 0.7950%	4,006,153.308 ± 0.9351%	4,0380,90.107 ± 1.1843%
Steady stats (hierarchic)	3,972,614.960 ± 0.6421%	3,979,528.765 ± 0.7714%	3,986,350.175 ± 0. 8781%	3,999442.957 ± 1. 0614%
Elitist generational (simultaneous)	3,999,566.762 ±0.7442%	4,007,250.052 ±0.8810%	4,021,509.314 ±1.1750%	4,057,448.273 ±1.5132%
Steady state (simultaneous)	3,989,635.184 ±0.7250%	3,993,724.920 +0.8905%	4,001,873.911 ±1.0833%	4,0149,65.078 ±1.3244%

TABLE 24.4 Flowtime Values for Two GA-Based Schedulers in a Static Setting

Flowtime ± % C.I (0.95)	Small	Medium	Large	Very Large
Elitist generational (hierarchic)	1,044,100,231.581 ±0.8697%	2,096,726,420.714 ±0.5659%	4,180,492,294.944 ±0.3958%	8,324,168,257.148 ±0.3812%
Steady state (hierarchic)	1,044,336,909.272 ±0.8623%	2,097,576,711.231 ±0.5738%	4,188,877,533.028 ±0.3651%	8,317,263,285.963 ±0.3415%
Elitist generational (simultaneous)	1,031,700,126.516 ±0.6638%	2,083,420,293.179 ±0.4290%	4,156,874,524.081 ±0.3465%	8,301,718,834.510 ±0.3522%
steady state (simultaneous)	1,034,719,380.878 ±0.7180%	2,083,408,328.335 ±0.4325%	4,155,161,769.915 ±0.3390%	8,296,717,932.193 ±0.3206%

resource utilization) values would be within the interval shown if the experiment were to run again.

24.7.3 Evaluation of the Results

As can be seen from Tables 24.3, 24.4, and 24.5, the steady-state GA in its hierarchic version shows the best performance for makespan values, and the confidence interval (CI) increases slowly with increase in the grid size. It can

TABLE 24.5 Resource Utilization Values for Two GA-Based Schedulers in a Static Setting

Resource Utilization ±% C.I(0.95)	Small	Medium	Large	Very Large
Elitist generational (hierarchic)	0.9991 ±0.0105%	0.9936 ±0.0392%	0.9907 ±0.0652%	0.9829 ±0.1021%
Steady state (hierarchic)	0.9995 ±0.0120%	0.9973 ±0.0273%	0.9954 ±0.0505%	0.9933 ±0.0861%
Elitist generational (simultaneous)	0.9885 ±0.0295%	0.9895 ±0.0424%	0.9802 ±0.0741%	0.9780 ±0.1958%
Steady state (simultaneous)	0.9945 ±0.0244%	0.9939 ±0.0410%	0.9905 ±0.0892%	0.9842 ±0.1108%

also be observed that the CI is larger for simultaneous optimization than for hierarchic optimization.

Regarding the flowtime, the steady-state GA in its simultaneous version shows the best performance. It should be noted that the value of flowtime is not affected by the average number of jobs assigned to a resource since it depends on the total number of jobs to be assigned. We also observed that for small grid size, the elitist generational GA in its simultaneous version obtained the best flowtime reductions.

Finally, regarding the resource utilization, we observed that the resource utilization is very good in general. The best values are obtained by the steady-state GA in its hierarchic version.

24.7.4 Parameters Used in a Dynamic Setting

In the dynamic setting the simulator is set using the parameter values in Table 24.6. Observe that in this case, the *initial* number of hosts and jobs is as follows: small (32 hosts, 384 jobs), medium (64 hosts, 768 jobs), large (128 hosts, 1536 jobs), and very large (256 hosts, 3072 jobs).

In Tables 24.7, 24.8, and 24.9 we show the results for makespan, flowtime, and resource utilization for both versions of GA-based scheduler and for each of them for both optimization modes.

Evaluation of the Results As can be seen from Tables 24.7, 24.8, and 24.9, for makespan values, steady-state GA again showed the best performance. However, for small and medium-sized grids, the makespan reductions are worse than in the static case. This could be explained by the fact that in the dynamic case changes in the available resources cause changes in the computational capacity, which in turn implies losts of execution time (due to the fluctuations). On the other hand, elitist generational GA showed to be more sensible to the grid size increment and does not make good use of the increment of the computing capacity of the grid. The simultaneous versions again showed worse performance; nonetheless, they performed

TABLE 24.6 Setup of the Simulator in the Dynamic Case

	Small	Medium	Large	Very Large
Init. hosts	32	64	128	256
Max. hosts	37	70	135	264
Min. hosts	27	58	121	248
MIPS	n(1000, 175)			
Add host	n(625,000, 93,750)	n(562,500, 84,375)	n(500,000, 75,000)	n(437,500, 65,625)
Delete host	n(625,000, 93,750)			
Total tasks	512	1024	2048	4096
Init. tasks	384	768	1536	3072
Workload	n(250,000,00, 43,750,000)			
Interrarival	e(7812.5)	e(3906.25)	e(1953.125)	e(976.5625)
Activation	resource_and_time_interval (250,000)			
Reschedule	true			
Host select	all			
Task select	all			
Local policy	sptf			
Number of runs	15			

TABLE 24.7 Makespan Values (Arbitrary Time Units) for Two GA-Based Schedulers in a Dynamic Setting

Makespan ± % C.I (0.95)	Small	Medium	Large	Very Large
Elitist generational (hierarchic)	4,0481,481,52.901 ±0.7560%	3,988,204.133 ±0.8501%	3,992,023.624 ±1.0724%	4,0481,16,478.205 ±1.7805%
Steady state (hierarchic)	4,051,858.600 ±0.8109%	3,979,957.581 ±0.7216%	3,981,408.541 ±0.9240%	3,978,104.733 ±1.4477%
Elitist generational (simultaneous)	4,062,331.154 ±0.8102%	3,999,261.610 ±0.9912%	4,015,066.055 ±1.4350%	4,041,820.132 ±1.9363%
Steady state (simultaneous)	4,063,425.465 ±0.8225%	3,994,804.904 ±0.8905%	3,995,161.988 ±1.2805%	4,0481,09,852.021 ±1.8390%

TABLE 24.8 Flowtime Values for Two GA-Based Schedulers in a Dynamic Setting

Flowtime ± % C.I (0.95)	Small	Medium	Large	Very Large
Elitist Generational (hierarhic)	1,044,100,231.581 ±0.8697%	2,096,726,420.714 ±0.5659%	4,180,492,294.944 ±0.3958%	8,324.168,257.148 ±0.3812%
Steady State (hierarchic)	1,044,336,909.272 ±0.8623%	2,097,576,711.231 ±0.5738%	4,188,877,533.028 ±0.3651%	8,317,263,285.963 ±0.3415%
Elitist generational (simultaneous)	1,031,700,126.516 ±0.6638%	2,083,420,293.179 ±0.4290%	4,156,874,524.081 ±0.3465%	8,301,718,834.510 ±0.3522%
Steady state (simultaneous)	1,0347,19,380.878 ±0.7180%	2,083,408,328.335 ±0.4325%	4,155,161,769.915 ±0.3390%	8,296,717,932.193 ±0.3206%

TABLE 24.9 Resource Utilization Values for Two GA-Based Schedulers in a Dynamic Setting

Resource Utilization ± % C.I (0.95)	Small	Medium	Large	Very Large
Elitist generational (hierarhic)	0.9991 ±0.0105%	0.9936 ±0.0392%	0.9907 ±0.0652%	0.9829 ±0.1021%
Steady state (hierarchic)	0.9995 ±0.0120%	0.9973 ±0.0273%	0.9954 ±0.0505%	0.9933 ±0.0861%
Elitist generational (simultaneous)	0.9885 ±0.0295%	0.9895 ±0.0424%	0.9802 ±0.0741%	0.9780 ±0.1958%
Steady state (simultaneous)	0.9945 ±0.0244%	0.9939 ±0.0410%	0.9905 ±0.0892%	0.9842 ±0.1108%

better than in the static case. Similar observations as in the makespan case hold for the flowtime. Finally, regarding resource utilization, it seems that GA-based schedulers for high dynamic environments are not able to obtain high resource utilization, which is due to not-so-good load balancing achieved by GAs in the dynamic setting. Again steady-state GA showed the best performance.

24.8 CONCLUSIONS

In this work we have presented an approach for experimental evaluation of metaheuristics-based schedulers for independent job scheduling using a grid simulator. In our approach, the metaheuristic schedulers can be developed independently from the grid simulator and can be coupled with the grid simulator using an interface, which allow communication between the scheduler and the simulator.

We have exemplified the approach for the experimental evaluation of four GA-based schedulers in a dynamic environment: elitist generational GA (in both hierarchic and simultaneous optimization mode) and steady-state GA (in both hierarchic and simultaneous optimization mode). The results of the study showed, on the one hand, the usefulness and feasibility of using the grid simulator for experimental evaluation of GA-based schedulers in a dynamic setting and, on the other hand, that steady-state GA performed best for both makespan and flowtime objectives in its hierarchic and simultaneous versions, respectively.

Acknowledgments

This research is partially supported by projects ASCE TIN2005-09198-C02-02, FP6-2004-ISO-FETPI (AEOLUS), MEC TIN2005-25859-E, and MEFOALDISI (Métodos formales y algoritmos para el diseño de sistemas) TIN2007-66523.

REFERENCES

1. A. Abraham, R. Buyya, and B. Nath. Nature's heuristics for scheduling jobs on computational grids. In *Proceedings of the 8th IEEE International Conference on Advanced Computing and Communications*, Cochin, India, 2000, pp. 45–52.
2. B. Adenso-Díaz and M. Laguna. Fine tuning of algorithms using fractional experimental designs and local search. *Operations Research*, 54(1):99–114, 2006.
3. E. Alba, F. Almeida, M. Blesa, C. Cotta, M. Díaz, I. Dorta, J. Gabarró, C. León, G. Luque, J. Petit, C. Rodríguez, A. Rojas, and F. Xhafa. Efficient parallel LAN/WAN algorithms for optimization: the MALLBA project. *Parallel Computing*, 32(5–6):415–440, 2006.
4. R. S. Barr, B. L. Golden, J. Kelly, W. R. Stewart, and M. G. C. Resende. Designing and reporting computational experiments with heuristic methods. *Journal of Heuristics*, 1(1):9–32, 2001.
5. T. Braun, H. Siegel, N. Beck, L. Boloni, M. Maheswaran, A. Reuther, J. Robertson, M. Theys, and B. Yao. A comparison of eleven static heuristics for mapping a class of independent tasks onto heterogeneous distributed computing systems. *Journal of Parallel and Distributed Computing*, 61(6):810–837, 2001.
6. R. E. Burkard, S. E. Karisch, and F. Rendl. QAPLIB: a quadratic assignment problem library. *Journal of Global Optimization*, 10(4):391–403 1997. http://www.opt.math.tu-graz.ac.at/qaplib/.
7. R. Buyya and M. M. Murshed. GridSim: a toolkit for the modeling and simulation of distributed resource management and scheduling for grid computing. *Concurrency and Computation: Practice and Experience*, 14(13-15):1175–1220, 2002.
8. H. Casanova. SimGrid: A toolkit for the simulation of application scheduling, In *Proceedings of the First IEEE/ACM International Symposium on Cluster Computing and the Grid (CCGrid'01)*, Brisbane, Australia, May 15–18, 2001, pp. 430–437.
9. S. P. Coy, B. L. Golden, G. C. Runer, and E. A. Wasil. Using experimental design to find effective parameter settings for heuristics. *Journal of Heuristics*, 7:77–97, 2000.
10. J. Carretero and F. Xhafa. Using genetic algorithms for scheduling jobs in large scale grid applications. *Journal of Technological and Economic Development: A Research Journal of Vilnius Gediminas Technical University*, 12(1):11–17, 2006.
11. C. Dumitrescu and I. Foster. GangSim: a simulator for grid scheduling studies. In *Proceedings of the 5th International Symposium on Cluster Computing and the Grid (CCGrid'05)*, Cardiff, UK. IEEE Computer Society Press, Los Alamitos, CA, 2005, pp. 1151–1158.
12. A. Freville and G. Plateau. Hard 0–1 multiknapsack test problems for size reduction methods. *Investigation Operativa*, 1:251–270, 1990.
13. J. H. Holland. *Adaptation in Natural and Artificial Systems*. University of Michigan Press, Ann Arbor, MI, 1975.
14. D. S. Johnson. A theoretician's guide to the experimental analysis of algorithms. In M. Goldwasser, D. S. Johnson, and C. C. McGeoch, eds., *Proceedings of the 5th and 6th DIMACS Implementation Challenges*. American Mathematical Society, Providence, RI, 2002, pp. 215–250.

15. L. Kang, A. Zhou, B. MacKay, Y. Li, and Zh. Kang. Benchmarking algorithms for dynamic travelling salesman problems. In *Proceedings of Evolutionary Computation (CEC'04)*, vol. 2, 2004, pp. 1286–1292.
16. D. Klusácek, L. Matyska, H. Rudová, R. Baraglia, and G. Capannini. Local search for grid scheduling. Presented at the Doctoral Consortium at the International Conference on Automated Planning and Scheduling (ICAPS'07), Providence, RI, 2007.
17. J. Kivijärvi, P. Fränti, and O. Nevalainen. Self-adaptive genetic algorithm for clustering. *Journal of Heuristics*, 9:113–129, 2003.
18. *k*-Tree benchmark. http://iridia.ulb.ac.be/~cblum/kctlib/.
19. A. Madureira, C. Ramos, and S. Silva. A genetic approach for dynamic job-shop scheduling problems. In *Proceedings of the 4th Metaheuristics International Conference (MIC'01)*, 2001, pp. 41–46.
20. M. Maheswaran, S. Ali, H. Siegel, D. Hensgen, and R. Freund, Dynamic mapping of a class of independent tasks onto heterogeneous computing systems. *Journal of Parallel and Distributed Computing*, 59(2):107–131, 1999.
21. K. Ranganathan and I. Foster. Simulation studies of computation and data scheduling algorithms for data grids. *Journal of Grid Computing*, 1(1):53–62, 2003.
22. R. L. Rardin and R. Uzsoy. Experimental evaluation of heuristic optimization algorithms: a tutorial. *Journal of Heuristics*, 7:261–304, 2001.
23. G. Ritchie. Static multi-processor scheduling with ant colony optimisation and local search. Master's thesis, School of Informatics, University of Edinburgh, UK, 2003.
24. G. Ritchie and J. Levine. A fast, effective local search for scheduling independent jobs in heterogeneous computing environments. *TechRep*. Centre for Intelligent Systems and Their Applications, University of Edinburgh, UK, 2003.
25. G. Ritchie and J. Levine. A hybrid ant algorithm for scheduling independent jobs in heterogeneous computing environments. Presented at the 23rd Workshop of the UK Planning and Scheduling Special Interest Group (*PLANSIG'04*), 2004.
26. H. J. Song, X. Liu, D. Jakobsen, R. Bhagwan, X. Zhang, K. Taura, and A. Chien. The microgrid: a scientific tool for modeling computational grids. *Journal of Science Programming*, 8(3):127–141, 2000.
27. A. Takefusa, S. Matsuoka, H. Nakada, K. Aida, and U. Nagashima. Overview of a performance evaluation system for global computing scheduling algorithms. In *Proceedings of the High Performance Distributed Conference*, IEEE Computer Society, Washington, DC, 1999, pp. 11–17.
28. Travelling salesman problem benchmark. http://elib.zib.de/pub/mp-testdata/tsp/tsplib/tsplib.html.
29. F. Xhafa, E. Alba, B. Dorronsoro, and B. Duran. Efficient batch job scheduling in grids using cellular memetic algorithms. *Journal of Mathematical Modelling and Algorithms* (online), Feb. 2008.
30. F. Xhafa, J. Carretero, and A. Abraham. Genetic algorithm based schedulers for grid computing systems. *International Journal of Innovative Computing, Information and Control*, 3(5):1–19, 2007.
31. F. Xhafa, J. Carretero, L. Barolli, and A. Durresi, A. Requirements for an event-based simulation package for grid systems. *Journal of Interconnection Networks*, 8(2):163-178, 2007.

32. A. YarKhan and J. Dongarra. Experiments with scheduling using simulated annealing in a grid environment. In *Proceedings of GRID'02*, 2002, pp. 232–242.

33. A. Y. Zomaya and Y. H. Teh. Observations on using genetic algorithms for dynamic load-balancing. *IEEE Transactions on Parallel and Distributed Systems*, 12(9):899–911, 2001.

CHAPTER 25

Remote Optimization Service

J. GARCÍA-NIETO, F. CHICANO, and E. ALBA

Universidad de Málaga, Spain

25.1 INTRODUCTION

Optimization problems are commonly tackled every day in the academic, industrial, and private domains, but only a few use optimization algorithms for solving their problems, especially in the industrial domain. The rest do not use optimization techniques because they do not know these techniques, they do not have enough knowledge to implement them or working with libraries, or they do not have the hardware resources required to work with the techniques.

In the literature we can find a plethora of libraries of optimization algorithms implemented in different programming languages [21], all with their own advantages and drawbacks. A common characteristic in almost all current related works is the use of *XML* [20], which facilitates the development of applications in such tasks as new algorithm creation, automatic code generation from XML specifications, integration of algorithms in services for its execution, and XML serialization of methods and objects. However, these libraries can only be used by people knowing the programming language in which the library is implemented, and even in this way, some time is required to be familiar with the algorithms.

In recent years, systems that integrate libraries of metaheuristic, exact, and hybrid algorithms have appeared. In addition, it is always interesting to offer methodologies and services for making the use of software repositories easier in order to make accessible algorithms, such as optimization algorithms, that can be beneficial for a lot of people in the world. In this context, ROS, the *remote optimization service* [3,10], ROS (Remote Optimization Service) provides a service for the remote execution of optimization algorithms through Internet. ROS is able to manage the access and execution of optimization algorithms, completely different themselves, and probably implemented in different programming languages.

Optimization Techniques for Solving Complex Problems, Edited by Enrique Alba, Christian Blum, Pedro Isasi, Coromoto León, and Juan Antonio Gómez

The rest of the chapter is organized as follows. The foundations of ROS and the state of the art of similar services are depicted in Section 25.2. Section 25.3 describes the ROS architecture, entities, components, and main features. In Section 25.4 the technology used in the communication layer of ROS is detailed, and the XML document specification used for the data exchange is introduced in Section 25.5. Section 25.6 covers the mechanism for encapsulating algorithms through *Wrappers*. Finally, in Section 25.7 the performance of ROS is evaluated using different optimization problems and algorithms, and Section 25.8 concludes the chapter.

25.2 BACKGROUND AND STATE OF THE ART

The objective of an optimization algorithm is to find the best of all possible solutions to a problem. Research in optimization has been very important in computer science from the beginning. Optimization is a very dynamic research field because new challenges appear every day: new engineering problems, new industrial situations, new services that need optimization, and a long list of other situations that constantly challenge the existing optimization techniques [11].

ROS was developed with the objective of establishing an optimization service on the Internet, that is, for easing the access and use of multiple optimization algorithms. In this way, different algorithms developed by different developers in different programming languages are grouped in a repository reachable from anywhere in the world through the Internet. In ROS we distinguish two types of actors: the *developer*, who provides algorithms and includes them in the system, and the *user*, who can interact with the available algorithms in a remote way for solving optimization problems.

Several recent initiatives exist presenting common features with ROS. In Table 25.1 we present these state-of-the-art systems together with some of their features. In particular, we show the operation mode (local or distributed) if they use XML for storing or exchanging information, the programming language, the way in which the algorithms are specified, and if they have a graphical user interface. In the last row we present the features of ROS in order to offer an early idea of its properties, allowing a fast comparison against existing proposals. In the following we are going to highlight the more interesting features of each system.

EAML is a language [1,20] based on XML that can be used for specifying evolutionary algorithms using tags defined in a dictionary (DTD [12]). From the EAML specification of the algorithm it is possible to generate C++ source code using a compiler called ECC (evolutionary computation compiler). EAML does not offer generic tags for extending the language and introducing new elements: When a new element has to be included in the language, its dictionary (DTD) must be modified. OPEAL is a library that includes evolutionary algorithms implemented using an extended version of EAML [14]. This library has the possibility of performing distributed executions using SOAP [15].

TABLE 25.1 Brief Summary of State-of-the-Art Systems Related to ROS

System	Year	Description	Operation	Use of XML	Prog. Lan.	Algorithm Specification	GUI
EAML + ECC [20]	2000	XML Specification and Compiler	Local	Yes	XML and C	Restricted to XML design	No
OPEAL [14]	2002	Library	Local and distributed	EAML SOAP	Perl	Restricted to prog. lang.	No
eaWeb [9]	2002	Web Service	Local	Yes	ASPX	Restricted to XML design	Yes
eaLib [16]	2002	Java API	Local	No	Java	Restricted to API classes	No
EA-Visualizer [7]	2001	Application	Local	No	Java	Restricted to available options	Yes
Evolvica [17]	2002	API and Code Editor	Local	No	Java	Restricted to API classes	Yes
ROS	2004	Distributed application	Local and distributed	Yes	Any	I/O format Wrapper	Yes

The web service *eaWeb* [9] offers an optimization service based on evolutionary algorithms. The user must specify the objective function and the parameters of the algorithm using XML documents from a web browser, and the service returns the results of the optimization process. The underlying idea of the system is similar to that of ROS, but unlike the latter, eaWeb restricts the types of algorithms, data structures, and parameters to use.

Another available tool is *eaLib*, a Java class library with which a developer with evolutionary computation knowledge can easily implement evolutionary algorithms. In addition, it includes a serialization mechanism for generating XML documents from solution collections. eaLib is not oriented to the remote execution of algorithms, but to the design of new evolutionary algorithms.

Evolvica is a successor of eaLib that provides a graphical user interface (GUI) for the development of evolutionary algorithms. This system is based on the *Eclipse* platform [2] and includes a Java editor, a debugger, and a visual results analyzer. It is aimed at expert users with knowledge of the Java language and evolutionary computation.

Finally, *EAVisualizer* [7] is another application implemented in Java. In this case the user does not need knowledge of Java language or of other programming languages, since in this system it is possible to design an evolutionary algorithm by interactively selecting the various components of the algorithm (operators, individual representation, etc.) from components defined previously.

Our proposal, ROS, has been designed to put together all the advantages of the systems mentioned. On the one hand, as in EAML and OPEAL, XML documents are used in ROS. However, unlike these systems, in this case they contain data involved in execution of the algorithm: parameters, results, execution orders, and so on. Instead of using languages for specifying the evolutionary algorithms, ROS incorporates algorithms already implemented by different developers. It perceives the algorithms as black boxes, where it introduces parameters and receives results, either locally or from a remote execution. Thanks to this feature, ROS is capable of working with many heterogeneous algorithms by different programmers and allocated remotely.

In all the systems presented (except *eaWeb*), the algorithms must be implemented in Java using the classes and components provided. This restriction is an important drawback for developers who want to include an algorithm implemented in a different programming language. The restriction to the use of Java is that the set of algorithms available in the system is a subset of those that could be included if no restriction would exist. To overcame this drawback, ROS provides a mechanism based on the *wrapper design pattern*, which allows it to include algorithms implemented in any programming language. All these algorithms can be used simultaneously in a transparent way for the client. For example, a genetic algorithm implemented in C++ can be solving an optimization problem for a user at the same time that an evolutionary strategy implemented in Modula 2 is solving another optimization problem for a second user.

In addition to the features mentioned, ROS also offers graphical user interfaces to ease the interaction with the system, as *EAVisualizer*, *Evolvica*, and *eaWeb* do.

25.3 ROS ARCHITECTURE

To provide a multiplatform and distributed nature to ROS, it was implemented in Java language and is composed of a set of entities interconnected using the client/server (C/S) paradigm. The architecture of ROS is based on four types of entity:

1. The *client* (see Figure 25.1) is an application that offers a graphical interface to users for interacting with the system. With this interface the users can carry out administrative operations (such as registration and authentication) as well as operations related to resolution of the optimization problems: selection of the algorithms to be applied, configuration of the algorithms selected (parameter setting, machines to run, definition of the objective function, etc.), execution control, and results collection.

2. The *primary server* is responsible for user authentication. It also offers information about algorithms and machines available to the user. By means of XML configuration documents, the primary server stores the information related to the algorithms implemented in each distribution server.

3. The *distribution server* (*server* in Figure 25.1) is a bridge between the client and the machine running the algorithm (worker). The distribution server is therefore a fundamental piece to make ROS scalable, flexible, and distributed in a network of workstations, since it hides the real location of the workers and controls the load balancing among them.

4. The *worker* houses the algorithms and executes them according to the orders specified by the client and routed through the distribution server. Once

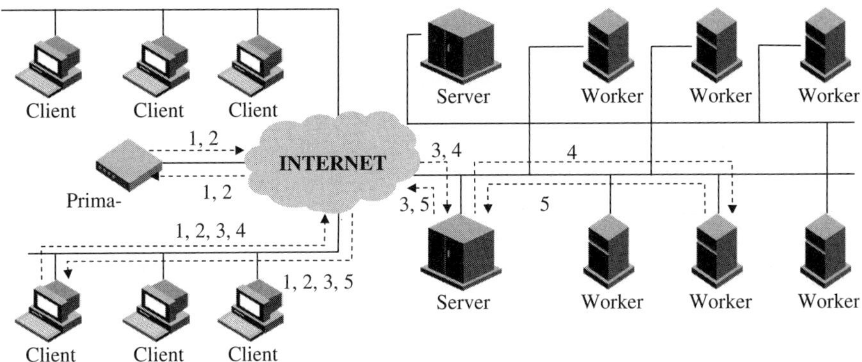

Figure 25.1 ROS architecture.

an execution is finished, it sends the results to the distribution server, which forwards them to the client.

Once we have presented the four entities of the system, we are going to detail the data flow sequence through the system in a normal operation. We follow the numeration and direction of dashed lines in Figure 25.1.

1. *User authentication*. First, the user fills the initial form of the ROS client GUI panel with his or her identification and password, and then it connects to the primary server. Once the user is authenticated in the system, the primary server returns a message of acceptance or rejection.

2. *Selection of algorithm, server, and programming language*. In this phase, the primary server sends a list of available algorithms, distribution server addresses, and programming languages. The client receives this information and shows it to the user, who makes a selection.

3. *Algorithm configuration*. The client, based on the algorithm, server, and programming language selected in the previous step, asks the distribution server the specific parameters and configuration options that the user must fill in order to run the algorithm. These parameters and options are specified in a XML document with empty tags that the client must fill with the data provided by the user. Thus, depending on the XML tags, the client generates a specific GUI panel view for each algorithm selected.

4. *Algorithm execution*. Once the user sets the parameters of the algorithm chosen by means of the visual panel, the XML document is filled by the client and subsequently, sent to the distribution server. This server sends the XML document to the suitable worker, which starts execution of the algorithm.

5. *Results collection*. After the end of algorithm execution, the results are sent in an XML document to the distribution server, which forwards them to

the client. Finally, the client shows the user the results of such execution in the visual panel.

One of the most important issues that the user is required to specify is the fitness function. To do this, the user must provide a function implemented in the same programming language as the algorithm selected and with a given name and signature determined by the algorithm. This function is updated together with the configuration options to the distribution server, which forwards it to the worker. In the worker, the function is compiled together with the algorithm, following the compiling instructions given by the developer of the algorithm.

25.4 INFORMATION EXCHANGE IN ROS

Due to the distributed nature of ROS, it needs to exchange information among its entities, which generally are deployed in different machines. Two versions of ROS are available, depending on the communication interface used: RPC with SOAP and sockets. In the SOAP version, information is exchanged by means of remote procedure calls between web services. The distribution server communicates with the process servers via RPC/SOAP, and the client communicates in the same way with the distribution server (see Figure 25.2). The advantage of using SOAP is that the different entities of ROS can run as web services and can be distributed geographically in a WAN (wide area network) with minimum risk that the communications will be cut by firewalls and other tools for implementing security policies.

The requirement to install web servers for remote object registry implies an additional complexity and computational charge in the system. For this reason,

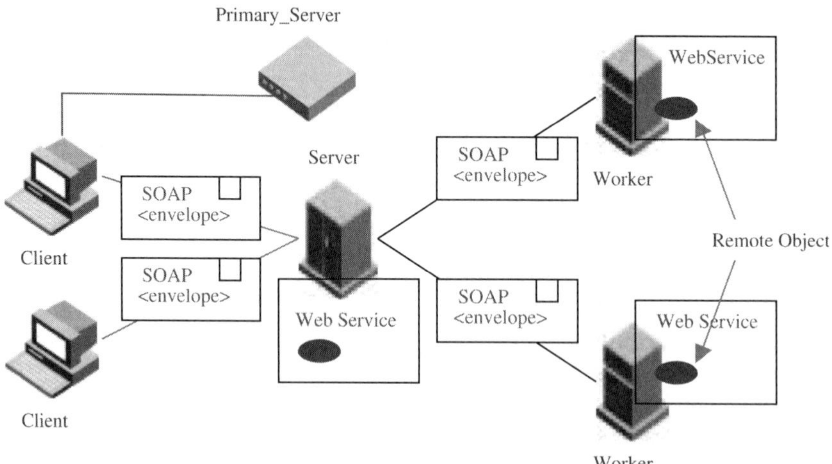

Figure 25.2 Summarized scheme of the SOAP version of ROS.

an implementation of ROS using Java sockets for the communication was also developed, making the communication lighter (since sockets are in a lower level with respect to RPC/SOAP) and avoiding an excessive computational charge. To avoid connectivity problems (due to firewalls, for example), this version of ROS should use the TCP 80 port. In the current implementation we use the TCP port 4000.

25.5 XML IN ROS

The main advantage of XML is that it is a standard for the specification of data. It makes the exchange of information among the different components of an application or even among different applications easy by imposing a set of rules on the data document elaboration. In ROS, XML is used to specify the configuration of a given algorithm and return the results to the client. The information contained in an XML document includes user information, features about the implementation of the algorithm, programming language, input parameters, and results.

All the XML documents that flow through the system are specified following the DTD (*data type document*) specification `optimization_algorithm.dtd` specially designed for this service (Figure 25.3, right). As we can see in Figure 25.3 (left), each document is organized hierarchically so that there is a root element `<optimization_algorithm>`, which contains four elements:

- The `<client>` element, containing all the user/client information (IP, user ID, etc.)
- The `<features>` element, containing information related to the features of the algorithm chosen (programming language, compilation/execution orders, etc.)
- The `<params>` element, which contains the parameters of the algorithm (mutation probability, number of individuals, fitness function, etc.)
- The `<results>` element, which contains the results of an execution (quality of solutions, computation time, number of evaluations, etc.)

There is an additional element that can be included inside the root element: `<others>`. This element allows us to specify additional features not included in the current DTD document. This is a mechanism to extend the expressive potential of the XML documents used in ROS.

Initially, an XML document is generated in the distribution server using the information collected from the user. In this phase, the user information and the features of the algorithm selected are included under the `<client>` and `<features>` tags of the XML document, respectively. When the user specifies the algorithm and the parameters, this information is included under the `<params>` element: The algorithm selected is specified by means of an attribute, `<params type = algorithm>`, and the parameters are included as subtags of the `<params>` tag. The resulting XML document is sent to the target worker

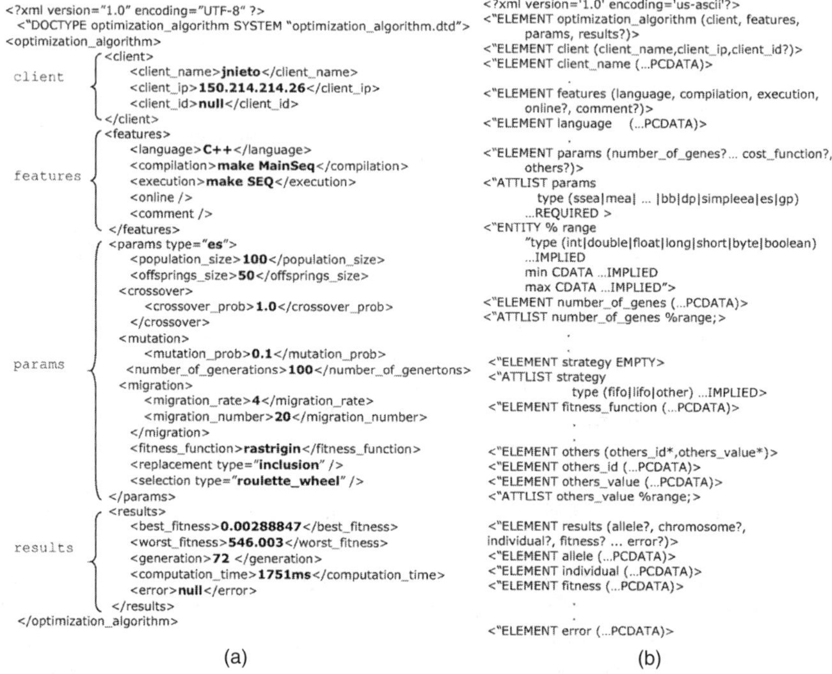

Figure 25.3 (a) XML document associated with an *evolutionary strategy* algorithm; (b) summarized DTD file employed to validate all XML documents in ROS.

through the distribution server. The worker reads the XML document in order to configure the algorithm and starts the execution. When the execution finishes, the worker collects all the results and includes it under the `<results>` tag of the XML document. This document is finally sent back to the user through the distribution server.

25.6 WRAPPERS

To be able to include in ROS a large number of algorithms implemented in different programming languages (C, C++, Java, Modula 2, etc.), we must use a general interface for dealing with the algorithms. We use the *wrapper design pattern* for this task. There is an abstract class called `Wrapper` that provides methods for generating configuration files from the XML document, launching compilation/execution orders, and writing the results in the XML document. In order to include a new algorithm in ROS, the developer must create a subclass of `Wrapper` and implement all previous methods of dealing with the algorithm. This subclass encapsulates all the details associated with the algorithm and allows ROS to communicate with the algorithm in a homogeneous way.

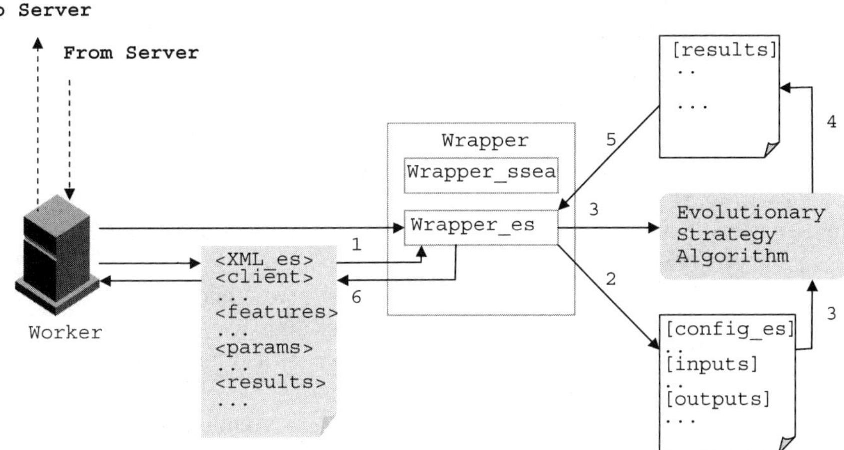

Figure 25.4 Algorithm encapsulation in ROS using the wrapper design pattern.

Figure 25.4 illustrates how the wrapper design pattern works in ROS. Let us suppose that we have an *evolutionary strategy* algorithm in ROS and that its associated `Wrapper` subclass is `Wrapper_es`. Initially (step 1 in Figure 25.4), the process server (worker) obtains the XML document with the configuration data of the algorithm from the distribution server. Then the worker invokes a method of `Wrapper_es` for generating the configuration file (specific to the available implementation of the evolutionary strategy) from the XML document (step 2). After that, the worker extracts the compilation command (if needed) in order to generate an executable file. The next step consists of running the algorithm using the execution command provided by `Wrapper_es` (step 3). When the execution finishes, `Wrapper_es` writes the results obtained into the XML document, which will be returned to the distribution server (steps 4, 5, and 6).

25.7 EVALUATION OF ROS

In this section we evaluate the performance of the system by solving a set of optimization problems with different optimization algorithms. We consider two different configurations of ROS: distributed in a LAN and deployed locally in a single machine. In the distributed configuration, the servers of ROS were executed in four machines with a Pentium-4 processor at 2.4 GHz and 512 Mbytes of RAM memory conncected by means of a 100-Mbps fast Ethernet network. These servers are the primary server, one distribution server, and two process servers. One of the process servers is running in a machine with Linux and the other one is running in a machine with Windows. There are some algorithms that can only be executed in a specific operating system (Linux or Windows). Using at least one worker running in each operating system, the user can select all these algorithms.

In addition, for those algorithms that can be executed in both operating systems, the user can compare the results obtained by the two implementations. The client application was executed in one machine with a Pentium-4 processor at 2.6 GHz and 512 Mbytes of RAM memory.

In the local configuration all the servers and the client application are running together on the same machine with a Pentium-4 processor at 2.6 GHz and 512 Mbytes of RAM memory. In this case the machine was first booted with Windows to evaluate the algorithms implemented in this operating system and after that we evaluated the Linux algorithms.

Tables 25.2 and 25.3 show the *execution time* ($e.t.$) and the *communication time* ($c.t.$) obtained after running different algorithms using ROS in both local and distributed configurations. We show the averages and the standard deviations over 50 independent runs.

For a given algorithm, we could expect lower execution times in the local configuration since the machine used is faster. However, we can notice that this does not happen in all cases. In particular, higher execution times in the local configuration are obtained for the cellular evolutionary algorithm (CEA) [5], the distributed evolutionary algorithm (DEA) [4], and the genetic algorithm for solving the problem called Venice Tides (VTGA) [8]. The reason of such an unexpected result could be the computational overload related to the base algorithm (implemented by different authors) and the operating system (previous results not shown confirm this hypothesis). For the remaining algorithms—steady-state genetic algorithm (SSGA) [19], evolutionary strategy (ES) [6], and simulated

TABLE 25.2 Experimental Results Using ROS with the Local Configuration

Alg.	Problem	O.S	Prog. Lan.	$\overline{e.t.}$ (ms)	$\sigma_{e.t.}$ (ms)	$\overline{c.t.}$ (ms)	$\sigma_{c.t.}$ (ms)
SSGA	OneMax	Linux	Java	201.76	18.08	995.96	61.35
CEA	Rastrigin	Win	Modula 2	1,336.40	42.13	1,838.20	165.73
DEA	Rastrigin	Win	Modula 2	1,311.70	23.96	1,630.40	70.52
ES	Rastrigin	Linux	C++	2,278.10	20.45	101.52	24.72
SA	OneMax	Linux	C++	70.74	5.08	99.90	40.25
VTGA	TSPr	Win	C++	276,620.00	49,361.00	760.96	151.02

TABLE 25.3 Experimental Results Using ROS with the Distributed Configuration

Alg.	Problem	O.S	Prog. Lan.	$\overline{e.t.}$ (ms)	$\sigma_{e.t.}$ (ms)	$\overline{c.t.}$ (ms)	$\sigma_{c.t.}$ (ms)
SSGA	OneMax	Linux	Java	296.28	68.94	1,510.00	549.91
CEA	Rastrigin	Win	Modula 2	1,104.20	7.76	1,778.10	39.53
DEA	Rastrigin	Win	Modula 2	1,124.90	4.73	1,744.60	57.64
ES	Rastrigin	Linux	C++	3,440.00	92.88	52.84	256.07
SA	OneMax	Linux	C++	126.00	43.53	264.64	28.91
VTGA	TSPr.	Win	C++	148,320.00	699.41	652.42	139.70

annealing (SA) [13]—the execution time in the local configuration is lower than in the distributed configuration.

In Figure 25.5 we plot the execution time of the 50 independent runs in the two configurations: distributed and local. We can observe that, as expected, the execution time is more stable in the local configuration. In the LAN configuration the execution time depends on the network traffic, which is unpredictable. If we focus on the communication time (see Figure 25.6), we can observe that more time is required in the LAN configuration since the processes are distributed in a network of workstations. However, in the genetic algorithm for the Venice lagoon problem, the evolutionary strategy, and the cellular genetic algorithm, this does not happen. The reason could be that the communication time includes preprocessing of XML and writing in files, which can slow communication down significantly in the local configuration. As happens with execution time, communication time is more stable in the local configuration.

One of the most important aspects in a study of the behavior of ROS in a network is the measurement of the data flow that arises during information

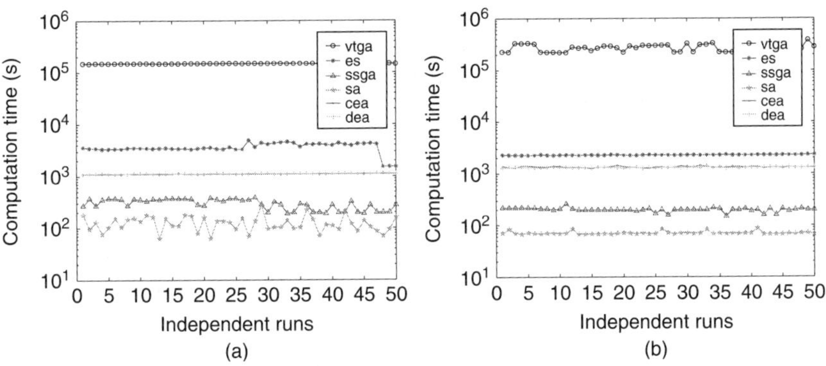

Figure 25.5 Execution time in the (a) distributed and (b) local configurations.

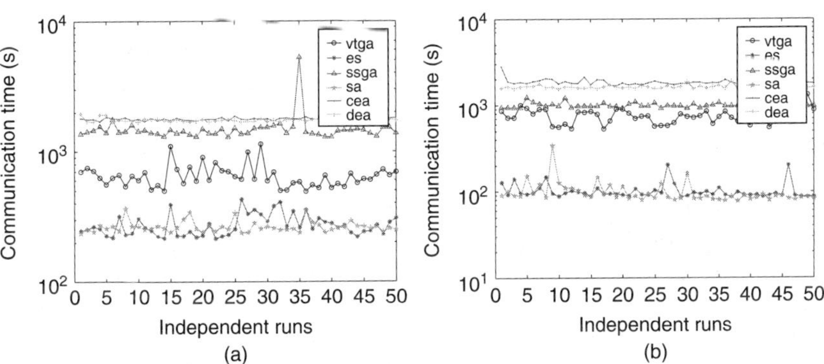

Figure 25.6 Communication time in the (a) distributed and (b) local configurations.

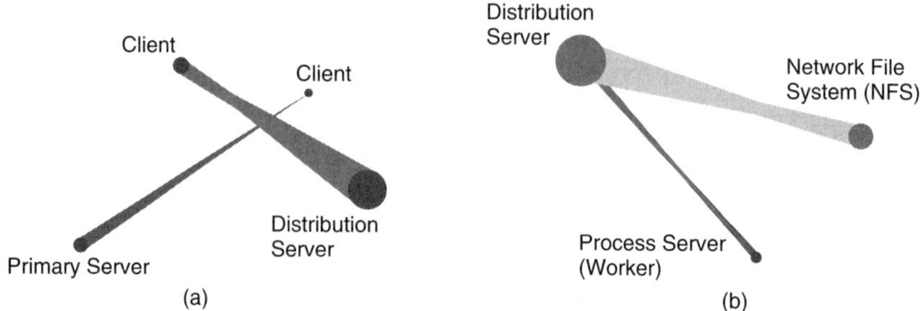

Figure 25.7 Data flow registered by Etherape in the network during a typical execution of ROS.

exchange between the various entities. Figure 25.7 consists of two diagrams (generated by means of the network application *Etherape* [18]), which show the traffic exchanged by the entities of ROS in a typical distributed configuration. Figure 25.7a reflects how the traffic trail generated by the interaction between the client and the primary server is thinner than that generated between the client and the distribution server, since in the second case the XML document is sent (more data). In Figure 25.7b the distribution server carries out a request to the process server. A file-reading operation using the network file system can also be observed.

25.8 CONCLUSIONS

In this chapter we have described ROS, a web service that provides access to a large set of optimization algorithms for solving optimization problems through the Internet. ROS is a platform-independent system based on client/server architecture that uses XML documents for information exchange. In addition, this service offers a graphical user interface for easing access to the system. From the point of view of the developers, the wrapper design pattern allows them to include new algorithms in any programming language with minimum effort.

As future work we plan to add new algorithms to the system and different implementations of the current algorithms. For dealing with the wide diversity of these algorithms and the different specific parameters that these algorithms require, it is necessary to generalize the XML specification (DTD).

We also defer for future work the study and development of methods for easing the incorporation of new algorithms in the system, imposing common design criteria for the development of these algorithms, developing classes for managing groups of algorithms, and including new graphical user interfaces.

The installation package for ROS is available at http://tracer.lcc.uma.es/ros/index.html. It is possible to evaluate the system by connecting it to active servers already working in our institution, whose IP addresses are configured in the client application.

Acknowledgments

The authors are partially supported by the Spanish Ministry of Science and Technology and FEDER under contract TIN2005-08818-C04-01 (the OPLINK project) and the Spanish Ministry of Industry and EUREKA-CELTIC by FIT-330225-2007-1 (the CARLINK project).

REFERENCES

1. EAML website. http://vision.fhg.de/ veenhuis/eaml/intro.html.
2. Eclipse website. http://www.eclipse.org.
3. E. Alba, J. García-Nieto, and F. Chicano. ROS: servicio de optimización remota. In J. Riquelme and P. Botella, eds., *Actas de las JISBD*, Barcelona, Spain, pp. 509–514, 2006.
4. E. Alba and J. M. Troya. An analysis of synchronous and asynchronous parallel distributed genetic algorithms with structured and panmictic islands. In *Proceedings of the 11 IPPS/SPDP'99*, pp. 248–256, 1999.
5. E. Alba and B. Dorronsoro. Cellular genetic algorithms. In *Operations Research*, vol. 42 *of computer Science Series*. Springer-Velag, New York, 2008.
6. T. Back, F. Hoffmeister, and H. Schewefel. A survey of evolution strategies. *Technical Report D-4600*, Department of Computer Science XI, University of Dortmund, Germany, 1991.
7. P. Bosman. Evolutionary algorithm visualization: EAVisualizer GUI. http://www.cs.uu.nl/people/peterb/computer/ea/eavisualizer, 2001.
8. J. C. Hernández C. Luque, and P. Isasi. Forecasting time series by means of evolutionary algorithms. In *Proceedings of the PPSN VIII*, vol. 3242 of *Lecture Notes in Computer Science*. Springer-Verlag, New York, pp. 1061–1070, 2004.
9. V. Filipovic. eaWeb: evolutionary algorithm web service. http://www.matf.bg.ac.yu/~vladaf/EaWeb, 2002.
10. J. García-Nieto, E. Alba, and F. Chicano. Using metaheuristic algorithms remotely via ROS. In *Proceedings of the Genetic and Evolutionary Computation Conference (GECCO'07)*, London, ACM Press, New York, p. 1510, 2007.
11. F. Glover and G. Kochenberger, editors. *Handbook of Metaheuristics*, vol. 57 of *International Series in Operations Research and Management Science*. Kluwer Academic, Norwell, MA, 2003.
12. E. R. Harold. *XML Bible*. IDG Books, Boston, MA 1999.
13. T. W. Manikas and J. T. Cain. Genetic algorithms vs. simulated annealing: a comparison of approaches for solving the circuit partitioning problem. *Technical Report 96-101*, Department of Electrical Engineering, University of Pittsburgh, Pittsburgh, PA, 1997.
14. J. J. Merelo-Guervós. OPEAL, Una librería de algoritmos evolutivos en Perl. In E. Alba et al. eds., *Actas del Primer Congreso Español sobre Algoritmos Evolutivos y Bioinspirados (AEB'02)*, pp. 54–59, 2002.
15. J. J. Merelo-Guervós, P. A. Castillo Valdivieso, and J. García Castellano. Algoritmos evolutivos P2P usando SOAP. In *Proceedings of the Primer Congreso Español sobre Algoritmos Evolutivos y Bioinspirados*, Mérida, Mexico, pp. 31–37, 2002.

16. A. Rummler. EALib. http://www.evolvica.org/ealib/index.html, 2002.
17. A. Rummler. Evolvica. http://www.evolvica.org, 2002.
18. J. Toledo, V. Adrighem, R. Ghetta, E. Mann, and F. Peters. Etherape, a graphical network monitor. http://etherape.sourceforge.net/.
19. F. Vavak and T. C. Fogarty. Comparison of steady state and generational genetic algorithms for use in nonstationary environments. *Technical Report BS16 1QY*. Faculty of Computer Studies and Mathematic, University of the West of England, Bristol, UK, 1996.
20. C. Veenhuis, K. Franke, and M. Köppen. A semantic model for evolutionary computation. In *Proceedings of the 6th International Conference on Soft Computing*, Iizuka, Japan, pp. 68–73, 2000.
21. S. Voß and D. Woodruff, eds. *Optimization Software Class Libraries*. Kluwer Academic, Norwell, MA, 2002.

CHAPTER 26

Remote Services for Advanced Problem Optimization

J. A. GÓMEZ, M. A. VEGA, and J. M. SÁNCHEZ
Universidad de Extremadura, Spain

J. L. GUISADO
Universidad de Sevilla, Spain

D. LOMBRAÑA and F. FERNÁNDEZ
Universidad de Extremadura, Spain

26.1 INTRODUCTION

In this chapter we present some remote services for advanced problem optimization. These tools permit us to get a set of resources useful to approach problems easily using the Internet as a channel of experimentation with the optimization algorithms involved.

The chapter is structured as follows. Section 26.2 covers the development of image-processing services via the Internet by means of algorithms implemented by software and reconfigurable hardware. In this way, advanced optimization techniques are used to implement the image-processing algorithms by custom-designed electronic devices, and an open platform is offered to the scientific community, providing the advantages of reconfigurable computing anywhere. Next, in Section 26.3 two web services using time series as optimization subjects are presented. The optimization algorithms are driven by the remote user, and the results are shown in graphical and tabular forms. Finally, Section 26.4 demonstrates the Abacus-GP tool: a web platform developed to provide access to a cluster to be used by researchers in the genetic programming area.

Optimization Techniques for Solving Complex Problems, Edited by Enrique Alba, Christian Blum, Pedro Isasi, Coromoto León, and Juan Antonio Gómez

26.2 SIRVA

In this section we present the results obtained inside the research line SIRVA (sistema reconfigurable de visión artificial) of the project TRACER (advanced optimization techniques for complex problems) [1]. The main goal of this research line is the development of image-processing services via the Internet by means of algorithms implemented by software and reconfigurable hardware with field-programmable gate arrays (FPGAs). In this way, advanced optimization techniques are used to implement image-processing algorithms by FPGAs and an open platform is offered to the scientific community, providing the advantages of reconfigurable computing anywhere (e.g., where these resources do not exist).

26.2.1 Background

At present, FPGAs are very popular devices in many different fields. In particular, FPGAs are a good alternative for many real applications in image processing. Several systems using programmable logic devices have been designed showing the utility of these devices for artificial-vision applications [2]. Whereas other papers display the results from implementing image-processing techniques by means of "standard" FPGAs or reconfigurable computing systems [3], in our work we have not only implemented image-processing operations using reconfigurable hardware and obtained very good results, but have also looked to add other services, thanks to the Internet. Basically, these services are:

- A client–server system: to access all the image-processing circuits we have developed in FPGAs. In this way, the user can send his or her artificial-vision problem, choose between hardware (FPGA) or software implementation, and then obtain a solution to the problem.
- A web repository of artificial-vision problems with interesting information about each problem: description, practical results, links, and so on. This repository can easily be extended with new problems or by adding new languages to existing descriptions.

These two additional services offer a set of benefits to the scientific community: They provide reconfigurable computing advantages where these resources are not physically available (without cost, and at any time), they promote the FPGA virtues because any researcher can compare the FPGA and software results (in fact, some of our FPGA-based implementations are even more than 149 times faster than the corresponding software implementation), and the easy extension of the open repository could convert it in a reference point.

26.2.2 Description

When a user accesses the web service (http://tracer.unex.es, then selects the research line SIRVA), the main page of the site is presented. The user can choose

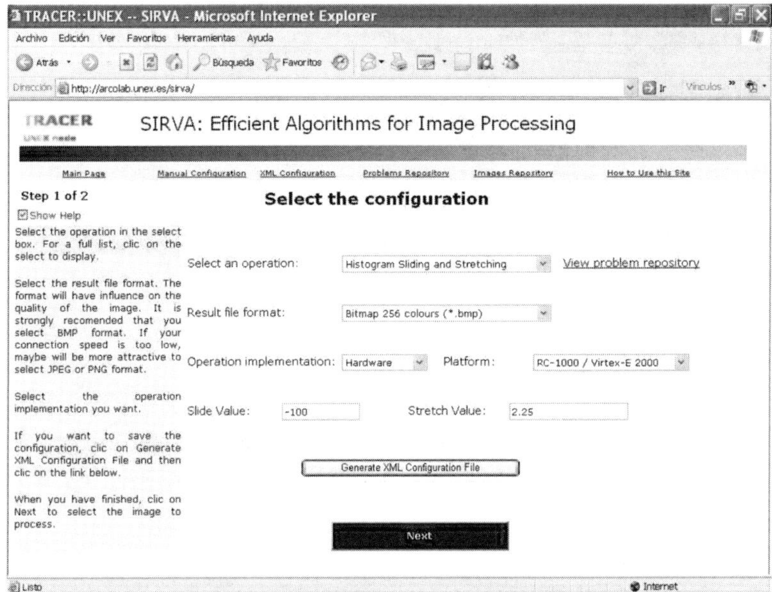

Figure 26.1 Page for manual configuration of SIRVA.

any of the options shown on this page. For space limitations we only explain the most important options of SIRVA. If the user selects *Manual Configuration* in the main page, he or she will go to the page shown in Figure 26.1. In this page the user can introduce, one by one, all the operation parameters: image-processing operation to perform, other configuration parameters, the format for the resulting image file (BMP, PNG, or JPEG), and the implementation he or she wants to use for performing the operation (software in a Pentium-4 or hardware in a FPGA).

After configuring the web service (using an XML file or manually), the following step (the Next button) is to select the image to process. After that, the user will see the result page. In this page (Figure 26.2), not only the original and processed images are shown, but also the main characteristics of the process performed: implementation selected (HW or SW), platform used (Virtex FPGA, Pentium-4), operation performed, image dimensions, operation time, and so on. On the other hand, the problem repository is based on an ASP page that shows the contents of the diverse XML files used to store the descriptions of different artificial-vision problems. Figure 26.3 presents an example (a more detailed explanation of SIRVA is included in ref. 4).

26.2.3 Technical Details

On the hardware side, we have used the Celoxica RC1000 board (http://www.celoxica.com) with a Xilinx Virtex XCV2000E-8 FPGA (http://www.xilinx.com). The circuit design has been carried out by means of the Handel-C language and

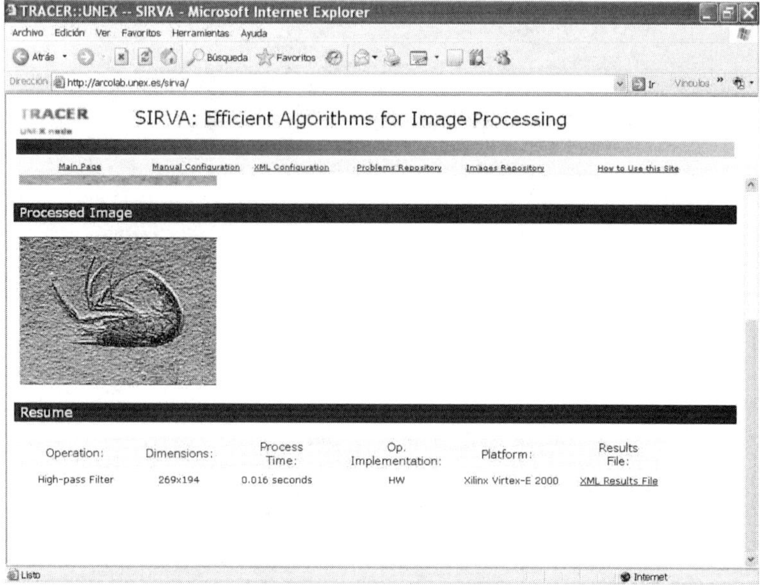

Figure 26.2 Result page (process summary) in SIRVA.

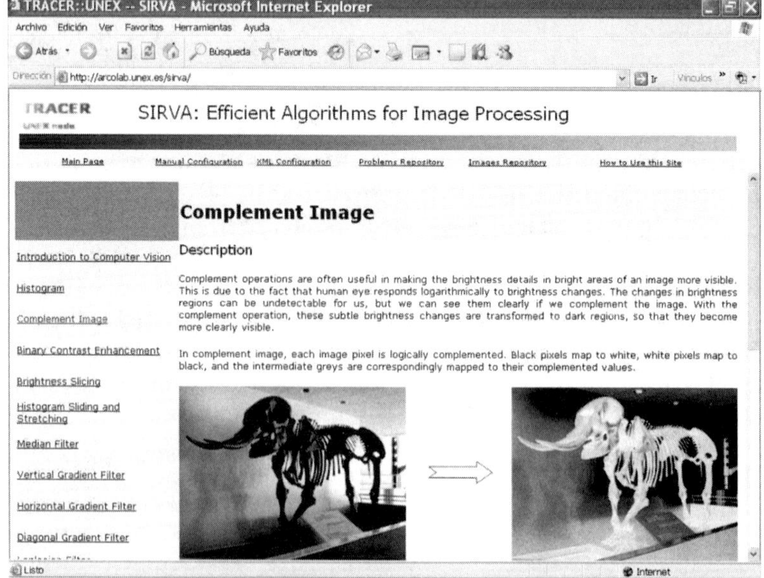

Figure 26.3 Problem repository of SIRVA.

the Celoxica DK platform. Thanks to DK the hardware description files have been obtained for the various circuits which are used later in the Xilinx ISE tool to generate the corresponding configuration files (.bit) for the FPGA.

For SIRVA system implementation we use Microsoft .NET, due to the facilities it offers for the creation of web services, and because the control library of the Virtex FPGA we use is implemented in C++ and the corresponding API provided by Celoxica advises the use of Microsoft Visual C++. In this platform the following technologies have been used for the implementation: ASP .NET, C#, ATL server, CGI, C++, and XML-XSLT.

26.2.4 Results and Conclusions

Table 26.1 presents several comparisons between the software and hardware (FPGA) versions of the various image-processing operations implemented in SIRVA. The software versions are executed in a 1.7-GHz Pentium-4 with 768 Mbytes of RAM, and the hardware versions in a Virtex-E 2000 FPGA. The software versions are written in C++ and compiled with the appropriate optimizations. All the results refer to images with 256 gray levels and are expressed in seconds. These times also include the communication time, and they are the average response times after 10 executions of each experiment. In order not to influence the results, all the measurements in Table 26.1 have been carried out with the same workload in the server.

As we can observe in Table 26.1, FPGA-based implementation clearly improves the performance obtained by software (some operations are more than 149 times faster). This is possible because our FPGA implementations use

TABLE 26.1 Software Versus FPGA for 2032 x 1524 Pixel Images

Image Operation	SW	HW	Speedup
Histogram	27.953	0.203	137.700
Complement image	0.078	0.001	78
Thresholding	0.078	0.001	78
Brightness slicing	0.078	0.001	78
Brightness/contrast adjustment	0.078	0.001	78
Median filter	34.328	5.536	6.201
Low-pass filter	26.195	0.461	56.822
High-pass filter	26.344	0.458	57.468
Laplacian filter	25.659	0.438	58.583
Vertical gradient filter	25.262	0.420	60.184
Horizontal gradient filter	25.203	0.417	60.475
Diagonal gradient filter	25.187	0.419	60.153
Binary erosion	26.188	0.175	149.219
Binary dilation	26.346	0.189	139.399
Gray-scale erosion	26.398	0.361	73.068
Gray-scale dilation	26.336	0.355	74.238

several parallelism techniques, include diverse optimizations, and have high operational frequencies (see ref. 4 for a more detailed explanation).

26.3 MOSET AND TIDESI

Here we present two web services that compute several algorithms to perform system identification (SI) for time series (TS). Although the main purpose of these developments was to experiment in optimization of the main parameters of the algorithms involved, it was found that due to their handling ease, configurability, and free online access, these services could be of great utility in teaching courses where SI is a subject: for example in the systems engineering field. The possible utility of these systems comes from their capacity to generate practical contents that support the theoretical issues of the courses.

26.3.1 Modeling and Prediction of Time Series

TSs are used to describe behaviors in many fields: astrophysics, meteorology, economy, and others. SI techniques [5] permit us to obtain the TS model. With this model a prediction can be made, but taking into account that the model precision depends on the values assigned to certain parameters. Basically, the identification consists of determining the ARMAX [6] model from measured samples, allowing us to compute the estimated signal and compare it with the real one, then reporting the error generated.

The recursive estimation updates the model at each time step, thus modeling the system. The more samples processed, the model's better the precision, because it has more information on the system's behavior history. We consider SI performed by the well-known recursive least squares (RLS) algorithm [6], which is very useful for prediction purposes. This identification is influenced strongly by the forgetting factor constant λ. There is no fixed value for λ, but a value between 0.97 and 0.995 is often used. Then the optimization problem consists basically of finding the λ value such that the error made in identification is minimal.

Recursive identification can be used to predict the following behavior of the TS from the data observed up to the moment. With the model identified in k time it is possible to predict the behavior in $k + 1$, $k + 2$, and so on. As identification advances over time, the predictions improve, using more precise models. If ks is the time until the model is elaborated and from which we carry out the prediction, the prediction will have a larger error and we will be farther away from ks. The value predicted for $ks + 1$ corresponds with the last value estimated before ks. When we have more data, the model has to be restarted to compute the new estimated values.

To find the optimum value of λ, we propose use of the parallel adaptive recursive least squares (PARLS) algorithm [7], where an optimization parameter λ is evolved for predicting new situations during iterations of the algorithm. This algorithm uses parallel processing units performing RLS identifications (Figure 26.4),

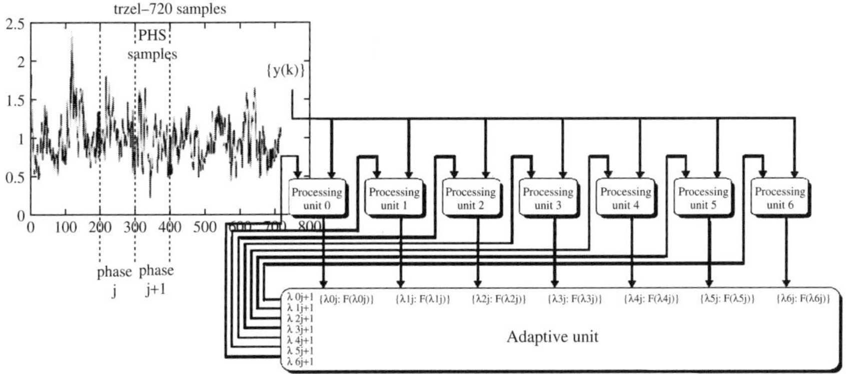

Figure 26.4 Parallel adaptive scheme used to optimize the time series problem.

and the results are processed in an adaptive unit to obtain the best solution. The goal of the iterations of the algorithm is that identifications performed by the processing units will converge to optimum λ values so that the final TS model will be a high-precision model.

26.3.2 MOSET

MOSET is an acronym for the Spanish term "modelacion de series temporales" (time series modeling). MOSET [Figure 26.5(a)] is an open-access web service [8] dedicated to the identification of TSs that use the RLS and PARLS algorithms.

All the experiments are made on a TS database obtained from several sources [7]. Each experiment consists of selecting a TS, the model size, a range of values for λ (several identifications are processed), and the algorithm to be used. According to the algorithm chosen, additional parameters must be selected (e.g., the number of parallel processing units if PARLS), as shown in Figure 26.5(b).

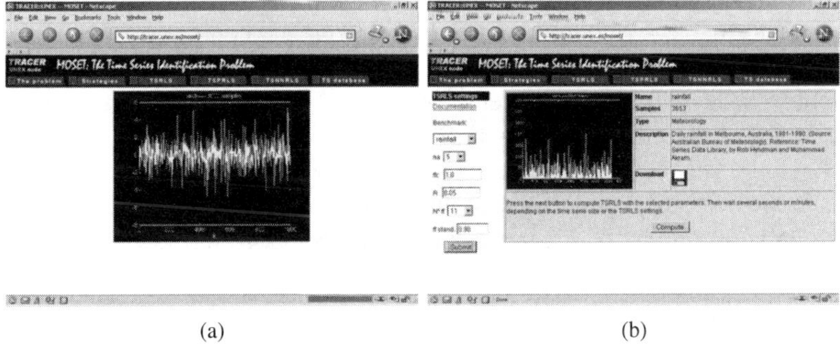

Figure 26.5 MOSET web service for identification of time series: (a) welcome page; (b) database and parameter selection.

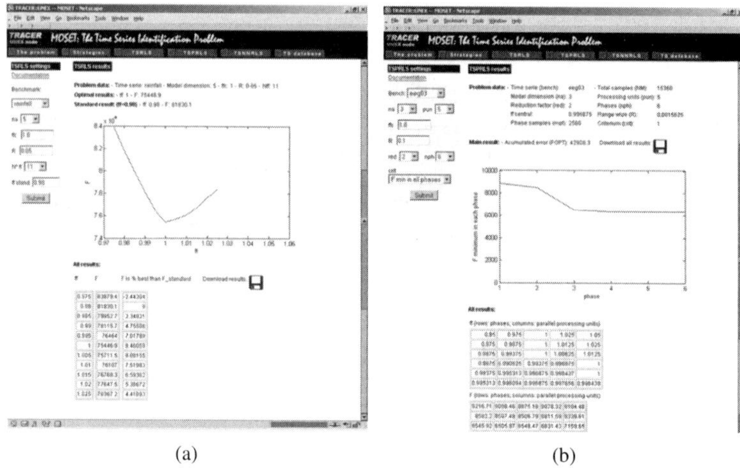

(a) (b)

Figure 26.6 MOSET results shown in graphics and tables: (a) RLS results; (b) PARLS results.

After a certain processing time (from seconds to minutes, depending on the TS selected), the downloadable results are shown by means of a curve whose minimum indicates the best value for λ (Figure 26.6).

26.3.3 TIDESI

TIDESI is an acronym for "tide system identification," a web service designed to model and predict a set of time series defining the tide levels in the city of Venice, Italy in recent decades. Although the database of TIDESI consists of this type of TS, the techniques are capable of supporting any other type of TS. The online access is public and free, and the prediction is programmed for the short, medium, and long term.

TIDESI has two options for processing the TS selected. The first option computes a simple RLS identification (calculating the estimated value, equal to prediction of the $k+1$ time). The second option computes a RLS identification whose model serves as a base to carry out future predictions for any term as a function of the user's requirements. In Figure 26.7 we can observe the selection of the TS, the identification/prediction method chosen, and the parameters that should be specified. In Figure 26.8 we observe how numerical and graphical results are shown, displaying signals with real, identified, and simulated values of the TS. In addition, the user can download the results for a more detailed study.

26.3.4 Technologies Used

To develop MOSET and TIDESI, a 2.4-GHz Pentium-4 PC with a Microsoft Windows 2003 Server operating system was used, running the Microsoft Internet

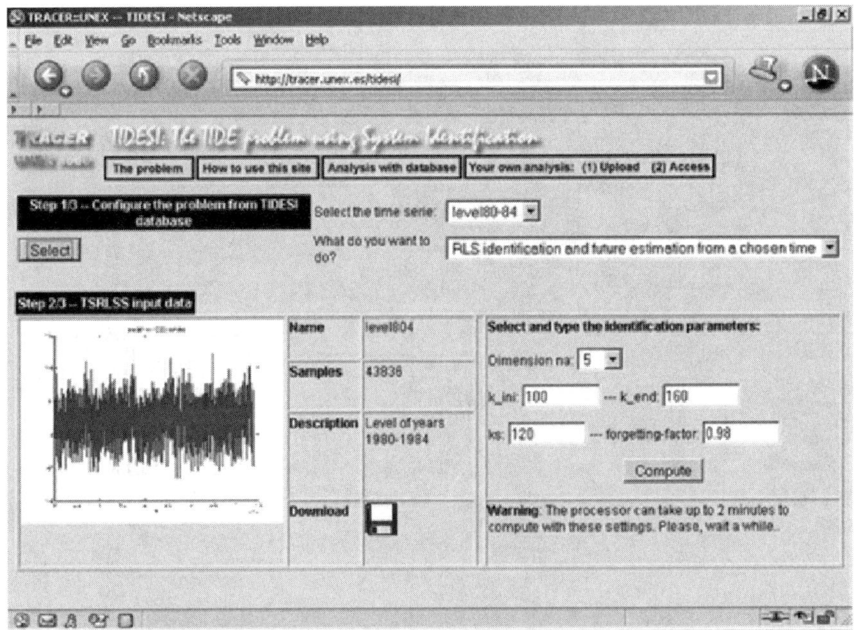

Figure 26.7 TIDESI web service to predict time series.

Information Server. To program the modeling, identification, and prediction algorithms, the Matlab language has been used [9], and the source codes are executed by means of the Matlab Webserver. Finally, TS database management has been carried out using Microsoft Active Server Pages and Microsoft Access.

26.3.5 Conclusions

Courses in the systems engineering area have enjoyed strong support in recent years thanks to the development of online teaching tools. The web services that we have described in this section can be used for teaching purposes, although they were developed with other experimental motivations. In addition, we want to allow the user the possibility of selecting external TS so that any researcher can use the TS of his or her interest.

26.4 ABACUS

Evolutionary algorithms (EAs) are one of the traditional techniques employed in optimization problems. EAs are based on natural evolution, where individuals represent candidate solutions for a given problem. The modus operandi of these algorithms is to crossover and mutate individuals to obtain a good enough solution for a given problem by repeating these operations in a fixed number of generations

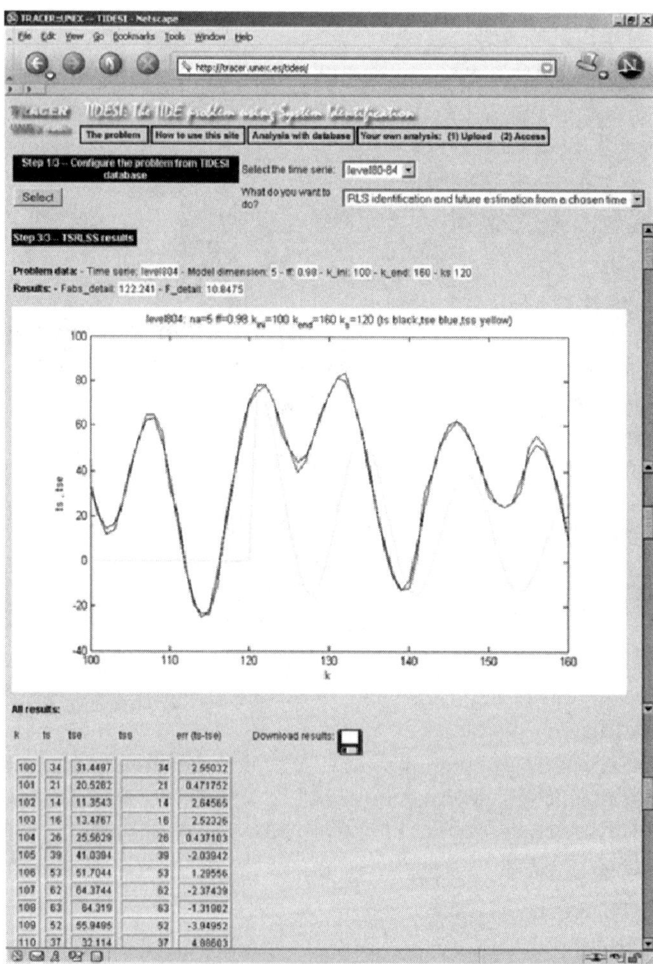

Figure 26.8 TIDESI results shown in graphics and tables.

(steps). However, when difficult problems are faced, these algorithms usually require a prohibitive amount of time and computing resources.

Two decades ago Koza developed a new EA called genetic programming (GP) [10]. GP is based on the creation of a population of programs considered as solutions for a given problem. Usually, GP individuals' structure is a tree, and they have different sizes. Therefore, different analysis times are needed to evaluate each one. This feature distinguish GPs from other EAs, such as genetic algorithms [11] or evolutionary strategies where individuals have a fixed size, so the evaluation time and computing resources (CPU and memory) required are always the same.

Several researchers have tried to reduce the computational time by means of parallelism. For instance, Andre and Koza employed a transputer network to improve GP performance [12]. Researchers usually employ two parallel models: fine and coarse grain (also known as an island model) [14,15]. As described by Cantu-Paz and Goldberg [16], two major benefits are obtained, thanks to parallel computers running EAs: first, several microprocessors improve the computational speedup; and second, the parallel-based coarse-grain model (another models are also possible) can produce improved performance.

Recently, new tools have been developed to employ the benefits of multiprocessor systems for EAs [17,18]), but there are few public multiprocessor systems that allow these parallel tools to be run. There are also several clusters where it is possible to run these tools, although they do not provide public access. Therefore, we introduce a new platform here, Abacus-GP [19], developed at the University of Extremadura (Spain) and publicly accessible (through a petition) to enable it to be used by GP researchers. Abacus-GP is a GNU/Linux-based parallel cluster comprising two modules: a web interface site in the cluster's firewall and an internal application running in the cluster master node. The Abacus-GP platform allows users to run GP experiments in the cluster by using a web interface. The GP-employed tool necessary to run jobs in Abacus is explained in ref. 20.

26.4.1 Cluster

The Abacus-GP platform is based on the high-performance computational cluster Abacus, set up by the research group on artificial evolution from the University of Extremadura in Spain, using the hardware components Commodity-Off-The-Self (COTS) and open-source software. Abacus is a parallel MPI/PVM-based Linux Beowulf cluster consisting of 10 nodes with Intel Pentium-4 microprocessors. The operating system is Rocks Linux Cluster (http://www.rocksclusters.org), a Linux-based distribution designed especially to create high-performance clusters.

The Abacus cluster consists of a master node and nine working nodes. The master node is in charge of running the jobs and scheduling the cluster, while the working nodes are focused exclusively on running jobs. Figure 26.9 shows the cluster structure. All the nodes are interconnected through a 100-Mbps fast Ethernet switch. Finally, the cluster is connected to the Internet using a GNU/Linux firewall which filters outside access to the cluster.

26.4.2 Platform

With the goal of achieving a more secure and easier access to the cluster and to help in the collaboration between researchers, we have developed a web tool that allows us to run experiments remotely and monitorize them later. When a job is finished, the system sends the results to the job owner by e-mail. Additionally, the user can access the final results (numerical and graphical modes) through the web interface. The website also has two additional tools: a finished experiment repository and a forum where web users can exchange information. Thanks to the web interface, this tool can be employed from any place in the world without knowing the underlying cluster structure.

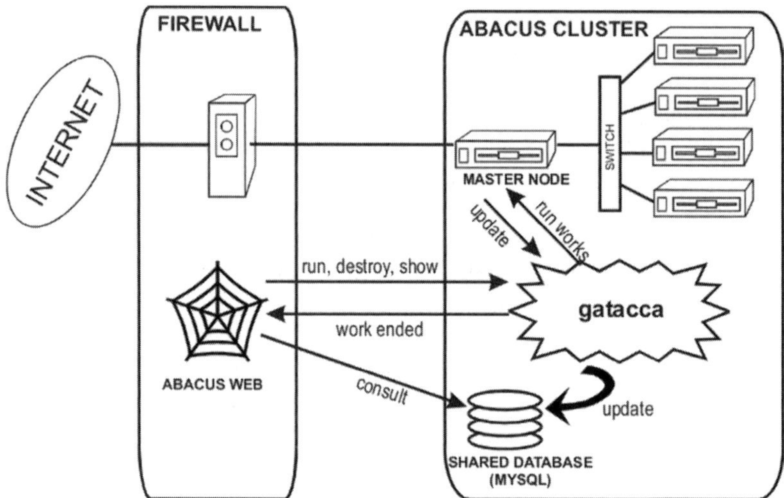

Figure 26.9 Communication between modules.

The cluster consists of two modules: a multiuser-authenticated website (Abacus web), which is the user interface and runs in the firewall, and *Gatacca*, a server that runs in the cluster and communicates uniquely with the web interface in order to launch new jobs in the cluster (Figure 26.9).

Both modules are coded in PHP and shell script and access the same MySQL data base (DB), which runs on the cluster. The website checks the job status in the DB and *gatacca* updates the information in the DB with the experimental results. Thus, *gatacca* has two main goals: first, to receive requests from the website in order to manage jobs and to know the ongoing status of the jobs launched, and second, to maintain the DB updated with the status of all the experiments running.

Figure 26.10 shows the web interface. This interface shows all the user-finished and still-running experiments, showing the startup time for each experiment and the estimated time to completion. When a user wants to launch a new experiment [Figure 26.10a], the system will ask the user to designate the necessary parameters to run it (population size, generations, etc.). Once the parameters are set up, the website sends a message to *gatacca* to launch the new job in the cluster. Once the message is received, *gataccca* creates a new folder in the cluster master node, copies in it all the necessary files, and finally, launches the job. Moreover, each completed job generation triggers *gatacca* to update the DB with the experiment information and estimates the time left to finish the job. Additionally, the user can consult the ongoing experiment status because *gatacca* provides the user with real-time numerical and graphical results. When the experiment is finished, *gatacca* provides the user with the results and charts through the web interface (Figure 26.11).

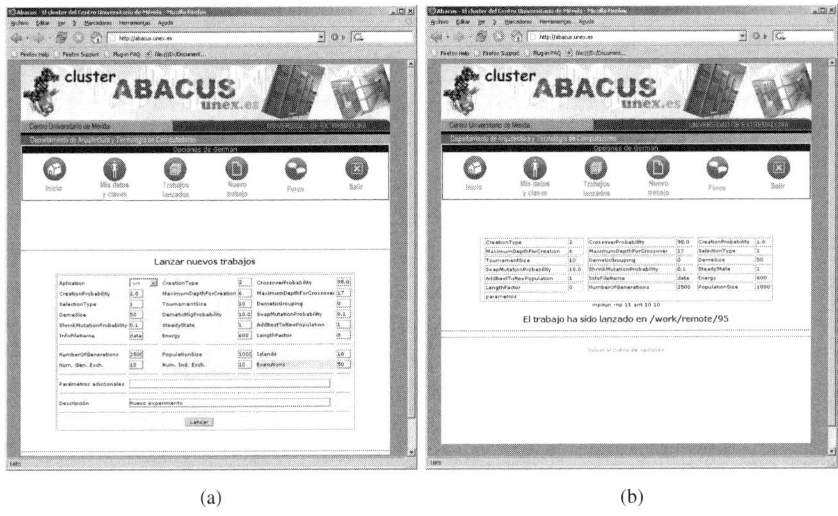

(a) (b)

Figure 26.10 Abacus-GP web interface: (a) parameter setup; (b) experiment launch.

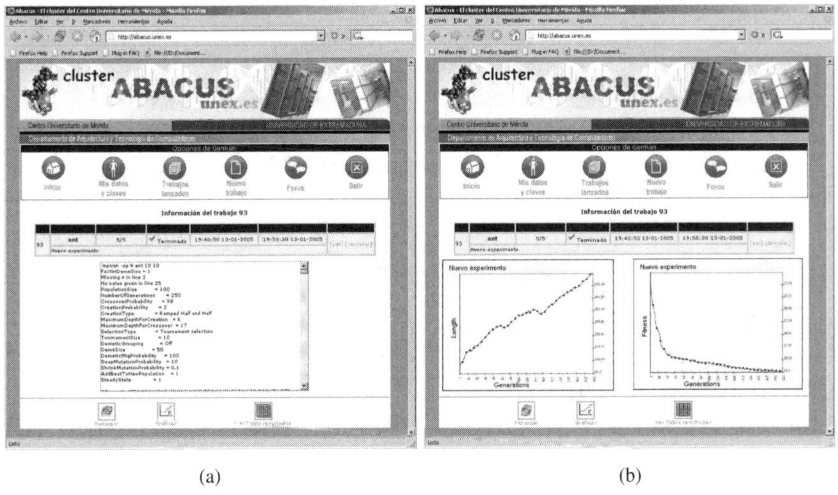

(a) (b)

Figure 26.11 Abacus-GP result representation: (a) data; (b) charts.

To test the operation of the cluster, we have developed two well-known GP problems in Abacus-GP: the Ant and the Even Parity 5 problems [10].

26.4.3 Conclusions

Abacus-GP is a public and secure platform for GP experimentation through the Internet that is useful to researchers since they can employ a cluster to run their experiments and share them with other colleagues easily by creating result repositories that can be shared. In the future, more bioinspired applications will be implemented and the web tool will also be improved to schedule several clusters using the same tool. This will allow us to distribute jobs between clusters and between zones, which will allow us to employ grid computing concepts.

Acknowledgments

The authors are partially supported by the Spanish MEC and FEDER under contracts TIC2002-04498-C05-01 (the TRACER project) and TIN2005-08818-C04-03 (the OPLINK project).

REFERENCES

1. J. A. Gomez-Pulido. TRACER::UNEX project. http://www.tracer.unex.es.
2. M. A. Vega-Rodríguez, J. M. Sanchez-Perez, and J. A. Gomez-Pulido. Guest editors' introduction—special issue on FPGAs: applications and designs. *Microprocessors and Microsystems*, 28(5–6):193–196, 2004.
3. M. A. Vega-Rodríguez, J. M. Sanchez-Perez, and J. A. Gomez-Pulido. Recent advances in computer vision and image processing using reconfigurable hardware. *Microprocessors and Microsystems*, 29(8–9):359–362, 2005.
4. M. A. Vega-Rodríguez, J. M. Sanchez-Perez, and J. A. Gomez-Pulido. Reconfigurable computing system for image processing via the Internet. *Microprocessors and Microsystems*, 31(8):498–515, 2007.
5. T. Soderstrom. *System Identification*. Prentice Hall, Englewood Cliffs, NJ, 1989.
6. L. Ljung. *System Identification*. Prentice Hall, Upper Saddle River, NJ, 1999.
7. J. A. Gomez-Pulido, M. A. Vega-Rodriguez, and J. M. Sanchez-Perez. Parametric identification of solar series based on an adaptive parallel methodology. *Journal of Astrophysics and Astronomy*, 26:1–13, 2005.
8. J. A. Gomez-Pulido. MOSET. http://www.tracer.unex.es/moset.
9. R. Pratap. *Getting Started with Matlab*. Oxford University Press, New York, 2005.
10. J. Koza. *Genetic Programming: On the Programming of Computers by Means of Natural Selection*. MIT Press, Cambridge, MA, 1992.
11. D. Goldberg. *Genetic Algorithms in Search, Optimization and Machine Learning*. Addison-Wesley, Reading, MA, 1989.
12. D. Andre and J. Koza. *Parallel Genetic Programming: A Scalable Implementation Using the Transputer Network Architecture*. MIT Press, Cambridge, MA, 1996.
13. D. Andre and J. Koza. Parallel genetic programming: a scalable implementation using the transputer network architecture. In *Advances in Genetic Programming*, vol. 2. MIT Press, Cambridge, MA, 1996, pp. 317–337.
14. F. Fernández, M. Tomassini, and L. Vanneschi. An empirical study of multipopulation genetic programming. *Genetic Programming and Evolvable Machines*, 4:21–51, 2003.

15. F. Fernandez de Vega. Parallel genetic programming: methodology, history, and application to real-life problems. In *Handbook of Bioinspired Algorithms and Applications*, Chapman & Hall, London, 2006, pp. 65–84.

16. E. Cantu-Paz and D. Goldberg. Predicting speedups of ideal bounding cases of parallel genetic algorithms. In *Proceedings of the 7th International Conference on Genetic Algorithms*. Morgan Kaufmann, San Francisco, CA, 2007.

17. F. Fernández. Parallel and distributed genetic programming models, with application to logic sintesis on FPGAs. Ph.D. thesis, University of Extremadura, Caceres, Spain, 2001.

18. F. Fernandez, M. Tomassini, W. F. Punch, and J. M. Sanchez. Experimental study of multipopulation parallel genetic programming. *Lecture Notes on Computer Sciences*, 1802:283–293, 2000.

19. F. Fernandez. ABACUS cluster. http://www.abacus.unex.es.

20. M. Tomassini, L. Vanneschi, L. Bucher, and F. Fernandez. An MPI-based tool for distributed genetic programming. In *Proceedings of the IEEE International Conference on Cluster Computing (Cluster'00)*. IEEE Computer Society Press, Los Alamitos, CA, 2000, pp. 209–216.

INDEX

Optimization Techniques for Solving Complex Problems, Edited by Enrique Alba, Christian Blum, Pedro Isasi, Coromoto León, and Juan Antonio Gómez

WILEY SERIES ON PARALLEL AND DISTRIBUTED COMPUTING

Series Editor: Albert Y. Zomaya

Parallel and Distributed Simulation Systems / Richard Fujimoto

Mobile Processing in Distributed and Open Environments / Peter Sapaty

Introduction to Parallel Algorithms / C. Xavier and S. S. Iyengar

Solutions to Parallel and Distributed Computing Problems: Lessons from Biological Sciences / Albert Y. Zomaya, Fikret Ercal, and Stephan Olariu (*Editors*)

Parallel and Distributed Computing: A Survey of Models, Paradigms, and Approaches / Claudia Leopold

Fundamentals of Distributed Object Systems: A CORBA Perspective / Zahir Tari and Omran Bukhres

Pipelined Processor Farms: Structured Design for Embedded Parallel Systems / Martin Fleury and Andrew Downton

Handbook of Wireless Networks and Mobile Computing / Ivan Stojmenović *(Editor)*

Internet-Based Workflow Management: Toward a Semantic Web / Dan C. Marinescu

Parallel Computing on Heterogeneous Networks / Alexey L. Lastovetsky

Performance Evaluation and Characteization of Parallel and Distributed Computing Tools / Salim Hariri and Manish Parashar

Distributed Computing: Fundamentals, Simulations and Advanced Topics, *Second Edition* / Hagit Attiya and Jennifer Welch

Smart Environments: Technology, Protocols, and Applications / Diane Cook and Sajal Das

Fundamentals of Computer Organization and Architecture / Mostafa Abd-El-Barr and Hesham El-Rewini

Advanced Computer Architecture and Parallel Processing / Hesham El-Rewini and Mostafa Abd-El-Barr

UPC: Distributed Shared Memory Programming / Tarek El-Ghazawi, William Carlson, Thomas Sterling, and Katherine Yelick

Handbook of Sensor Networks: Algorithms and Architectures / Ivan Stojmenović *(Editor)*

Parallel Metaheuristics: A New Class of Algorithms / Enrique Alba (*Editor*)

Design and Analysis of Distributed Algorithms / Nicola Santoro

Task Scheduling for Parallel Systems / Oliver Sinnen

Computing for Numerical Methods Using Visual C++ / Shaharuddin Salleh, Albert Y. Zomaya, and Sakhinah A. Bakar

Architecture-Independent Programming for Wireless Sensor Networks / Amol B. Bakshi and Viktor K. Prasanna

High-Performance Parallel Database Processing and Grid Databases / David Taniar, Clement Leung, Wenny Rahayu, and Sushant Goel

Algorithms and Protocols for Wireless and Mobile Ad Hoc Networks / Azzedine Boukerche (*Editor*)

Algorithms and Protocols for Wireless Sensor Networks / Azzedine Boukerche (*Editor*)

Optimization Techniques for Solving Complex Problems / Enrique Alba, Christian Blum, Pedro Isasi, Coromoto León, and Juan Antonio Gómez (*Editors*)